ELECTROMAGNETIC VIBRATIONS, WAVES, AND RADIATION

George Bekefi

and

Alan H. Barrett

Department of Physics
Massachusetts Institute of Technology

The MIT Press

Cambridge, Massachusetts, and London, England

PUBLISHER'S NOTE

This format is intended to reduce the cost of publishing certain works in book form and to shorten the gap between editorial preparation and final publication. Detailed editing and composition have been avoided by photographing the text of this book directly from the authors' typescript.

This book was printed and bound in the United States of America

Library of Congress Cataloging in Publication Data

Bekefi, George.
Electromagnetic vibrations, waves, and radiation.

Includes index.
1. Electromagnetic fields. 2. Electrodynamics.
I. Barrett, Alan Hildreth, joint author. II. Title.
QC665.E4B44 537 77-10421
ISBN 13: 978-0-262-52047-8
ISBN 10: 0-262-52047-8

20 19 18 17 16 15 14 13

To Claire and Emerich Bekefi

PREFACE

In this book is presented an introduction to classical electrodynamics, with emphasis on the oscillatory aspects of the electromagnetic field – that is, on vibrations, waves, radiation, and the interaction of waves with matter. The content is designed primarily for the use of second or third year students of physics and engineering who have had a semester of mechanics and a semester of electricity and magnetism. The aim throughout has been to provide a mathematically unsophisticated treatment of the subject, but one that stresses modern applications of the principles involved.

The book is an outgrowth of a one-semester course taught by us during the past five years at the Massachusetts Institute of Technology. And although, traditionally, it has been a course for sophomores in the Department of Physics, it has been well attended by students from other departments in the Schools of Science and Engineering. During our involvement in the course we have not used a prescribed text; instead we distributed notes which were revised and extended each year. This book is the culmination of four successive editions which have been exposed to the scrutiny of some 2000 students.

There are eight chapters. In the first chapter, the reader is introduced to the physics of oscillators. Mechanical oscillations are often more easily visualized than electrical ones, and for that reason examples from mechanical systems are freely used. But the stress is on electrical systems and on the motion of charged particles in particular. For example, to give the student as much exposure as possible, we prepared (Appendix 3) a Fortran computer code that allows him to watch the motion of an electron subjected simultaneously to a dc magnetic field and a crossed RF electric field. Nonlinear and coupled oscillations are also included in this chapter. Chapter 2 deals with waves; for the sake of simplicity only the scalar wave equation is discussed at this stage, and examples are drawn from waves on strings, sound waves in pipes, etc. Fourier analysis of periodic disturbances and analysis of wave packets are introduced at the end of the chapter. After a brief review of Maxwell's equations, Chapter 3 is taken up with the propagation of electromagnetic waves in vacuum. Wave polarization, linear and angular momentum associated with a wave, and radiation pressure are discussed. The whole of Chapter 4 is devoted to the classical problem of radiation from accelerated charges. The radiation field is calculated by the method of J. J. Thomson. This approach is physically very revealing and for that reason is preferred, in an introductory text such as this one, to the formal mathematical techniques used in more advanced books. The latter part of the chapter is taken up with applications: to antennas, to bremsstrahlung, to cyclotron radiation. A section on black-body radiation is also included.

Guided waves and resonant cavities are discussed in Chapter 5. Traditionally, these subjects were limited in their usefulness to radio and microwave frequencies. Today, lasers and integrated optics make this subject relevant also to optical wavelengths, a fact which we tried to bring out in the text. Chapter 6 is devoted to a discussion of the interaction of electromagnetic waves with matter – dielectrics, metals, and plasmas. On the microscopic level the subject is developed along the lines of the classical electron theory of Lorentz. On the macroscopic level, Maxwell's field equations are formulated in terms of "physical" quantities, namely $\vec{E}$, $\vec{B}$, $\vec{P}$, and $\vec{M}$. The traditional use of $\vec{D}$ and $\vec{H}$ is purposely avoided in this book since at this level of a student's experience these quantities tend to confuse. Chapter 7 deals with boundary value problems: reflection and refraction at dielectric and metal surfaces. This is followed by a discussion of Thomson and Rayleigh scattering and the problem of radiation reaction. The chapter ends with an account of stimulated emission and of the free electron laser. A nonquantum mechanical approach to this problem is adopted. Chapter 8 deals with interference and diffraction. Here we revert to scalar waves and the approximate theories of Huygens, Fresnel, and Kirchhoff. This is forced upon us by the extreme mathematical difficulties encountered in a rigorous electromagnetic treatment of this subject.

Short Appendices on the addition of periodic vibrations and on the use of complex numbers are included. The book ends with some 70 problems for home assignment.

M. K. S. units are used throughout.

The book contains more material than can be covered in one semester. The individual instructor must pick his own balance between the topics and examples to be covered. Furthermore, the material presented lends itself to a wide range of lecture demonstrations which can be both instructive and fun, for the students as well as the lecturer, but that further limits the available time. We have made no attempt to prejudge those chapters or sections which might be appropriate for a one semester course as there are many variables to be considered in such a selection.

Finally we wish to acknowledge our debt to the many students and M. I. T. faculty who have suffered through previous versions of our notes and offered their constructive criticisms which have led to the final version.

George Bekefi and Alan H. Barrett
Cambridge, Massachusetts, 1977

CONTENTS

CHAPTER 1

THE OSCILLATOR

In the restless universe around us all things are in a state of incessant motion. Some is pretty much random: for example, the seemingly senseless jiggling of atoms in a bottle of gas or the meandering of stars in their galaxies. In contradistinction, and at the other end of a scale of "orderliness" are the periodic phenomena which are the subject matter of this book. The swinging pendulum, the vibrations of the ammonia molecule, the tuned circuit of a radio, and the laser beam all have in common an aesthetically most satisfying property – periodicity. It is a phenomenon of rejuvenation in which the pattern of movement repeats over and over again. When the movement occurs about an equilibrium position fixed in space, we speak of oscillators and of standing modes. When the periodic oscillations travel from place to place, we speak of waves and propagation.

Many systems in nature closely approximate what is known as simple harmonic motion – the oscillation of a system about an equilibrium position *ad infinitum*. But notice the words "closely approximate"; simple harmonic motion is not physically realizable because such a motion would have had to begin an infinitely long time ago and would continue for an infinite time into the future. All systems are subject to external influences which perturb the motion in one way or another, causing it to lose (or gain) energy, thereby destroying the simple oscillatory, or harmonic, character of its motion. Why then is simple harmonic motion so important in physics and engineering, and why do we spend time and effort studying it? The answer lies not only in the fact that simple harmonic motion is a good approximation to many physical processes but also in the fact that more complicated processes can be thought of as being made up of several harmonic motions acting independently. This is a powerful concept which we shall have occasion to use as we study more complex systems, such as two oscillators coupled together in some manner, the vibrations of a taut wire given an arbitrary initial distortion, or the propagation of electromagnetic waves in dielectric media, to name just a few.

1.1 THE SIMPLE HARMONIC OSCILLATOR

The primary characteristics of simple harmonic motion are:

The motion is about an equilibrium position at which position no net force acts on the system.

A restoring force, proportional to the displacement from the equilibrium position, acts to restore the system to its equilibrium position.

The motion is periodic, i.e., the motion repeats itself after a time T_0, known as the period.

It may seem surprising but we need no more facts than these to derive and solve the "equation of motion" of a simple harmonic oscillator. Such is the universality of simple harmonic motion!

If some parameter of the system designating its displacement from the equilibrium position is characterized by a time-dependent coordinate $\psi(t)$, then Newton's law tells us that

$$F = Ma = M \frac{d^2\psi(t)}{dt^2}$$

where F is the net force on the system of mass M and a is the acceleration which equals $d^2\psi(t)/dt^2$. But for simple harmonic motion, the net force is the restoring force and so must be given by

$$F = -\beta\psi(t).$$

This equation combines all the properties of the restoring force stated above: it is proportional to the displacement from the equilibrium $\psi(t)$ and the minus sign shows that F acts to "turn back" $\psi(t)$ as it tries to increase away from equilibrium. The coefficient β will depend on the specific problem being done but it will not be a function of either time or $\psi(t)$. Combining our two equations we have

$$M \frac{d^2\psi(t)}{dt^2} = -\beta\psi(t)$$

or

$$\boxed{\frac{d^2\psi(t)}{dt^2} + \omega_o^2\psi(t) = 0} \tag{1.1}$$

where we have written

$$\omega_o^2 = \frac{\beta}{M}.$$

Equation 1.1 is the basic equation of simple harmonic motion; it exhibits the most fundamental of all oscillatory motions and has as one possible solution

$$\psi(t) = \cos(\omega_o t). \tag{1.2}$$

Indeed, this motion can be viewed as the building block out of which ever more complex oscillatory phenomena can be assembled. That Eq. 1.2 is a solution of Eq. 1.1 can be checked by differentiation and direct substitution: $d\psi/dt = -\omega_o \sin \omega_o t$; $d^2\psi/dt^2 = -\omega_o^2 \cos \omega_o t$. Adding to this the second term of Eq. 1.1, that is, $\omega_o^2\psi(t) = \omega_o^2 \cos \omega_o t$, correctly yields the value of zero, and thus balances the right-hand side of this equation.

We shall see that the symbol $\psi(t)$ can represent any number of physical quantities, as, for example, the angular displacement of a swinging pendulum, the twist of a torsion fiber supporting a galvanometer mirror, the current in a tuned oscillator or even the electric or magnetic fields sloshing back and forth in a resonant microwave cavity. Nonetheless, the physical contents of Eqs. 1.1 and 1.2 become

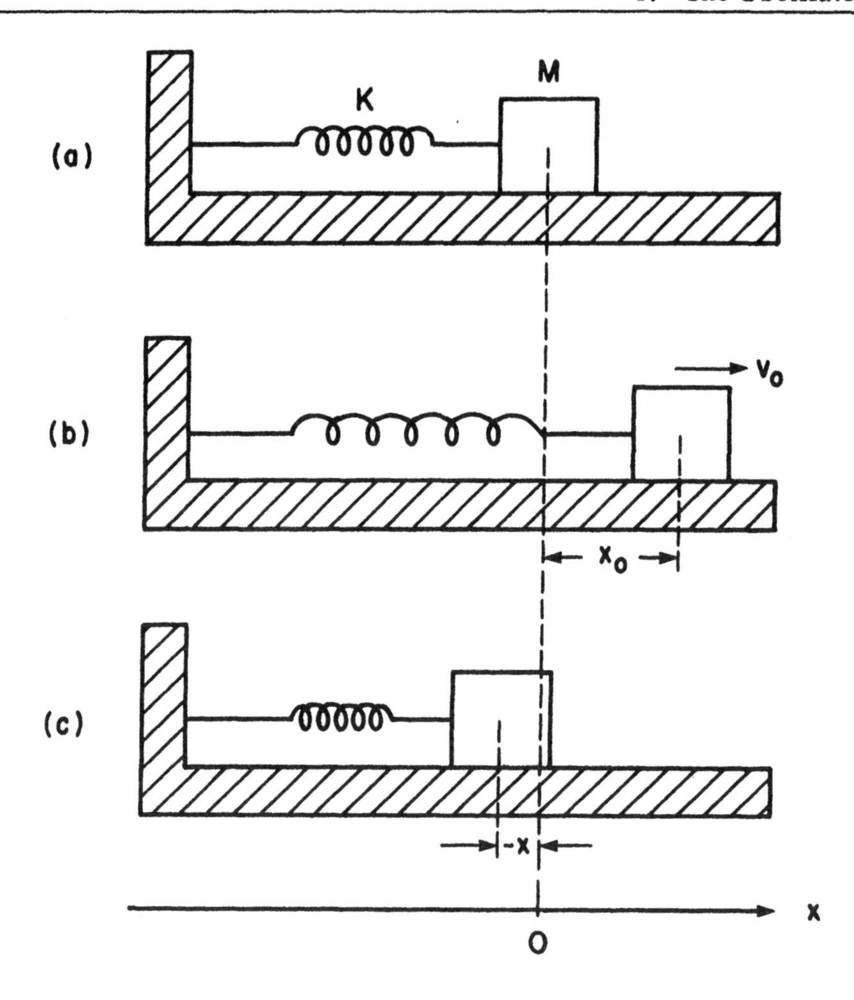

Fig. 1.1 Horizontal oscillations of a mass on a spring. (a) Equilibrium position with spring in a relaxed state; (b) configuration at time $t = 0$ when the mass is displaced distance x_0 and given an initial push with velocity v_0; (c) general configuration at some later time t.

most transparent in the context of some simple mechanical motion; thus, without lack of generality and with no further ado we refer to Fig. 1.1 which is meant to represent an object on a frictionless table, oscillating at the end of a light spring. Here, therefore, $\psi(t)$ represents the displacement $x(t)$ of the mass M with $d^2\psi/dt^2 = d^2x/dt^2$ as its acceleration a. The acceleration is caused by a force which is just the pull (or push) of the spring. Now, experiments show that springs, bending beams, and many other elastic substances, if not excessively abused by too much stretching, bending, and so on, have a restoring force proportional to the amount of displacement. For the spring this means that

$$F(x) = -Kx \tag{1.3}$$

and is a statement of Hooke's law. We note that the force described by the equation is always in a direction toward the origin and its tendency is to restore the initial state of equilibrium. Therefore K, the "spring constant," is of necessity a positive constant.

The equation of motion of the mass at the end of the spring is obtained by substituting Eq. 1.3 into Newton's equation $\vec{F} = M\vec{a}$ with the result that

$$\boxed{M\frac{d^2x}{dt^2} = -Kx} \tag{1.4}$$

or

$$\frac{d^2x}{dt^2} + \omega_o^2 x = 0 \tag{1.5}$$

where, by analogy with Eq. 1.1 we have written that

$$\boxed{\omega_o^2 = \frac{K}{M}.} \tag{1.6}$$

Observe that the quantity ω_o which as we shall see plays a crucial role in describing the characteristics of the oscillations, is uniquely determined by the static properties of the system, in our case the mass M and the spring constant K.

The solution

Our task is to determine the complete time history, past, present, and future, of the mass-spring system. This is what we mean by a solution of the problem. Now, as has already been demonstrated at the beginning of the section, a solution of Eq. 1.5 (or of the equivalent Eq. 1.1) is

$$x(t) = (1 \text{ meter}) \times \cos(\omega_o t), \tag{1.7}$$

a result which possesses the sought-for periodicity of motion: Suppose at some time t_1 the mass is plus x_1 units from the origin, and has reached this point, having traveled, say, from left to right in our Fig. 1.1; then $x_1(t_1) = \cos(\omega_o t_1)$. When at some later times t_2, t_3, etc. the mass again passes the plus x_1 point, it follows that $\cos(\omega_o t_1) = \cos(\omega_o t_2) = \cos(\omega_o t_3)$, etc. For this to happen, and with the proviso that in traversing x_1 the mass once again proceeded from left to right, we must have that $\omega_o t_2 = \omega_o t_1 + 2\pi$, $\omega_o t_3 = \omega_o t_2 + 2\pi$, and so on. Thus the time intervals $(t_2 - t_1)$, $(t_3 - t_2)$, etc. between successive identical displacements are precisely the same and have the value,

$$T_o = \frac{2\pi}{\omega_o} \text{ seconds} \tag{1.8}$$

where T_o is called the period of the motion; ω_o is called the angular frequency and it denotes the object's repetition rate in radians per second. The frequency of vibration or just "frequency," usually denoted by symbols f_o (or ν_o), is related to ω_o through $f_o = \omega_o/2\pi$; its unit is one vibration per second and in modern usage is called the Hertz (Hz). In summary, then, the three quantities that we shall most often encounter are related as follows:

$$\omega_o = 2\pi f_o (2\pi\nu_o) = \frac{2\pi}{T_o} \tag{1.9}$$

ω_o ↑	$2\pi f_o(2\pi\nu_o)$ ↑	$\frac{2\pi}{T_o}$ ↑
angular frequency in radians/sec	frequency in hertz (Hz)	period in seconds

For reasons of brevity, however, we shall often refer to ω_o as the frequency, although it will be understood that "angular frequency" is actually implied.

The frequency range of naturally occurring oscillators is vast indeed. From Eq. 1.6 we see that the frequency is the higher the lower the mass and the larger the force per unit displacement. The earth shaken by an earthquake may well oscillate (for a short time at least) at frequencies of 1 Hz or lower. The ammonia molecule, which is the basic ingredient of certain very precise clocks, has its lowest frequency of oscillation at approximately 2×10^{10} Hz; and an atomic electron bound to the nucleus by the electric coulomb force exhibits oscillations at frequencies that are typically 10^{15} Hz and greater. Many additional examples could well be given.

Whether the frequency is high or low, there is one important property shared by all simple harmonic oscillators: the frequency is independent of the size of the excursions. Galileo discovered this fact in 1581 by watching the swinging of a lamp in the cathedral in Pisa. Indeed, when we look at Eqs. 1.2 or 1.7, the maximum distance of excursion (the so-called amplitude of the motion) enters nowhere in our expressions. This is a little disconcerting and one may wonder what is wrong. The answer is simply this: whereas Eq. 1.7 is a solution of the differential equation (1.5) it is an incomplete solution. One can readily check by direct substitution that $x_1(t) = A\cos(\omega_0 t)$ where A is a constant, is also a solution. But are we finished, now that we have incorporated the requisite amplitude A in our result? The answer is no, because $x_2(t) = B\sin(\omega_0 t)$ likewise proves to be a perfectly satisfactory result. The dilemma caused by this apparent nonuniqueness of the results is fully resolved in the theory of differential equations where we find that for any linear homogeneous second-order equation, of which Eqs. 1.1 and 1.7 are special examples, the general and fully unique solution is given by the sum of both solutions, $x(t) = x_1(t) + x_2(t)$, or written out in full,

$$\boxed{x(t) = A\cos(\omega_0 t) + B\sin(\omega_0 t)} \tag{1.10}$$

with A and B as two (and no more or less than two) arbitrary coefficients. This principle of superposition obeyed by all linear differential equations is a statement of the fact that the displacement produced by two disturbances acting together is just the addition of the two individual events acting on the system separately.* The continued application of this most fundamental of principles so permeates much of our thinking that we are likely to forget that in the real physical world the principle of superposition is sometimes no more than a first good approximation to reality; and that when the medium being considered is stressed too much by too violent jiggling (e. g., Hooke's law fails when $x(t)$ is too large), the simple addition of disturbances can become inapplicable. Despite these reservations, we shall confine ourselves in this book almost exclusively to systems that obey the superposition principle — we call them linear systems. There are two reasons for this. First, we know how to solve them and we don't know how to go about solving the majority of nonlinear problems. Second, the laws of electricity and magnetism, in regions of empty space at least, can always be described by linear differential equations.

The so-called arbitrary coefficients A and B appearing in Eq. 1.10 are not arbitrary in the sense that they serve to nail down the initial state of the oscillating system. For example, one may wish to specify at time $t = 0$ the initial position and the initial velocity of the oscillating mass. If we denote the former by x_o and the latter by v_o (see Figs. 1.1 and 1.2), then Eq. 1.10 gives: $x(t=0) = x_o = A\cos(0) + B\sin(0)$ which yields the result that $A = x_o$. Similarly, we have for the velocity, $v = dx/dt = -A\omega_o \sin(\omega_o t) + B\omega_o \cos(\omega_o t)$. Thus, setting $v = v_o$ at $t = 0$ gives $B = v_o/\omega$. Inserting these values in Eq. 1.10 gives

$$x(t) = x_o \cos(\omega_o t) + \frac{v_o}{\omega_o} \sin(\omega_o t) \tag{1.11}$$

which fully determines the position of the mass at all times, $t \geq 0$.

Equation 1.10 may be cast into the form

$$\boxed{x(t) = C\cos(\omega_o t + \phi)} \tag{1.12}$$

*See Appendix 1 for the mathematics of superposition.

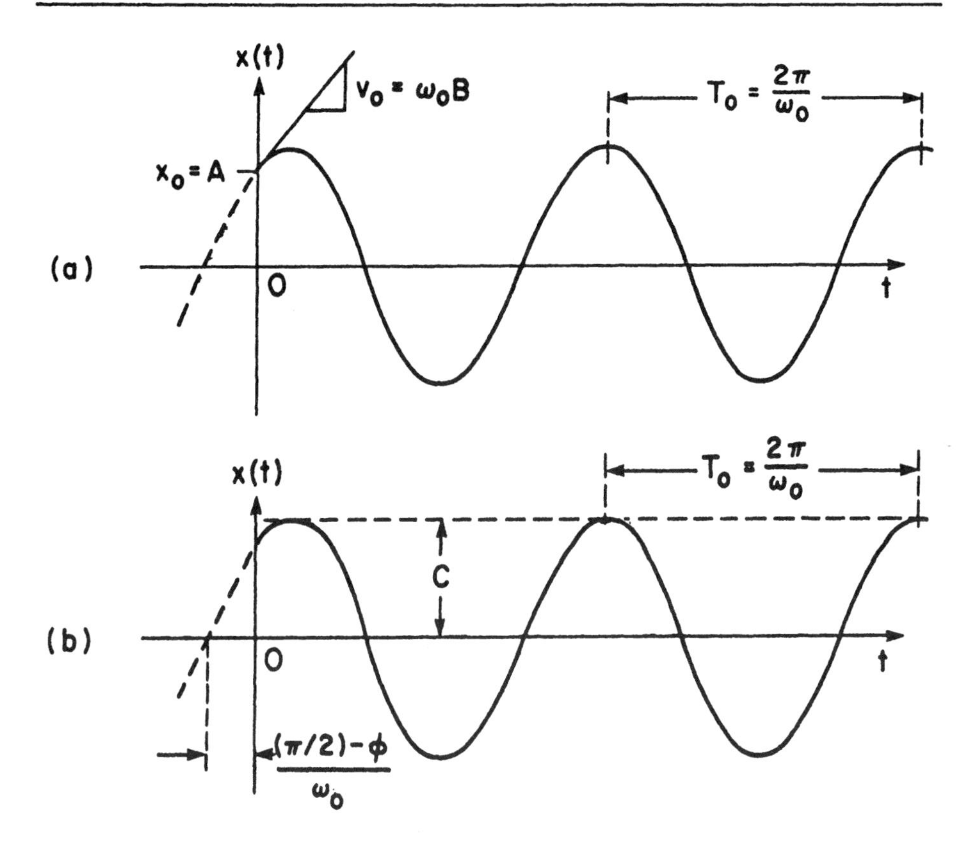

Fig. 1.2 Displacement as a function of time of a simple harmonic oscillator. In (a) are specified the initial conditions $x = x_0$, $v = v_0$ at $t = 0$ as per Eq. 1.11. In (b) are specified the amplitude C and the phase ϕ as per Eq. 1.12. The constants A, B, C, and ϕ are connected through Eq. 1.13. Note that the initial displacement $x_0 = A$ is not the amplitude C of oscillation.

which often proves to be more convenient for purposes of computations. We see once again that the equation contains two arbitrary coefficients, C and ϕ, as it must in order to be a complete solution of our second-order differential equation (1.5). In order to make contact between Eqs. 1.12 and 1.10 we first expand the cosine of the latter through use of the identity $C \cos(\omega_0 t + \phi) = C [\cos(\omega_0 t) \cos(\phi) - \sin(\omega_0 t) \sin(\phi)]$, and then compare this result with Eq. 1.10. Since the two equations are required to apply at all times it follows that

$A = C \cos\phi$ and $B = -C \sin\phi$ with the result that

$$C = \sqrt{A^2 + B^2}$$
$$\phi = \tan^{-1}\left(-\frac{B}{A}\right). \tag{1.13}$$

Figure 1.2 shows that while C prescribes the amplitude, ϕ specifies the initial phase angle of the motion and thus determines our choice of the zero of time. Often (but not always!) we are not especially interested in ϕ and in such cases we can conveniently reset our clock so as to make $\phi = 0$.

Yet another form for x(t) is its representation in the complex plane. We now know that $x(t) = C\cos(\omega_0 t + \phi)$ is a solution of Eq. 1.5. An equally good solution is $j\,C\sin(\omega_0 t + \phi)$ where $j = \sqrt{-1}$. Thus, by the principle of superposition

$$x(t) = C\cos(\omega_0 t + \phi) + j\,C\sin(\omega_0 t + \phi) \tag{1.14}$$

is likewise a solution. We invoke the mathematical identity

$$\cos\theta + j\sin\theta = e^{j\theta} \tag{1.15}$$

due to Euler, and obtain the important result

$$\boxed{x(t) = C\,e^{j(\omega_0 t + \phi)}} \tag{1.16}$$

Much use will be made of this complex representation of oscillatory phenomena.* Its utility lies in the fact that exponentials are more readily manipulated than sines or cosines. However, when we come to the end of a given problem, we must remember that only the real part of the right-hand side of Eq. 1.16 is physically significant and that it is this real part which is the true representation of the oscillator's displacement x(t).

*See Appendix 2 which discusses some of the properties of complex quantities.

Energy

At any one time of the oscillatory cycle, the energy in the spring-mass system is contained in part in the form of kinetic energy $(1/2)\,Mv^2$ of the mass, and in part in the form of potential energy $(1/2)\,Kx^2$ of the spring. From Eq. 1.12 it follows that the kinetic energy is

$$\left.\begin{aligned} K \cdot E &= \frac{1}{2} M \left(\frac{dx}{dt}\right)^2 \\ &= \frac{1}{2} M\omega_o^2 C^2 \sin^2(\omega_o t + \phi) \\ &= \frac{1}{2} KC^2 \sin^2(\omega_o t + \phi) \end{aligned}\right\} \tag{1.17}$$

where the third form of the equation is obtained from the second with the aid of the result that $\omega_o^2 = K/M$. The potential energy is

$$\begin{aligned} P \cdot E &= \frac{1}{2} Kx^2 \\ &= \frac{1}{2} KC^2 \cos^2(\omega_o t + \phi). \end{aligned} \tag{1.18}$$

Thus we see that as time progresses the energy flows back and forth between kinetic and potential. Suppose the phase $\phi = \frac{\pi}{2}$; then at zero time all the energy resides in the kinetic energy of motion. A quarter cycle later it is all in the form of potential energy, and so on. The total energy, however,

$$\begin{aligned} U &= K \cdot E + P \cdot E \\ &= \frac{1}{2} KC^2 \\ &= \frac{1}{2} M\omega_o^2 C^2 \end{aligned} \tag{1.19}$$

is independent of time in accord with our expectations, since we have considered a system with no frictional loss. Note that the energy in the oscillation is proportional to the square of its amplitude C — a result that is of considerable generality.

Often, when the oscillations of a system are too rapid it is not possible to detect the instantaneous values associated with the motion, but only certain time-averaged values. For example, one may well ask what the oscillatory kinetic and oscillatory potential energy is averaged over a period of oscillation T. To perform such a computation we invoke the fact that the time average of any quantity f(t) is given by

$$\langle f(t) \rangle = \frac{1}{T} \int_{t}^{t+T} f(t)\, dt \tag{1.20}$$

where the angular brackets $\langle\ \rangle$ signify time averaging. Now substituting Eq. 1.17 in Eq. 1.20 we obtain for the time-averaged kinetic energy

$$\begin{aligned} \langle K \cdot E \rangle &= \frac{1}{2} KC^2 \frac{1}{T} \int_{t}^{t+T} \sin^2(\omega_o t + \phi)\, dt \\ &= \frac{1}{4} KC^2 \end{aligned} \tag{1.21}$$

where the last result follows from evaluating the integral and finding its value to be T/2. A similar calculation for the potential energy yields

$$\langle P \cdot E \rangle = \frac{1}{4} KC^2, \tag{1.22}$$

showing that the energy is shared equally between kinetic and potential. The total energy adds up once again to the value of $U = (1/2)\, KC^2$.

<u>Example</u>

A massless spring of spring constant K has an unstretched length ℓ_o equal to 15 cm. When a mass M, equal to 1.5 kg, is suspended from the spring, under the influence of gravity, the spring is extended an additional distance ℓ_e of 10 cm. If the mass is now displaced 2 cm downward from its equilibrium position and released with an upward velocity of 5 cm/sec, find the following quantities:

a) The frequency of oscillation, ω_o.
b) An equation describing its displacement from equilibrium as a function of time.
c) The velocity of the mass at zero displacement, i.e., the equilibrium position.
d) The total energy of the system.
e) The time average potential energy.

Solution

Static equilibrium is obtained when the upward force exerted by the spring, $K\ell_e$, balances the gravitational force, Mg. Thus

$$K\ell_e = Mg$$

$$K = \frac{Mg}{\ell_e}.$$

The frequency of oscillation ω_o is given by Eq. 1.5, hence

a) $$\omega_o = \sqrt{\frac{K}{M}} = \sqrt{\frac{g}{\ell_e}} = \sqrt{\frac{9.8}{0.1}} = 9.9 \text{ rad/sec.}$$

The equation describing the general motion of the mass M is Eq. 1.10. The coefficients A and B have been evaluated in terms of the initial displacement and velocity in Eq. 1.11. Thus we have

$$x_o = 2 \text{ cm} \qquad v_o = -5 \text{ cm/sec}$$

where we have taken the positive x direction downward.

b) $$x(t) = 2\cos(9.9\,t) - \frac{5}{9.9}\sin(9.9\,t) \text{ cm.}$$

An equivalent way of writing this result is

$$x(t) = 2.06\cos(9.9\,t + 0.248) \text{ cm}$$

where Eqs. 1.12 and 1.13 have been used.

The velocity of M is given by

$$v = \frac{dx}{dt} = -(2.06)(9.9)\sin(9.9\,t + 0.248).$$

When the mass M is at the equilibrium position (zero displacement) we have

$$9.9\,t + 0.248 = \pi/2$$

(or any odd multiple of $\pi/2$). Thus the velocity at this position is

c) $$v = \pm 20.4 \text{ cm/sec.}$$

The plus or minus sign is to be chosen depending on whether M is moving upward (-) or downward (+) as it passes through zero. The speed is the same in either case.

The total energy U is the sum of the kinetic energy KE and the potential energy PE. If we adopt the zero of energy as the energy of mass M at rest at the equilibrium position, then

$$KE = \frac{1}{2} Mv^2$$

$$PE = \frac{1}{2} K(\ell_e + x)^2 - Mgx - \frac{1}{2} K\ell_e^2 = \frac{1}{2} Kx^2.$$

The total energy can be evaluated in several ways: (1) at $x = 0$ where all the energy is kinetic, (2) at $x = x_0$ where all the energy is potential because the velocity is zero, or (3) at an arbitrary x. We shall adopt the last possibility.

$$KE = \frac{1}{2}(1.5)(0.204)^2 \sin^2(9.9\,t + 0.248) \text{ joules}$$

$$PE = \frac{1}{2}\,\frac{(1.5)(9.8)}{0.1}\,(0.0206)^2 \cos^2(9.9\,t + 0.248) \text{ joules}$$

d) $$U = KE + PE = 0.0312 \text{ joules.}$$

The time-dependent potential energy is as given above

$$PE = 0.0312 \cos^2(9.9\,t + 0.248) \text{ joules.}$$

Using Eq. 1.20 we find the time-average potential energy

$$\langle PE \rangle = \frac{9.9}{2\pi}(0.0312)\int_0^{2\pi/9.9} \cos^2(9.9\,t + 0.248)\,dt$$

e) $\langle PE \rangle = (0.0312)\left(\frac{1}{2}\right) = 0.0156$ joules.

Phase-space trajectory

Equation 1.12 tells us that the position x(t) and the momentum $p(t) = M(dx/dt)$ of the oscillating mass at any instant of time are

$$x = C\cos(\omega_0 t + \phi)$$
$$p = -M\omega_0 C\sin(\omega_0 t + \phi). \tag{1.23}$$

Sometimes the explicit dependence on time is not of primary interest. Rather, one would like to know, for a given position x, the velocity or momentum. To determine the value of p corresponding to a given value of x, we eliminate time between the two foregoing expressions by squaring them and then using trigonometric relation $\cos^2\theta + \sin^2\theta = 1$. The result is

$$\frac{x^2}{C^2} + \frac{p^2}{M^2\omega_0^2C^2} = 1. \tag{1.24}$$

Since the total energy U is connected to C through $U = (1/2)\,M\omega_0^2C^2$ (see Eq. 1.19) it follows that

$$\frac{x^2}{\left(2U/M\omega_0^2\right)} + \frac{p^2}{(2MU)} = 1. \tag{1.25}$$

This is the equation of an ellipse with a major semi-axis $a = \sqrt{2U/M\omega_0^2}$ and a minor semi-axis $b = \sqrt{2MU}$. A plot of p as a function of x is known as the particle's phase-space diagram and is illustrated in Fig. 1.3a. The arrows indicate the direction of the particle's trajectory. We note that as x increases, the momentum and thus the velocity, decrease, which is just as it should be for the case of a freely oscillating body. When one changes the initial condition of the oscillation, namely U, another ellipse results with different sized major and minor axes (see Fig. 1.3b). The ellipses never

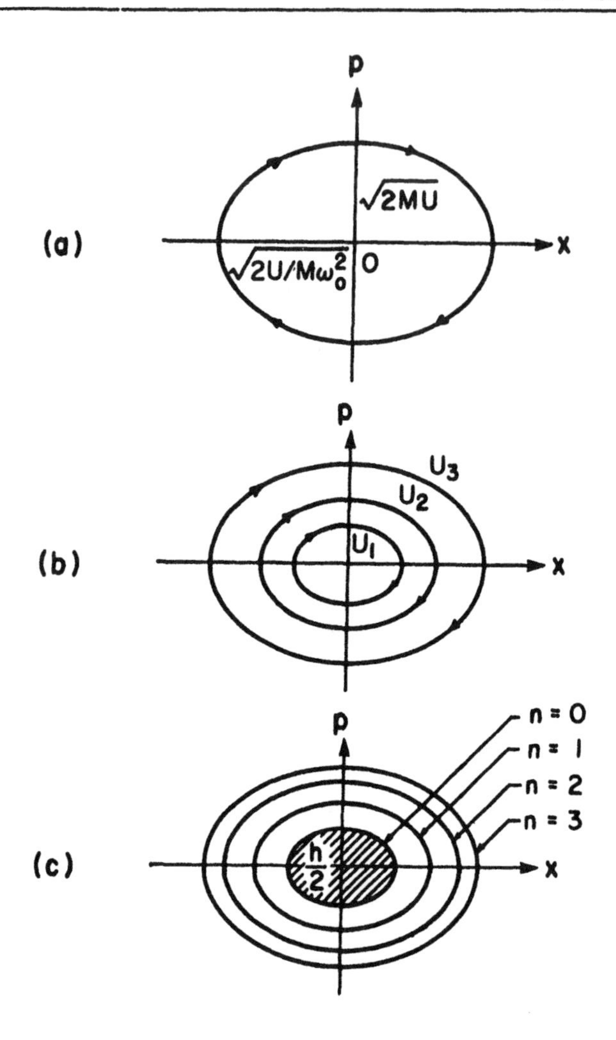

Fig. 1.3 Phase-space trajectories of a simple harmonic oscillator. (a) An oscillator with an energy U; (b) an oscillator with three different initial conditions U_1, U_2, U_3; (c) lowest four states of a quantum-mechanical harmonic oscillator.

intersect; if they did, the oscillator would have the possibility of choosing one of two or more paths and its motion would not be unique. (Do not confuse paths in the p-x plane with the object's path in real space.)

Now, the area of an ellipse with axes of lengths a and b is given by πab. Therefore the phase-space area A(U) for an oscillator of energy U is

$$\left.\begin{aligned} A(U) &= \int p\,dx = \pi ab \\ &= \frac{2\pi U}{\omega_o} \\ &= \frac{U}{\nu_o} \end{aligned}\right\} \qquad (1.26)$$

where ν_o is the natural frequency of oscillation in hertz. As the energy of the oscillator increases, the area enclosed by its phase-space trajectory increases. The p-x plane can thus be covered with a continuous distribution of ellipses, infinite in number, since U can be varied continuously through all possible values. We may mention that according to quantum mechanics U cannot in fact be continuous but has discrete values given by

$$U_n = \left(n + \frac{1}{2}\right) h\nu_o \qquad (n = 0, 1, 2, 3, \ldots.) \qquad (1.27)$$

with h as Planck's constant. This signifies that out of all the classically possible ellipses only certain discrete ones are allowed. These discrete "quantized ellipses" representing the quantum states of the harmonic oscillator have areas given by

$$\left.\begin{aligned} A(U_n) &= \frac{U_n}{\nu_o} \\ &= \left(n + \frac{1}{2}\right) h \\ &= \frac{1}{2}h, \frac{3}{2}h, \frac{5}{2}h, \ldots \end{aligned}\right\} \qquad (1.28)$$

They are illustrated in Fig. 1.3c. Of course, for very large values of n, that is, for highly excited oscillators of large energy (and long oscillatory periods) the adjacent ellipses are so closely spaced that the classical picture is well approximated.

1.2 OSCILLATORS AND SOME OF THEIR USES

The basic mass-spring system discussed in the previous section serves as a prototype for the understanding of a variety of natural and man-made simple harmonic oscillators. A few of these, together with their applications will now be described.

Mechanical oscillations – the pendulum

The pendulum, whether it is one from a grandfather clock, or a high precision instrument used in geodesy, consists of a heavy, rigid mass able to swing freely about a knife edge acting as the pivot (see Fig. 1.4). Its equation of motion is readily derived. Suppose 0 is the horizontal axis perpendicular to the plane of the paper about which the pendulum swings back and forth; its mass is M and its center of gravity is located at C. At some instant of time t, C has an angular displacement θ relative to the vertical. At this instant, the torque about the axis through 0 equals $-Mgh \sin\theta$ where h is the distance between 0 and C. The torque produces in the body an angular acceleration given by $d^2\theta/dt^2$, and if I_0 is the body's moment of inertia about 0, the resultant equation of motion is

$$I_0 \frac{d^2\theta}{dt^2} = -Mgh \sin\theta. \tag{1.29}$$

Note that the angle θ is in radians. If this angle is sufficiently small so that the Taylor's expansion, $\sin\theta = \theta - \theta^3/3! + \theta^5/5! - \ldots.$ can be cut off after the first term, the *nonlinear* equation (1.29) reduces to the by now familiar, linear homogeneous equation

$$\frac{d^2\theta}{dt^2} + \frac{Mgh}{I_0}\theta = 0. \tag{1.30}$$

Then, by comparison with Eqs. 1.5 and 1.12 it has the solution

$$\theta = \theta_0 \cos(\omega_0 t + \phi) \tag{1.31}$$

where θ_0 is the amplitude and where

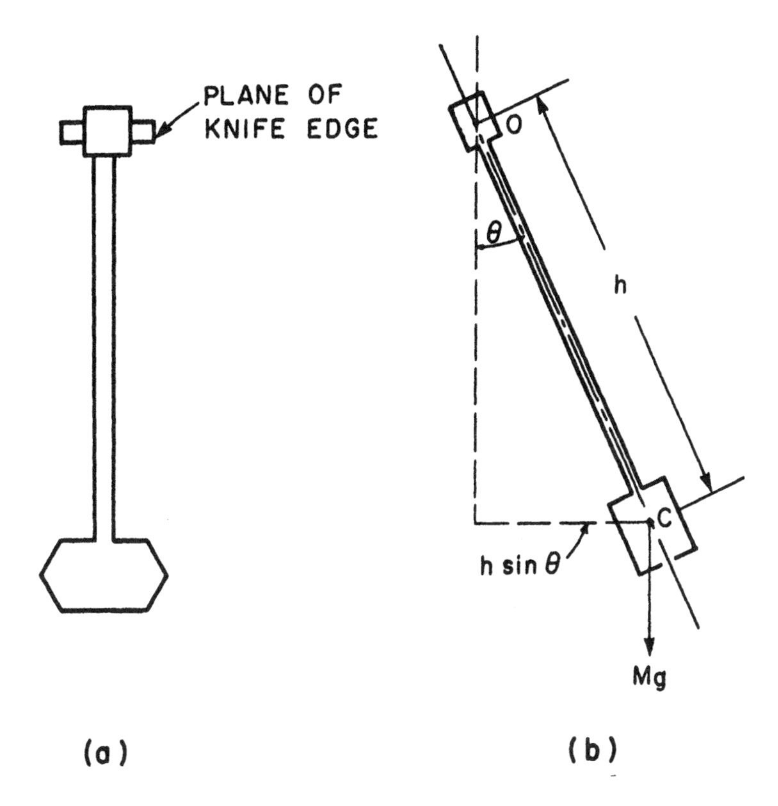

Fig. 1.4 (a) Typical shape of precision pendulum often made from invar because of low thermal expansion of this alloy of iron and nickel. (b) Parameters that enter into pendulum's motion. 0 is the point of suspension, C the center of gravity. The pendulum like that shown in (a) was used in an accurate determination of g. [See text and J. E. Jackson, Geophysical Journal 4, 375 (1961).]

$$\omega_0 = \sqrt{\frac{Mgh}{I_0}} \tag{1.32}$$

is the angular frequency. The period of oscillation is given by

$$T = 2\pi \sqrt{\frac{I_0}{Mgh}}. \tag{1.33}$$

If the pendulum were an ideal point mass supported by a weightless rod or string of length $h = \ell$, the moment of inertia I_0 would simply be $M\ell^2$ and the period would take on the value

$$T = 2\pi \sqrt{\frac{\ell}{g}}. \tag{1.34}$$

Such a pendulum of length, say, equal to 25 cm has a period of approximately 1 second ($g \approx 9.81$ m/sec^2). The motion being simple harmonic, the period is independent of amplitude θ_0, a fact that was stressed already in section 1.1. However, the exact nonlinear equation (1.29) does not have this property and T must, in fact, depend somewhat on θ_0. We can find out how much since Eq. 1.29 is one of the few nonlinear equations that can be solved exactly. The answer is

$$T(\text{exact}) = 4\sqrt{\frac{I_0}{Mgh}}\, F(\theta_0) \tag{1.35}$$

$$= 2\pi \sqrt{\frac{I_0}{Mgh}} \left[1 + \frac{1}{4}\sin^2\left(\frac{\theta_0}{2}\right) + \frac{9}{64}\sin^4\left(\frac{\theta_0}{2}\right) + \ldots\right]$$

where $F(\theta_0)$ is known as the "complete elliptic integral of the first kind"; the second line of the equation is a power series expansion of this function. From this we can determine the error in timing caused by finite amplitude: suppose $\theta = 0.1$ radian ($\approx 6°$ of arc), then $T(\text{exact}) \approx 2\pi[I_0/Mgh]^{1/2}\,[1 + (1/1600) + \ldots]$. The approximate result (1.33) gives the slightly shorter period, $T = 2\pi[I_0/Mgh]^{1/2}$. The error in using the latter is therefore approximately 0.05 percent.

Until some two decades ago the pendulum was the only precise way of determining g absolutely (in 1952 Volet was the first to make a modern, accurate determination of g by timing a freely falling body; since that time the free-fall technique has surpassed the pendulum method). The basis of the pendulum determination is the accurate measurement of its period followed by application of Eq. 1.33 or some modification thereof. Since neither I_0 nor h can be measured with precision, this expression is not suitable for the calculation of g as it stands. Kater (1818) showed how this difficulty can be circumvented, but we shall not pursue the details of this clever idea here. Let us just say that with extreme care, where one may go as far as swinging the pendulum in an evacuated vessel to eliminate air drag and viscous air resistance, compensating for temperature variations and bending of the pendulum etc., an absolute measurement of g to approximately 1 part in a million can be achieved. The best accepted value of g reduced to what it would be on the equator (at sea level) is

$$\begin{aligned} g &= 9.78049 \text{ m/sec}^2 \\ &= 978049 \text{ milligalileo} \end{aligned} \qquad (1.36)$$

where the milligalileo (or mgal) is the unit commonly used in geophysics (1 mgal = 10^{-5} m/sec^2). Because of the difficulties not many absolute measurements of g have been made at different places of the earth's surface. For that reason relative measurements of g at various stations are of great importance since they can be used to fill in gaps where absolute measurements are lacking. The pendulum method of obtaining relative measurements of g is, in principle, very simple. Referring to Eq. 1.33 we see that if the same pendulum is swung, under identical conditions at two places where the values of gravity are g_1 and g_2, the corresponding periods T_1 and T_2 are related by the equation

$$\frac{g_2}{g_1} = \frac{T_1^2}{T_2^2}. \qquad (1.37)$$

In this way gravity maps of the entire earth's surface can be made. An important application of such gravity surveying has been in the search for minerals, oil and gas. For example, the last two often occur in sedimentary basins and tend to accumulate in porous formations near structural deformations, such as salt domes in the Gulf Coast of the U.S.A. Salt with a density of approximately 2.2 g/cm^3 is normally less dense than other sedimentary rock; thus a measurement of g along the earth's surface above such a dome manifests itself as an "anomaly" in the value of g; that is, g departs from the normal average value for the terrain in question. Figure 1.5 illustrates the profile of g across a salt dome found in Russia. The variation in g is seen to compare favorably with the profile obtained by means of seismic measurements. It is interesting to note that theoretical modeling of the dome (dashed line) based on the gravity

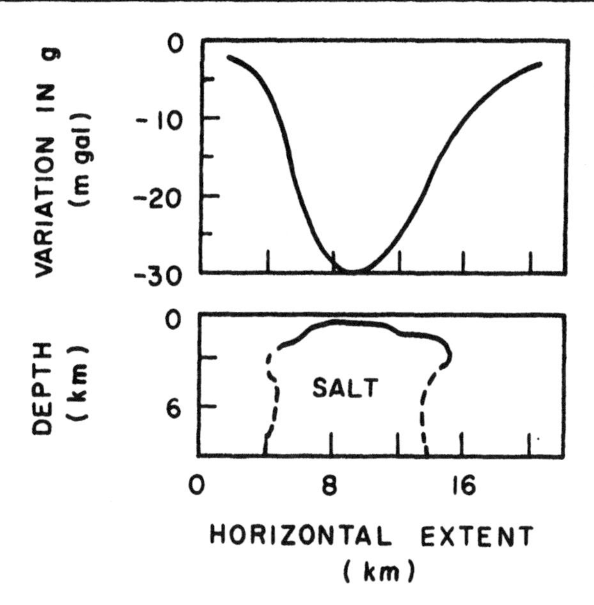

Fig. 1.5 Variation of g across a large salt dome is shown in upper curve. The solid line in the bottom curve shows profile of dome obtained from seismic measurements. The dashed part was determined from theoretical modeling based on g measurements. [After The Earth's Shape and Gravity by G. D. Garland (Pergamon Press, 1965); see also Gravity and the Earth by A. H. Cook (Wykeham Publications, London, 1969); and Elementary Gravity and Magnetics by L. L. Nettleton, Monograph Series No. 1, 1971 Soc. Exploration Geophysicists.]

measurements also tells us a great deal about the dome's profile in the vertical direction (i.e., depth). The salt dome depicted in Fig. 1.5 is an exceptionally large one. Even then the maximum variation in g is only 30 mgal. Domes on the Gulf Coast are much smaller and the typical changes in g are approximately 5 mgal. Thus we see that a gravity meter needs to measure variations of g to better than 0.1 mgal, which is only one part in ten million of the total gravitational field in which the measurements are made.

Sound oscillations – the Helmholtz resonator

In the course of analyzing musical sounds, Helmholtz made use of air resonators. They were made from metal or glass and had two apertures, a small one that was stuck in one's ear and a wider one,

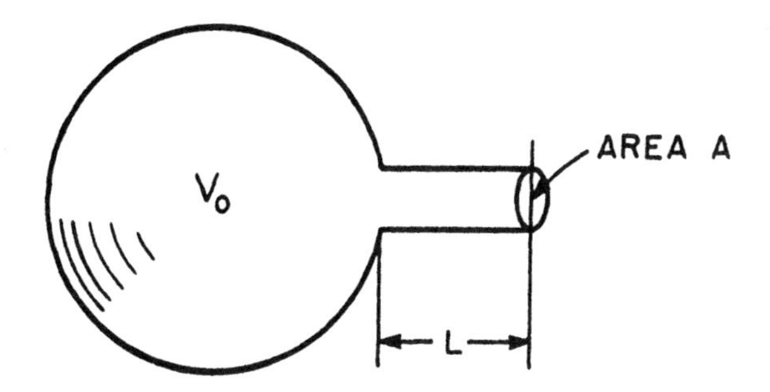

Fig. 1.6 Schematic of a Helmholtz resonator.

which was held in the direction of the sound. Whenever the sound contained a frequency equal to the natural frequency of oscillation of the resonator, the latter responded and this response was detected by the ear. An analogous effect is obtained by blowing gently across the opening of a bottle, causing it to resonate.

Consider then a simplified version of a resonator shown in Fig. 1.6. Let V_o be its volume, A the cross-sectional area of its neck (assumed to be uniform) and L the length of its neck. When the opening to the bottle is excited by blowing across it, or otherwise, the motion is confined mainly to the air in the neck which acts as a piston causing compressions and rarefactions of the air in the volume V_o. Thus if ρ_o is the density of the (unperturbed) air, the mass times the acceleration of the air piston is $AL\rho_o(d^2x/dt^2)$ where x is its displacement from equilibrium.

Now the return force F_x is due to the pressure differential dp = $(p-p_o)$ with p as the pressure in V_o and p_o the equilibrium pressure. Since pressure is defined as the force per unit area,

$$\begin{aligned} F_x &= A\,dp \\ &= A\left(\frac{dp}{dV}\right)_o Ax \end{aligned} \tag{1.38}$$

where the subscript zero signifies that the differential dp/dV is to be evaluated around the equilibrium volume. Observe, by comparison with Eq. 1.3, that the quantity $A^2(dp/dV)_o$ plays the same role in sound oscillations as the spring constant K in the vibrations of the mass-spring system. Indeed, the parameter

$$\kappa \equiv -\left(\frac{dp}{dV/V}\right)_o = -\left(V\frac{dp}{dV}\right)_o \tag{1.39}$$

is a measure of the stress-to-strain relationship of material in bulk and is called its bulk modulus. Stated in words, it says that the force exerted per unit area is proportional to the fractional change in the material's volume, with κ as the constant of proportionality. Returning now to Eq. 1.38 we see that we still have to evaluate (dp/dV), the rate at which the gas pressure varies with volume. To this end we assume that air behaves pretty much as an ideal gas and that the compressions and rarefactions are so rapid that there is insufficient time for heat to be able to flow from the former to the latter. Such adiabatic behavior is known to be well satisfied by virtually all sound vibrations with the result that

$$pV^{\gamma} = \text{constant}, \tag{1.40}$$

which is the appropriate "equation of state" of these systems. Here γ is the ratio of the specific heat at constant pressure to the specific heat at constant volume. The differentiation of p with respect to V is now readily carried out with the result that $(dp/dV)_o = -\gamma p_o/V_o$. The restoring force of Eq. 1.38 then becomes

$$F_x = -\left(\gamma\frac{A^2 p_o}{V_o}\right)x. \tag{1.41}$$

Recall that p_o is the equilibrium pressure in the resonator. The application of F = ma yields the sought-for equation of motion

$$AL\rho_o\frac{d^2x}{dt^2} = -\left(\gamma\frac{A^2 p_o}{V_o}\right)x$$

from which we see that the air in the neck executes simple harmonic oscillations giving, for the frequency of the Helmholtz resonator, the value

$$\omega_o = \sqrt{\left(\gamma \frac{p_o}{\rho_o}\right)\left(\frac{A}{V_o L}\right)}. \tag{1.42}$$

The combination of terms $(\gamma p_o/\rho_o)^{1/2}$ turns out to be the expression for the speed of sound v. We can easily find its value; at standard temperature and pressure we have for air $\gamma = 1.40$, $\rho_o = 1.29\ \text{kg/m}^3$ and p_o (1 atmosphere) $= 1.013 \times 10^5\ \text{newtons/m}^2$ with the result that $v = 332$ m/sec. Hence Eq. 1.42 becomes

$$\begin{aligned} \omega_o &= v\sqrt{\frac{A}{V_o L}} \\ &= 332\sqrt{\frac{A}{V_o L}}\ \text{rad/sec.} \end{aligned} \tag{1.43}$$

Take a one-liter wine bottle having a 5 cm long neck of radius 1 cm; then in accordance with Eq. 1.43 its lowest resonant frequency $\nu_o = \omega_o/2\pi = 133$ Hz. We point out that expression (1.43) is not fully correct since the motion of the air is not confined exactly to the length of the neck only, but extends beyond it. Certain "end corrections" require that L of Eq. 1.43 be replaced by L (effective) where

$$L(\text{effective}) \approx L + 1.2\ r \tag{1.44}$$

with r as the neck radius.

Electrical oscillations – the LC circuit

The most basic of all electrical resonators is the circuit of Fig. 1.7 comprising a capacitance C and the inductance L. Suppose the capacitor is charged up to some value and at a time $t = 0$ the switch S is closed, thus completing the electrical circuit. Then, at a later time t, the charge on the capacitor will have the value Q(t), the current flowing in the circuit will be I(t), and the potential difference (i.e., the voltage) developed across the capacitor will be V(t). Taking V(t) to be positive when the upper capacitor plate is positively charged and taking the current flow at that instant to be in the direction shown, the

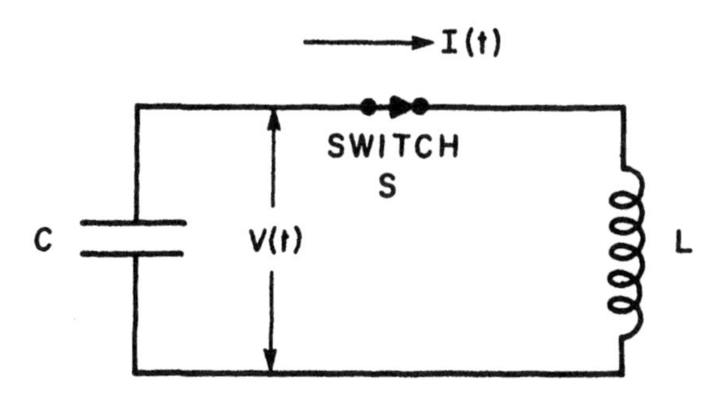

Fig. 1.7 An LC circuit.

relations connecting Q, I, and V are given by

$$I = -\frac{dQ}{dt} \text{ amp} \tag{1.45}$$

and

$$Q = CV \text{ coulomb} \tag{1.46}$$

so that

$$I = -C\frac{dV}{dt}. \tag{1.47}$$

Now, the current flow through the inductance causes a potential difference to develop across its terminals in accordance with the relation

$$V = L\frac{dI}{dt} \text{ volt} \tag{1.48}$$

where V must be the <u>same</u> voltage difference that develops across the capacity C. Equations 1.47 and 1.48 are first-order differential equations containing the two variables V and I in mixed form. To eliminate, say, V we proceed as follows: differentiate Eq. 1.48 with respect to time with the result that $dV/dt = L\, d^2I/dt^2$ and then substitute for dV/dt from Eq. 1.47. The outcome is that

$$\boxed{\frac{d^2I}{dt^2} + \frac{1}{LC} I = 0} \tag{1.49}$$

showing that the current varies sinusoidally as

$$I = I_o \cos(\omega_o t + \phi) \tag{1.50}$$

and oscillates at the frequency

$$\omega_o = \sqrt{\frac{1}{LC}} \text{ rad/sec.} \tag{1.51}$$

The voltage therefore varies as

$$V = -L\omega_o I_o \sin(\omega_o t + \phi) \tag{1.52}$$

a result that follows after substituting Eq. 1.50 in Eq. 1.48. From this we see that the voltage and current are out of phase by 90°.

The oscillations involve a continuous flow of energy back and forth between the capacitor and the inductance, or from the electric field in C to the magnetic field in L. Suppose at some instance in time I is zero. Then the magnetic energy $U_M = (1/2)\,LI^2$ is zero, whereas the electrical energy $U_E = (1/2)\,CV^2$ is at its peak value. A quarter cycle later the capacitor is discharged and all the energy that was contained in it now resides in the magnetic field of the coil. The instantaneous power flowing in the circuit is given by

$$\left.\begin{aligned} P &= VI \\ &= -L\omega_o I_o^2 \cos(\omega_o t + \phi)\sin(\omega_o t + \phi) \\ &= -\frac{1}{2} L\omega_o I_o^2 \sin(2\omega_o t + 2\phi) \end{aligned}\right\} \tag{1.53}$$

The structure of this equation reminds one of the power dissipated in a resistance, $P = I^2R$. Indeed, the quantity $L\omega_o$ has the dimensions of ohms and is known as the impedance of the inductor. However, in contradistinction to the power flow in a resistor, here there is no energy loss from the circuit. We see this right away by calculating the rate of energy dissipation $\langle P \rangle$ averaged over a period of the oscillation: we insert Eq. 1.53 in Eq. 1.20 and obtain $\langle P \rangle = 0$.

There are innumerable ways in which the basic L-C circuit is

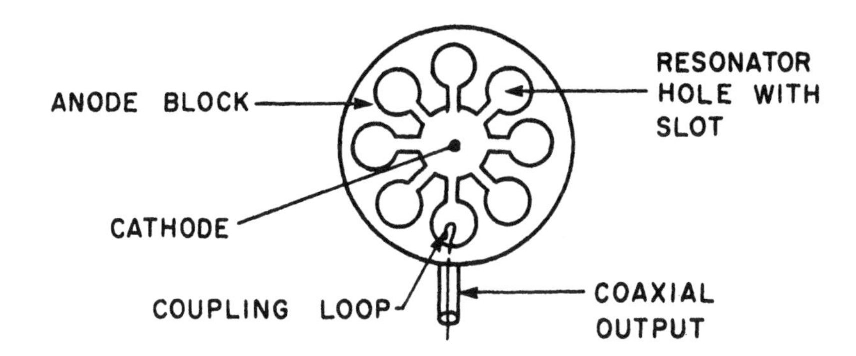

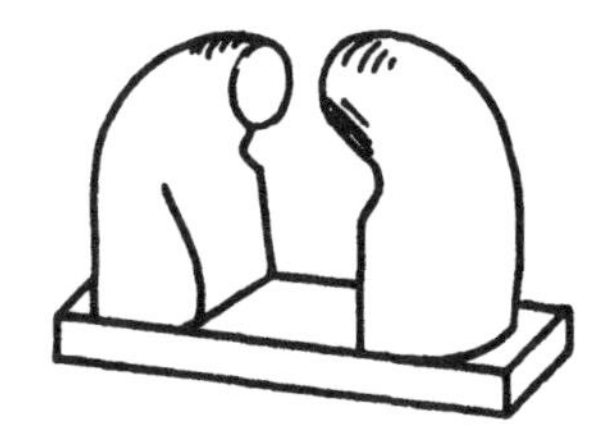

Fig. 1.8 A famous type of magnetron developed during World War II and the horseshoe magnet used with it. Oscillation frequency is approximately 3 GHz (3×10^9 Hz). [For details of this and other magnetrons see Principles of Radar by J. F. Reintjes and G. T. Coate (McGraw Hill, New York, 1952); also Microwave Magnetrons by G. B. Collins (McGraw Hill, New York, 1948).]

employed. One nice example is the way in which it is integrated into the magnetron. This is a device capable of generating high-power microwaves such as are used in radar or in the familiar microwave ovens used for cooking. One form of magnetron is comprised of eight resonators spaced uniformly in a circle (Fig. 1.8). Each resonator has the shape of a cylindrical hole and a slot both of which are cut out of a solid block of copper. Electrons boil off from a heated cathode wire located at the center of the structure. They are accelerated radially by a steady voltage applied between the cathode wire and the copper block which acts as the anode. A steady magnetic field of a thousand or more gauss is supplied by a horseshoe magnet and is made to act parallel to the cathode wire (that is, into the plane of the paper). The combined action of the radial voltage and the axial magnetic field cause the electrons to take up a roughly circular path concentric with the cathode. The purpose of these orbiting electrons is twofold. First, to induce charges and current flows at each successive hole and slot resonator, which in turn generate the requisite electric and magnetic fields within the resonator; and second, once these fields have been established, to maintain the electromagnetic oscillations which would otherwise die out as a result of the joule heat losses in the copper block and other losses. Thus, the oscillations are maintained at the expense of the energy of the orbiting electrons. A small coupling antenna inserted into one of the resonators then extracts the electromagnetic energy from the magnetron device.

Each hole and slot resonator can be viewed as an L-C circuit. In first approximation we can treat the slot as a parallel plate capacitor and assume that only electrical energy is stored there. If the slot has a width d, a length s, and a depth b as is shown in Fig. 1.9, then the capacity is

$$\begin{aligned} C &= \frac{\epsilon_o \times \text{area of capacitor}}{\text{spacing between plates}} \text{ farad} \\ &= \frac{\epsilon_o sb}{d} \end{aligned} \tag{1.54}$$

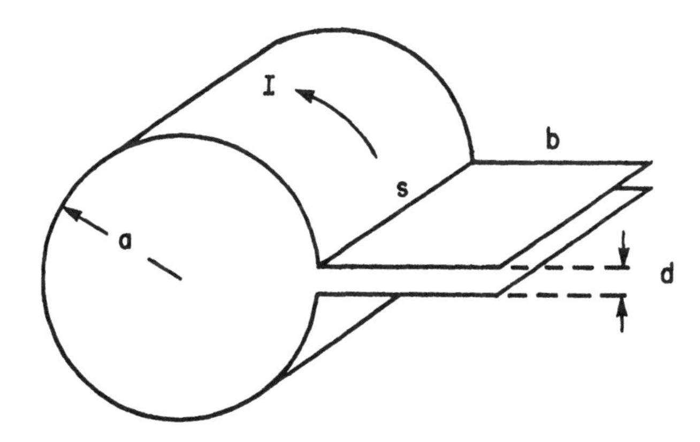

Fig. 1.9 Diagram of the magnetron hole and slot resonator. The slot can be approximated by a parallel plate capacitor of capacitance C, and the hole by a long one-turn solenoidal coil with a current sheet flowing as shown. The self-inductance of the coil is L.

where $\epsilon_0 = 8.854 \times 10^{-12}$ coulomb^2/meter^2-newton. The hole, on the other hand, can be viewed as a single-turn solenoid and thus as the repository of the magnetic energy. If we neglect the fringing fields at the ends of the solenoid, we obtain for the magnetic field within its hole the value

$$B = \frac{\mu_0 I}{s} \text{ webers/m}^2. \tag{1.55}$$

Here $\mu_0 = 4\pi \times 10^{-7}$ newton/amp^2 and I is the current flowing around the hole. The self-inductance L is obtained from the relation L = magnetic flux/current with the result that

$$\begin{aligned} L &= \frac{\pi a^2 B}{I} \\ &= \frac{\mu_0 \pi a^2}{s} \text{ henry} \end{aligned} \tag{1.56}$$

where a is the hole radius. Finally, substituting for C and L in Eq. 1.51 we arrive at the resonant frequency of the magnetron

$$\begin{aligned} \omega_0 &= \sqrt{\frac{1}{LC}} \\ &\approx \sqrt{\frac{d}{\epsilon_0 \mu_0 \pi a^2 b}}. \end{aligned} \tag{1.57}$$

(The fact that there are seven more hole and slot resonators does not affect the result for ω_0 appreciably.) We see that ω_0 is independent of the length s of the hole and slot structure, but is a function only of d, a, and b. Typical values for these quantities are $a \approx 0.5$ cm, $d/b \approx 0.5$ and on substituting these into Eq. 1.57 we find that $\nu_0 = \omega/2\pi \approx 3.8 \times 10^9$ Hz. Today magnetrons are manufactured over a wide range from $\sim 10^9$ Hz to $\sim 3 \times 10^{10}$ Hz.

Electron oscillations – the plasma frequency

Electron oscillations in atoms and molecules provide an almost limitless wealth of examples concerning oscillators on the atomic scale of things. Their full appreciation and understanding is furnished by quantum theory, a subject that lies outside the scope of this book. However, there exists an important situation concerning the oscillation of free electrons which can be treated fully by classical theory. This then is the subject matter to which we shall address ourselves here.

The ionosphere, the solar corona, large chunks of interstellar gas, and the common fluorescent lamp are a few of the places where electrons exist in a free state. They are intermingled with positive ions in such proportion so as to make the entire mixture electrically neutral (atoms or molecules are sometimes present, but we will ignore their existence). This gas composed of electrons and ions is

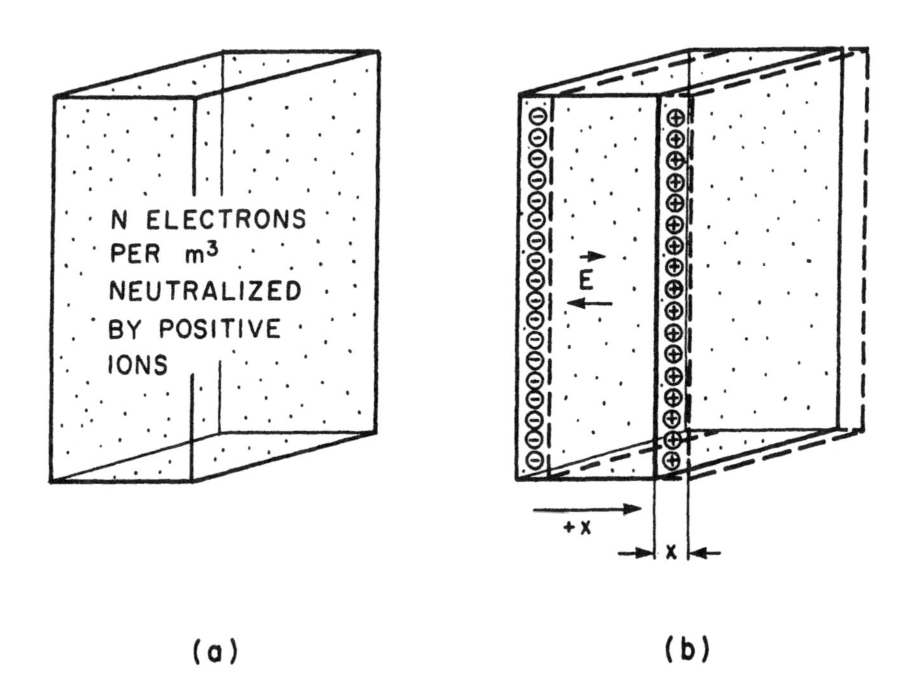

Fig. 1.10 (a) A plasma slab where the electrons and ions are in equilibrium with one another. The ions provide a neutralizing background for the electrons. (b) The electrons are pushed to the left relative to the ions by a distance x, creating two charged sheets with electric field $\vec{E}$ between them.

called a plasma and it exhibits some interesting oscillatory properties as we shall now demonstrate.

For simplicity consider a uniform slab of plasma as illustrated in Fig. 1.10a. Initially the whole system is assumed to be in equilibrium. Now picture a perturbation caused by an external agency in which each electron is displaced from its initial position by a distance x (the ions being so much more massive than the electrons that they will hardly move and we assume that they remain fixed). As a result of the displacement, there will be an excess of electrons at the left plasma-vacuum boundary and an excess of ions at the right boundary (Fig. 1.10b). Note, however, that except for these two charged sheets, the bulk of the plasma remains electrically neutral. The formation of the charged sheets has an important consequence — the creation of an electric field within the plasma volume. Now, it is known that a uniform sheet of charge of surface density σ produces an electric field distributed symmetrically on each side, of magnitude

$$E = \frac{\sigma}{2\epsilon_0} \text{ volt/m.} \tag{1.58}$$

Thus the total field at any point within the plasma, due to both charged sheets, is the sum of the two fields (1.58), namely

$$E = \frac{\sigma}{\epsilon_0}. \tag{1.59}$$

If we have N electrons per unit volume of plasma, uniformly distributed throughout, $\sigma = Nqx$ (where q is the elementary charge) and

$$\vec{E} = -\frac{Nq}{\epsilon_0}\vec{x}. \tag{1.60}$$

It is precisely this electric field acting on each plasma electron which tends to restore the initial equilibrium of the system. A given electron of mass m experiences a restoring force of magnitude qE and its equation of motion is given by

$$m\frac{d^2x}{dt^2} = -\frac{Nq^2}{\epsilon_0}x. \tag{1.61}$$

Therefore, every electron oscillates about its equilibrium position as

$$x = x_o \cos(\omega_p t + \phi) \tag{1.62}$$

with a frequency given by

$$\omega_p = \sqrt{\frac{Nq^2}{m\epsilon_o}} \text{ rad/sec.} \tag{1.63}$$

This frequency is known as the plasma frequency and is normally denoted as ω_p. Since the electronic charge q and mass m are known constants, ω_p is solely a function of the density of charges N. Using the values $q = 1.6 \times 10^{-19}$ coulomb, and $m = 9.01 \times 10^{-31}$ kg, we find that

$$\begin{aligned} \nu_p &= \omega_p/2\pi \\ &= 8.979\sqrt{N} \text{ Hz} \end{aligned} \tag{1.64}$$

where N is in units of particles per cubic meter. A typical value of density in the ionosphere (it varies greatly with height and time of day) is $\sim 10^{11}$ m^{-3} and thus $\nu_p \sim 3 \times 10^6$ Hz. In the fluorescent lamp N ~ 10^{17} m^{-3} and $\nu_p \sim 3 \times 10^9$ Hz. It is clear that a single measurement of ν_p can establish for us the density of plasma electrons. For this reason it has become a widely used technique of determining N.

It may well be asked how one goes about exciting the electrons to oscillate in the manner indicated above. One way is to impinge radio or microwaves of the right frequency at the plasma. Another is to shoot charged particles or neutral matter into the ionized interior. In nature such processes manifest themselves quite dramatically in the solar corona which is an ionized plasma surrounding the visible solar disk (the photosphere). At various sporadic intervals, flares get ejected from the photosphere and travel through the corona. As the flare plows its way through the corona, it excites local plasma oscillations and these can be detected as high-frequency radio signals arriving at Earth stations. It is interesting to note that N decreases with distance r from the sun and thus, by Eq. 1.64, so does the frequency ν_p. Since N(r) is known from other measurements, a

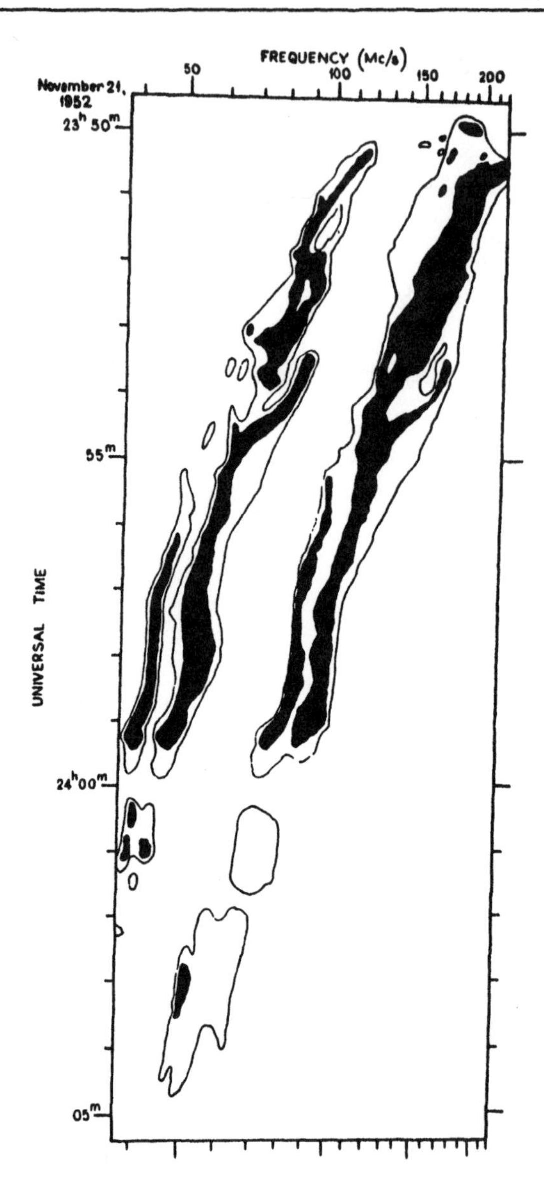

Fig. 1.11 Intensity contours of radio signals observed from a solar burst plotted on a map of signal frequency ν versus the time at which it was received. The left-hand dark trace corresponds to $\nu \approx \nu_p$ and the right-hand trace to $\nu \approx 2\nu_p$ where ν_p is the local plasma frequency of the solar corona. [From The Sun by G. P. Kuiper (University of Chicago Press, 1953).]

determination of the time history of ν_p tells the observer the velocity (and position) of the solar flare. The graph illustrated in Fig. 1.11 represents such an observation. The heavy, lower frequency trace represents oscillations at ν_p. The other parallel trace occurs at

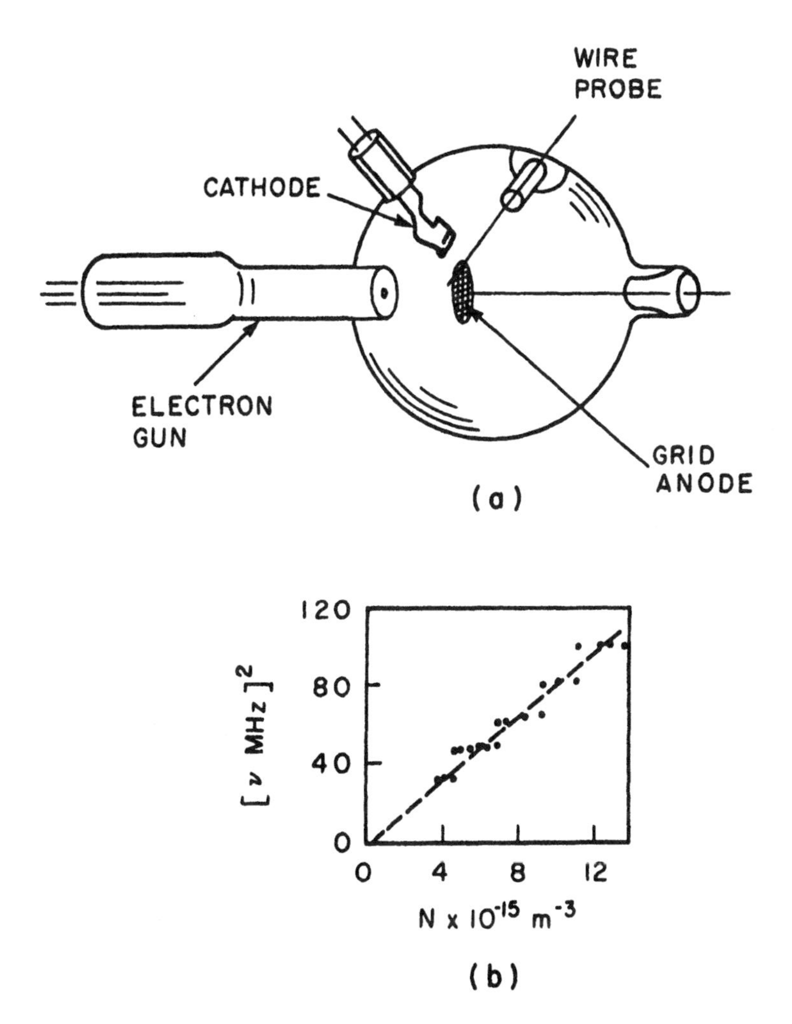

Fig. 1.12 The plasma is generated by electrical breakdown between a hot oxide-coated cathode and grid anode. Electron oscillations are excited by firing ~300 eV electrons from gun into plasma. The plasma oscillations are detected with a fine wire probe and measured as function of the density N. [After D. H. Looney and S. C. Brown, Phys. Rev. 93, 965 (1954).]

approximately $2\nu_p$. The generation of signals at the second harmonic is a most intriguing phenomenon, and is not fully understood. Let us just remark that generation of higher harmonics in any system is typical of the presence of nonlinear effects (see section 1.6).

Figure 1.12a represents a laboratory technique of exciting plasma oscillation. First, a plasma is generated in a large pyrex bowl by applying a sufficiently high voltage between a cathode and anode, thereby breaking down the mercury vapor contained in the bowl. Then an electron beam from a gun is fired into the ionized gas. The oscillations are picked up by an antennalike device. As shown in Fig. 1.12b, the square of the oscillation frequency is proportional to the electron density, in agreement with Eq. 1.64.

We have seen that the oscillatory behavior of all the systems discussed in this section is the result of the interplay of two opposing tendencies, a "restoring force" that attempts to restore the system to its initial equilibrium; and the inertia that tends to preserve the existing motion and causes the system to overshoot. For the mass-spring combination the characteristic quantities are K, the restoring force per unit displacement, and the inertial mass M. The corresponding parameters describing the other oscillators are listed in Table 1.1.

Table 1.1 Parameters describing the restoring force and inertia of several simple harmonic oscillators.

System	"Restoring Force"	"Inertia"
Mass on a spring	Spring constant K	Mass M
Pendulum	Torque	Moment of inertia I_o
Acoustic oscillations	Bulk modulus κ	Mass density ρ
Electrical oscillations	[Capacitance C]$^{-1}$	Inductance L
Plasma oscillations	Nq^2/ϵ_o	Electron mass m

1.3 DAMPED OSCILLATIONS

The world is not as simple as we tried to make it appear in earlier sections. There is friction, and an oscillator, once started by some means or other, does not continue to vibrate indefinitely with a constant amplitude. Ordinarily the inclusion of friction in the differential equation for $\psi(t)$ (Eq. 1.1) make the problem difficult to solve because of the complex character of the frictional force. There are, however, circumstances in which the friction is proportional to the first derivative with respect to time of the "moving" quantity ψ, and a solution is readily attained. For example, an object moving sufficiently slowly through a viscous medium experiences a resistive force proportional to its speed; another example is electrical charge flowing through a fixed resistance, a situation which we shall study shortly. If, on the other hand, the resistance varies in a more complicated manner (say, $F(\text{friction}) \propto (\text{velocity})^2$), one must usually resort to a numerical solution of the differential equation.

Our main concern in this book is with oscillations in electromagnetic systems, and for that reason we shall illustrate the mathematical procedures by turning to the important case of the RLC circuit.

The series RLC circuit

The series RLC circuit is illustrated in Fig. 1.13; it differs from Fig. 1.7 only in that a resistance R is incorporated in the system.

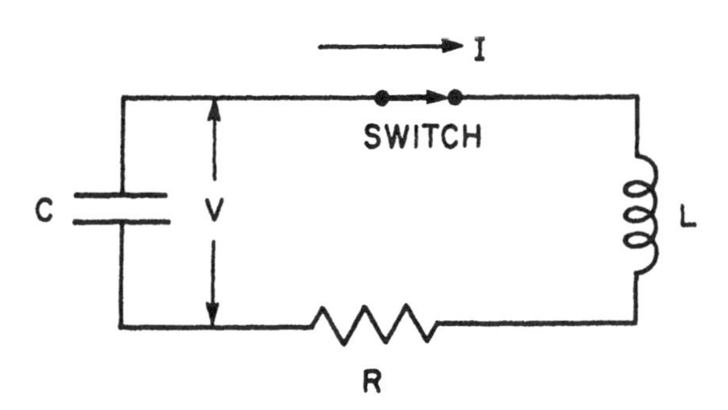

Fig. 1.13 Series RLC circuit.

The resistance R may be an actual resistor purposely inserted into the circuit, or it could represent the intrinsic resistance of the coil or leads. Equations 1.45, 1.46, and 1.47 still apply to the problem at hand; but now the voltage V of Eq. 1.48 acts across both R and L with the result that

$$V = RI + L\frac{dI}{dt} \text{ volt.} \tag{1.65}$$

We eliminate V between Eqs. 1.47 and 1.65, using the method described earlier, and find that

$$\frac{d^2I}{dt^2} + \frac{R}{L}\frac{dI}{dt} + \frac{1}{LC}I = 0 \tag{1.66}$$

where the new, second term describes the presence of damping. Recalling that $\omega_o = 1/\sqrt{LC}$ is the resonant frequency of the undamped circuit, and defining the effect of the resistance through

$$\beta = \frac{R}{L}, \tag{1.67}$$

we can write Eq. 1.66 in the more concise form

$$\boxed{\frac{d^2I}{dt^2} + \beta\frac{dI}{dt} + \omega_o^2 I = 0.} \tag{1.68}$$

This equation serves not only as the basic equation of the RLC circuit, but is also the "master equation" for any simple harmonic system in which the friction is proportional to the first derivative of the oscillating quantity. To solve it let us try the elegant technique of complex exponentials and use as the trial function

$$I = I_o\, e^{j(pt+\phi)} \tag{1.69}$$

where I_o is the amplitude and ϕ the phase — both taken to be real quantities. Inserting Eq. 1.69 in 1.68 we find that

$$\left(p^2 - jp\beta - \omega_o^2\right) I_o\, e^{j(pt+\phi)} = 0 \tag{1.70}$$

which is satisfied by the trivial result $I_o = 0$. If the amplitude is to be finite, the nontrivial and necessary requirement is that

$$p^2 - jp\beta - \omega_o^2 = 0 \qquad (1.71)$$

which has the solution

$$2p = \pm\sqrt{4\omega_o^2 - \beta^2} + j\beta. \qquad (1.72)$$

Two important cases emerge from this result, giving an entirely different physical behavior to the system. One is where p is complex, which occurs when $4\omega_o^2 > \beta^2$ or equivalently, $4/LC > (R/L)^2$. This is the case of "light damping." The other is the situation where $4\omega_o^2$ equals or is less than β^2; here p is purely imaginary and we refer to this as "heavy damping." Let us begin with the former.

Light damping

Taking the positive solution of Eq. 1.72 (the negative would serve equally well with no change in the answer), substituting it in Eq. 1.69 and taking the real part of all this (since only the real part has physical meaning for us) we arrive at the result that

$$I = I_o\, e^{-(\beta/2)t} \cos(\omega t + \phi) \qquad (1.73)$$

where for the sake of brevity we have written

$$\omega = \sqrt{\omega_o^2 - \frac{\beta^2}{4}} = \sqrt{\frac{1}{(LC)} - \frac{R^2}{4L^2}}. \qquad (1.74)$$

This, then, is the general solution with its two "arbitrary" parameters I_o and ϕ. The essential feature of the result is the exponential decrease of the amplitude of the current. Also, the oscillatory frequency ω differs from ω_o, the frequency of undamped oscillations. However, when R is not excessively large, ω does not depart

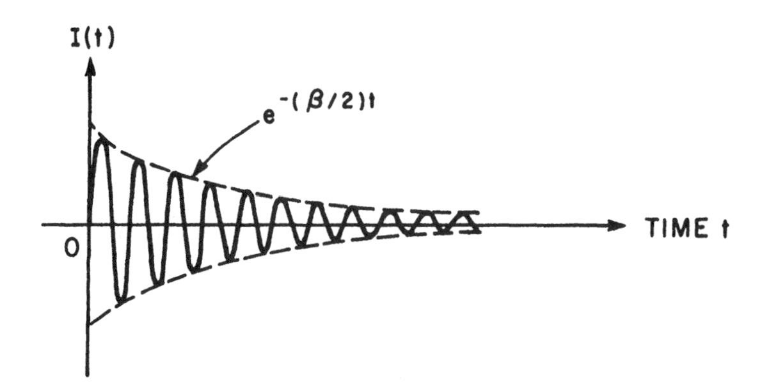

Fig. 1.14 Damped oscillations of the current in a series RLC circuit. The phase ϕ of Eq. 1.73 is taken to be $-\pi/2$ so that $I = 0$ at $t = 0$.

significantly from ω_0. Figure 1.14 illustrates the variations of current with time; $t = 0$ is the time at which the switch is thrown to complete the circuit. The dashed line in the figure shows how the amplitude falls off as a function of time. We see that it decreases by $(1/e)^{th}$ of its original value in a time of $(2/\beta)$ seconds. Of course, the voltage across the capacitor likewise falls off with time; inserting Eq. 1.73 in Eq. 1.65, and performing the differentiation dI/dt, we find that

$$V(t) = I_0 R\, e^{-(\beta/2)t}\left[\frac{1}{2}\cos(\omega t+\phi) - \frac{\omega}{\beta}\sin(\omega t+\phi)\right]. \tag{1.75}$$

In contrast to the pure simple harmonic oscillations discussed in sections 1.1 and 1.2, the energy in the damped oscillation is not constant in time. Rather, the energy is continuously given up to the damping medium where it is dissipated as heat (joule heating, I^2R, in the case of the electrical resistor). The total energy stored in the system at any one time is

$$U(t) = \frac{1}{2}LI^2(t) + \frac{1}{2}CV^2(t) \text{ joule}. \tag{1.76}$$

Now, we know $I(t)$ and $V(t)$; they are given by expressions (1.73) and (1.75). We insert these in our formula, and after some rearranging of terms we get

$$U(t) = \left(\frac{1}{2}LI_o^2\right) e^{-\beta t} \times \left\{1 + LC\left[\left(\frac{\beta}{2}\right)^2 \cos(2\omega t + 2\phi) - \left(\frac{\beta\omega}{2}\right)\sin(2\omega t + 2\phi)\right]\right\}. \quad (1.77)$$

The important point of this result is that except for the oscillating part, the energy decays exponentially as

$$\boxed{U = U_o\, e^{-\beta t}.} \quad (1.78)$$

The oscillations show up as wiggles around the exponentially falling trace of Eq. 1.78, but they are quite small when the damping is light (see Fig. 1.15). This equation tells us that the oscillator's energy falls to $(1/e)^{th}$ of its initial value U_o in a time

$$\boxed{\tau = \frac{1}{\beta}.} \quad (1.79)$$

This time is known as the lifetime of the oscillations, and is a

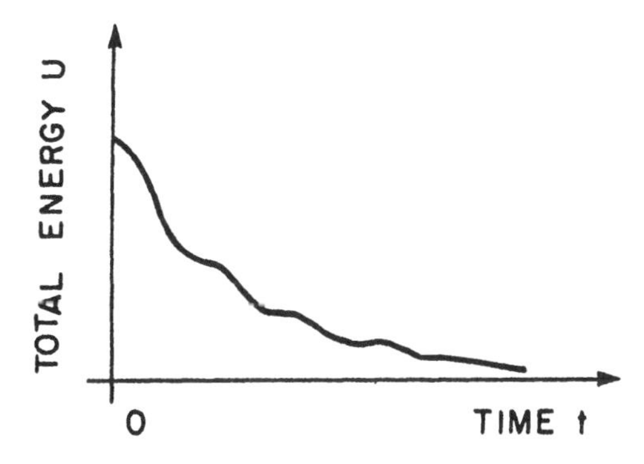

Fig. 1.15 Decrease in the energy with time of a damped harmonic oscillator. The small wiggles about the exponentially decreasing function are caused by the second term [---] of Eq. 1.77. The wiggles come from the fact that the energy loss rate has maxima and zeros. The zeros in dU/dt come whenever I or V go through zero.

quantity that appears time and again in many fields of physics.

The rate at which energy keeps decreasing is easily found: differentiating Eq. 1.78 with respect to time we obtain that $dU/dt = -\beta[U_o \exp(-\beta t)]$. But the term in brackets is just the instantaneous value of the energy U; so $dU/dt = -\beta U$, which says that the rate at which energy is dissipated by the system is proportional to the energy that is present at the time; this, of course, is the very property of the exponential function which changes by the same fractional amount during equal intervals of time. The power dissipated in damping is dU/dt; therefore, if one would want to maintain the oscillations indefinitely, a source with this amount of power would have to keep feeding the oscillator. This subject will be discussed in the next section.

Recall that in section 1.1 we showed that the phase-space trajectory of a perfect harmonic oscillator is a closed ellipse (Fig. 1.3a). In a damped oscillator the radius vector drawn in the p-x plane is continuously decreasing – the spiral phase path shown in Fig. 1.16 is typical of all damped harmonic motion.

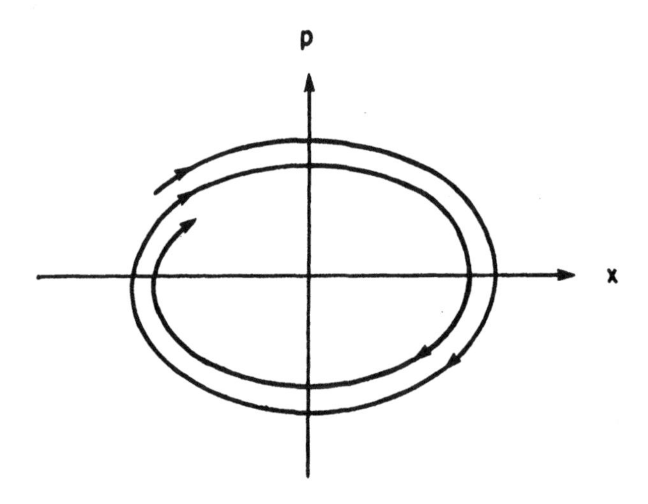

Fig. 1.16 Phase-space trajectory of a damped harmonic oscillator; p is its instantaneous momentum and x its coordinate. In the RLC circuit p is proportional to the voltage across the inductance and x the current through the circuit.

Heavy damping

The oscillatory motion of a damped oscillator ceases when p of Eqs. 1.69 and 1.72 becomes purely imaginary, because then the real parts of complex exponentials like exp[jp] become monotonically decreasing functions. A glance at Eq. 1.72 shows that the onset of non-oscillatory motion occurs when the damping has increased to the point where

$$4\omega_o^2 = \beta^2 \tag{1.80}$$

or

$$R = 2\sqrt{\frac{L}{C}}. \tag{1.81}$$

When this condition prevails the system is said to be critically damped. The solution of Eq. 1.68 now becomes

$$I = (A+Bt)\, e^{-(\beta/2)t}, \tag{1.82}$$

as is readily proven by substituting this formula in Eq. 1.68. The two constants A and B are determined from initial conditions. The corresponding voltage V on the capacitor comes from inserting Eq. 1.82 in Eq. 1.65 with the result that

$$V = \left[LB + \frac{1}{2}R(A+Bt)\right] e^{-(\beta/2)t}. \tag{1.83}$$

We shall apply these findings to the following practical application.

Discharge of a capacitor bank

A bank of capacitors provides a most useful and versatile way of storing electrical energy. Its great virtue is that it is capable of delivering large amounts of stored energy in very short times, of the order of microseconds or less. Figure 1.17 shows a schematic of a typical system. A direct current power supply charges the capacitors of the bank with a charging resistor and a closed charging switch. The charging resistor limits to acceptable levels the current being drawn from the power supply. This charging process may take many

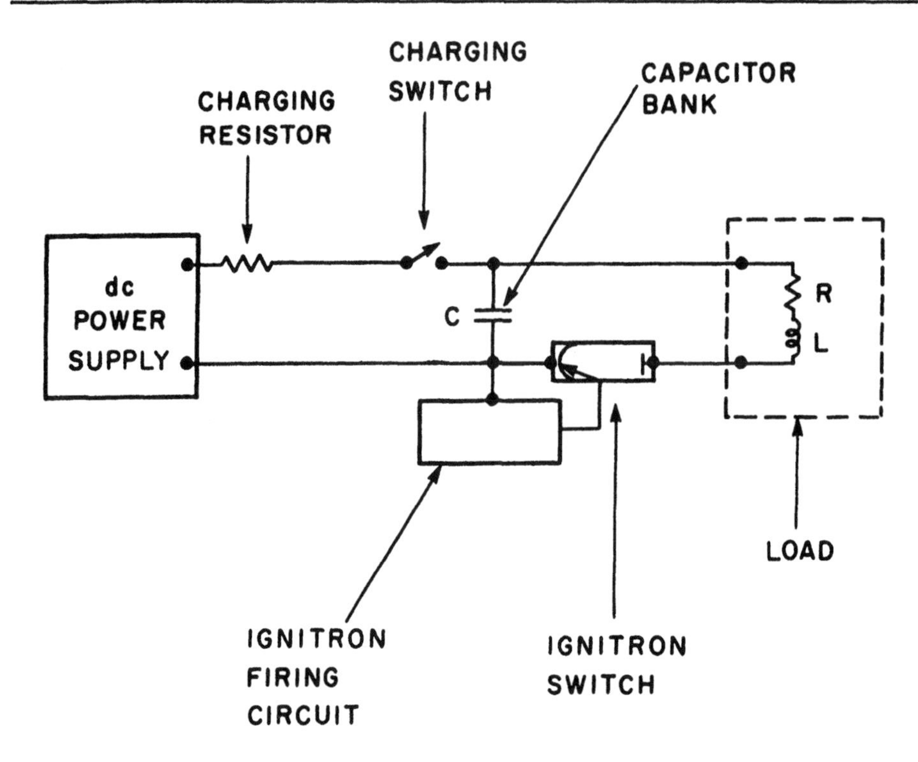

Fig. 1.17 Schematic of capacitor bank. The RLC circuit is completed by firing the ignitron after charging capacitor bank from dc supply. When closed, a big ignitron can carry as much as 100 kA peak current and pass as much as 30 coulombs of charge per shot.

seconds and even minutes. Suppose we want to store a maximum of 20,000 joules and we have a dc supply capable of delivering 1000 watts of power; hence the charging time is about 20 seconds.

During the charging process, the "discharge" switch is open. One commercial form of switch is the ignitron. This is quite an ingenious device in which a command trigger from a firing circuit causes ionization in a diode containing mercury vapor. The ionized mercury atoms filling the space between cathode and anode cause the tube to conduct, thus closing the switch in times less than one microsecond. Once the switch is closed, current flows to the load. The load may be virtually anything — a solenoid in which we wish to create large magnetic fields for a short time, an electron gun from which a burst of energetic electrons is desired, or even a fine wire that explodes as a result of the deposition of large amounts of power in it. Ordinarily the load has resistance R and nearly always has some inductance L (the capacitor bank itself and the leads connecting the load have inductance). Therefore we have an RLC circuit.

Let us now take the following specific situation. A capacitor bank having a total C equal to 100 μfarads is charged to a peak voltage V_o of 20 kV. The resistance of the load, leads, etc., is 0.045 ohm and their inductance is 10.1 microhenries. In summary, then,

$$\left.\begin{aligned} C &= 10^{-4} \text{ farad} \\ R &= 0.045 \text{ ohm} \\ L &= 1.01 \times 10^{-5} \text{ henry} \\ V_o &= 2 \times 10^{4} \text{ volt} \\ U_o &= (1/2)\, CV_o^2 = 20{,}000 \text{ joule} \end{aligned}\right\} \tag{1.84}$$

With these values R is less than $2\sqrt{L/C}$ and thus the system is lightly damped and oscillatory. The frequency of oscillation (Eq. 1.74) is 5000 Hz. This means that since initially the current must be zero (because when the ignitron is fired at t = 0, the current cannot jump discontinuously), the time the current reaches its first peak value is $t = T/4 = 50$ μsec. How big is its peak value? Since $|I| = C\, dV/dt$, and V is given by Eq. 1.75, we can compute I at any time after the switch is closed. We find the value of 56,500 amperes. This current, of course, keeps diminishing as a result of damping; for example, the next peak current in the reversed direction has only the value of 45,200 amperes, and so on. The oscillatory aspects of V and I are often undesirable features in supplying energy to the load. In addition, the voltage oscillations experienced by the capacitors are damaging to them and shorten their useful life; but this can be remedied as will be shown below.

Now, let us keep all the values listed under Eqs. 1.84 the same with one exception: increase R to the new value

$$R = 0.64 \text{ ohm} \tag{1.85}$$

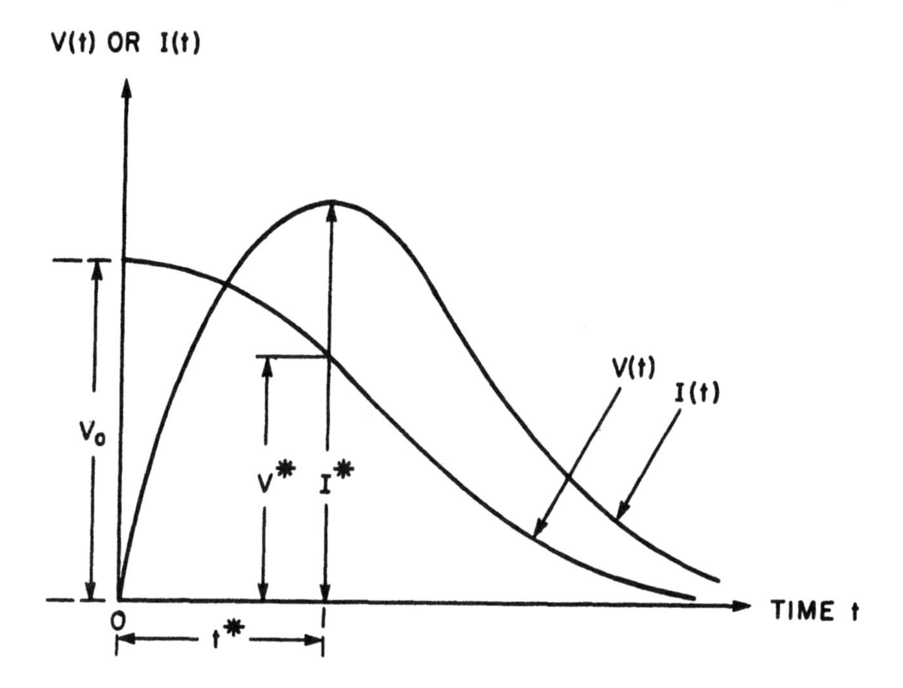

Fig. 1.18 Current and voltage as a function of time in a critically damped RLC circuit with initial conditions: at $t = 0$, $V = V_o$, $I = 0$. I^* is the peak current attained at time t^*. The voltage is V^*.

which in accordance with relation (1.81) corresponds to a critically damped circuit. To find how I and V vary, the appropriate initial condition needs to be inserted in Eqs. 1.82 and 1.83. As before they are: $V = V_o$, $I = 0$ at $t = 0$. This yields

$$I = \frac{V_o}{L}\, t\, e^{-(\beta/2)t}$$

$$V = V_o\left[1 + \frac{1}{2}\frac{R}{L}t\right] e^{-(\beta/2)t} \tag{1.86}$$

and Fig. 1.18 shows how I and V vary in time. Neither quantity is seen to be oscillatory. The voltage across the capacitor decreases almost exponentially; the current rises from its initially zero value and reaches a maximum current I^* in a time $t = t^*$. At this time the voltage has fallen to a value $V = V^*$. We shall leave to the reader to show, using Eqs. 1.86 that, in fact,

$$\left.\begin{aligned} I^* &= 0.736 \frac{V_o}{R} \\ t^* &= \frac{2L}{R} = \frac{RC}{2} \\ V^* &= 0.736\, V_o \end{aligned}\right\} \qquad (1.87)$$

With our values of V_o, R, and L, these give $I^* = 23,000$ amperes, $t^* = 32\ \mu$sec, $V^* = 14,720$ volts. When we compare these values with those obtained for the underdamped case we see that the peak current I^* is less than can be obtained with the lightly damped circuit, but it is reached in a shorter time.

When the resistance is further increased, beyond the value of 0.64 Ω, such that $R > 2\sqrt{L/C}$, the broad features illustrated in Fig. 1.18 remain, but the voltage decay is slower, the current maximum is smaller and the time to reach it takes longer. This so-called "overdamped" situation is neither very interesting nor very important and we shall not pursue the subject further.

Example

What is the lifetime of an oscillator whose natural frequency is ω_o and whose frequency when lightly damped is less than ω_o by one part in ten thousand?

Solution

The damped frequency ω is

$$\omega = \omega_o\left(1 - \frac{1}{10^4}\right) = 0.9999\,\omega_o.$$

From Eq. 1.74

$$\beta^2 = 4\left(\omega_o^2 - \omega^2\right) = 4\,\omega_o^2(1-0.9998)$$

$$\beta = 2\omega_o\sqrt{2 \times 10^{-4}} = 2.8 \times 10^{-2}\,\omega_o.$$

From Eq. 1.79 the lifetime τ is

$$\tau = \frac{1}{\beta} = \frac{36}{\omega_o} = 5.7\ T_o$$

where T_o is the period of the natural oscillations.

1.4 RESONANCE

To get an oscillator to oscillate, we must feed energy to it to first build up its oscillations; and once they are built up, we must keep feeding it energy to compensate for the inevitable frictional losses. How to do this efficiently, and the outcome of such excitations are subject matters of this section. Clearly, to provide energy in any haphazard manner is not the way to go. Consider the mass-spring of Fig. 1.1; if we tap the mass at random intervals of time, then we sometimes tap it when it goes the right way and sometimes when it goes the wrong way. But if we keep tapping just right we can get the mass to go like wild, and if there were no damping to limit the amplitude of the oscillation, the mass and spring would eventually break away from their mooring. This is what we call resonance. One of the simplest things we can think of is to excite the resonator with a harmonically varying driver — this is also the problem which requires the least mathematics — and so we now proceed to examine it.

The sinusoidally driven oscillator

Let us, then, return to the series RLC circuit of Fig. 1.13 and incorporate in it an energy source represented in Fig. 1.19 as the time-varying voltage generator $\mathcal{E}(t)$. The sum of the voltages RI, $L \times dI/dt$ and Q/C across the resistance, inductance, and capacitance, respectively, must be made up by the electromotive force $\mathcal{E}(t)$ of the generator, with the result that

$$\boxed{L\frac{d^2Q}{dt^2} + R\frac{dQ}{dt} + \frac{1}{C}Q = \mathcal{E}(t).} \tag{1.88}$$

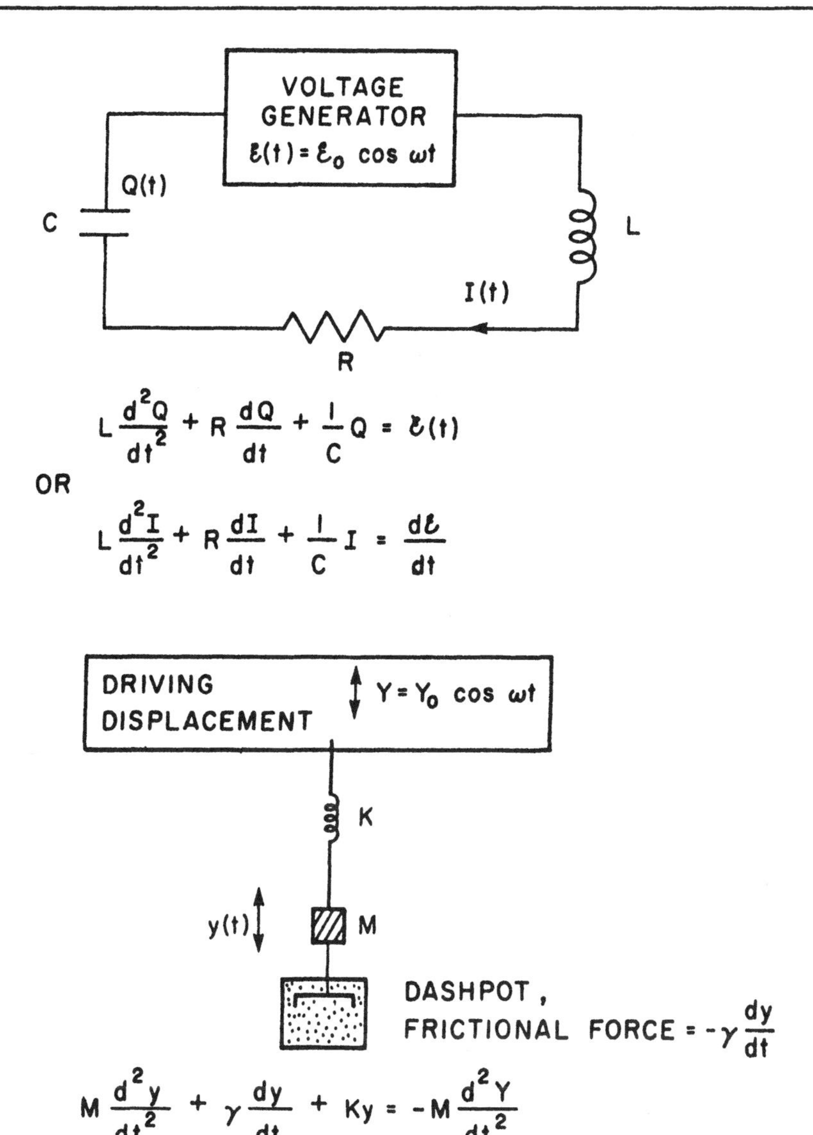

Fig. 1.19 Series RLC circuit driven by a voltage source (above) and a mechanical analogue (below). Notice that the left-hand sides of the equations are all of identical form, but the right-hand sides representing the driving terms have somewhat different forms depending on the physics of the problem. Since the drivers are taken to be sinusoidal, the differences are only in the phase and in a constant of proportionality [$d \cos \omega t/dt = -\omega \sin \omega t$; $d^2 \cos \omega t/dt^2 = -\omega^2 \cos \omega t$ and so on.]

Since the current I rather than the charge Q is of more general interest, we differentiate the equation with respect to time and obtain

$$L\frac{d^2I}{dt^2} + R\frac{dI}{dt} + \frac{1}{C}I = \frac{d}{dt}\mathscr{E}(t) \qquad (1.89)$$

which, except for the new source term $d\mathscr{E}/dt$, is identical with Eq. 1.66 of the previous section. One or the other of Eqs. 1.88 or 1.89 will now serve us as the basis for computing the characteristics of driven, damped harmonic oscillators. The equations are second-order, linear, but inhomogeneous differential equations, with the source term providing the inhomogeneity, and thus adding somewhat to the mathematical complexities.

Let us apply a sinusoidal electromotive force $\mathscr{E} = \mathscr{E}_0 \cos \omega t$ of frequency ω assumed to be generated by some agency external to the circuit. For purposes of computation we write this force in complex notation as $\mathscr{E} = \mathscr{E}_0 \exp(j\omega t)$, but with the realization that eventually the real part is to be taken of all quantities in order to find physical results. Inserting this form in Eq. 1.89 and rearranging terms we obtain

$$\frac{d^2I}{dt^2} + \beta\frac{dI}{dt} + \omega_0^2 I = \left(\frac{j\omega\mathscr{E}_0}{L}\right)e^{j\omega t} \qquad (1.90)$$

where once again $\beta = R/L$ and $\omega_0^2 = 1/LC$.

When the oscillator is first excited by the driver, it wants to oscillate in its customary mode and with its customary frequency $\left[\omega_0^2 - (\beta/2)^2\right]^{1/2}$ described in the previous section. It has not had time to get acquainted with the driver, so to speak, and to be forced by the driver to follow its detailed instructions. Thus, during the initial period of time, fairly complex transient effects take place which consist of a superposition of the free (but damped) oscillator motions and the driver's imposed motions. However, this transient period can be fairly short-lived. In times $\tau > 1/\beta = L/R$, the oscillator's natural tendencies have pretty much died out as a result of the frictional forces. It will then settle down and execute constant amplitude vibrations of the type

$$I = I_0 \, e^{j(\omega t + \delta)} \tag{1.91}$$

which is the form of trial function that we shall now use to solve Eq. 1.90. The important point is that in the results of this subsection the assumption is implicit that all transient behavior has damped out by the time we take up our calculations. To solve for I_0 and δ, we substitute Eq. 1.91 in Eq. 1.90 and find that

$$\left[\left(\omega_0^2 - \omega^2\right) + j\beta\omega\right] I_0 = \left(\frac{j\omega \mathscr{E}_0}{L}\right) e^{-j\delta} \tag{1.92}$$

which is the equation to be satisfied. But this is really not one equation, but two — one for real and one for imaginary quantities. Thus

$$\begin{aligned} \left(\omega_0^2 - \omega^2\right) I_0 &= \left(\frac{\omega \mathscr{E}_0}{L}\right) \sin \delta \\ \beta\omega I_0 &= \left(\frac{\omega \mathscr{E}_0}{L}\right) \cos \delta \end{aligned} \tag{1.93}$$

where, in obtaining this result we used the identity $\exp(-j\delta) = \cos\delta - j\sin\delta$. The two simultaneous equations in the two unknowns I_0 and δ are easily solved, giving

$$\begin{aligned} I_0 &= \frac{(\omega \mathscr{E}_0/L)}{\sqrt{\left(\omega_0^2 - \omega^2\right)^2 + (\omega\beta)^2}} \\ \tan \delta &= \frac{\omega_0^2 - \omega^2}{\beta\omega}, \end{aligned} \tag{1.94}$$

with the result that (on taking the real part of Eq. 1.91)

$$\boxed{I = \frac{(\omega \mathscr{E}_0/L)}{\sqrt{\left(\omega_0^2 - \omega^2\right)^2 + (\omega\beta)^2}} \cos(\omega t + \delta).} \tag{1.95}$$

This, then, is the sought-for expression* for the current driven in the circuit by the electromotive force $\mathscr{E} = \mathscr{E}_o \cos \omega t$. It exhibits a number of very interesting characteristics. The current is purely simple harmonic. Maintaining the driver amplitude $\mathscr{E}_o$ constant and varying its frequency ω, the current is greatest when

$$\omega = \omega_o = \frac{1}{\sqrt{LC}}, \tag{1.96}$$

which is precisely the natural frequency of the undamped LC circuit. Also, at this frequency, $\tan \delta = 0$, $\delta = 0$ and the current is exactly in phase with the driving voltage $\mathscr{E}$; its value is

$$I = \frac{\mathscr{E}_o}{R} \cos(\omega t) \tag{1.97}$$

which is the same current that would flow through the resistance R alone, with L and C removed. The plot of the current amplitude as a function of frequency ω shown in Fig. 1.20a illustrates convincingly the resonant nature of the system. If the resistance R is reduced, the peak current at $\omega = \omega_o$ becomes larger and larger, and the curve skinnier and skinnier. The frequency dependence of the phase angle δ is depicted in Fig. 1.20b. We observe that at high frequencies $\omega > \omega_o$, δ is negative and approaches the value of $-\pi/2$. In the parlance of electronics, "the current lags the voltage"; by this is meant that the current reaches its peak value after the voltage does. This is reminiscent of how an inductance behaves when the voltage is suddenly turned on. Looking at Eq. 1.89 we see that at high frequency the first term, that is, the inductance term $L\, d^2I/dt^2$ indeed dominates,

---

*Another way of arriving at this result is to use the trial function $I = A \exp(j\omega t)$ where A is now a complex constant (see Appendix 2). Inserting this in Eq. 1.90 yields

$$I = (j\omega \mathscr{E}_o/L) \left[\omega_o^2 - \omega^2 + j\omega\beta\right]^{-1} \exp(j\omega t)$$

which, on taking the real part, can be cast into the form given in the text.

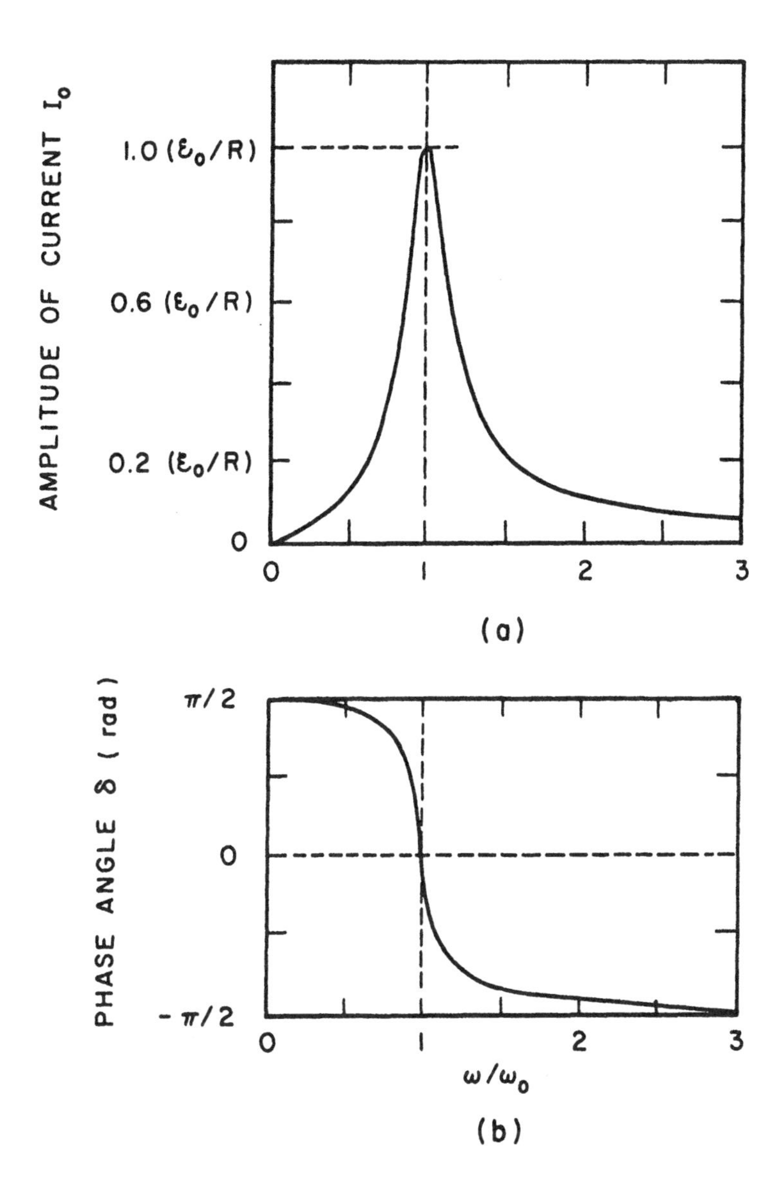

Fig. 1.20 The current amplitude I_0 and the phase angle δ of Eqs. 1.94 as a function of driver frequency ω of a series RLC circuit. The damping coefficient β was chosen so that $\beta/\omega_0 = 0.2$, or equivalently, that $R\sqrt{C/L} = 0.2$.

because d^2I/dt^2 is proportional to ω^2. At low frequency, on the other hand, the capacitive term $1/C$ takes over, δ approaches $+\pi/2$ and now the "current leads the voltage" in accordance with the behavior of a capacitive circuit. (In a purely resistive circuit the current is exactly in phase with the driving voltage.) The phase transition from $+\pi/2$ to $-\pi/2$ becomes increasingly abrupt the smaller the damping.

Energy and power dissipation

In the RLC circuit energy is alternately stored in the capacitor, released from it and then stored in the inductance. Simultaneously energy is being dissipated by the resistor. Let us calculate these quantities but let us average them over a period of the oscillation, since it is the averaged quantities which prove to be more useful. The instantaneous rate at which energy is delivered to the circuit is given by the product $\mathscr{E}I$, so that

$$P(t) = \mathscr{E}_o I_o \cos(\omega t)\cos(\omega t + \delta). \tag{1.98}$$

The power P(t) averaged over a period of the oscillation T is obtained with the aid of Eq. 1.20:

$$\langle P\rangle = \frac{1}{T}\,\mathscr{E}_o I_o \int_0^T \cos(\omega t)\cos(\omega t + \delta)\,dt. \tag{1.99}$$

By writing that $\cos(\omega t+\delta) = \cos(\omega t)\cos(\delta) - \sin(\omega t)\sin(\delta)$, and evaluating the ensuing integrals we get

$$\langle P\rangle = \frac{1}{2}\,\mathscr{E}_o I_o \cos\delta. \tag{1.100}$$

Note that when there is no dissipation, $\beta = 0$, $\tan\delta$ of Eq. 1.94 goes to infinity, $\delta = \pi/2$, $\cos\delta = 0$ and consequently $\langle P\rangle = 0$, which just confirms our expectations that with no dissipation, no power needs to be supplied to the circuit. It is instructive to look at the detailed structure of Eq. 1.100 and substitute in it the known values of I_o and δ as obtained from Eqs. 1.94. After some algebraic manipulations we find that

$$\langle P\rangle = \frac{1}{2}\left(\frac{\mathscr{E}_o^2}{R}\right)\frac{\beta^2\omega^2}{\left(\omega_o^2-\omega^2\right)^2 + \beta^2\omega^2}. \tag{1.101}$$

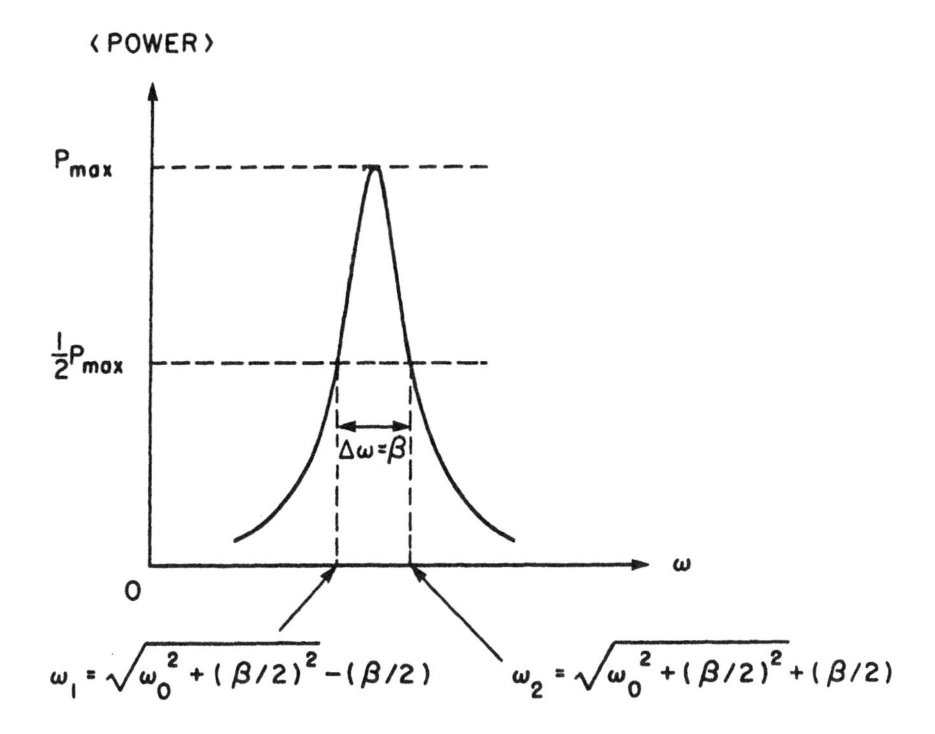

Fig. 1.21 Time-average power dissipated in a series RLC circuit driven at different frequencies ω. $P_{max} = \mathscr{E}_0^2/2R$.

This time-averaged power $\langle P \rangle$ dissipated in the circuit, just like the current I, exhibits a pronounced resonance behavior. For fixed values of $\mathscr{E}_0$ and R, the dissipation attains its highest value $\left(\mathscr{E}_0^2/2R\right)$ at the resonant frequency ω_0. The shape of the curve is illustrated in Fig. 1.21. As the resistance R is decreased, curves are obtained similar to the one shown, but they become taller and narrower. The customary way of defining the width of a resonant curve is the following: read off the peak value of $\langle P \rangle$, in our case $\left(\mathscr{E}_0^2/2R\right)$, and find the frequencies ω_1 and ω_2 on either side of the peak, where $\langle P \rangle$ has fallen to half its value. The width $\Delta\omega = \omega_2 - \omega_1$ is then known as the "full linewidth at half maximum power" or, in short, the "resonance width." To find ω_1 and ω_2 one simply sets the left side of Eq. 1.101 equal to $\left(\mathscr{E}_0^2/4R\right)$ and solves the resultant quadratic equation in the variable ω^2. The outcome is

$$\omega_1 = \sqrt{\omega_0^2 + (\beta/2)^2} - (\beta/2)$$
$$\omega_2 = \sqrt{\omega_0^2 + (\beta/2)^2} + (\beta/2). \qquad (1.102)$$

Therefore

$$\boxed{\begin{aligned}\Delta\omega &= \beta \\ &= \frac{R}{L}\end{aligned}} \qquad (1.103)$$

showing that the width of the power resonance curve is just equal to the damping constant β. But we recall (Eq. 1.79) that β is also equal to the reciprocal of the energy lifetime τ of the freely decaying oscillator. Thus we have arrived at a very important result, the connection between the frequency width of a driven oscillator and its energy lifetime, when the same oscillator is excited and then just left to its own devices:

$$\boxed{\Delta\omega = \frac{1}{\tau}.} \qquad (1.104)$$

Therefore, if one of these quantities is determined from experiment, for example, the other is immediately known through the use of Eq. 1.104.

When the dissipation constant β is fairly light, the resonance curve of Fig. 1.21 is quite narrow and most of the important line structure (that lies, say, within a few $\Delta\omega$'s on either side of the resonance) is not far removed from ω_0. Therefore, the term $\omega_0^2 - \omega^2 = (\omega_0 + \omega)(\omega_0 - \omega)$ of Eq. 1.101 can be approximated by $2\omega(\omega_0 - \omega)$ with the result that

$$\langle P \rangle \approx \frac{1}{2}\left(\frac{\mathcal{E}_0^2}{R}\right)\frac{(\beta/2)^2}{(\omega_0 - \omega)^2 + (\beta/2)^2} \qquad (1.105)$$

which is a much simpler looking equation than the full-blown resonance formula (1.101), and has enjoyed widespread use. For example, optical line shapes can often be well approximated by an equation of this form, and are then referred to as having a Lorentzian profile.

Having discussed energy dissipation in some detail, we must now turn our attention, if only briefly, to the equation of energy storage. At any instant of time, the energy residing in the RLC circuit is the sum of the energies stored in the inductance and capacitance,

$$U(t) = \frac{1}{2} LI^2(t) + \frac{Q^2(t)}{2C}. \tag{1.106}$$

Now, we know I(t) from Eq. 1.95 and we can find Q(t) from $-\int I(t)\,dt$. When we time-average the resulting expression for U(t) we find that

$$\langle U \rangle = \frac{1}{4} I_o^2 L \left\{ 1 + \left[\frac{\omega_o^2}{\omega^2} \right] \right\} \tag{1.107}$$

where, once again, I_o is given by Eq. 1.94 and where once again we discover, not surprisingly, that $\langle U \rangle$ has a resonant structure similar (but not of exactly the same shape) to that obtained earlier for $\langle P \rangle$.

The "Q" of an oscillator

The damping in an oscillator determines the relative amounts of energy being dissipated and the amounts being stored in its components. The "Q" (quality) is a dimensionless parameter designed to give a quantitative measure of this. For an oscillator driven at the frequency ω, the Q is defined as

$$\begin{aligned} Q &= \omega \frac{\text{average energy stored in a cycle}}{\text{average power dissipated in a cycle}} \\ &= \omega \frac{\langle U \rangle}{\langle P \rangle} \end{aligned} \tag{1.108}$$

from which we see that the ideal, frictionless system has an infinite Q. At the other end of the "quality" scale would be the critically damped oscillator (for which $\beta = 2\omega_0$) which as it turns out would have a Q of approximately 0.5. But, when the system is as heavily damped as this, the Q is no longer a very meaningful quantity. The resonant nature of the oscillator hardly manifests itself unless the Q is greater than one or two (see Fig. 1.22).

To see how Q depends on the damping parameter β, we substitute Eqs. 1.100 and 1.107 in 1.108, we insert the values of I_0 and δ from Eq. 1.94 and obtain the pleasantly simple result

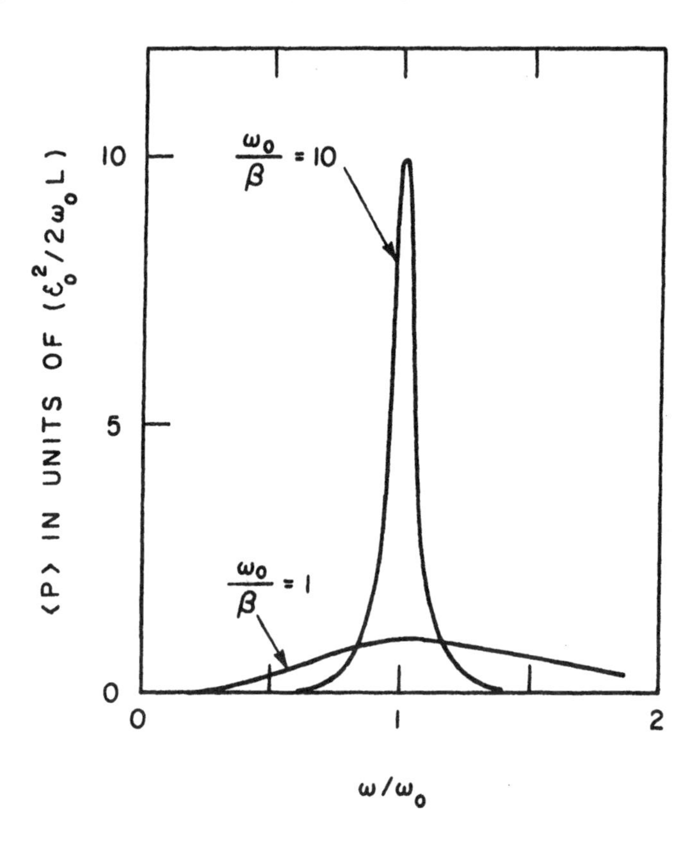

Fig. 1.22 Time-average power dissipated in an RLC circuit as a function of frequency. The power is expressed in units of $(\mathcal{E}_0^2/2\omega_0 L)$. When $\omega_0/\beta = 10$, the Q value ≈ 10 (see Eq. 1.110). When $\omega_0/\beta = 1$, $Q \approx 1$ but here the formula (1.110) is rather poor and is best not used. With all quantities held constant except for β, the peak power dissipated at $\omega = \omega_0$ is inversely proportional to the damping constant β.

$$Q = \frac{\omega^2 + \omega_o^2}{2\beta\omega}. \tag{1.109}$$

A further simplification can be instituted by noting that in low loss systems ω is close to ω_o so that to good approximation we can write that $\omega \approx \omega_o$ and

$$Q \approx \frac{\omega_o}{\beta}. \tag{1.110}$$

This equation, in fact, provides us with two alternative definitions of Q, one in terms of the half power resonance width $\Delta\omega$ (Eq. 1.103) of a driven oscillator,

$$\boxed{Q = \frac{\omega_o}{\Delta\omega}} \tag{1.111}$$

and the other in terms of the lifetime τ of the decaying oscillator (Eq. 1.79)

$$Q = \omega_o\tau. \tag{1.112}$$

Thus we have three ways open to us of finding the Q of a system either theoretically or experimentally. Determine $\langle U \rangle$ and $\langle P \rangle$ and derive Q from Eq. 1.108; find the resonance line shape of the driven oscillator, measure up the frequency width at the half power point and use Eq. 1.111; or determine the lifetime τ of the dying oscillator and use Eq. 1.112. Ordinarily, of the three methods, the second is the one most easily carried out. Table 1.2 lists the Q's of a few interesting systems.

The effect of transients

Everything in Eq. 1.95 is fully determined and specified. Therefore this equation cannot be the complete solution of the second-order differential equation (1.90). If it were a complete solution, where then are the two so-called arbitrary coefficients? But we know what

Table 1.2 Order of magnitude Q values of some interesting systems; ν_o is the approximate resonance frequency of the oscillator.

System	ν_o (Hz)	Q
Seismograph (see next subsection)	0.1	1
Moon [obtained from impact of Apollo 12 lunar module; see Science 167 (1970)]	1 to 6	10^4
Resonant microwave cavity made from superconducting material [lead alloy walls at cryogenic temperature; see E. Maxwell, Progress in Cryogenics, IV (Heywood, London), p. 123; also section 5.4]	10^9 to 10^{10}	10^6 to 10^7
Typical optical line emitted from atomic oscillators of a low-pressure discharge tube	5×10^{14}	10^7
Laser cavity formed between two mirrors facing each other (see section 5.4)	10^{14} to 10^{15}	10^8 to 10^9

is missing; it is the transients which we conveniently allowed to die out in all the calculations of the previous subsection. To reinstate them, the theory of linear differential equations tells us to proceed as follows: Solve Eq. 1.90 with the right-hand driving term equal to zero; the solution is $I_1(t)$, and is called the "complementary function." To this add the "particular solution" $I_2(t)$ that reproduces the right-hand side of Eq. 1.90, and you are done. Since we have solved for I_1 in section 1.3, and for I_2 in this section, we have the full result,

$$I(t) = A\, e^{-(\beta/2)t} \cos(\Omega t + \phi) + I_o \cos(\omega t + \delta). \tag{1.113}$$

Here A and ϕ are the "arbitrary" amplitude and phase of the transient part of the solution which will eventually be determined from initial conditions; $\Omega = \sqrt{\omega_o^2 - (\beta/2)^2}$ is the frequency of the transient oscillations; I_o and δ are the current amplitude and phase, respectively, of the steady oscillations defined by Eqs. 1.94; and finally ω

is the frequency of the driver.

An interesting situation is one in which the oscillations are allowed to build up from zero, that is, at time $t = 0$ nothing is moving. In the RLC circuit we may then ask for the condition that $I = 0$ at $t = 0$. Of course, this single initial condition does not suffice to solve the problem because we have two constants to evaluate, A and ϕ. We could then ask that the charge on the capacitor $Q = \int I\,dt$ be zero at $t = 0$, so that initially the total energy stored in the system be zero. The computations are straightforward but tedious and we shall not pursue them further. Rather, we show pictorially in Fig. 1.23 what can be expected. The type of results we obtain (assuming that the

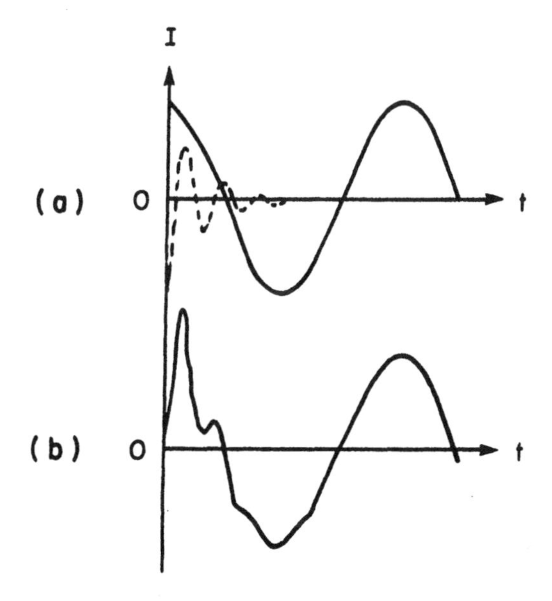

Fig. 1.23 The effect of the transient on a driven oscillator. (a) Solid curve is the sinusoidal driven oscillation of the current and the dashed curve is the transient generated by the act of turning on the driver at time $t = 0$. The frequency ω of the driven oscillations is taken to be $(1/7)^{th}$ the frequency Ω of the transient oscillations. The damping constant $\beta = 0.30\,\Omega$. (b) Net current obtained by superposition of the two oscillations. Observe the large distortions of the sinusoidal shape at early times. [After J. B. Marion, Classical Dynamics of Particles and Systems (Academic Press, New York, 1970).]

damping coefficient β is fairly small) depends critically on the relative frequency ω of the driver and the natural frequency of oscillation ω_0 of the freely oscillating system (we assume that β is sufficiently small so that $\Omega = \sqrt{\omega_0^2 - (\beta/2)^2}$ can be well represented by ω_0).

Figure 1.23 illustrates the situation where the driving frequency ω is very much less than the natural frequency ω_0. At the top of the figure we show the individual currents that make up the complete solution: the transient current I_1 and the steady driven current I_2. Observe that we have arranged matters so that at $t = 0$, $I_1 = -I_2$ with the result that the net current I is zero. At the bottom of the figure is shown the superposition of the two currents. Observe the serious distortion of the sinusoidal shape experienced by the oscillator at early times.

In Fig. 1.24 we give sketches of the oscillations as one changes the frequency of ω relative to ω_0. Sketch (a) is the same as that depicted in the previous figure, namely for $\omega \ll \omega_0$. Sketch (b) shows what happens when ω is close to ω_0. We now see that the amplitude of the driven current exhibits a slow almost period oscillation. This "modulation" of the amplitude is typical whenever two oscillations of nearly equal frequencies are allowed to interact with one another. These so-called "beats" have a frequency equal to $\omega - \omega_0$ and thus a period of $T = 2\pi/(\omega - \omega_0)$. If it were not for the damping term $\exp[-(\beta/2)t]$ in Eq. 1.113, the beats would persist forever; in the presence of damping, however, the beats gradually die out and the oscillator settles into a steady state; the sum and difference frequencies come about as a result of the addition of the two cosines of Eq. 1.113. In sketch (c) we have the case where ω is equal to ω_0, showing that now the oscillations build up in a gradual uniform way. And, finally, the last picture represents an oscillation being driven at a frequency very much greater than its natural frequency. A kind of modulation is again apparent.

Example

The circuit of Fig. 1.19 has components whose values are $R = 0.2$ ohms, $L = 10^{-4}$ henries, $C = 10^{-2}$ microfarads, and

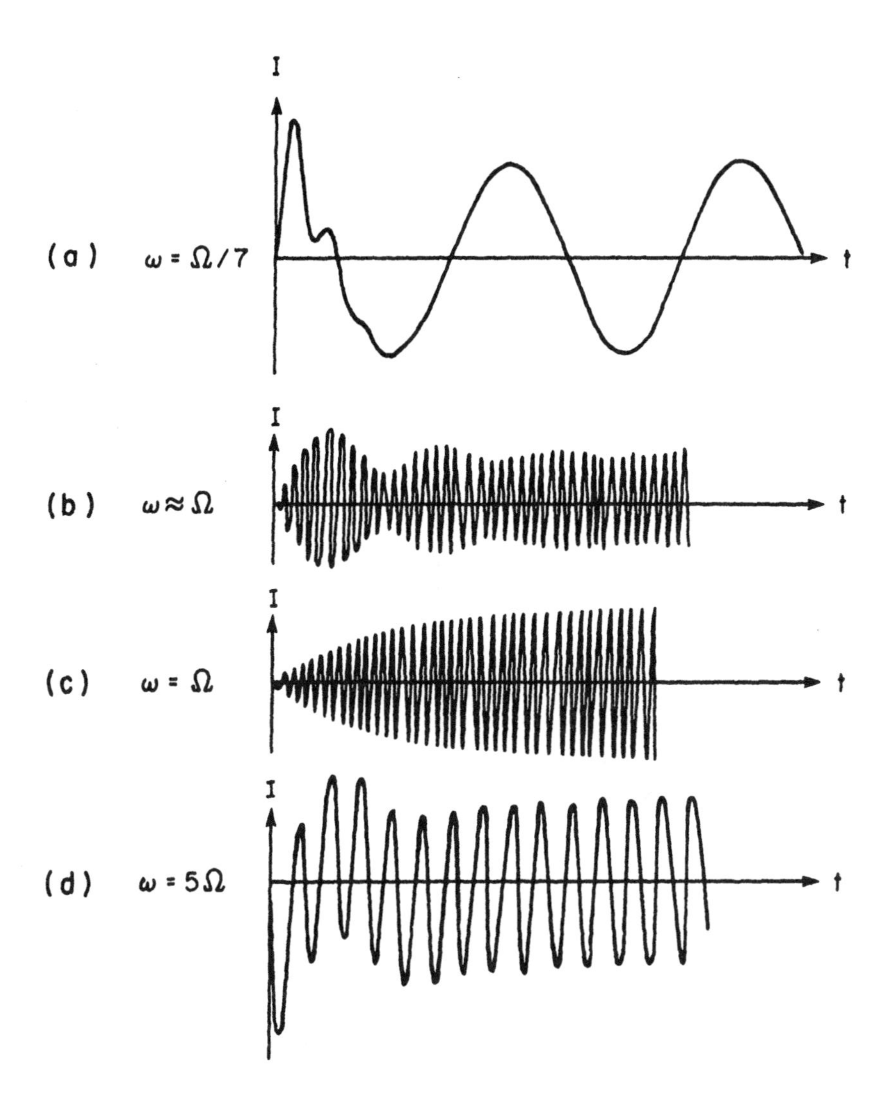

Fig. 1.24 Transient response of an oscillator with a periodic driving force of different frequencies ω relative to the frequency $\Omega = \sqrt{\omega_o^2 - (\beta/2)^2}$; ω_o is the natural frequency of oscillation of the undamped oscillator. In all these cases, $\beta \ll \omega_o$ so that Ω is almost equal to ω_o. At time $t = 0$, $I = 0$.

$\mathcal{E}_o$ = 1.5 volts. Give a general expression for the current in the circuit as a function of frequency and time after the transients have died out. Compute the current flowing in the circuit at resonance, the Q of the circuit, and the lifetime of the oscillations when the driving voltage is replaced by a short circuit.

Solution

The resonant frequency ω_o is given by Eq. 1.96 as

$$\omega_o = \frac{1}{\sqrt{LC}} = \frac{1}{\sqrt{(10^{-4})(10^{-8})}} = 10^6 \text{ sec}^{-1}$$

and the β of Eq. 1.90 is

$$\beta = \frac{R}{L} = \frac{0.2}{10^{-4}} = 2 \times 10^3 \text{ sec}^{-1}.$$

The general expression for the current is given in Eq. 1.95 which becomes

$$I(t) = \frac{1.5 \times 10^4 \omega}{\sqrt{(10^{12} - \omega^2)^2 + 4 \times 10^6 \omega^2}} \cos(\omega t + \delta) \text{ amp.}$$

where δ is (Eq. 1.94)

$$\tan \delta = \frac{10^{12} - \omega^2}{2 \times 10^3 \omega}.$$

At resonance, $\omega = \omega_o$ and the current becomes

$$I(t) = 7.5 \cos \omega_o t \text{ amperes.}$$

The Q of the circuit is found from Eq. 1.110:

$$Q = \frac{\omega_o}{\beta} = \frac{10^6}{2 \times 10^3} = 500$$

and the lifetime of the oscillations with the driving voltage removed is

$$Q = \omega_o \tau$$

$$\tau = \frac{Q}{\omega_o} = \frac{1}{\beta} = 5 \times 10^{-4} \text{ sec.}$$

1.5 OTHER EXAMPLES OF DRIVEN OSCILLATORS

The unified aspects of science – that is, the representation of a diversity of apparently unrelated physical behavior by a single idea or equation is nowhere better exemplified than by the case of the oscillator. Indeed, the RLC circuit and its workings have become so much a part of the scientists' and engineers' know-how that, for example, it has become quite common practice to adopt the language of electronics and to draw pictures of equivalent electrical circuits for nonelectrical systems. The purpose of this section is to acquaint the reader with a few other examples of driven oscillators and at the same time apply to them what we have learned in the previous section.

The seismometer

The seismometer and its sister instrument, the seismograph, are units placed on or into the ground for the purpose of detecting motions of the earth. These motions are due to earthquakes or they can be induced by explosions as is the case in seismic explorations. It is necessary to record motions as small as 10^{-8} inches (a few angstroms!).

The basis of all the instruments is the simple harmonic oscillator (pendulum, mass on a spring, etc.) attached to a support that rests firmly on the ground. The motion of the support due to earth vibrations then drives the oscillator. A simple and widely used type is illustrated in Fig. 1.25. The instrument has a coil and a magnet; the coil is rigidly fixed to the support while the heavy magnet is supported by a spring or a system of springs. Any relative motion between the coil and the spring causes a change in magnetic flux and this, by

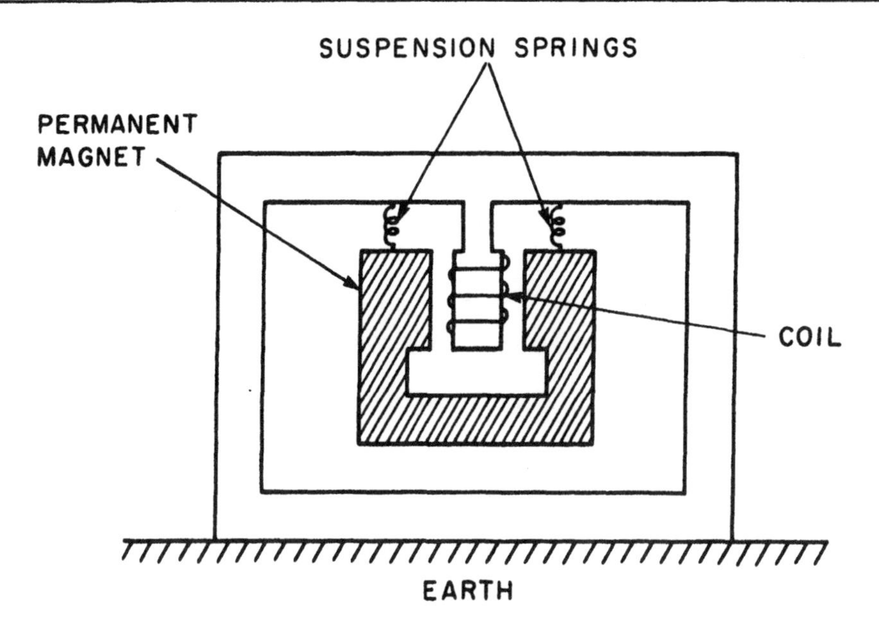

Fig. 1.25 Schematic of electromagnetic seismometer. When the heavy spring-suspended magnet moves relative to the coil, an electromotive force is generated across the coil winding proportional to the relative velocity of the coil and magnet. The signal is amplified electronically and plotted as a function of time on a chart recorder or magnetic tape. [See Introduction to Geophysical Prospecting by M. B. Dobrin (McGraw Hill Book Company, New York, 1952); also Elementary Seismology by C. F. Richter (Freeman & Co., San Francisco, 1958).]

Faraday's law, induces a voltage across the coil's terminals which is proportional to the velocity of motion. Thus we see that the coil moves with the earth while the magnet acts as the inertial element. What then is the equation of motion? Referring to Fig. 1.26, let y be the displacement of the mass at some instant of time t measured relative to a fixed point in space 0; and let y_o be its equilibrium position. The restoring force of the spring acting on the mass M is $K(y-y_o)$ with K as the spring constant. However, if the motion of the earth causes the support to move a distance Y relative to 0, the restoring force becomes $K(y-y_o-Y)$ and the equation of motion of M takes on the form

$$M\frac{d^2y}{dt^2} = -K(y-y_o-Y) - \gamma\frac{d}{dt}(y-y_o-Y) \qquad (1.114)$$

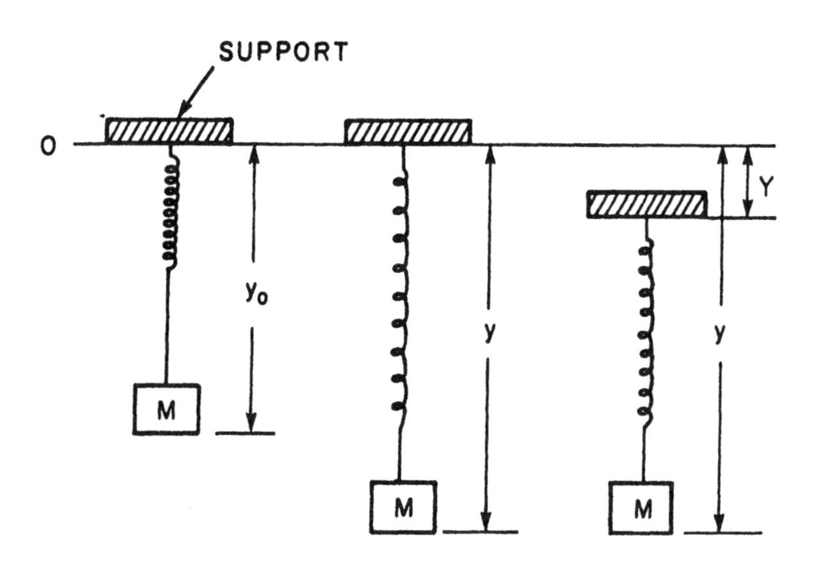

Fig. 1.26 Principle of the seismograph. A mass M and a spring hang vertically from a support. As a result of motion of the earth, the support moves a distance Y relative to a fixed point in space 0. This motion drives the mass-spring system: y_0 is the equilibrium position of the mass and y is its displaced position. Such an arrangement measures vertical motions of the earth. Other types of seismographs are designed to be sensitive to X and Z, the two horizontal displacements.

where we have included a damping force proportional to the velocity of M. Since the seismogram is a measure of the quantity $(y-y_0-Y)$ we define a new variable $p \equiv (y-y_0-Y)$ and obtain

$$M\frac{d^2p}{dt^2} + \gamma\frac{dp}{dt} + Kp = -M\frac{d^2Y}{dt^2} \tag{1.115}$$

which relates the seismogram deflections p to the actual vertical ground displacement Y. The differential equation is seen to be of precisely the same type as Eqs. 1.88 and 1.89 for the RLC circuit. Of course, seismic signals are a far cry from the nice sinusoidal oscillations we assigned earlier to our electrical energy source driving the circuit. But no matter; we know, and we shall learn in greater detail later, that pretty much any physical quantity like Y(t) can be decomposed (Fourier-analyzed, as it is called) into a sum of sine and cosine functions. So we might just as well take one of these, say,

$$Y = Y_0 \cos(\omega t), \tag{1.116}$$

learn what its consequences are, and then by the principle of superposition obtain the net outcome of all such sinusoidal perturbations. When we insert Eq. 1.116 in Eq. 1.115 and solve by the method described in section 1.4, we find that

$$p = \text{transient "complementary" function}$$

$$\text{plus}$$

$$Y_0 \frac{\omega^2}{\sqrt{\left(\omega_0^2 - \omega^2\right)^2 + (\beta\omega)^2}} \cos(\omega t - \delta) \tag{1.117}$$

where $\omega_0^2 = K/M$, $\beta = \gamma/M$, and $\tan\delta = \beta\omega/\left(\omega_0^2 - \omega^2\right)$.

Equation 1.117 tells us a great deal about how to design a seismometer. First, we must get rid of the transient part of the solution because it tells nothing concerning the motion of the earth — its characteristics are dominated by properties of the freely decaying oscillations. To eliminate this term sufficient damping must be provided. The remaining term, that is, the particular solution, is now proportional to Y_0, the amplitude of the earth's motion, and this is just what we are looking for. Since the seismic signal is composed of many frequencies ω, there are some that will be at and near the resonance frequency ω_0. Such signals, being near resonance, will be greatly amplified (see Fig. 1.20) all out of proportion to signals at other frequencies. This distortion is bad and is remedied by making seismometers with low Q values. It is quite customary to choose critical or near-critical damping ($\beta = 2\omega_0$) with the result that

$$p = Y_0 \frac{\omega^2}{\omega_0^2 + \omega^2} \cos(\omega t - \delta). \tag{1.118}$$

We now have to decide on a suitable value of ω_0 compared with the important range of frequencies ω expected in the seismic signal. Should ω_o be small or large compared with ω? If $\omega_o \ll \omega$, then the amplitude of p equals Y_o and is independent of frequency ω. The seismograph is then a direct reading instrument of the earth displacements. If, on the other hand, $\omega_o \gg \omega$, p becomes proportional to $Y_o\omega^2$ which is not a displacement but an acceleration. Such an instrument is known as an accelerometer. Earthquakes have periods T in the range ~0.1 to ~10 sec and seismographs whose natural period is of the order of 15-30 sec are commonly used. Thus, $\omega_o < \omega$ and the instruments act approximately as displacement meters. In seismic prospecting the dominant frequencies are higher ($\nu = \omega/2\pi \sim 30$ Hz) and therefore ω_o can be made correspondingly larger.

Electrical polarizability of atoms

The response of an atom (or molecule) to an applied electric field is a subject that, strictly speaking, would require us to go deeply into the quantum theory of atomic structure. However, it is possible to obtain a simplified but nonetheless quite adequate model by making use of one or two basic results concerning the structure of atoms.

A simple atom such as a hydrogen atom can be pictured comprised of a tiny, almost pointlike nucleus carrying an elementary charge +q, and a revolving electron of charge −q. The instantaneous positions of the electron are fairly uncertain, but when looked at over a sufficiently long time (greater than $\sim 10^{-16}$ seconds) the electronic structure appears as a quasi-stationary, smeared out, charged cloud surrounding the nucleus. The clcud is close to being spherically symmetric, dense in the center and falling off in density at the edges. Most of the charge resides within a sphere of $\sim 5 \times 10^{-9}$ cm radius (~0.5 Å) and we call this the atomic radius (Fig. 1.27). Note that in this picture the centers of the positive and negative charges coincide; the atom does not have an intrinsic dipole moment of its own. Such classes of atoms are called nonpolar. However, the centers of positive and negative charges may not coincide as is the case for the water molecule, but for the sake of simplicity we shall exclude these so-called polar substances here.

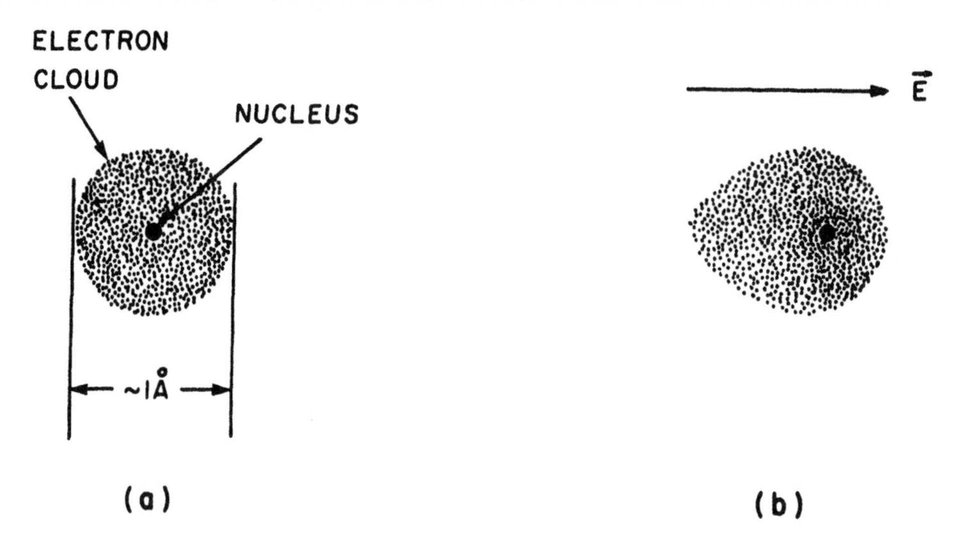

Fig. 1.27 (a) The spherically symmetric electron cloud concentric with the nucleus of an unperturbed hydrogen atom. (b) In the electric field $\vec{E}$, the negative electron cloud is displaced relative to the positively charged nucleus and the atom is said to be polarized.

If we now subject a nonpolar atom or nonpolar molecule to an electric field $\vec{E}$, the electron cloud becomes displaced relative to the nucleus and a dipole moment is generated. A rigorous determination of the effective displacement $\vec{r}$ is a complicated problem in quantum mechanics. It is, however, plausible and actually confirmed by rigorous theory, that the electron behaves to good approximation as if it were bound to its equilibrium position by a quasi-elastic, springlike restoring force $\vec{F}$(restoring) = $-K\vec{r}$. Now, suppose that the atom is acted upon by a sinusoidal electric field

$$\vec{E} = \vec{E}_o\, e^{j\omega t} \tag{1.119}$$

which could be the alternating field generated between a pair of capacitor plates, or it could be the electric field of a radio signal, or even of a light signal. The equation of motion of the atomic electron of mass m and charge q is then given by*

$$m\frac{d^2\vec{r}}{dt^2} + \gamma\frac{d\vec{r}}{dt} + K\vec{r} = q\vec{E}_o\, e^{j\omega t} \tag{1.120}$$

where $\gamma(d\vec{r}/dt)$ represents a phenomenological damping force. Dividing through by m gives

$$\frac{d^2\vec{r}}{dt^2} + \beta\frac{d\vec{r}}{dt} + \omega_o^2\vec{r} = \left(\frac{q}{m}\right)\vec{E}_o\, e^{j\omega t} \tag{1.121}$$

with $\beta = \gamma/m$ and $\omega_o = \sqrt{K/m}$. As we already know, the steady-state solution is

$$\vec{r} = \frac{(q/m)}{\sqrt{\left(\omega_o^2 - \omega^2\right)^2 + (\beta\omega)^2}}\,\vec{E}_o\cos(\omega t - \delta) \tag{1.122}$$

and the time-varying dipole moment $\vec{p} = q\vec{r}$ is

$$\vec{p} = \frac{(q^2/m)}{\sqrt{\left(\omega_o^2 - \omega^2\right)^2 + (\beta\omega)^2}}\,\vec{E}_o\cos(\omega t - \delta). \tag{1.123}$$

Thus the net outcome of applying an electric field is the creation of atomic dipoles with each electron contributing the amount given by Eq. 1.123. We see that $\vec{p}$ is proportional to the electric field; the constant of proportionality is known as the electrical polarizability of the atom and is an important parameter in the analysis of atomic structure. It is generally denoted by the letter α. Since, as we shall see in a later chapter, α can be readily deduced from measurements, chemists and physicists use it quite extensively to help them unravel the atomic structure of more complex atoms and molecules.

The net polarization experienced by the substance in question is the sum total of contributions from each electron. If we have a system of identical molecules of density N per cubic meter, it is

*For an electron, q = −e; for a proton, q = +e. Using q instead of +e or −e makes the equations general for any charge and eliminates any chance of making errors in sign.

convenient to define the total "density of polarization" $\vec{P} = N\alpha\vec{E}$ so that

$$\vec{P} = \frac{(Nq^2/m)}{\sqrt{\left(\omega_o^2 - \omega^2\right)^2 + (\beta\omega)^2}} \vec{E}_o \cos(\omega t - \delta). \tag{1.124}$$

But, this is just a definition — the physics is all in Eq. 1.123.

It is clear that a substance is likely to exhibit the effects of polarization most dramatically at a frequency at and near the resonant frequency ω_o. One of these effects is the phenomenon of absorption where electrical energy will be removed from the driver and dissipated by the atoms. If the electric field is that of an electromagnetic wave passing over the ensemble of atoms the wave amplitude will be diminished. After passage through the substance, moreover, when the incident electromagnetic wave contains many frequencies, the ones corresponding to $\omega = \omega_o$ will be absorbed while the other frequency components will remain virtually unaffected. Dark absorption lines will appear. The solar spectrum, for example, shows thousands upon thousands of such features known as Fraunhofer lines. In the foregoing analysis we assumed implicitly that there was only one resonance frequency. In general there are many such frequencies even in a system of the same kind of molecules. Equation 1.124 must then be replaced by a more general expression,

$$\vec{P} = \left[\left(\frac{Nq^2}{m}\right) \sum_k \frac{f_k}{\left(\omega_k^2 - \omega^2\right)}\right] \vec{E}_o \cos \omega t \tag{1.125}$$

where for simplicity we have neglected damping. The parameter f_k is defined in such a way that Nf_k is the total number of electrons per unit volume having the resonance frequency ω_k. For most gases and for many solids (as, for example, the various glasses) the ω_k's are at frequencies corresponding to the ultraviolet. For example, hydrogen, oxygen, and air have their important resonant frequencies at 3.40×10^{15} Hz, 3.55×10^{15} Hz, and 3.98×10^{15} Hz, respectively. There are none in the visible region of the spectrum — if there were any, we would see absorption bands in air and glass would be neither transparent nor colorless!

Observe that so far the motion of the nuclei in the applied electric field has been neglected. The reason for this can be easily understood. The light electrons follow the field almost instantaneously up to very high frequencies, including the visible spectrum. The nuclear masses, on the other hand, are so heavy that they cannot follow the field in the high frequency region and their polarizability for visible light frequencies is negligibly small compared with the polarizability for the electrons. Therefore, when, if ever, are nuclear contributions important? To make an estimate, remember that the resonant frequency ω_o is proportional to (spring constant K/mass)$^{1/2}$. Now, it is reasonable to assume that the quasi-elastic restoring forces are about the same for electrons and nuclei, so that K is the same for both, and $\omega_o \propto 1/\sqrt{\text{mass}}$. For hydrogen the nuclear mass is 1840 times larger than the electron mass so that ω_o for the nucleus is approximately 1/43 that for the electron. This then brings the resonant frequency from 3.40×10^{15} Hz down to about 8×10^{13} Hz which is in the infrared regime of the spectrum. It is here that nuclear contributions to $\vec{P}$ are expected to be felt. In the limit of zero frequency, that is, for the static, time-independent electric perturbations, $\vec{p}$ from Eq. 1.123 reduces to

$$\begin{aligned} \vec{p} &= \frac{q^2}{m\omega_o^2}\vec{E}_o \\ &= \frac{q^2}{K}\vec{E}_o \end{aligned} \tag{1.126}$$

which shows clearly that now contributions to $\vec{p}$ from nuclear motion certainly cannot be omitted.

Equation 1.122 has the nice property that the electron displacement $|\vec{r}|$ is directly proportional to the strength $|\vec{E}_o|$ of the electric field. This makes for simplicity — the system is linear, superposition holds and the summation of effects like that shown by Eq. 1.125 is entirely justified. But, is $|\vec{r}|$ always going to be proportional to E_o? It is unlikely, if the excursion of the electron is too large. Take $|\vec{E}_o|$

as high as 10^6 volts/meter, substitute this in Eq. 1.122 and assume for simplicity that $\omega \ll \omega_0$ and $\beta = 0$. Using the value $\omega_0/2\pi = 3.4 \times 10^{15}$ Hz applicable to hydrogen, we find that $|\vec{r}| \approx 5 \times 10^{-16}$ meters, which is a very tiny displacement indeed, corresponding to about one millionth of the atomic radius. Linearity is likely to be a good approximation. To displace the electron through a full atomic radius, which in effect rips it out and ionizes the atom, requires (according to Eq. 1.122) a field of approximately 10^{11} volts/meter. In actual practice a field less than one hundredth as big suffices to do the job. However, well before onset of such a destructive event the linearity between $|\vec{r}|$ and $|\vec{E}_0|$ becomes invalid and nonlinear aspects of the material's structure come to the fore. High-power lasers can generate electric fields of these magnitudes when focused into the material in question and they are being used in very exciting studies of the nonlinear properties of matter.

Electron cyclotron heating

Some interesting things happen when, in addition to the oscillating electric field $\vec{E}(t)$ of the previous subsection, the electron is simultaneously subjected to a static magnetic field $\vec{B}_0$. For simplicity, let the electron be a free electron as in a plasma, rather than being bound to the nucleus. Also, let the electric field act along the x axis and the magnetic field along the z axis, as is shown in Fig. 1.28a. Suppose for the moment that $\vec{E}(t)$ is zero and only $\vec{B}_0$ acts. When an electron is injected into the magnetic field with an initial velocity v along, say, the x axis, the magnetic force

$$\vec{F}(\text{magnetic}) = q(\vec{v} \times \vec{B}_0) \tag{1.127}$$

will bend its path into the form of a circle of radius $v(qB_0/m)^{-1}$. Now, let us put back the electric field; it will give the electron periodic kicks with a force

$$\begin{aligned} \vec{F}(\text{electric}) &= q\vec{E}(t) \\ &= q\vec{E}_0 \cos(\omega t) \end{aligned} \tag{1.128}$$

and these kicks will sometimes aid the electron's motion when the phase and direction are right, and sometimes it will hinder the motion when they are wrong. By and large the detailed trajectory is complicated and difficult to show pictorially.

Let us then write the equations of motion along the three coordinate axes, using $\vec{F} = m\vec{a}$ and Eqs. 1.127 and 1.128:

$$\left.\begin{aligned} m\frac{d^2x}{dt^2} &= qE_o \cos(\omega t) + qB_o \frac{dy}{dt} \\ m\frac{d^2y}{dt^2} &= 0 \qquad\qquad - qB_o \frac{dx}{dt} \\ m\frac{d^2z}{dt^2} &= 0 \qquad\qquad + \quad 0 \end{aligned}\right\} \tag{1.129}$$

Assuming for the present that the electron has no velocity along the z axis, that is, $v_z = dz/dt = 0$, the last of the three equations drops out and need not concern us further. We differentiate the first equation once with respect to time and obtain that

$$m\frac{d^2v_x}{dt^2} = -qE_o\omega \sin(\omega t) + qB_o \frac{d^2y}{dt^2} \tag{1.130}$$

where we have written $v_x = dx/dt$ as the x component of the particle's velocity. We now insert the second equation (1.129) into Eq. 1.130 and thus eliminate d^2y/dt^2 in favor of dx/dt. The result is that

$$\frac{d^2v_x}{dt^2} + \omega_c^2 v_x = -\left(\frac{q}{m}\right) E_o\omega \sin(\omega t) \tag{1.131}$$

where

$$\omega_c = \frac{qB_o}{m} \tag{1.132}$$

is called the electron cyclotron frequency and, in fact, represents the rate at which the electron revolves around the magnetic field line when E(t) equals zero. The form of Eq. 1.131 is by now entirely familiar to us. It is once again the equation of the driven oscillator in which v_x is the time-varying quantity, ω_c the natural frequency oscillation of the freely oscillating system, and $-(q/m)\,E_0\omega\,\sin(\omega t)$ is the driving force. A similar equation can be found for v_y, the y component of the velocity of the charge. This is given by

$$\frac{d^2v_y}{dt^2} + \omega_c^2 v_y = -\left(\frac{q}{m}\right) E_0\omega_c \cos(\omega t). \tag{1.133}$$

That we have two equations should not be surprising; after all, motion in two dimensions can be thought of as a superposition of two simple harmonic oscillations at right angles to one another.

To derive the particle orbits we must solve these two equations for v_x and v_y (the method is described in Appendix A2 for $\omega \neq \omega_c$) and then integrate the results to obtain the positions x and y for any time t. The results are illustrated in Fig. 1.28b.

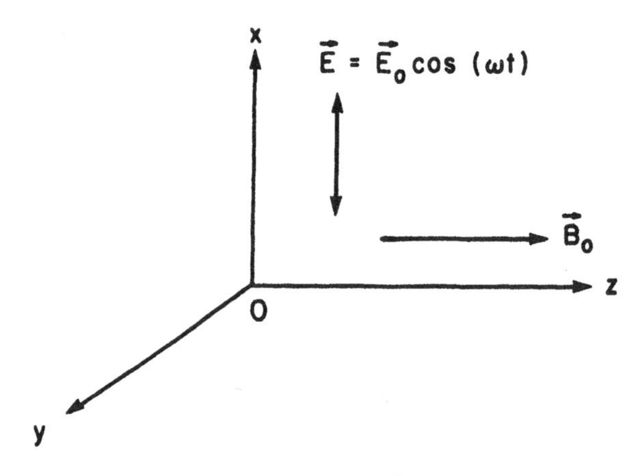

Fig. 1.28 (a) A static magnetic field $\vec{B}_0$ oriented along the z axis and an oscillating electric field $\vec{E}$ oriented along the x axis simultaneously act on a charge q (electron). In the absence of $\vec{E}$, the electron would move in a helical path along the z axis; if it had no initial velocity along z, it would just go in a circular orbit around a magnetic field line. Its speed around the circle is $v = (qB_0/m)R$ where R is the radius of its circular motion.

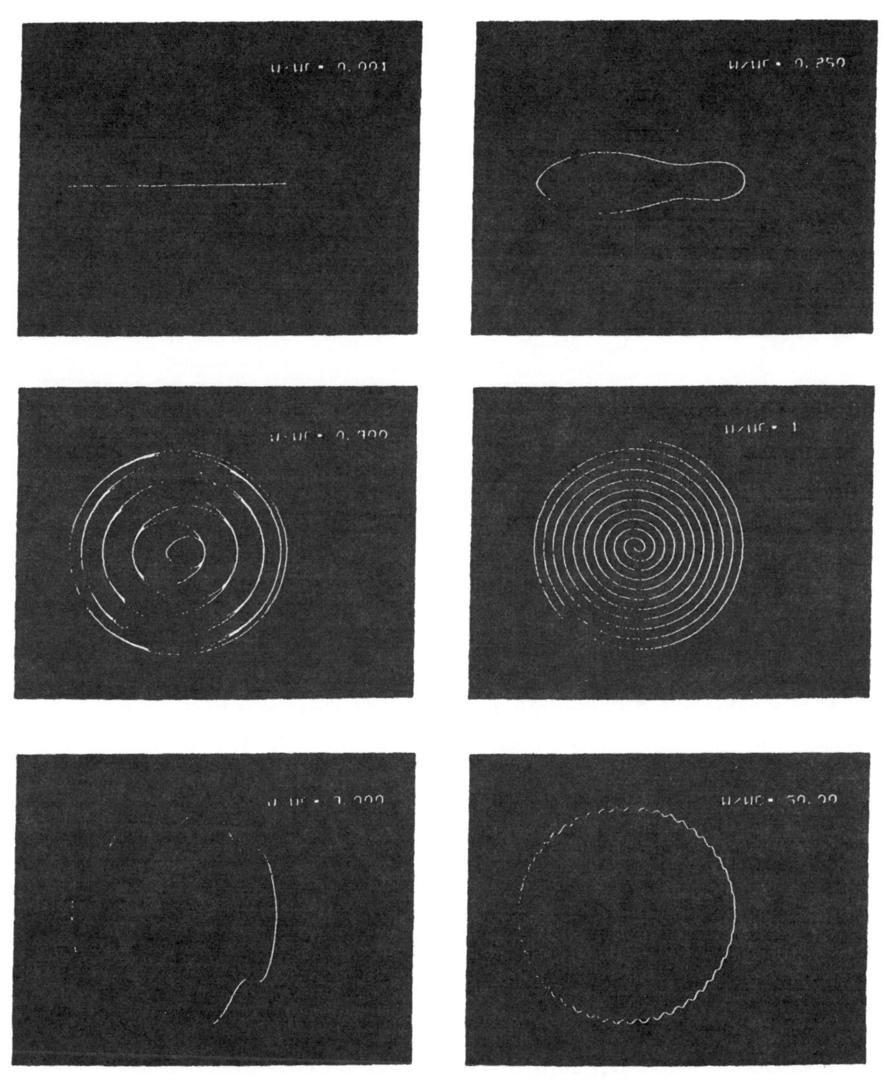

Fig. 1.28b Computer-generated orbits of an electron in a steady magnetic field B_o (pointing out of the page) and an RF electric field at right angles to B_o; ω is the frequency of the RF field and $\omega_c = qB_o/m$ is the cyclotron frequency. The computer program in FORTRAN is given in Appendix 3. When $\omega \ll \omega_c$ the electron drifts slowly in a quasi-elliptical path with a drift velocity $\vec{v} = (\vec{E} \times \vec{B}_o)/B_o^2$. When $\omega \gg \omega_c$ the motion is a circular gyration dominated by $\vec{B}_o$ and modulated by the high frequency electric field. When $\omega = \omega_c$ the motion is no longer periodic (see text).

We note that when $\omega \neq \omega_c$ the orbits are periodic. The one exception is the important case $\omega = \omega_c$ which shows that the orbit is continuously increasing.

When $\omega = \omega_c$ the motion is not periodic and the customary periodic trial solutions with which we are so familiar can no longer be used to solve these differential equations. Suppose, for example, that initially (when $t = 0$) $v_x = v_y = 0$. Then the solution of Eqs. 1.131 and 1.133 is of the form

$$v_x = A \sin(\omega_c t) + Bt \cos(\omega_c t)$$
$$v_y = A' \sin(\omega_c t) + B't \sin(\omega_c t) \qquad (1.133a)$$

showing that the second term on the right-hand side is aperiodic and the amplitude of the motion grows linearly with time. If the particle is initially at the origin of coordinates ($x = y = 0$ at $t = 0$) and if it has initially zero velocity $v_x = v_y = 0$ at $t = 0$ and accelerations $a_x = qE_o/m$, $a_y = 0$ at $t = 0$, the subsequent position coordinates are

$$x = \left(\frac{qE_o}{2m\omega_c^2}\right)[(\omega_c t) \sin(\omega_c t)]$$

$$y = \left(\frac{qE_o}{2m\omega_c^2}\right)[(\omega_c t) \cos(\omega_c t) - \sin(\omega_c t)],$$

a result which a patient reader can easily derive using (1.133a) as a trial solution of the differential equations (1.131) and (1.133). The particle trajectory corresponding to these solutions is shown in the center of Fig. 1.28b.

Consider an electron initially at rest, subjected to the electric and magnetic fields under discussion. At resonance, $\omega = \omega_c$, the amplitudes can be extraordinarily large and the initially cold, immobile electron will have gained much kinetic energy. This is the concept of electron cyclotron heating. In practice (see Fig. 1.29) a resonant microwave cavity with a volume of several liters is filled with unionized gas like hydrogen maintained at a pressure of $\sim 10^{-3}$ Torr.

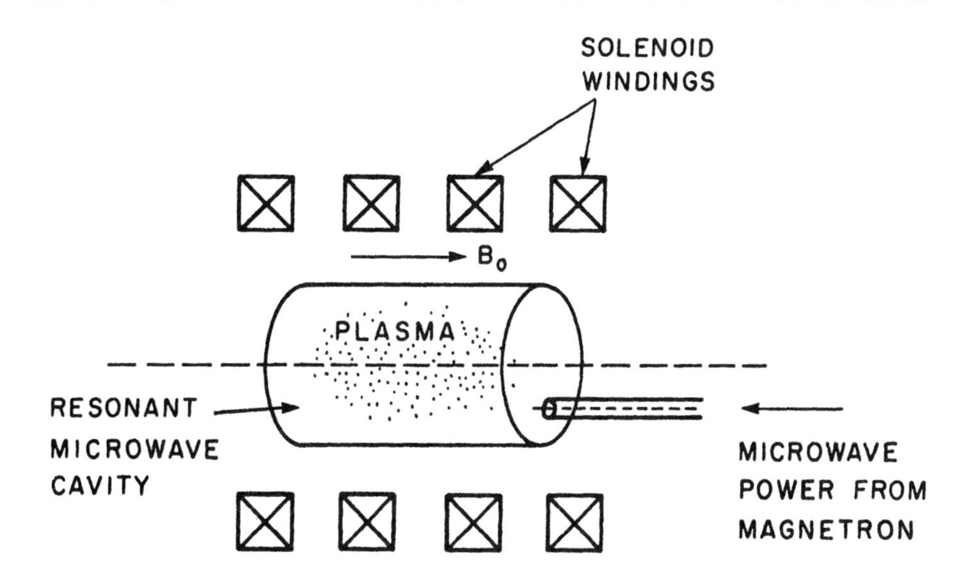

Fig. 1.29 Sketch of apparatus in an electron cyclotron heating experiment of a plasma. The frequency ω of the microwaves and the electron cyclotron frequency $\omega_c = (qB_0/m)$ are adjusted to be nearly the same so that maximum microwave power can be delivered to the plasma. The cylindrical microwave cavity has many modes of oscillation with different frequencies and orientations of the electric vector. For cyclotron resonance to occur, $\vec{E}$ must be orthogonal to $\vec{B}_0$. [See R. A. Dandl et al., Nuclear Fusion **4**, 344 (1964).]

Tens of kilowatts of microwave power at a frequency $\omega/2\pi$ of, say, 10^{10} Hz is pumped into the resonator where rf electric fields of thousands of volts per centimeter are created. The static magnetic field is provided by a solenoid surrounding the cavity. For resonance to occur $\omega_c = qB_0/m = \omega$, from which it is deduced that the magnetic field must have a value equal to approximately 3500 gauss. Fine tuning on $|\vec{B}_0|$ is provided so that the exact resonance conditions can be attained. Initially, the gas is un-ionized, but the cavity may contain a stray electron or so from a cosmic ray. The electron is heated by the cyclotron resonance mechanism, and gains large speeds. When it collides with a neutral atom, it has enough energy to ionize it (it requires at least 13.5 eV) and an electron-ion pair is born.

The new electron gets heated, collides, and so on. Eventually the entire gas becomes virtually fully ionized. Then the microwave energy goes into heating the electrons further and further, and since electrons collide with ions, the latter likewise become hot. In this manner it is found possible to heat electrons to energies of several thousand electron volts and ions to energies of several hundred volts. At one time it was hoped that electron and ion temperatures of approximately 10 keV could be achieved, which is the energy necessary for controlled thermonuclear fusion to occur. Unfortunately, a number of problems have arisen. Just to mention two, there is the problem of electron motion along magnetic field lines. A hot electron can move freely along $\vec{B}_0$, leave the active heating region, strike a wall and thus be lost from the plasma pool. So-called "magnetic bottles" partially alleviate this difficulty, but only partially. There is also the fact that as the electron becomes more and more energetic, its mass increases as a result of relativity by an amount $[1-(v/c)^2]^{-1/2}$; the cyclotron frequency ω_c decreases and that particular electron gets out of resonance with the applied microwave field.

1.6 NONLINEARITIES

When a spring is stretched too far, a pendulum swung too hard, when too much electrical current is passed through a resistor or when too strong an electric field acts on the atoms of a dielectric, the system will probably respond in quite a complicated way. No longer is it true to say that the response r is proportional to the excitation e — as was assumed implicitly hitherto. We have before us the domain of nonlinear phenomena where the response could well be of the form

$$r = ae + be^2 + ce^3 + \dots \qquad (1.134)$$

where a is the coefficient associated with the familiar linear response, and b, c, etc., are coefficients of the nonlinear terms. Since a is often much larger than b, c ... the nonlinearities contribute noticeably only during strong excitation e. This section will discuss the importance of nonlinearities and how they manifest themselves. That they are important and have widespread applications goes without question. Radio and television communication the way we know it today could not exist without nonlinear elements. Indeed, the way we see and hear is based on nonlinearities in the eye and ear.

Harmonic generation

Suppose the nonlinear system is subjected to a sinusoidal excitation

$$e = e_o \sin(\omega t). \tag{1.135}$$

Then the response is

$$r = ae_o \sin(\omega t) + be_o^2 \sin^2(\omega t) + ce_o^3 \sin^3(\omega t) + \ldots. \tag{1.136}$$

which is a result obtained by inserting Eq. 1.135 in Eq. 1.134. Using standard trigonometric identities like $\cos(2\theta) = (1 - 2\sin^2\theta)$ this equation can be rewritten as

$$\begin{aligned} r = ae_o \sin(\omega t) &+ \tfrac{1}{2} be_o^2[1 - \cos(2\omega t)] \\ &+ \tfrac{1}{4} ce_o^3[3\sin(\omega t) - \sin(3\omega t)] \\ &+ \ldots. \end{aligned} \tag{1.137}$$

The first term on the right-hand side tells us nothing that we did not know already: that a pure sinusoidal excitation gives a pure sinusoidal response at the <u>same</u> frequency, as long as all nonlinearities are absent. The second term contains $\cos(2\omega t)$ showing that a quadratic nonlinearity contributes a signal at twice the excitation frequency, a tertiary nonlinearity contributes a signal at three times the excitation frequency, etc. Such harmonic generation is of great significance since it permits us to manufacture from low-frequency signals ever higher frequency signals albeit with greatly diminished intensity. This has been known in electronics for decades, but a beautiful modern example of this comes to us from optics, in an experiment first carried out in 1961 by P. A. Franken et al. In this experiment (Fig. 1.30) light from a pulsed high-power ruby laser delivering ~3 kW of power was focused onto a crystal of quartz. The red incident light at a

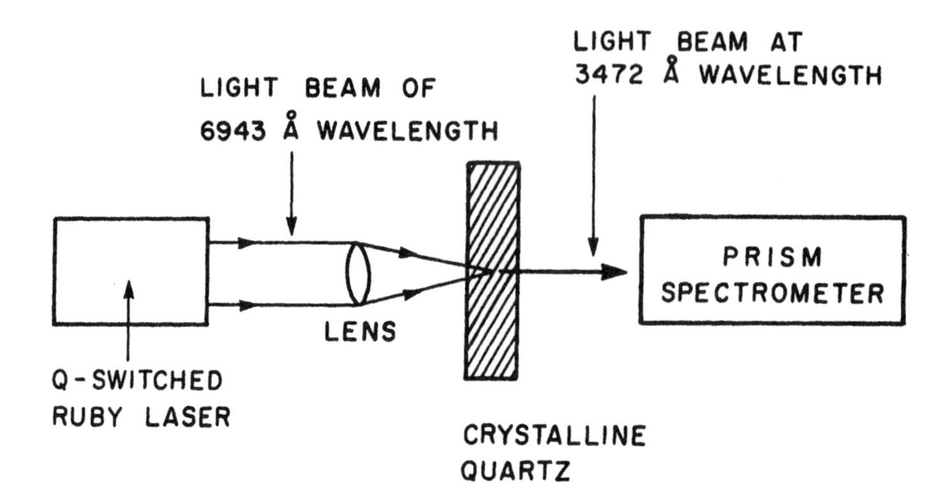

Fig. 1.30 Generation of an optical harmonic. The laser produced approximately 3 joules of light in a pulse 1 msec long. [See P. A. Franken et al., Phys. Rev. 7, 118 (1961); for later, more dramatic results see Scientific American, July 1963; also article by J. A. Giordmaine in Scientific American, April 1964.]

wavelength of 6943 Å emerged as a beam of green light(!) of wavelength 3471.5 Å (recall that halving the wavelength is equivalent to doubling the frequency). The intensity of the green radiation was found to be approximately 10^{-8} times weaker than the intensity of the incident ruby light. The cause of this second harmonic generation lies in the form of the polarizability of quartz. This material (and others, as for example KDP, potassium dihydrogen phosphate) have an induced dipole moment p which varies with the electric field E of the light wave as (cf. section 1.5)

$$p = \alpha E + \alpha_2 E^2 \tag{1.138}$$

thus exhibiting the quadratic nonlinearity required for second harmonic generation. Materials like calcite, on the other hand, have their lowest nonlinear term equal to $\alpha_3 E^3$ and hence generate the third harmonic of the incident light.

Let us now examine a little more quantitatively the oscillatory motion in the presence of the quadratic nonlinearity. For this purpose we take the simplest situation we can picture, that of a freely oscillating, undamped particle of mass M experiencing a restoring force

$$F(\text{restoring}) = -Kx + \epsilon Mx^2. \tag{1.139}$$

Its equation of motion is then given by

$$\frac{d^2x}{dt^2} + \omega_o^2 x - \epsilon x^2 = 0 \tag{1.140}$$

where $\omega_o = \sqrt{K/M}$ is the natural frequency of linear oscillation. There is no simple analytic solution of Eq. 1.140. However, if ϵ is taken to be a very small quantity, then it is plausible to assume that the solution of Eq. 1.140 will not differ appreciably from that of the simple harmonic oscillation, $x_o \cos(\omega_o t)$. We therefore write that

$$x = x_o \cos(\omega_o t) + \epsilon x_1 \tag{1.141}$$

where our task is to determine the function x_1 caused by the small nonlinearity present in the oscillation. In the so-called "perturbation" procedure we substitute Eq. 1.141 in Eq. 1.140 and then discard all terms with ϵ^2, ϵ^3, etc. as being of higher order of smallness, and therefore negligible, compared with terms proportional to ϵ itself.*
The result is that

$$\epsilon\left(\frac{d^2x_1}{dt^2} + \omega_o^2 x_1 - x_o^2 \cos^2(\omega_o t)\right) = 0 \tag{1.142}$$

which, for finite ϵ can only be satisfied when the term in brackets is zero. We therefore have a differential equation for the sought-for quantity x_1. To solve it we write $\cos^2(\omega_o t)$ as $[1 + \cos(2\omega_o t)]/2$,

$$\frac{d^2x_1}{dt^2} + \omega_o^2 x_1 = \frac{x_o^2}{2}[1 + \cos(2\omega_o t)] \tag{1.143}$$

*This perturbation procedure is not valid universally. For example, it breaks down completely when the equation of motion is

$$\frac{d^2x}{dt^2} + \omega_o^2 x - \epsilon x^3 = 0.$$

and use for a trial solution the form

$$x_1 = A\cos(2\omega_0 t) + B \tag{1.144}$$

with A and B as two coefficients to be determined. (Note that the form of Eq. 1.143 and the trial solution (1.144) are familiar to us from the discussion of the driven oscillator of section 1.4.) Inserting (1.144) in (1.143) gives

$$-3\omega_0^2 A\cos(2\omega_0 t) + \omega_0^2 B = \frac{x_0^2}{2}\cos(2\omega_0 t) + \frac{x_0^2}{2}. \tag{1.145}$$

But this equation must be true at all times. For this to be so, the first term on the left must equal the first term on the right, and the second term on the left must equal the second term on the right, so that

$$A = -\frac{x_0^2}{6\omega_0^2}$$
$$B = \frac{x_0^2}{2\omega_0^2}. \tag{1.146}$$

We thus arrive at the (approximate) solution of the nonlinear oscillator equation (1.140), correct to first order in the coefficient ϵ:

$$x(t) \approx x_0\cos(\omega t) + \epsilon\,\frac{x_0^2}{2\omega_0^2}\left[1 - \frac{1}{3}\cos(2\omega_0 t)\right] \tag{1.147}$$

and find, in accord with our expectations, harmonic displacements both at fundamental frequency ω_0 and the second harmonic frequency $2\omega_0$ (if we proceed to solve for terms involving ϵ^2, ϵ^3, etc., additional harmonics at $3\omega_0$, $4\omega_0$ ... would appear in the solution).

Rectification

Many systems, by virtue of their nonlinear characteristics possess the ability to convert an oscillatory excitation e into a response r which contains a steady component in addition to the oscillatory signals just discussed. The term $\left(be_o^2/2\right)$ of Eq. 1.137 is just such a component, as is the term $\left(\epsilon x_o^2/2\omega_o^2\right)$ of Eq. 1.147. We observe that both came from a quadratic nonlinearity, and both are proportional to the square of the excitation amplitude e_o.

This ability to convert is known as rectification and finds widespread use in electrical circuits where it is often desired to change ac currents to dc currents. In fact, any electrical device that has a high resistance to current in one direction and low resistance to current in the other direction will effect rectification. The important devices used for this purpose include the thermionic diode, certain crystals which have been incorporated into diodes, and junctions between semiconductor materials. The operation of the thermionic diode as a rectifier is particularly easy to understand. One of the two electrodes, the cathode, is coated with a material which emits electrons when heated to high temperatures. The other electrode (the anode) is kept uncoated and cold and cannot emit electrons. Thus, when the anode is positive, current can flow and when it is negative, the current is zero. Figure 1.31 illustrates a simple rectifier circuit, together with the input and output waveforms. We point out that the output waveform shown in the figure requires that during the conduction part of the ac cycle (when the anode is positive) the current is proportional to the applied voltage. In general this turns out to be a rather poor assumption. Under one set of conditions when the diode is said to be "space-charge limited" the current is, in fact, proportional* to $(\text{voltage})^{3/2}$.

Detection

Some of the most important detectors of oscillatory behavior constructed by nature and by man have incorporated in them a nonlinear element as an essential ingredient. Such is the case of the light sensors in our eyes and the sound sensors in our ears. Man-made

*This is known as the Child-Langmuir law.

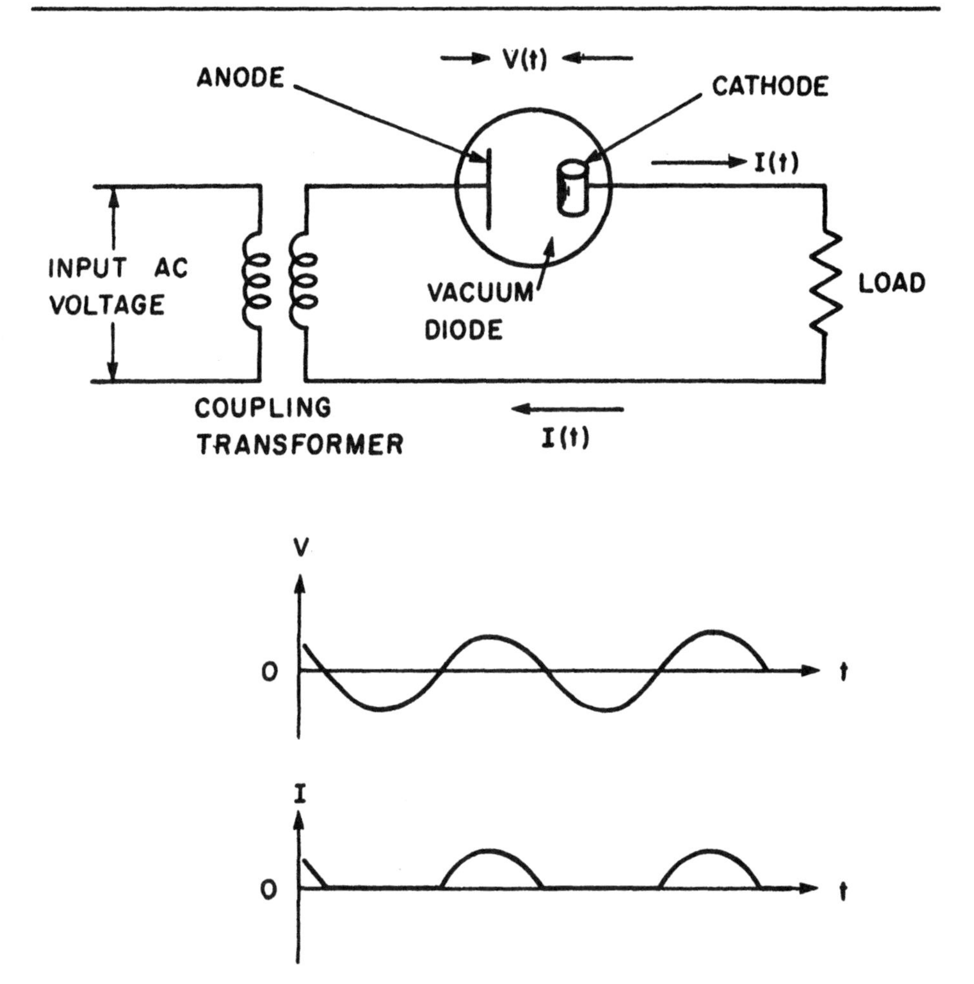

Fig. 1.31 Simple half-wave rectifier circuit and the input and output waveforms. The coated cathode cylinder is heated by a filament within it (not shown). It is assumed that when V is positive, I is proportional to V and when V is negative, I = 0.

devices include the crystal diode detector to detect microwave oscillations, the photodiode or photomultiplier to detect light oscillations, and calorimetric devices of all sorts to detect either sound or electromagnetic vibrations.

These instruments all share one common property — they respond not to the instantaneous positive and negative excursions of the oscillations, but rather to some time-averaged quantity associated with the oscillatory motions. And it is well that they do so because it is difficult to imagine how, for example, the relatively slow photochemical-neural response of the human eye could possibly follow the 10^{15} or so oscillations per second of visible light. The photodiode using the highly mobile electrons can do better than the human eye but it still cannot follow oscillations as rapid as this; the highest frequency that has been achieved is several times 10^9 Hz which is still almost a million times too slow.

A very desirable class of instruments are the so-called "square-law detectors" which are characterized by a response r to an excitation e given by

$$r = Ae^2 \tag{1.148}$$

where A is some constant determined by the detailed properties of the instrument. If e is a sinusoidal excitation $e = e_0 \cos(\omega t + \phi)$, substitution in Eq. 1.148 and use of the time-averaging procedure given by Eq. 1.20 yields

$$\langle r \rangle = \frac{1}{2} Ae_0^2, \tag{1.149}$$

showing that the measured observable quantity $\langle r \rangle$ is proportional to the square of the oscillation amplitude. In the photodiode, for example, the response r is simply the diode current generated by the incident light; e is proportional to the electric field of the light wave and e^2 is proportional to the light intensity. Not all detectors are precisely square-law. Some approach this condition well over large ranges of excitation levels e from the very smallest to the very largest. Others, like the crystal diode mentioned earlier, may have a linear response at small e, a square-law response for intermediate e and some quite complex response at very large e. In such cases, the instrument must be carefully calibrated before it can be used for quantitative studies of the oscillatory properties in question.

Modulation

The sending of information, speech, music, etc., via electromagnetic waves is man's principal means of transmitting intelligence rapidly over large distances. It is advantageous to do this at high frequencies. The reason is that at a high frequency level the transmission problems are easier. More important, it is possible to transmit a number of messages simultaneously without interference if the frequency band is different for each message.

A high-frequency perfectly sinusoidal oscillation $e = e_o \cos(\omega t + \phi)$ carries in itself no information other than the trivial statement that the oscillation in fact exists. To send information, the wave must somehow be tagged by the message itself, and this can be accomplished by varying either e_o, ω, or ϕ in some prescribed way. This is known as modulation, and more specifically, it is called amplitude, frequency, or phase modulation according to whether e_o, ω, or ϕ is tagged. Here we shall concern ourselves only with amplitude modulation.

Whereas the high-frequency unperturbed oscillation (known as the carrier)

$$E_c = E_{oc} \cos(\omega_c t) \tag{1.150}$$

may well be perfectly sinusoidal, the perturbation (i.e., message) will normally be a very complex function of time. Nevertheless, we can decompose it into a large number of sinusoidal oscillations, and we choose one of these (referred to as the modulation) and represent it by

$$E_m = E_{om} \cos(\omega_m t). \tag{1.151}$$

The desired amplitude-modulated carrier wave then has the form

$$E = [E_{oc} + a E_{om} \cos(\omega_m t)] \cos(\omega_c t), \tag{1.152}$$

which is just a sinusoidal oscillation $\cos(\omega_c t)$ with a time-varying amplitude given by the terms enclosed in brackets; the dimensionless coefficient a determines the maximum variation in amplitude for a given modulating signal E_{om}. Equation 1.152 in the form written is not very informative; but it can be expanded to read

$$\left.\begin{aligned} E = {} & E_{oc}\cos(\omega_c t) \\ & + \frac{1}{2}\alpha E_{om}\cos(\omega_c + \omega_m)t \\ & + \frac{1}{2}\alpha E_{om}\cos(\omega_c - \omega_m)t \end{aligned}\right\} \tag{1.153}$$

This expanded version shows that Eq. 1.152 is no more than a superposition of three perfectly harmonic oscillations: the carrier at frequency ω_c, an oscillation at the sum frequency $(\omega_c + \omega_m)$, and an oscillation at the difference frequency $(\omega_c - \omega_m)$. Figure 1.32 illustrates the variation of E with time as determined by inserting numerical values into Eq. 1.152 or 1.153. Note that in general the carrier frequency ω_c (which may be a radio wave at ~100 kHz or a microwave at hundreds of megahertz) is very much greater than the signal frequency ω_m (which, if it were speech or music, would have a frequency typically in the range from 20 Hz to 20 kHz). A plot of the frequency spectrum, that is, of the oscillation intensity as a function of frequency, is shown in Fig. 1.33a. In general, as we have said, the modulating signal is not sinusoidal but is comprised of a bunch of frequencies of different amplitudes. The spectrum of such a more general situation is sketched in Fig. 1.33b.

The question we have not answered yet is how one goes about achieving amplitude modulation in a practical system. The answer is, by impressing two sinusoidal voltages (the carrier and modulation) on a nonlinear electrical element. The element may be the triode vacuum tube illustrated in Fig. 1.34. The carrier and modulation voltages are both applied to the grid circuit of the triode and the modulated oscillation appears as a voltage across, say, a resistive load. To examine the modulation analytically, we shall suppose that the current I_p flowing in the plate circuit is related to the input grid voltage V through the nonlinear relation,

$$I_p = aV + bV^2 \tag{1.154}$$

where a and b are constants characteristic of the vacuum tube in

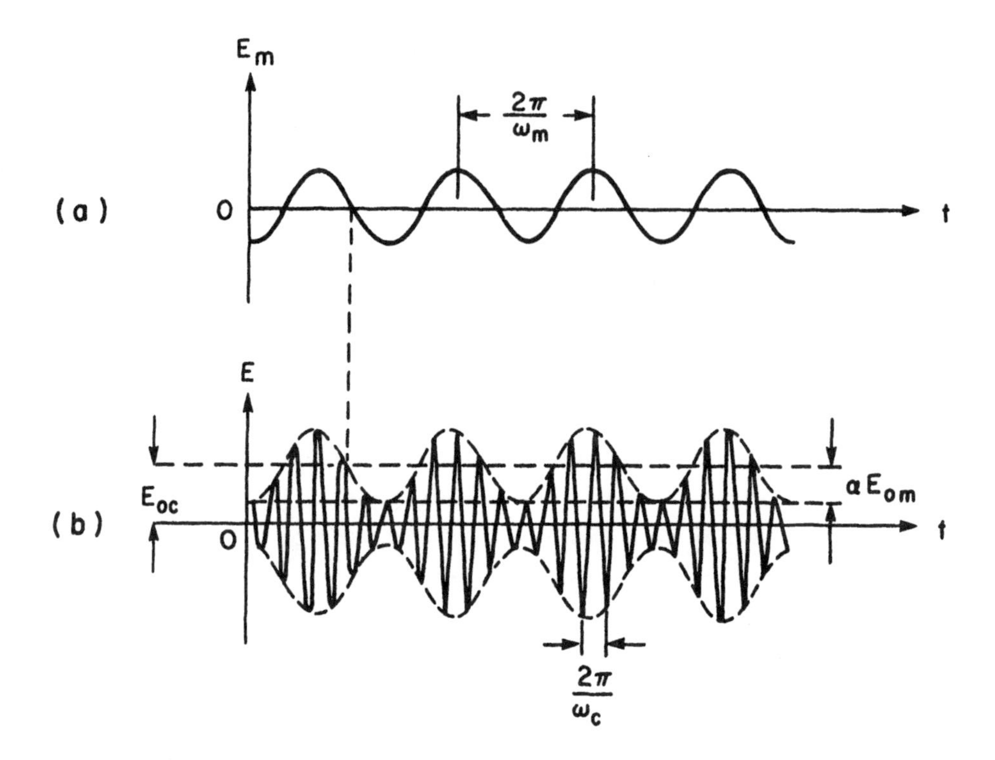

Fig. 1.32 (a) Low-frequency modulating signal, (b) modulated carrier as per Eq. 1.152. The envelope shown by the dashed, curved line is defined by the equations $\pm[E_{oc} + aE_{om} \cos(\omega_m t)]$. Observe the simple construction from which E_{oc} and aE_{om} are derived. To obtain E_{oc}, take a time when $E_m = 0$ and drop a perpendicular from the top to the bottom figure (shown dashed). The perpendicular then intersects the lower curve at the point where E equals E_{oc}.

question. Imposing the excitation potential

$$V = V_{om} \cos(\omega_m t) + V_{oc} \cos(\omega_c t) \tag{1.155}$$

onto Eq. 1.154 one finds, after some trigonometric manipulations, that the output plate current is given by

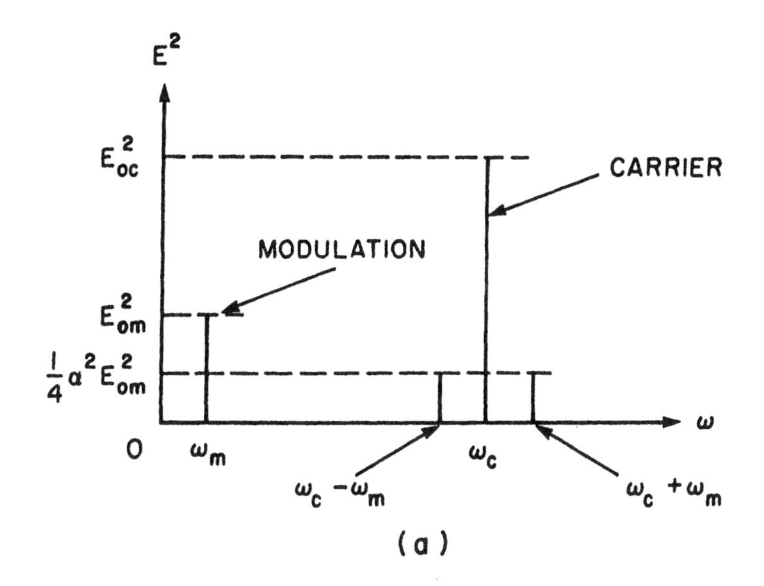

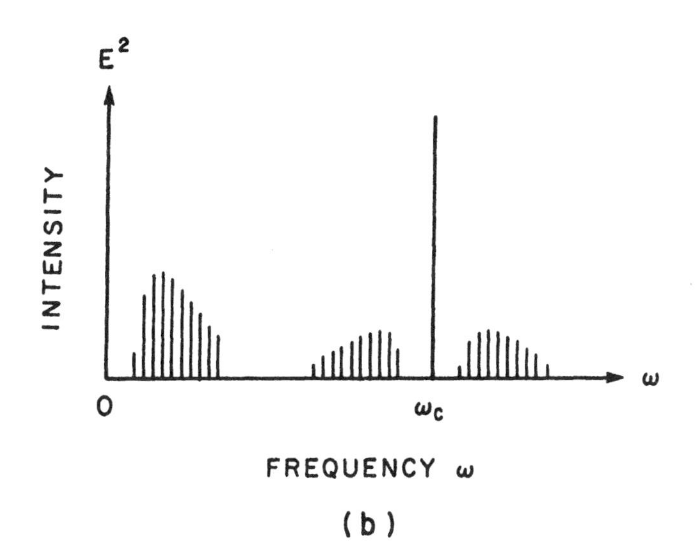

Fig. 1.33 Frequency spectrum of a modulated carrier. In (a) it is modulated by a sinusoidal wave of frequency ω_m; in (b) it is modulated by a complicated (but more realistic) distribution of waves of different amplitudes. Along the vertical axes are plotted the squares of the amplitudes, which are proportional to the intensity.

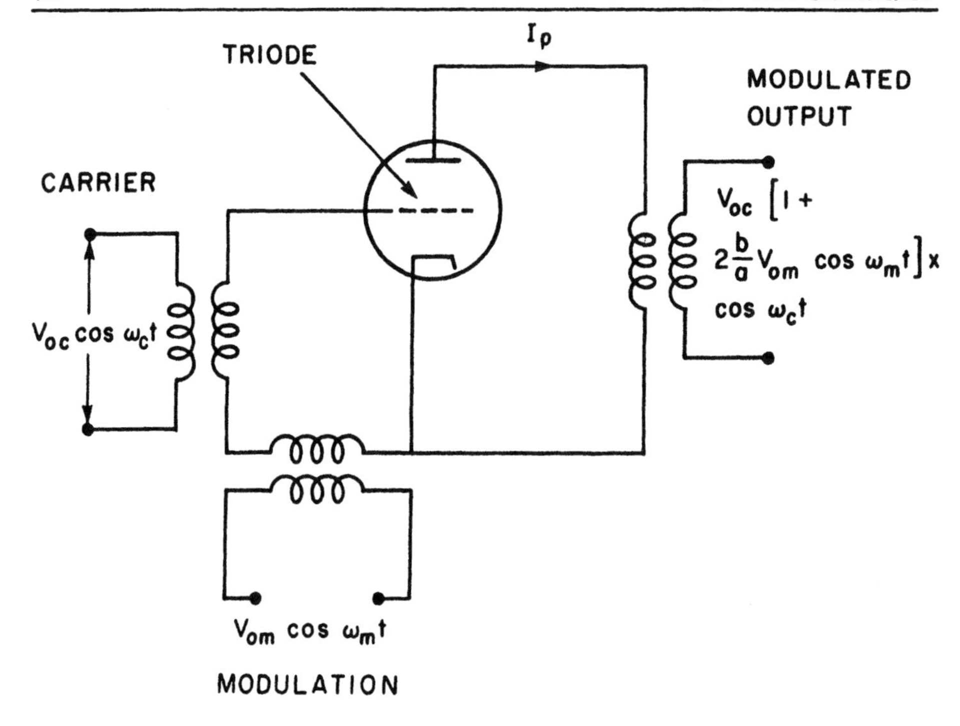

Fig. 1.34 The circuit of a simple modulator, exploiting the nonlinear current-voltage characteristics of the triode. The transformers couple the external circuits to the triode circuit.

$$\left.\begin{aligned} I_p = {} & aV_{oc} \cos(\omega_c t) + aV_{om} \cos(\omega_m t) \\ & + bV_{oc}V_{om} \cos(\omega_c + \omega_m)t \\ & + bV_{oc}V_{om} \cos(\omega_c - \omega_m)t \\ & + \frac{1}{2} bV_{oc}^2 + \frac{1}{2} bV_{om}^2 \\ & + \frac{1}{2} bV_{oc}^2 \cos(2\omega_c t) + \frac{1}{2} bV_{om}^2 \cos(2\omega_m t) \end{aligned}\right\} \tag{1.156}$$

The plate current contains, in addition to a steady dc component, oscillations at six separate frequencies, ω_c, ω_m, $(\omega_c \pm \omega_m)$, $2\omega_c$, and $2\omega_m$. If, however, $\omega_c \gg \omega_m$, as is generally the case, and if, in addition, all those frequencies that are not near ω_c are suitably filtered out, then the only components of current flowing through the load are

$$I_p \approx aV_{oc} \cos(\omega_c t) + bV_{oc}V_{om} \cos(\omega_c + \omega_m)t$$
$$+ bV_{oc}V_{om} \cos(\omega_c - \omega_m)t \tag{1.157}$$

which can be rearranged to read

$$I_p = aV_{oc}\left[1 + 2\left(\frac{b}{a}\right) V_{om} \cos(\omega_m t)\right] \cos \omega_c t. \tag{1.158}$$

This is precisely of the form given by Eq. 1.152. An electromagnetic signal proportional to the right-hand side of Eq. 1.158 is radiated out by an antenna. In the receiver, this procedure is reversed (the signal is demodulated) and an audio signal is delivered to the listener.

The human ear behaves in a similar way. When two tones are sounded together, the two original tones plus others that arise from the nonlinearity of the ear will be audible. They are called combination tones. We may describe the human ear reasonably well by including in the driven oscillator equations of sections 1.4 and 1.5 a nonlinear term proportional to the square of the displacement. Thus the basilar membrane or ear drum are pictured as springs with a restoring force of the form F(restoring) = $-Kx + bx^2$. The resulting equation of motion is

$$\frac{d^2x}{dt^2} + \omega_o^2 x - \beta x^2 = A_1 \cos(\omega_1 t) + A_2 \cos(\omega_2 t) \tag{1.159}$$

where A_1 and A_2 are the amplitudes of the incident tones of frequencies ω_1 and ω_2; and β is a constant assumed to be small. We shall not attempt to solve this equation here. But it is fairly clear, on the basis of our previous discussion, that the solution for x contains terms with the following frequencies: ω_1, ω_2, $2\omega_1$, $2\omega_2$, $(\omega_1 + \omega_2)$, and $(\omega_1 - \omega_2)$. Of these six frequencies, only the last one is lower than the original tones and is therefore easiest to identify.

Parametric excitation

We learned in sections 1.4 and 1.5 that large oscillatory motions are built up by a periodic external force acting on a vibrating system.

Here we will show that a periodic variation of a suitable parameter of the system itself can accomplish the same thing. This is known as parametric excitation. As an example, there is the well-known building up of oscillations by a child on a swing who rhythmically stands up and sits down, thereby changing the center of gravity of the system. A simple analogue of this is a pendulum whose length is being varied by pulling and releasing the string as is indicated in Fig. 1.35. Suppose that during each passage through the vertical position the bob is raised a small distance $\Delta\ell$ and at each extreme displacement, the bob is lowered by the same distance $\Delta\ell$. Thus, during each period T_o, the pendulum is lengthened twice and shortened twice; that is, the periodic variations of the parameter are executed at twice the natural frequency of oscillation $\omega_o = 2\pi/T_o$. Note that this differs from the

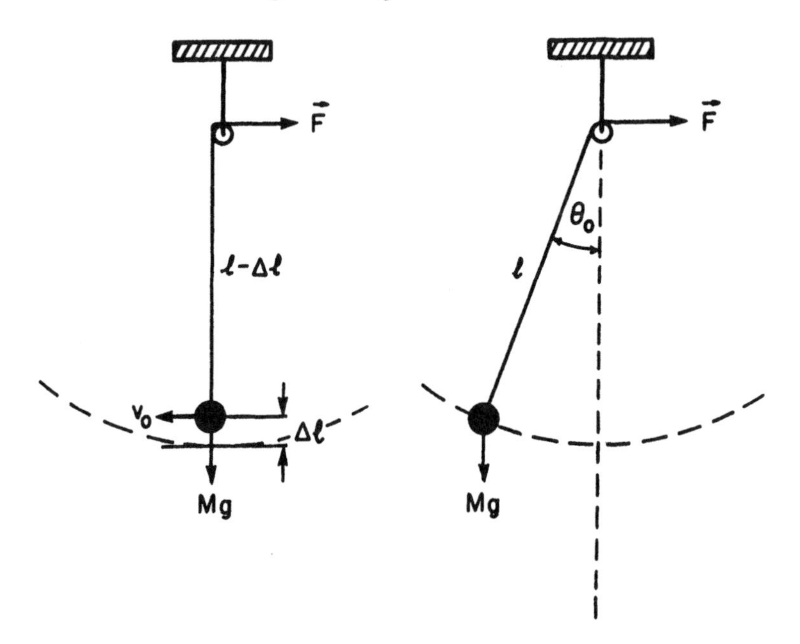

Fig. 1.35 Parametric excitation of a swinging pendulum, "pumped" by raising and lowering the bob; θ_o is the maximum angular excursion, and v_o the maximum velocity. The pumping frequency ω_p is twice the natural frequency $\omega_o = \sqrt{g/\ell}$. The dashed arc of the circle represents the trajectory of the bob in the absence of parametric pumping.

case of the externally driven oscillator where we found that the most efficient way of building up the oscillations occurred when the driver frequency ω equaled ω_o. The fact that in the parametric excitation scheme the driving frequency equals $2\omega_o$ suggests that here we are dealing with a nonlinear system, as indeed we are.

Any buildup of oscillations must come from the work done by the external force in raising and lowering the bob of the pendulum. In the vertical position the pendulum gains $2Mg\Delta\ell$ of potential energy per oscillation (the factor of two comes from the fact that we raise the bob twice during each period), and at its farthest deflection θ_o it loses $2Mg(\cos\theta_o)\,\Delta\ell$ of energy. In addition, F does work against the centripetal force Mv_o^2/ℓ which is highest when the pendulum is vertical and zero when it reaches its farthest deflection, where it has no velocity (v_o is the maximum velocity of the bob). Adding up all the gains and losses, we find that the total work done by the external force equals

$$W = 2Mg\Delta\ell\left[(1-\cos\theta_o) + \frac{v_o^2}{\ell g}\right]. \tag{1.160}$$

Since $Mv_o^2/2 = Mg\ell(1-\cos\theta_o)$, we can eliminate the speed v_o from the above equation (1.160) and obtain

$$W = 6Mg\Delta\ell(1-\cos\theta_o), \tag{1.161}$$

from which we see that the work done by F is positive so that the amplitude of the swings will keep growing with each period of oscillation. Note that the phase of the acting force relative to the swinging pendulum is very important. If the phase is changed so that the bob is lowered when it is in its vertical position, energy will be supplied to the engine, which changes the length of the string, and the amplitude of the pendulum's oscillations will damp rather than grow.

What is fun in mechanics becomes a serious matter in electronics which abounds with parametric devices. When we compare the equation of motion of the pendulum (1.29) with the "equation of motion" of the LC circuit (1.49), we see that the length of the pendulum has the same relation to the first equation as C has to the second equation.

From this we infer that by varying the capacity of an LC circuit it should be possible to build up its current and voltage oscillations. Figure 1.36 illustrates the circuit in question. The time-varying capacitance

$$C(t) = C_0 + C_1(t) \tag{1.162}$$

is assumed to be made up of a static capacity C_0 and a small time-varying part $C_1(t)$ in parallel with it. We can imagine that the variations $C_1(t)$ are caused by pulling and pushing together the plates of a parallel plate capacitor. Suppose that at the instant of time $t = 0$ the capacitor is fully charged. At this instant pull the plates apart a small amount. Work is required to do this because the plates carry opposite charges which attract one another. The work done goes into increasing the electric energy between plates since the charge on the capacitor cannot change instantaneously. Thus the voltage $V = Q/C$ increases. A quarter of a period later push the plates together to their initial separation. At this time ($t = T_0/4$) there is no charge on the plates and no work is needed to do this. Still another quarter period later, pull the plates apart once again, and so on. In practice, of course, the change of capacitance is not this abrupt, but sinusoidal

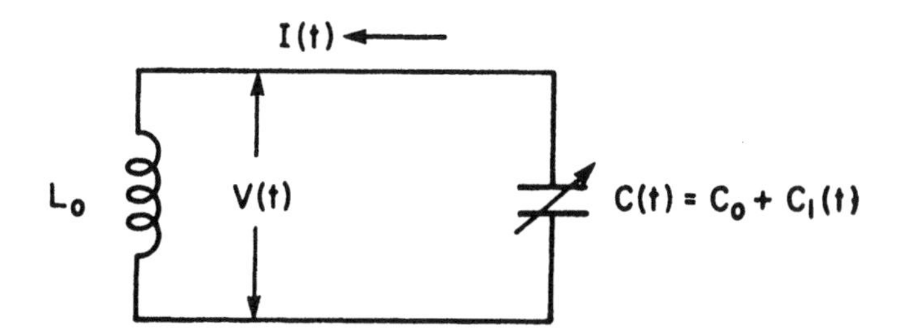

Fig. 1.36 Circuit of a parametric oscillator. When the time-varying capacitance is "pumped" at the right frequency and in the correct phase (see text), the voltage and current oscillations can be made to grow exponentially. In actual practice the pump power must exceed a certain minimum threshold value before growth occurs, to overcome the inevitable loss of energy in the circuit resistance R (not shown). [For details and fairly complicated mathematics of problem, see Coupled Mode and Parametric Electronics by W. H. Louisell (John Wiley & Sons, New York, 1960).]

and Eq. 1.162 has the form

$$C(t) = C_o + C_{o1} \cos(2\omega_o + \phi)t. \tag{1.163}$$

Note that the so-called "pumping" frequency $2\omega_o$ is twice the natural frequency of oscillation of the LC circuit, $\omega_o = 1/\sqrt{L_o C_o}$.

The circuit equations can be obtained easily. Using Eqs. 1.45 through 1.49 we have for the voltage that

$$V = L_o \frac{dI}{dt} \tag{1.164}$$

and for the current that

$$I = -\frac{d}{dt}(CV). \tag{1.165}$$

But in accordance with Eq. 1.162 $C = C_o + C_1(t)$, so that

$$I = -C_o \frac{dV}{dt} - C_1 \frac{dV}{dt} - V \frac{dC_1}{dt}. \tag{1.166}$$

We can now eliminate V between Eqs. 1.164 and 1.166 with the result that

$$\frac{d^2I}{dt^2} + \frac{1}{L_o C_o} I = -\frac{C_1}{C_o}\frac{d^2I}{dt^2} - \frac{1}{C_o}\left(\frac{dI}{dt}\right)\left(\frac{dC_1}{dt}\right) \tag{1.167}$$

where we have written the familiar terms on the left of the equation and the new terms on the right.* We shall not solve this equation, but only note that in some respects it can be likened to the driven oscillator equation (1.89) with the right-hand side playing the role of the generator of electromotive force. But now the generator is no longer a simple external sinusoidal source of emf but a complex nonlinear element which is internal to the system itself.

*When the complete "dynamics" of the time-varying capacitor is solved it is found that C_1 is not only a function of t but also of the current I, making Eq. 1.167 nonlinear.

Parametric amplifiers are used at radio and microwave frequencies where extremely low noise operation is desired. A semiconductor junction such as a silicon junction diode functions as the voltage variable capacitance. This element is placed in a circuit in such a manner that it can receive two voltages, one at the signal frequency ω_o and the other at the pump frequency $2\omega_o$. A thousand-fold gain in intensity of the input signal is typical for these instruments.

In conclusion we would like to point out that the parametric system we have described is just one special type amongst a large variety of ideas. The nonlinear capacitance could be replaced by a nonlinear inductance. Also, pumping at twice the signal frequency is not absolutely mandatory. Imagine a system composed of two resonant circuits having frequencies ω_{o1} and ω_{o2} and coupled to one another by a nonlinear capacitance (or inductance). We can then expect growth of the signals provided that the frequency of the pump ω_p obeys the relation

$$\omega_p = \omega_{o1} + \omega_{o2}. \tag{1.168}$$

(The combination $\omega_p = \omega_{o1} - \omega_{o2}$ looks equally promising, but it turns out that no growing oscillations occur.) The situation we have been treating corresponds to a special "degenerate" case where the signal at the one frequency ω_o interacts with itself so that $\omega_p = \omega_o + \omega_o = 2\omega_o$.

1.7 COUPLED OSCILLATORS

Thus far we have confined ourselves to the study of a single, isolated oscillator. To be sure, in section 1.4 we have discussed how the oscillator behaves when it is driven by a periodic force, but we did not consider its possible back-reaction upon the driver. In many instances this reaction is sufficiently weak that its neglect is not serious. In other cases, however, the incessant exchange of energy between two or more connected oscillators profoundly affects their respective motions, and it is these motions that we shall study in the present section.

The vibrations of a system of two or more coupled oscillators are in general neither simple harmonic, nor are they necessarily periodic in time. This is true even though the system is comprised of "ideal" elements, namely undamped simple harmonic oscillators connected by perfect spring-like forces obeying Hooke's law. Figure 1.37 illustrates what appears to be a particularly simple system; nonetheless, the resulting motion is seen to be quite complex. The important point that will emerge is that despite the complexity, the motion can be analyzed into what are known as "normal modes" which themselves are simple harmonic. Any one of these modes can exist independently of the others, and by proper choice of initial conditions the system can be excited to oscillate in that mode only. When this is achieved, each individual oscillator of the system vibrates at the same well-defined frequency; in addition, the displacements from equilibrium of the separate parts retain constant ratios throughout all times. Figure 1.38 illustrates the configuration of the normal modes when two identical pendula are coupled by a spring, and the normal modes of oscillation of a light, stretched string supporting equal, equidistant masses. It is fairly obvious that for simple systems like those shown, symmetry considerations and a little intuition is all that is required to guess the mode configurations. However, the oscillation frequencies of the individual modes are not so easily arrived at, and to obtain them the equations of motion of the component oscillators must be derived and solved. As a rule of thumb, when a system has two modes, the antisymmetrical mode has the higher frequency and the symmetrical mode the lower frequency. In a more complex system consisting of more than two coupled oscillators, the mode that has the highest degree of symmetry has the lowest frequency. If this symmetry is destroyed, the springs must "work harder" in the antisymmetrical modes and the frequency is raised (recall from Eq. 1.6 that the frequency squared is proportional to the restoring force per unit mass).

Two coupled oscillators

The equations of motion of two coupled oscillators A and B usually have the form

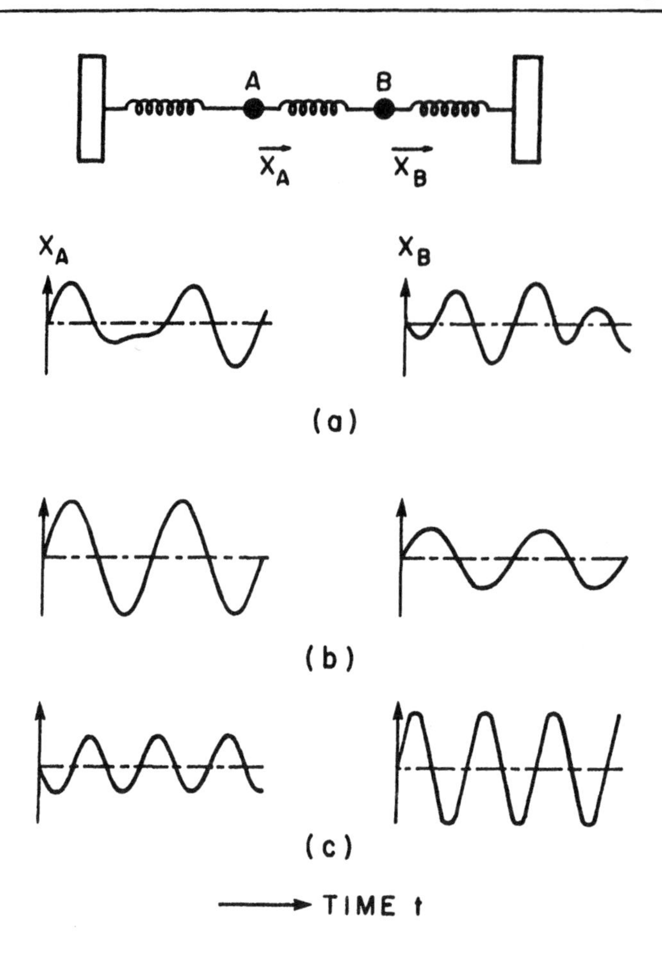

Fig. 1.37 Two objects A and B connected by massless springs and oscillating on a frictionless horizontal table. (a) Displacements of A and B as a function of time when the masses are set in motion in some arbitrary way; (b) and (c) show two normal modes from which the general motion can be synthesized; (b) represents the lower frequency symmetrical mode ω_1, and (c) the higher frequency antisymmetrical mode ω_2. If the system is started "right" the two masses will oscillate indefinitely in one or the other normal mode. In the case shown, $\omega_2 = \sqrt{3}\,\omega_1$. In this situation, when ω_1 and ω_2 are not commensurable, the general motion shown in (a) never repeats and thus is not periodic. The condition for any sort of periodicity is that $n\omega_1 = m\omega_2$ where n and m are any two integers (explain why).

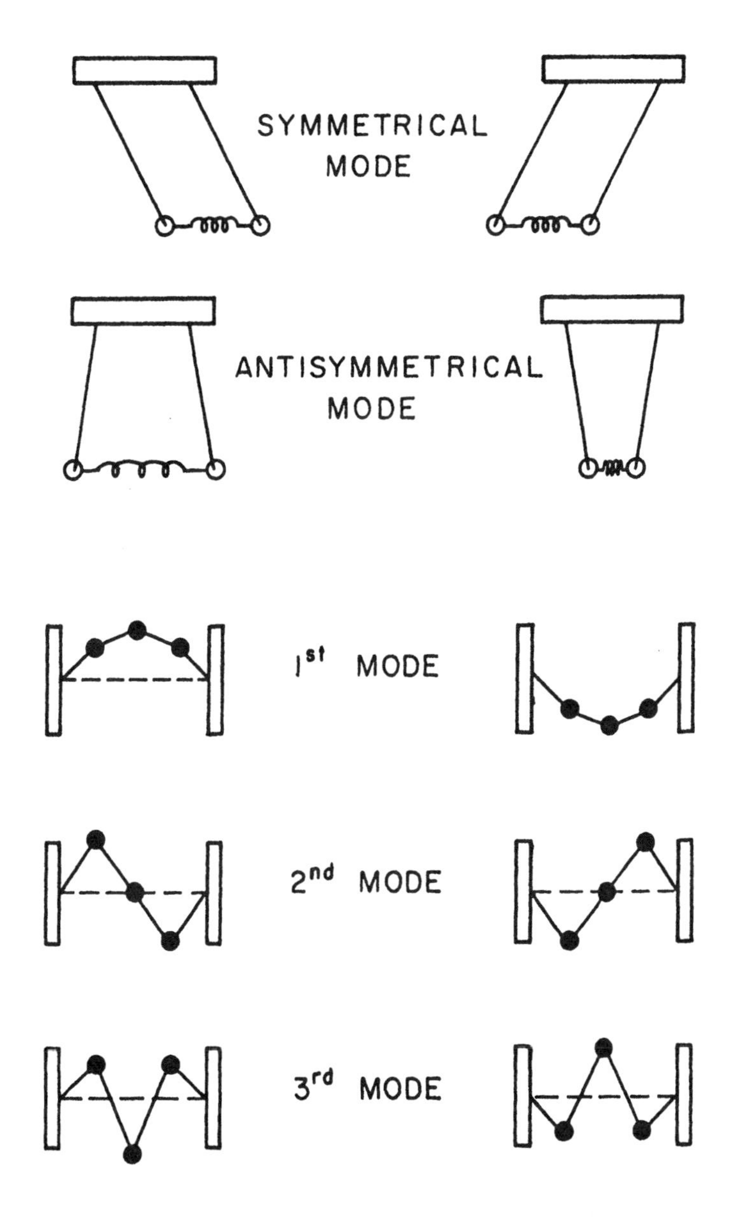

Fig. 1.38 Normal modes of two coupled pendula and three coupled masses.

$$\frac{d^2x_A}{dt^2} = -a_{11}x_A - a_{12}x_B$$
$$\frac{d^2x_B}{dt^2} = -a_{21}x_A - a_{22}x_B. \tag{1.169}$$

The differentials on the left represent the accelerations. The terms $-a_{11}x_A$ and $-a_{22}x_B$ are the self-restoring forces (per unit mass) acting on the individual oscillators; and the terms $-a_{12}x_B$ and $-a_{21}x_A$ are the coupling forces. The coefficients a are numerical constants with dimensions of $(\text{time})^{-2}$ and have values that depend on the problem at hand. To show that the two equations have indeed the appropriate form, consider as an example, the system shown in Fig. 1.39. Each oscillator has a mass M and a spring of spring constant K. The spring constant of the coupler is κ. If, at some instant of time, mass A is displaced from its equilibrium position through a distance x_A, the force it experiences equals $-Kx_A - \kappa(x_A-x_B)$. Similarly, mass B experiences a force equal to $-Kx_B - \kappa(x_B-x_A)$. Thus, the equations of motion are

$$M\frac{d^2x_A}{dt^2} = -(K+\kappa)\, x_A + \kappa x_B$$
$$M\frac{d^2x_B}{dt} = +\kappa x_A - (K+\kappa)\, x_B \tag{1.170}$$

and are precisely of the type given by Eqs. (1.169) with

$$a_{11} = \frac{K+\kappa}{M} \quad ; \quad a_{12} = -\frac{\kappa}{M}$$
$$a_{21} = -\frac{\kappa}{M} \quad ; \quad a_{22} = \frac{K+\kappa}{M}. \tag{1.171}$$

Equations (1.169) are dependent and must be solved simultaneously because each contains the two unknown displacements x_A and X_B. The solution is not difficult if we adopt the normal mode concept introduced

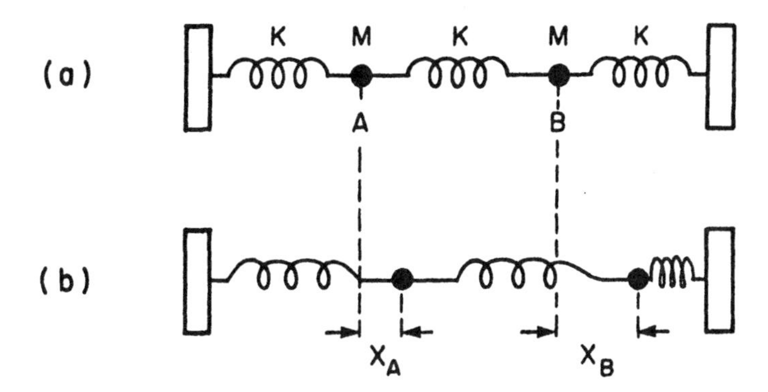

Fig. 1.39 Two coupled mass-spring systems oscillating on a horizontal frictionless table. In (a) the masses are in their equilibrium positions. In (b) they are shown in some general configuration after displacement from equilibrium.

earlier. We simply assume that the system oscillates in one of its normal modes, which in mathematical terms

$$\begin{aligned} x_A &= A\cos(\omega t+\phi) \\ x_B &= B\cos(\omega t+\phi) \end{aligned} \tag{1.172}$$

means that the oscillators have the same frequency ω, the same phase ϕ and a constant, but as yet unknown amplitude ratio A/B. Inserting Eqs. (1.172) in (1.169) and performing the differentiations with respect to time yields two homogeneous linear equations in x_A and x_B:

$$\begin{aligned} (a_{11}-\omega^2)x_A + a_{12}x_B &= 0 \\ a_{21}x_A + (a_{22}-\omega^2)x_B &= 0. \end{aligned} \tag{1.173}$$

They have a trivial solution $x_A = x_B = 0$, and a nontrivial solution given by setting the determinant of its coefficients equal to zero:

$$\begin{vmatrix} a_{11}-\omega^2 & a_{12} \\ a_{21} & a_{22}-\omega^2 \end{vmatrix} = 0 \tag{1.174}$$

or equivalently,

$$(a_{11}-\omega^2)(a_{22}-\omega^2) - a_{12}a_{21} = 0$$

This is a quadratic equation in the variable ω^2. Its two roots, which we denote ω_1 and ω_2, are:

$$\omega_1^2 = \frac{1}{2}\left[(a_{11}+a_{22}) - \sqrt{(a_{11}-a_{22})^2 + 4a_{12}a_{21}}\right]$$
$$\omega_2^2 = \frac{1}{2}\left[(a_{11}+a_{22}) + \sqrt{(a_{11}-a_{22})^2 + 4a_{12}a_{21}}\right]. \qquad (1.175)$$

They are the frequencies of the two normal modes (denoted 1 and 2) which the system is capable of supporting. Evidently ω_1 is the lower frequency and is associated with the symmetrical mode, while ω_2 is the higher frequency and is associated with the antisymmetrical mode.

The shapes of the modes are specified by the amplitude ratio $R \equiv (A/B)$. From the first equation (1.173) we see that $(x_A/x_B) = -a_{12}/(a_{11}-\omega^2)$ and since $x_A/x_B = A/B$, it follows that configurations of the two characteristic oscillations are

$$R_1 = \left(\frac{A}{B}\right)_1 = -\frac{a_{12}}{a_{11}-\omega_1^2}$$
$$R_2 = \left(\frac{A}{B}\right)_2 = -\frac{a_{12}}{a_{11}-\omega_2^2}. \qquad (1.176)$$

Now that we have found the mode frequencies and their amplitude ratios, we have obtained all there is to know about the coupled two-oscillator system. What remains is to write down the most general solution which is merely a superposition of the normal mode solutions:

$$x_A = A_1 \cos(\omega_1 t + \phi_1) + A_2 \cos(\omega_2 t + \phi_2)$$
$$x_B = \frac{A_1}{R_1}\cos(\omega_1 t + \phi_1) + \frac{A_2}{R_2}\cos(\omega_2 t + \phi_2). \qquad (1.177)$$

Observe that there is a total of four and no more than four arbitrary coefficients in the equations. They are A_1, A_2, ϕ_1, and ϕ_2. Their existence allows us to fit the foregoing solutions to any desired initial conditions such as initial displacements $x_A(t=0)$ and $x_B(t=0)$, and initial velocities $(dx_A/dt)_{t=0}$ and $(dx_B/dt)_{t=0}$.

Let us apply these results to the mass-spring problem illustrated in Fig. 1.39. We insert Eqs. 1.171 in Eqs. 1.175 and obtain for the mode frequencies the values

$$\begin{aligned} \omega_1 &= \sqrt{\frac{K}{M}} \\ \omega_2 &= \sqrt{\frac{K + 2\kappa}{M}}. \end{aligned} \tag{1.178}$$

We insert Eqs. 1.171 and Eqs. 1.178 in Eqs. 1.176 and obtain the amplitude ratios:

$$\begin{aligned} R_1 &= \left(\frac{A}{B}\right)_1 = 1 \\ R_2 &= \left(\frac{A}{B}\right)_2 = -1. \end{aligned} \tag{1.179}$$

The configurations are shown schematically in Fig. 1.40. Now suppose the system is set into oscillation in a way that initially mass B is at

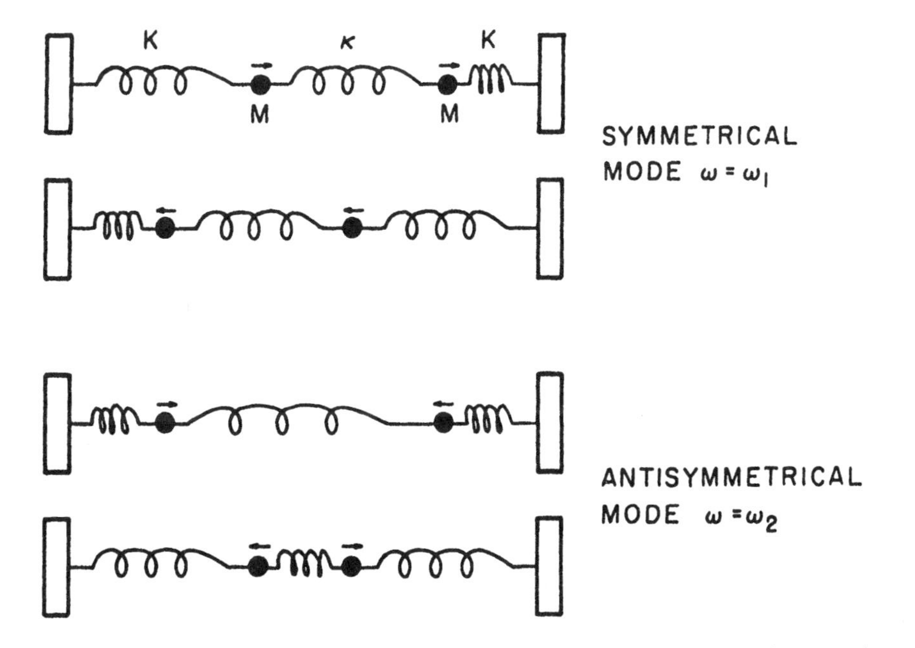

Fig. 1.40 The two normal modes of the coupled mass-spring system of Fig. 1.39.

rest in its equilibrium position $x_B = 0$, but mass A is displaced through a distance $x_A = x_o$ and then released from rest. What is the subsequent motion of the two masses? Inserting the aforementioned boundary conditions $[x_A(t=0) = x_o,\ x_B(t=0) = 0,\ (dx_A/dt)_{t=0} = (dx_B/dt)_{t=0} = 0]$ into Eqs. 1.177 gives $\phi_1 = \phi_2 = 0$ and $A_1 = A_2 = x_o/2$. Also $R_1 = 1$ and $R_2 = -1$, with the result that

$$
\begin{aligned}
x_A &= \frac{x_o}{2}\cos(\omega_1 t) + \frac{x_o}{2}\cos(\omega_2 t) \\
&= x_o \cos\left(\frac{\omega_1 - \omega_2}{2}\,t\right)\cos\left(\frac{\omega_1 + \omega_2}{2}\,t\right)
\end{aligned}
\tag{1.180}
$$

$$
\begin{aligned}
x_B &= \frac{x_o}{2}\cos(\omega_1 t) - \frac{x_o}{2}\cos(\omega_2 t) \\
&= -x_o \sin\left(\frac{\omega_1 - \omega_2}{2}\,t\right)\sin\left(\frac{\omega_1 + \omega_2}{2}\,t\right)
\end{aligned}
\tag{1.181}
$$

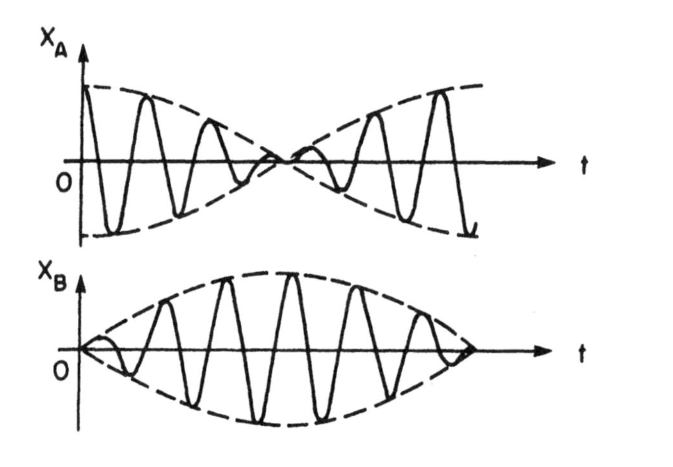

Fig. 1.41 The motions of two identical, weakly coupled oscillators like those shown in Fig. 1.39 with $\kappa \ll K$. Initially only mass A is displaced from the origin. As time increases, the amplitude of A decreases slowly whereas B increases from zero. Hence energy is transferred from A to B. When all the energy has been transferred to B it begins to flow back to A, and this keeps repeating as time goes on. The dashed curves represent the envelopes of the amplitude modulations $x_o\cos[(\omega_1-\omega_2)t/2]$ and $-x_o\sin[(\omega_1-\omega_2)t/2]$ of Eqs. 1.180 and 1.181, respectively. [cf. Fig. 1.32.]

where the frequencies ω_1 and ω_2 are given by Eqs. 1.178. These then are the equations giving the complete time history of the masses' displacements. An interesting special case arises when the coupling between the masses is weak so that $\kappa \ll K$. In this situation the frequencies of the two modes are nearly the same. As a consequence, the displacements x_A and x_B are almost simple harmonic with slowly varying amplitudes given by $x_o \cos[(\omega_1 - \omega_2)t/2]$ and $-x_o \sin[(\omega_1 - \omega_2)t/2]$. This is illustrated in Fig. 1.41. Observe the occurence of so-called beats — a phenomenon which we met already in section 1.6 in connection with another problem, that of modulation in a nonlinear circuit [see also Appendix 1].

Example

Many problems of coupled oscillators display a high degree of symmetry but it would be misleading to believe that this is always the case. Consider two identical masses A and B of mass M hanging vertically, as shown, and coupled with identical springs of constant K to each other and the horizontal support. Analyzing vertical motion only, derive the coupled differential equations of motion and solve them to find the most general solutions for X_A and X_B.

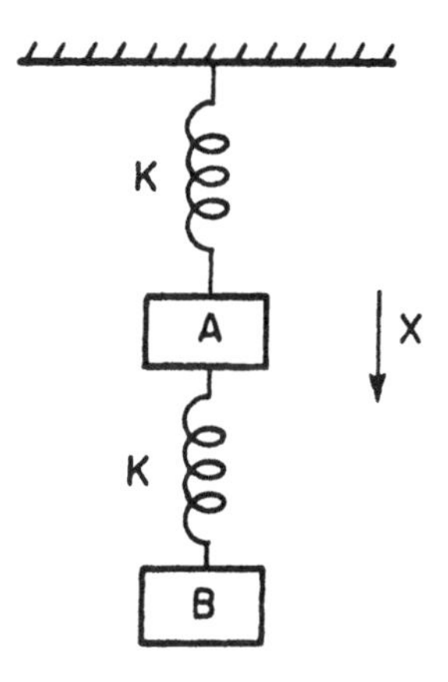

Solution

The derivation of the coupled differential equations of motion is often simplified by imagining one mass (B) fixed and evaluating the forces on A as it is moved and then considering A fixed, letting B move, and again evaluating the forces on A. This yields one of the equations, i.e., the equation involving $\ddot{X}_A$. Then repeat the process for mass B and $\ddot{X}_B$.

For example, if X_A and X_B are measured positive downward from the equilibrium positions of A and B, then we have

$$M\ddot{X}_A = F_A = \underbrace{-KX_A - KX_A}_{\substack{\text{Force on A}\\ \text{when A is}\\ \text{moved and B}\\ \text{held fixed.}\\ \text{Force acts to}\\ \text{decrease } X_A.}} + \underbrace{KX_B}_{\substack{\text{Force on A}\\ \text{when B is}\\ \text{moved and A}\\ \text{held fixed.}\\ \text{Force acts to}\\ \text{increase } X_A.}}$$

Similarly,

$$M\ddot{X}_B = F_B = -KX_B + KX_A.$$

Thus by writing

$$\omega_o^2 = \frac{K}{M}$$

the coupled equations of motion can be written as

$$\ddot{X}_A + 2\omega_o^2 X_A - \omega_o^2 X_B = 0$$

$$\ddot{X}_B + \omega_o^2 X_B - \omega_o^2 X_A = 0.$$

Note that these equations are of the same form as Eqs. 1.169. Note, also, the lack of symmetry between A and B is already apparent at this stage of the problem, as of course it must be.

As a trial solution we take

$$X_A = A\cos(\omega t + a)$$

$$X_B = B\cos(\omega t + a)$$

and substitute in the differential equations. This yields

$$\left(-\omega^2 + 2\omega_o^2\right) A - \omega_o^2 B = 0$$

$$-\omega_o^2 A + \left(-\omega^2 + \omega_o^2\right) B = 0.$$

Each of these equations gives a ratio A/B:

$$\frac{A}{B} = \frac{\omega_o^2}{2\omega_o^2 - \omega^2} \qquad \frac{A}{B} = \frac{\omega_o^2 - \omega^2}{\omega_o^2}.$$

Now equate these two expressions and solve for the normal frequencies. The result is

$$\omega^2 = \frac{\omega_o^2}{2}\{3 \pm \sqrt{5}\}.$$

So the normal frequencies are

$$\omega_- = \omega_o \left\{\frac{3 - \sqrt{5}}{2}\right\}^{1/2}$$

$$\omega_+ = \omega_o \left\{\frac{3 + \sqrt{5}}{2}\right\}^{1/2}.$$

Each value of ω may now be substituted in either equation for A/B to give the ratio of the amplitudes of X_A and X_B for that particular normal frequency. Then we have

$$\frac{A_-}{B_-} = \frac{1}{2}(\sqrt{5} - 1) \quad \text{using } \omega_-.$$

$$\frac{A_+}{B_+} = -\frac{1}{2}(\sqrt{5} + 1) \quad \text{using } \omega_+.$$

We can now write our general solution as

$$X_A = \frac{\sqrt{5} - 1}{2} B_- \cos(\omega_- t + \alpha_-)$$

$$- \frac{\sqrt{5} + 1}{2} B_+ \cos(\omega_+ t + \alpha_+)$$

$$X_B = B_- \cos(\omega_- t + \alpha_-) + B_+ \cos(\omega_+ t + \alpha_+).$$

Note our solution has four undetermined constants, B_-, B_+, α_-, and α_+ which become determined when the positions and velocities of each mass A and B are given at some instant of time.

Two coupled electrical circuits

When an LC circuit is brought near another similar electrical circuit, oscillations in the one can be induced in the other. Different coupling schemes will be found in Fig. 1.42. For purposes of illustration we shall analyze the capacitively coupled system shown in Fig. 1.42a, but, for the sake of simplicity we make the two resonators identical, that is, $C_A = C_B = C$ and $L_A = L_B = L$. Kirchhoff's circuital relation demends that, in traversing each circuit completely, the total

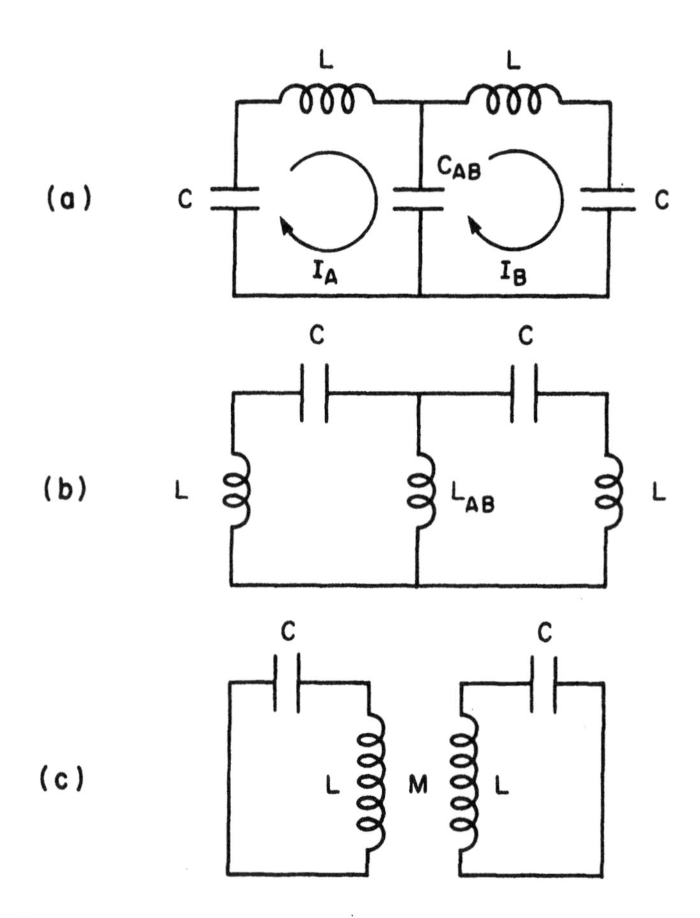

Fig. 1.42 Coupled electrical oscillators. In (a) the coupling is capacitive via C_{AB}; in (b) the coupling is inductive via L_{AB}; and in (c) the coupling is via a mutual inductance M as is typical in a transformer.

voltage drop be zero, so that [cf. Eqs. 1.49 and 1.89]

$$L\frac{d^2I_A}{dt^2} + \frac{1}{C}I_A + \frac{1}{C_{AB}}[I_A - I_B] = 0$$
$$L\frac{d^2I_B}{dt^2} + \frac{1}{C}I_B + \frac{1}{C_{AB}}[I_B - I_A] = 0 \tag{1.182}$$

a result which we can rearrange to read

$$\frac{d^2I_A}{dt^2} = -\left[\frac{1}{LC} + \frac{1}{LC_{AB}}\right]I_A + \frac{1}{LC_{AB}}I_B$$
$$\frac{d^2I_B}{dt^2} = -\left[\frac{1}{LC} + \frac{1}{LC_{AB}}\right]I_B + \frac{1}{LC_{AB}}I_A. \tag{1.183}$$

These equations are just equations (1.169) with coefficients given by

$$a_{11} = \left[\frac{1}{LC} + \frac{1}{LC_{AB}}\right] \quad ; \quad a_{12} = -\frac{1}{LC_{AB}}$$
$$a_{22} = \left[\frac{1}{LC} + \frac{1}{LC_{AB}}\right] \quad ; \quad a_{21} = -\frac{1}{LC_{AB}}. \tag{1.184}$$

The characteristic frequencies of the two normal modes are then found by inserting Eqs. 1.84 in Eqs. 1.175 with the result that

$$\omega_1 = \sqrt{\frac{1}{LC}}$$
$$\omega_2 = \sqrt{\frac{1}{LC} + \frac{2}{LC_{AB}}}. \tag{1.185}$$

Observe that if one of the oscillators (say B) were stopped by cutting the circuit and reducing I_B to zero, the oscillation frequency of circuit A would then equal $\omega_o \equiv \sqrt{(1/LC) + (1/LC_{AB})}$. Similarly, if A were stopped and B were allowed to oscillate, it would do so at the frequency ω_o. Thus we see that the effect of coupling is to split the common frequency ω_o into two frequencies ω_1 and ω_2, the one being lower than ω_o and the other higher than ω_o.

Many coupled oscillators – electrical filters

The general problem of many coupled oscillators lies beyond the scope of this book and the interested reader is referred to one or the other of the following texts: A. P. French, Vibrations and Waves (W. W. Norton, 1971); J. B. Marion, Classical Dynamics of Particles and Systems (Academic Press, 1970). The task we set ourselves here is to examine, somewhat cursorily, the special case of a ladder network of identical elements, in our case inductances L and capacitors C arranged as shown in Fig. 1.43. This ladder is an extension of the two-element system shown in Fig. 1.42a and discussed in the previous subsection. If, then, we consider the current I_n in the nth element we have [cf. Eqs. 1.182] that

$$L\frac{d^2 I_n}{dt^2} + \frac{1}{C}[I_n - I_{n-1}] + \frac{1}{C}[I_n - I_{n+1}] = 0. \qquad (1.186)$$

Assuming sinusoidal current oscillations $I_n = I_n^o e^{j\omega t}$ of frequency ω, we differentiate twice the first term of the above equation with respect to time and obtain $-L\omega^2 I_n$. Inserting this result and rearranging terms it follows that the amplitudes denoted by superscripts o are related by

$$\frac{I_{n-1}^o}{I_n^o} + \frac{I_{n+1}^o}{I_n^o} = 2 - \omega^2 LC. \qquad (1.187)$$

To solve this difference equation we note that all the elements of the ladder are exactly similar, and intuition suggests that therefore the current ratio in successive elements is constant, that is, $\left(I_{n-1}^o / I_n^o\right) =$

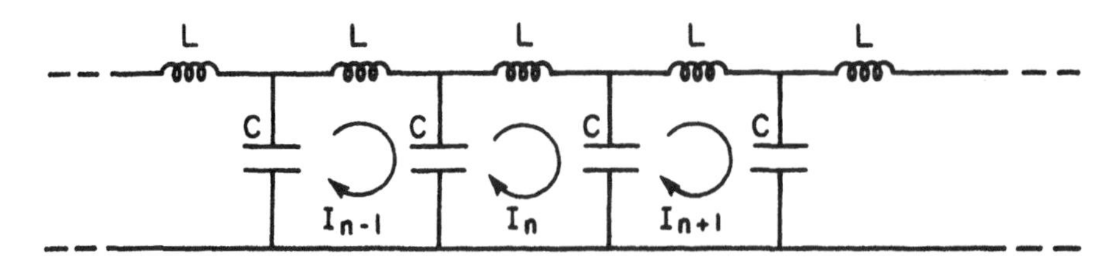

Fig. 1.43 A ladder of identical coupled LC circuits. The coupling is capacitive.

$I_n^o/I_{n+1}^o = \ldots =$ constant. Setting this constant equal to, say, $\pm \exp(-p)$ where p is some yet-to-be-determined coefficient, we have that

$$\frac{I_{n-1}^o}{I_n^o} = \frac{I_n^o}{I_{n+1}^o} = \ldots . = \pm e^{-p}. \qquad (1.188)$$

Substitution of Eq. 1.188 in Eq. 1.187 determines p; that is, $\exp(-p) + \exp(+p) = \pm(2 - \omega^2 LC)$. But $\exp(p) + \exp(-p) = 2 \cosh p$ with the result that

$$\cosh p = \pm\left[1 - \frac{\omega^2 LC}{2}\right]. \qquad (1.189)$$

This ends the mathematics; let us see what it all means.

Knowledge of p yields the ratio of currents in successive elements of the ladder [Eq. 1.188]. And p is given in terms of physical quantities by Eq. 1.189. It is a transcendental equation which may be solved graphically by plotting the functions $y = \cosh p$ and $y = \pm[1 - (\omega^2 LC/2)]$ as is indicated in Fig. 1.44; the points of intersection of the curve and the straight line give the roots of the equation. The important thing to notice is that a root exists only if $y > 1$ or equivalently if $(\omega^2 LC/2) > 2$. This means that there is a critical frequency associated with the system whose value is given by

$$\omega_{crit} = \frac{2}{\sqrt{LC}}. \qquad (1.190)$$

The result suggests that when the ladder is excited at its input end (located at the far left) by a signal of frequency $\omega > \omega_{crit}$, the signal will, in traversing the ladder, keep decreasing exponentially as $I_n/I_{n+1} = \exp(-p)$ (mathematically the signal could keep increasing as $\exp(+p)$ but that would mean an infinite current as $n \to \infty$, which is nonphysical).

Have we exhausted all solutions of Eq. 1.189; and what is happening at frequencies $\omega < \omega_{crit}$? The answer is no, we have not found all solutions because there is no reason for us to exclude the possibility that p is imaginary. And, as we shall find momentarily, imaginary p will also tell us what is taking place at low frequencies $\omega < \omega_{crit}$.

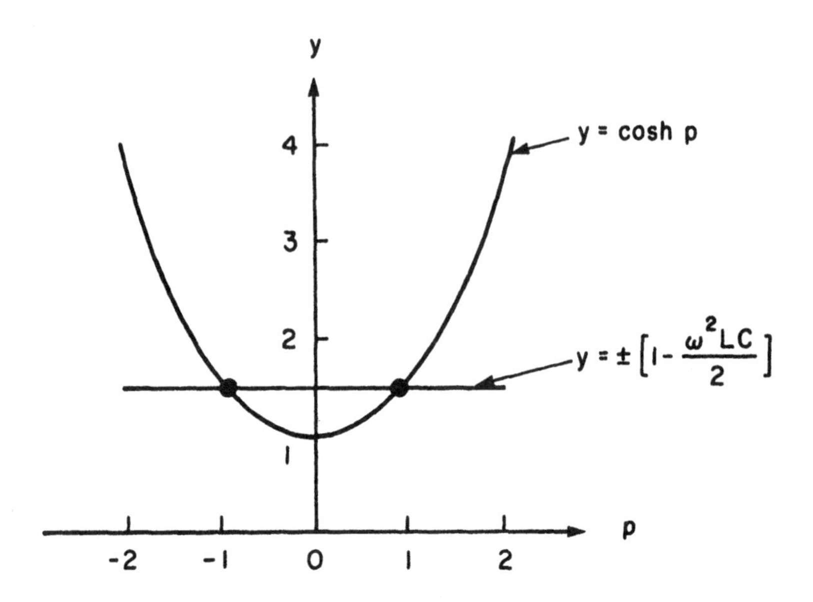

Fig. 1.44 Graphical solution of the transcendental equation cosh p = $\pm[1-(\omega^2 LC)/2]$. For p real, intersection points (i.e., solutions) can exist only if $(\omega^2 LC/2)$ is greater or equal to 2. Thus $\omega^2 \geqslant 4/LC$. If p is purely imaginary (p = jq), then the transcendental equation has a solution only if $\omega^2 \leqslant 4/LC$. In the last-named regime of frequencies, the current oscillations can propagate along the ladder suffering phase shift but no decrease in amplitude.

Therefore, setting p = jq in Eq. 1.188 yields

$$\frac{I^o_{n-1}}{I^o_n} = \frac{I^o_n}{I^o_{n+1}} = \ldots\ldots = \pm e^{jq} \tag{1.191}$$

where q is given by Eq. 1.189 with p replaced by jq:

$$\cos q = \pm\left[1 - \frac{\omega^2 LC}{2}\right]. \tag{1.192}$$

Equation 1.191 tells us that now the current suffers no degradation in amplitude as it proceeds from stage to stage, but merely a phase shift given by q. And Eq. 1.192 tells us that in order to be in this "nondissipative" regime the signal frequency must be between zero and

$\omega_{crit} = 2/\sqrt{LC}$.

It is now obvious that what we have created is a lowpass filter circuit capable of transmitting signals virtually unattenuated (there is always some resistive loss) for all frequencies from dc to ω_{crit}. Signals with frequencies greater than ω_{crit} are rejected exponentially; the degree of rejection increases as the number of stages increases and the larger ω is relative to ω_{crit}. In practice four to five stages are usually quite sufficient.

When the inductances and capacitances are interchanged (Fig. 1.45) the ladder network changes from a lowpass filter to a highpass filter. Using the methods just described, we leave it to the reader to show that the circuit of Fig. 1.45 passes all frequencies above a critical frequency given by

$$\omega_{crit} = \frac{1}{2\sqrt{LC}} \quad \text{(highpass).} \tag{1.193}$$

Lowpass, highpass (and bandpass) filters like those we have been discussing are important in virtually all electronic systems. Mechanical analogues used in acoustical filtering are also widely used.

Until this section we have given no thought to the possibility that different parts of an oscillating system may be out of phase with one another. When the bob of a pendulum passes through its maximum displacement, so does every portion of the rod (see Fig. 1.4) from which it is suspended; when the bob passes through its equilibrium

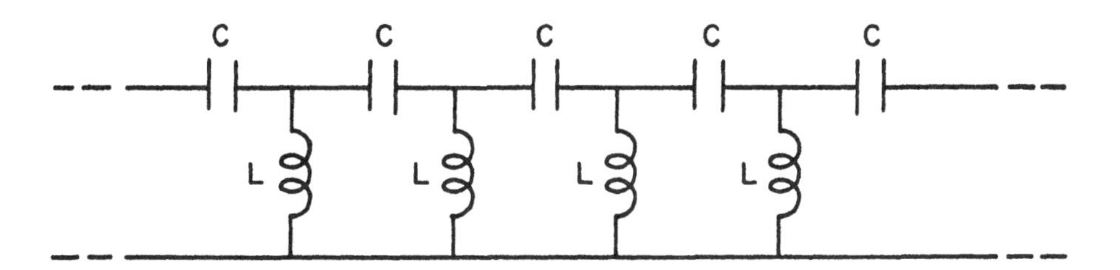

Fig. 1.45 A ladder of inductively coupled LC circuits. This acts as a highpass filter and allows signals to pass whose frequencies are greater than $1/2\sqrt{LC}$.

position, so does the rod. All pieces of the system are in exact phase with one another. But now we know that this is not a necessary state of motion of an oscillating system. The two coupled oscillators shown in Fig. 1.38 can be in or out of phase, depending on the mode of their vibration. In a ladder of identical oscillators, Eq. 1.191 tells us that we can even have a continuous progression of phases in which the oscillating currents in successive elements are of the form

$$\left.\begin{aligned} I_1 &= I_1^o\, e^{j\omega t} \\ I_2 &= I_2^o\, e^{j\omega t} \\ &= I_1^o\, e^{j(\omega t + q)} \\ I_3 &= I_3^o\, e^{j\omega t} \\ &= I_2^o\, e^{j(\omega t + q)} \\ &= I_1^o\, e^{j(\omega t + 2q)} \end{aligned}\right\} \qquad (1.194)$$

$$\vdots$$

etc.

The phase progression in fact turns out to be a phase regression, that is, a delay relative to the free end of the ladder at which the signal is being injected. The implication is that it takes time for the signal to travel up the ladder, a result that is both plausible and of profound importance. It is the basis of all wave propagation, the subject of the next chapter.

CHAPTER 2

THE WAVE

The gentle sway of the ocean surface, the burst of triumphant music from the orchestra, the sight of the rising Sun, or the rhythmic beats of a pulsating object somewhere in outer space would not be known to us if it were not for the existence of the all-pervasive phenomenon – the wave – which fills the greater part of the physical universe and brings such glorious knowledge to us. What then is this thing we call the wave? It is really just another form of oscillation such as we have just studied, but with one major difference. Whatever the quantity was that moved in Chapter 1, whether it was the oscillating mass on a spring or the voltage in an LC circuit, the oscillations were strictly localized and periodic in time alone. In a wave the moving "thing" is physically spread out over large, yes, even infinite domains, and the oscillating quantity can be periodic both in time and in space. In Chapter 1 we had the satisfaction of illustrating to the reader what a vast variety of phenomena could rest on a single physical thought and on a single mathematical expression, Eq. 1.1. To be sure, we modified and refined Eq. 1.1 as our insight deepened. We added damping because dissipation is almost inevitable. Then, to overcome the damping and keep the oscillations going we added an engine – the driver of oscillations. And finally, knowing that objects are not usually perfectly elastic we inquired about the role of nonlinearities. The wave, just like the single oscillator, is describable by a differential equation

$$\boxed{\frac{\partial^2 \psi}{\partial t^2} = v^2 \frac{\partial^2 \psi}{\partial z^2}} \qquad (2.1)$$

which tells us immediately that the moving quantity ψ is now a function of both time t and position z. This formula, known as the one-dimensional wave equation is one of the most important in all of physics and we shall study its implications in much detail.

2.1 TRAVELING WAVES

To begin with a familiar example, consider a very, very long, flexible (but inextensible) string held at one end, the origin 0, and kept under constant tension, which could be exerted for example by a spring fastened at the other end (see Fig. 2.1). At 0 the string is attached to an oscillating device (the "transmitter") which shakes the string periodically by displacing the end point 0 in a direction perpendicular to z, the direction of the string. If we denote by s(0, t') the string's displacements at the origin 0, and if we assume that the displacements are simple harmonic, we can write that

$$s(0, t') = s_0 \cos(\omega t' + \phi) \tag{2.2}$$

where s_0 is the amplitude and ϕ the phase relative to some choice of the origin of time. What then is the displacement s(z, t) of an arbitrary point P on the string, distance z from 0 and at a time t? Now, from common experience of watching traveling water waves, for example, we are fairly confident of two things; first, that the oscillations appear pretty much the same wherever we look along z (that

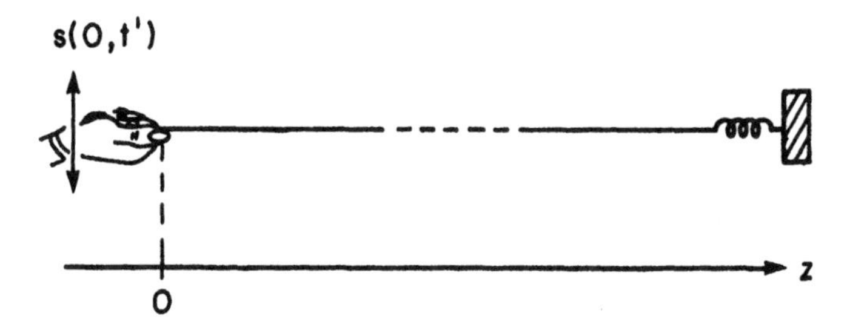

Fig. 2.1 Excitation of a traveling wave on a long string kept under constant tension.

is, the string propagates the oscillation without changing its shape); and second, that whatever is happening at position z and time t is delayed relative to what is happening at z = 0 and an earlier time t', by the fact that it takes time for the disturbance to reach a given point on the string. If v is the velocity, then

$$t' = t - \frac{z}{v} \tag{2.3}$$

and

$$\begin{aligned} s(z, t) &= s(0, t') \\ &= s_o \cos\left[\omega\left(t - \frac{z}{v}\right) + \phi\right] \end{aligned} \tag{2.4}$$

where the second form of Eq. 2.4 was obtained by inserting Eq. 2.3 in Eq. 2.2. (We assume that the far end of the string is so remote that the disturbance s(z, t) has not reached it during times of interest here; hence we don't have to bother just yet about complications occurring where the string is attached to the spring.)

The quantity v represents the velocity at which the disturbance propagates along the string and is usually referred to as the <u>phase</u> velocity or the propagation velocity of the wave. It must not be confused with the velocity acquired by any one point on the string. A given mass element of string oscillates with simple harmonic motion about an equilibrium position in a direction perpendicular to z and has an instantaneous velocity given by differentiating Eq. 2.4 with respect to time: $ds/dt = -\omega s_o \sin[\omega(t-(z/v)) + \phi]$. The phase velocity v, on the other hand, is directed along z and has a magnitude as yet unknown to us. The shape of the string at different time intervals is illustrated in Fig. 2.2. The heavy dot shows that the motion of an arbitrary mass element is indeed always transverse to z. It is important to realize that neither in this wave or in any other wave is there any net transport of "shaking matter" traveling with the wave.

When we scan Fig. 2.2 vertically and concentrate our attention on any one specific element of the string, we see that it executes simple harmonic motion in time with a period $T = 2\pi/\omega$, which is precisely how a simple harmonic oscillator behaves, as was described in

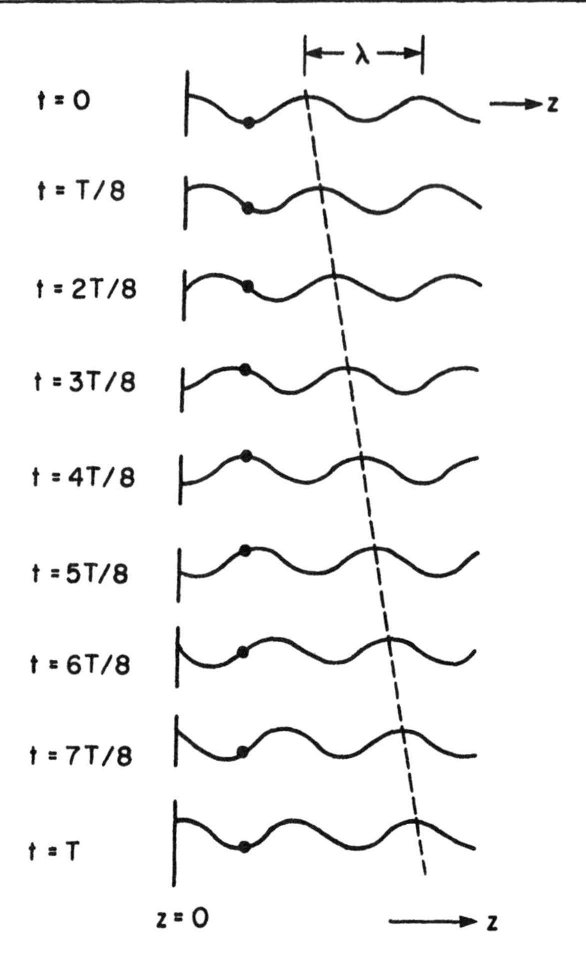

Fig. 2.2 A sinusoidal transverse wave on a string traveling toward the right, shown at intervals of one eighth of a period. The transmitter is at $z = 0$. The black dot shows the position of an element of string. The dashed line shows how the wave form advances steadily to the right. The oscillations are assumed to have gone on for such a long time that all transients have died out.

Chapter 1. If, on the other hand, we freeze time and examine the spatial configuration of the string, we observe that it is also periodic in space. This double periodicity is characteristic of wave motion. It is easy to see from Eq. 2.4 that the spatial pattern repeats whenever z changes by an amount $2\pi v/\omega$. This repetition length is called the wavelength and is denoted by λ, that is,

$$\boxed{\begin{aligned}\lambda &= \frac{2\pi v}{\omega} \text{ meters} \\ &= \frac{v}{\nu}\end{aligned}} \tag{2.5}$$

where the second version of the relation follows from the connection, $\omega = 2\pi\nu$, which relates the angular frequency ω in radians/sec to the frequency ν in hertz. Therefore, if we so wish, we can eliminate the velocity v from Eq. 2.4 in favor of the wavelength λ and write

$$s = s_o \cos\left(\omega t - \frac{2\pi}{\lambda} z + \phi\right) \tag{2.6}$$

which is a perfectly acceptable and often used form of representing the traveling sinusoidal wave. However, in modern usage one rarely finds equations written with the "2π divided by the wavelength," because it is both inconvenient and a little clumsy. Instead, one defines a propagation constant k as

$$\boxed{k = \frac{2\pi}{\lambda}} \tag{2.7}$$

and obtains

$$\boxed{s = s_o \cos(\omega t - kz + \phi)} \tag{2.8}$$

a form which we shall use exclusively in this book. In terms of k, then, the phase velocity and angular frequency are connected by

$$\boxed{v = \frac{\omega}{k}.} \tag{2.9}$$

As mentioned already, we know nothing as yet concerning the properties of v; we do not know its magnitude, neither do we know whether it is a function of the frequency of excitation ω. To discuss

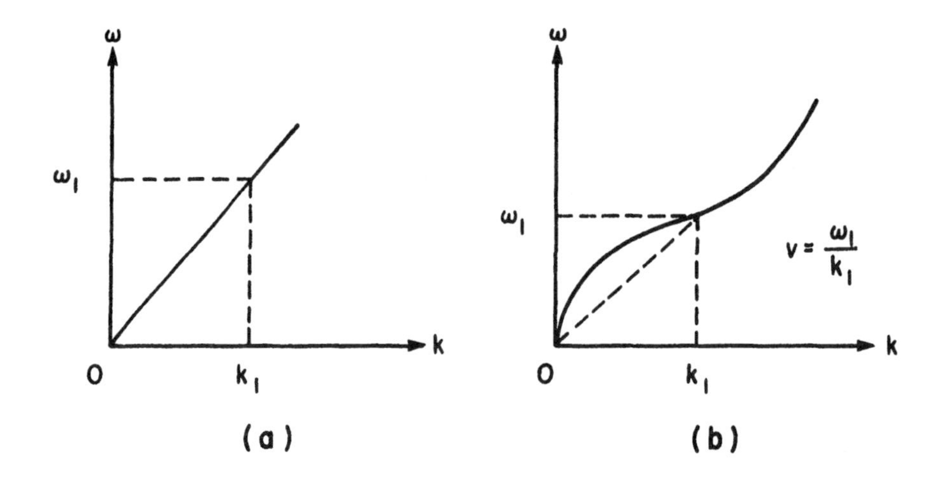

Fig. 2.3 Dispersion diagrams. In (a) the phase velocity $v = \omega/k$ is a constant, independent of frequency of excitation. The wave is dispersionless as, for example, in transverse waves on a perfectly flexible string, or in sound waves. In (b) the phase velocity is a function of frequency, and the wave is said to exhibit dispersion. The phase velocity first decreases and then increases as is known to be the characteristic of deep-water waves, where $\omega^2 = ak + bk^3$ with a and b as constants.

these facts one must set up the detailed equations of motion for the system in question. Indeed, the prime objective in solving any wave propagation problem is to determine v and thereby find the relationship between the excitation frequency ω and the wave number k. This is known as finding the dispersion relationship and a plot thereof is known as the dispersion diagram. The latter is illustrated in Fig. 2.3. Observe how easy it is to determine the phase velocity v of the wave at any desired frequency ω. Draw a straight line from the point on the dispersion curve in question to the origin, and measure off the distances ω_1 and k_1; the phase velocity corresponding to the frequency ω_1 is then given by $v = \omega_1/k_1$. Note that when the wave propagates along the negative z direction, k is in effect negative and the equation for the wave is now given by

$$s = s_o \cos(\omega t + kz + \phi). \tag{2.10}$$

On the next few pages we shall illustrate how one goes about

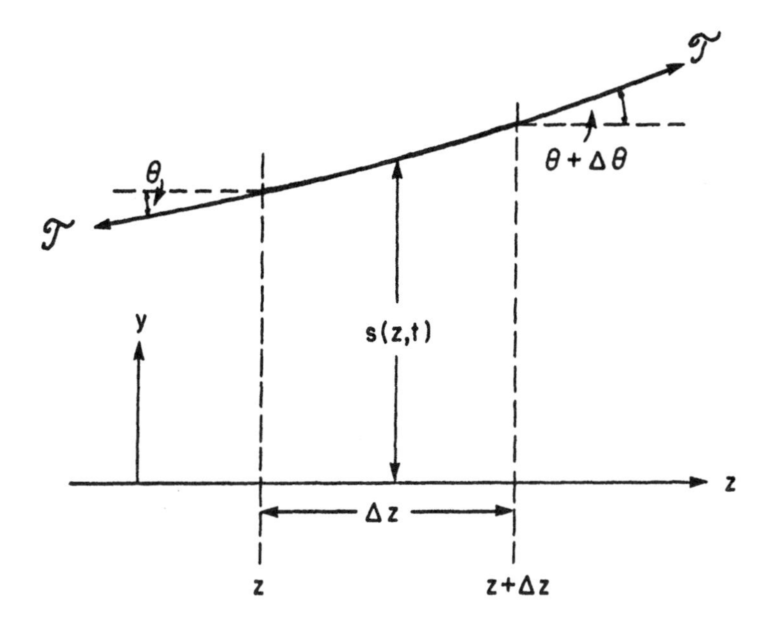

Fig. 2.4
Section of vibrating string displaced a distance s(z, t) from its equilibrium position; $\mathscr{T}$ is the tension in the string. At (z+dz), $\mathscr{T}_y(z+dz) \approx \mathscr{T}(\partial s/\partial z)_{z+\Delta z}$ and at z, $\mathscr{T}_y \approx \mathscr{T}(\partial s/\partial z)_z$. The net force on the element Δz is $F_y = \mathscr{T}[(\partial s/\partial z)_{z+\Delta z} - (\partial s/\partial z)_z]$. A Taylor's expansion of $(\partial s/\partial z)_{z+\Delta z} = (\partial s/\partial z)_z + (\partial^2 s/\partial z^2)_z \Delta z$ + etc. Hence, $F_y = \mathscr{T}(\partial^2 s/\partial z^2)\Delta z$ which is an alternate way of deriving Eq. 2.13.

calculating v in terms of the physical properties of the medium which supports the wave motion.

Transverse waves on a string

Consider a flexible and inextensible string maintained under constant tension $\mathscr{T}$ provided by, say, the spring shown in Fig. 2.1. To find how the displacement s varies with position z and time t, we isolate a short section of the string as indicated in Fig. 2.4 and find the net force acting on it. It is given by

$$F_y = \mathcal{T} \sin(\theta + \Delta\theta) - \mathcal{T} \sin \theta$$
$$F_z = \mathcal{T} \cos(\theta + \Delta\theta) - \mathcal{T} \cos \theta \qquad (2.11)$$

where θ and $\theta + \Delta\theta$ are the angles subtended by the tangents to the string at the two ends, respectively, as measured relative to the z axis. We now assume that the displacements are everywhere sufficiently small so that the angles formed by the various segments are small fractions of a radian. Subject to this, a number of approximations can be made, notably that $\sin \alpha \approx \alpha - (\alpha^3/3!)$, $\cos \alpha \approx 1 - (\alpha^2/2!)$, $\tan \alpha \approx \alpha + (\alpha^3/3)$. Then retaining only those terms containing the first power in θ and neglecting terms with higher power, Eqs. 2.11 become

$$F_y \approx \mathcal{T}\Delta\theta$$
$$= \mathcal{T}\left(\frac{\partial\theta}{\partial z}\right)\Delta z \qquad (2.12)$$
$$F_z = 0 \text{ (to first order in } \Delta\theta\text{)},$$

showing that only the y-directed force is of importance. Observe that when we write $\Delta\theta = (\partial\theta/\partial z)\Delta z$ the use of the partial derivative $(\partial\theta/\partial z)$ signifies that in differentiating with respect to z, the value of time is held constant. For small displacements, θ is approximately equal to $\tan \theta$ and $\tan \theta$ is (exactly) equal to $(\partial s/\partial z)$ (it is just the average slope of the string) with the result that

$$F_y \approx \mathcal{T}\left(\frac{\partial^2 s}{\partial z^2}\right)\Delta z. \qquad (2.13)$$

But the string is inextensible and since we are considering only small movements from equilibrium, our segment of string has only one component of acceleration, which is in the y component, and its value is $\partial^2 s/\partial t^2$. Here partial derivatives denote that the acceleration is to be computed at a fixed position z. Taking μ as the mass per unit length of string and neglecting gravity, Newton's second law $F_y = (\mu\Delta z)\, \partial^2 s/\partial t^2$ results in

$$\frac{\partial^2 s}{\partial t^2} = \left(\frac{\mathscr{T}}{\mu}\right)\frac{\partial^2 s}{\partial z^2}. \tag{2.14}$$

We recognize this immediately as the wave equation, and comparison with Eq. 2.1 yields the sought-after phase velocity

$$\begin{aligned} v &= \frac{\omega}{k} \\ &= \sqrt{\frac{\mathscr{T}}{\mu}}. \end{aligned} \tag{2.15}$$

It is a constant number and is a function of the static properties $\mathscr{T}$ and μ of the string alone. Since it is independent of frequency, ω varies linearly with k and the dispersion diagram for waves on the string is like that shown in Fig. 2.3a. The waves are then said to be nondispersive, which is the case of the perfectly flexible string assumed here. In actual practice an object like a piano string is less than perfect and we then find a somewhat more complicated dispersion relation which has the form

$$\frac{\omega^2}{k^2} = \frac{\mathscr{T}}{\mu} + a^2k^2 \tag{2.16}$$

where a is a small constant of dimensions of length. This tells us that waves of higher frequency have a somewhat higher phase velocity than waves of lower frequency; for the perfectly flexible string $a = 0$ and Eq. 2.16 reduces to Eq. 2.15. Take as an example a steel piano string of 1-mm diameter (10^{-3} meters) held taut with a tension $\mathscr{T}$ = 500 newtons (this corresponds to hanging a weight of about 50 kg from its free end). The density of steel is approximately 9 gm/cm^3 or 9×10^3 kg/m^3. The mass per unit length μ is then 9×10^3 multiplied by the cross-sectional area of the wire, or $\mu = 7.07 \times 10^{-3}$ kg/meter. Using this value of μ and the value of $\mathscr{T}$ given above in expression (2.15) for the phase velocity, we obtain that $v \approx 266$ m/sec. If a transmitter fastened to one end of the string oscillates the string transversely with a frequency of, say, 256 Hz (which is the middle C on the piano), the wavelength imposed by this frequency is given by $\lambda = v/\nu$ and thus equals about 1.04 meters.

Sound waves in a pipe

As our second example we consider the propagation of sound waves in a fluid which could be air or some other gas, or for that matter, it could be a liquid. We picture the fluid as contained in a very long pipe extending along the z axis and driven by a piston oscillating back and forth in a direction parallel to the z axis, as in Fig. 2.5. This causes each element of fluid to move back and forth along z about an equilibrium position. It also causes the pressure in the fluid to oscillate and these pressure oscillations are accompanied by compressions and rarefactions in the fluid density. Note that all of the

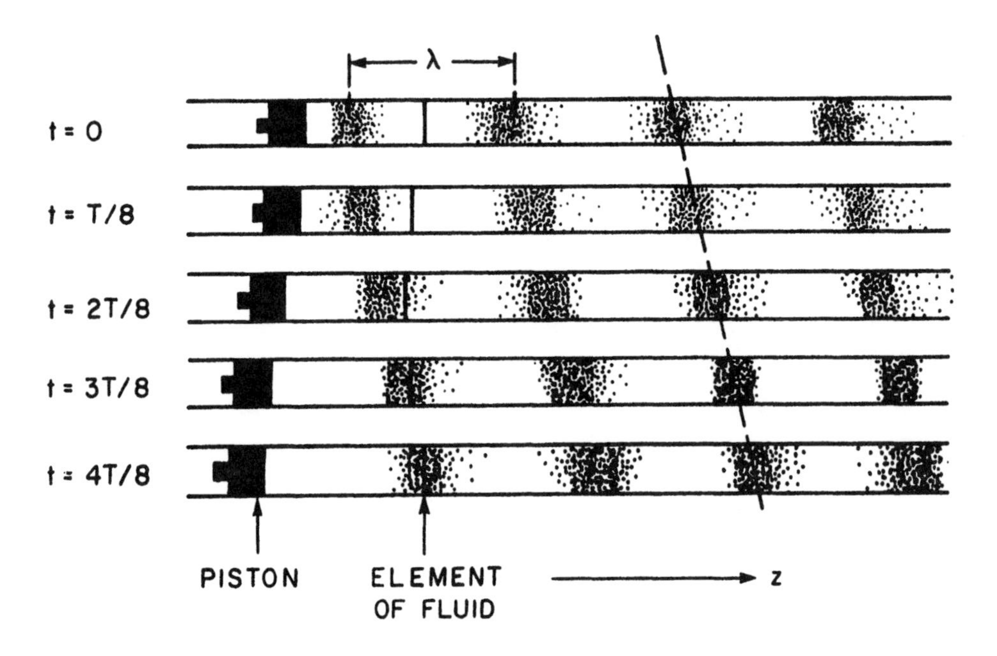

Fig. 2.5 Longitudinal sound wave in a pipe, driven by a piston executing harmonic oscillations. Dark areas represent condensations of the fluid; light areas, regions of rarefactions. The thin vertical line shows the instantaneous position of a given element of fluid; it is executing simple harmonic vibrations about a fixed equilibrium point. The slanted dashed line shows the progress of a particular condensation; it moves along the z axis with the phase velocity v.

oscillations take place along the propagation axis rather than transverse to it, as was the case of the vibrating string. For this reason we call the sound oscillations longitudinal as opposed to the transverse oscillations on the string. Indeed, the entirety of all types of wave phenomena in nature can be divided into these two major classes, transverse and longitudinal (but on occasion we shall meet waves that are a mixture of both, being partly longitudinal and partly transverse).*

The purpose of the pipe is mainly one of mathematical convenience. It makes the problem one-dimensional with z as the axis along which any changes are taken to occur. It is assumed that all points that lie on a given section perpendicular to z undergo identical displacements. We could make the pipe diameter go to infinity but to maintain one-dimensionality the oscillating piston would likewise have to have an infinite cross section. Such an abstraction of reality has many uses. It has a name — it is called a plane wave. And there is no harm in using it as long as we keep in mind that a plane wave in wide open space is an approximation to what actually happens, namely that the finite sized transmitter excites something closer to a spherical wave, but when we are sufficiently far from the transmitter and when we look at a restricted portion of space, it looks like a plane wave.

Let us then go back to the pipe but neglect any possible frictional or viscous forces which the wall may exert on the fluid. Isolate a section of fluid, which in the undisturbed condition lies between z and $(z+\Delta z)$. If A is the cross-sectional area of the pipe, the elementary volume of fluid lying between z and $(z+\Delta z)$ is $V_o = A\Delta z$. As a result of the piston displacement, the front and back ends of the fluid element move through distances $s(z)$ and $s(z+\Delta z)$ respectively (see Fig. 2.6). The change in volume is given by $\Delta V = [s(z+\Delta z) - s(z)]A$ and the

*There are more complicated systems where the motions need not be in a straight line. In the case of waves on the surface of deep water elements of water make circular orbits in a vertical plane lying parallel to the direction of propagation. Torsional vibrations in rods give circular motions perpendicular to the direction of propagation.

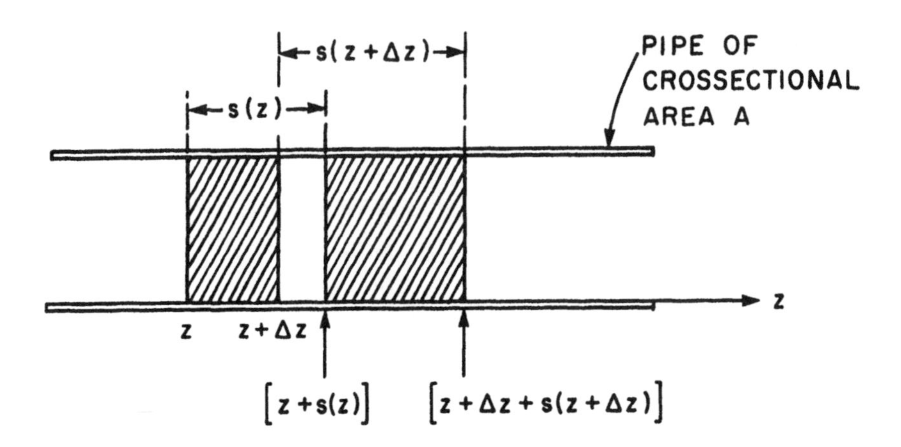

Fig. 2.6 The motion of an element of fluid of length Δz induced by a longitudinal sound wave traveling down the pipe; $s(z)$ is the displacement of the front surface of the element, and $s(z+\Delta z)$ is the displacement of the back surface.

fractional change in volume by

$$\frac{dV}{V_o} = \frac{s(z+\Delta z) - s(z)}{\Delta z} = \frac{\partial s}{\partial z} \tag{2.17}$$

where $V_o = A\Delta z$ is the volume of the undisturbed element. The second form of the equation comes about as a result of taking a very tiny slice Δz and proceeding to the limit dz. Accompanying this change in volume is a change in pressure p (measured relative to the equilibrium pressure p_o of the undisturbed system). But we know already the relation between p and ΔV; we discussed it in conjunction with the Helmholtz resonator and the result is given by formula (1.39). Inserting (2.17) in (1.39), it follows that

$$p = -\kappa \frac{\partial s}{\partial z} \tag{2.18}$$

where κ is the elasticity (or bulk modulus) of the fluid in question. The net force acting on the element Δz is $F = -A\Delta p$ where Δp is the difference between the values of p at z and $(z+\Delta z)$. For small displacements this difference is easily shown to equal $(\partial p/\partial z)\Delta z$ with the result that

$$F = -A\left(\frac{\partial p}{\partial z}\right)\Delta z. \qquad (2.19)$$

This is the force causing an acceleration $\partial^2 s/\partial t^2$ in a mass of fluid equal to $\rho_o A\Delta z$ with ρ_o as its density. Use of $\vec{F} = m\vec{a}$ then gives

$$\frac{\partial^2 s}{\partial t^2} = -\frac{1}{\rho_o}\frac{\partial p}{\partial z}. \qquad (2.20)$$

The differential pressure p (above or below p_o) is eliminated by use of Eq. 2.18 yielding the wave equation

$$\frac{\partial^2 s}{\partial t^2} = \frac{\kappa}{\rho_o}\frac{\partial^2 s}{\partial z^2} \qquad (2.21)$$

with

$$v = \sqrt{\frac{\kappa}{\rho_o}} \qquad (2.22)$$

as the wave's phase velocity. This result bears a strong resemblance to Eq. 2.15 for the wave velocity on the vibrating string. The bulk modulus κ plays the role of the tension $\mathscr{T}$ and the mass density ρ_o(kg/m^3) plays the role of the line density μ(kg/m).

The magnitude of the sound velocity varies greatly from substance to substance and in particular depending on whether the substance is gaseous, liquid or solid. In gases the pressure and volume changes are known to follow a particularly simple law, pV^γ = constant (see Eq. 1.40). When this statement is inserted in Eq. 1.39 and the differentiation performed, the bulk modulus κ takes on the value γp_o and the expression for the sound velocity becomes

$$v = \sqrt{\gamma\frac{p_o}{\rho_o}}\ \text{m/sec} \quad \text{(for gases)}. \qquad (2.23)$$

In the case of air $\gamma = 1.40$, $\rho_o = 1.29$ kg/m^3 and p_o (at one atmosphere) $= 1.01 \times 10^5$ N/m^2, with the result that v at NTP equals 331 m/sec. The measured value of 331.45 m/sec is in very gratifying agreement with this prediction from the theory. Satisfactory but somewhat less good agreement is found for solids and liquids when the measured v are compared with predictions from Eq. 2.22. Take, for example, the case of aluminum which has a density $\rho_0 = 2.7 \times 10^3$ kg/m^3 and a bulk modulus κ equal to 7.46×10^{10} n/m^2. This gives a velocity v = 5256 m/sec which is to be compared with the directly measured value of 5100 m/sec. The accompanying table lists the velocities in several other materials. It is interesting to observe the large difference in the sound of speed in wood when this is measured along two different directions. Materials which exhibit different propagation characteristics along different axes are said to be anisotropic. One of the most notable examples of this is light propagation along different principal axes of a crystalline substance.

Table 2.1. Sound velocity in different materials.

Material	v (m/sec)
Helium (NTP)	965
Water (distilled at 25°C)	1498
Wood (ash, along the fiber)	4670
Wood (ash, along the rings)	1260
Silver	2680
Brick	~3650

In the preceding calculation the molecular nature of the fluid in the pipe was ignored and the medium was treated as though it were a continuous, homogeneous substance. This may be a fairly good picture for solids and liquids but one becomes a little worried in the case of gases. We know that a gas is composed of molecules in random motion, separated by distances that are huge compared with their size. The random velocities for air molecules at room temperature are on an average equal to about 300 m/sec and this is comparable with the wave velocity of 331 m/sec. Thus each gas molecule has, superposed on its large random velocity, the oscillatory velocity which constitutes the wave. Might the random motion spoil the organized oscillatory property of the wave, scramble the phases and disrupt the works? This could very well happen if a bunch of molecules in, say, the crest of the wave carried the "crest" oscillatory velocity into the neighboring trough where they would add their organized motion to the velocities of the "trough" molecules. What saves the situation are the numerous collisions between molecules; in one or two collisions a molecule loses all knowledge of its earlier motion and is thus incapable of carrying the information very far. What is the criterion that scrambling (or "phase mixing" as it is called) does not occur? Simply, that the mean-free path between collisions be short compared with the wavelength λ of the oscillation. For air at atmospheric pressure the mean-free path L is approximately 10^{-7} meters. In speech or music the highest frequency of interest is ~20 kHz which corresponds to a wavelength $\lambda = (332/20\times10^3) \approx 1.7\times10^{-2}$ meters, and our criterion is therefore well satisfied. There can, however, be situations (studies at low gas pressures and high frequencies) where phase mixing due to the large random motions of atoms may well be of importance.

2.2. THE ENERGY OF WAVES

Waves carry energy. If this were not so, messages could not be sent by them. Also, writers of science fiction (and more down-to-Earth applied scientists) would not talk about death rays and similar appliances. Let us then go back to section 2.1 where we discussed the pipe filled with fluid, and furnished at one end with an oscillating piston (Fig. 2.5). This piston moves back and forth about its equilibrium position $z = 0$ and thus generates a wave. The force exerted by the piston on the fluid equals $(p_0 + p)A$ where A is the area of the pipe, p_0A is the equilibrium force and pA is the differential force, with p given by Eq. 2.18. The instantaneous power P_0 exerted by the piston on the fluid is $(p_0+p)A$, multiplied by the instantaneous velocity $(\partial s/\partial t)_0$ of the piston. Subtracting the equilibrium contribution which is of no interest here, we get

$$P_o = pA\left(\frac{\partial s}{\partial t}\right)_o \quad \text{watt}$$
$$= -A\kappa\left(\frac{\partial s}{\partial z}\right)_o \left(\frac{\partial s}{\partial t}\right)_o . \tag{2.24}$$

The subscripts "o" denote that this is the power exerted on the fluid at the origin $z = 0$ where the transmitter (piston) is located. If there is no damping and if the wave indeed transports energy from place to place, the power given by Eq. 2.24 must be the same at any general point z down the pipe and is thus equal to the energy passing any downstream point per second. The energy flux, that is, the energy crossing unit area per second is then given by the equation

$$S(z, t) = -\kappa\left(\frac{\partial s}{\partial z}\right)\left(\frac{\partial s}{\partial t}\right) \quad \text{watt/m}^2. \tag{2.25}$$

This is a quite general result for any kind of periodic perturbations by the piston.* Let us apply it, however, to the case of a sinusoidal wave propagating along the positive z axis,

$$s(z, t) = s_o \cos(\omega t - kz + \phi). \tag{2.26}$$

Inserting this equation in Eq. 2.25 gives

$$S(z, t) = \kappa\omega k s_o^2 \sin^2(\omega t - kz + \phi). \tag{2.27}$$

The right-hand side of the equation is seen to be positive at all times, showing that energy is indeed traveling from left to right (along $+z$). But is energy, perhaps, flowing from the piston into the wave during one half of the cycle and back into the piston during the second half of the cycle, so that no net energy is imparted to the wave? The answer is no, because on time averaging Eq. 2.27 over one period of the oscillation (as per instructions given by Eq. 1.20), we obtain

$$\langle S(z, t)\rangle = \frac{1}{2}\kappa\omega k s_o^2 \tag{2.28}$$

*The result for small transverse oscillations on a string is very similar to Eq. 2.25. That is, the power traveling along the string equals $-\mathcal{T}(\partial s/\partial z)(\partial s/\partial t)$ watts where $\mathcal{T}$ is the tension in the string and s its transverse displacement. Prove this result.

which is clearly not zero. Note that the energy flux is proportional to the square of the amplitude, a result that applies to all types of wave motion.

It is instructive to cast Eq. 2.25 into a slightly different form by noting that $k(\partial s/\partial t) = -\omega(\partial s/\partial z)$ with the result that $S(z, t) = \kappa(k/\omega) \times (\partial s/\partial t)^2$. But (ω/k) is just the phase velocity v so that

$$S(z, t) = \frac{\kappa}{v}\left(\frac{\partial s}{\partial t}\right)^2. \tag{2.29}$$

We may now compare this, (by what may appear to be a somewhat far-fetched analogy), with the power in an electrical circuit, $P = RI^2 = R(dQ/dt)^2$ where R, I, and Q are the resistance, current and charge, respectively. Thus, the quantity (κ/v) plays the role of the resistance R and is termed the "characteristic impedance" of the propagating medium, denoted by the letter Z_o:

$$\begin{aligned} Z_o &= \frac{\kappa}{v} \\ &= \sqrt{\gamma p_o \rho_o}. \end{aligned} \tag{2.30}$$

The second form of this equation is written for the specific case of a gaseous medium for which $\kappa = \gamma p_o$ and $v = \sqrt{\gamma p_o/\rho_o}$. Observe that Z_o is a function of the static properties of the medium alone, and it will prove to be a useful quantity in our later discussions of more complex wave phenomena. However, a word of warning: Whereas the resistance R is a measure of the amount of damping in the electrical circuit, Z is not; rather, it is a measure of the amount of radiant energy abstracted from the transmitter and radiated by the wave.

The time-average energy $\langle S(z, t)\rangle A\,dt$ transported at time t across the area A situated at z must at an earlier time have filled a cylinder of length $v dt$ and cross-sectional area A (see Fig. 2.7), where v is the wave velocity. If $\langle U(z, t)\rangle$ is the time-average energy density in the wave (in joules/m^3), the total energy contained within the cylinder is $\langle U(z, t)\rangle Av\,dt$. Since this energy must balance the energy crossing the area A, it follows that

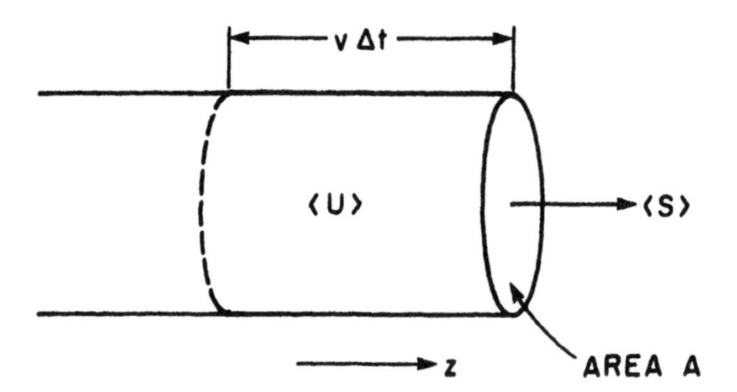

Fig. 2.7 The time-averaged flux S(watt/m^2) of radiant energy crossing the area A of a cylindrical pipe containing the waves. If $\langle U \rangle$ is the energy density of radiation (joule/m^3) in the pipe, then $\langle S \rangle = v\langle U \rangle$ where v is the wave speed.

$$\boxed{\langle S(z, t) \rangle = v \langle U(z, t) \rangle} \tag{2.31}$$

which is a result of considerable generality; it relates the flow of radiant energy to its density. Our introduction of the phase velocity v in the foregoing computation sounds plausible but is subject to qualification. Energy travels at the phase velocity only as long as v is not a function of frequency. When the medium is dispersive, v of Eq. 2.31 must be replaced by the so-called group velocity which we shall discuss in section 5.3.

The intensity and the decibel

The intensity, or as it is sometimes called, "the irradiance," is the radiation flux averaged over a period of the oscillation T;

$$\begin{aligned} I &= \frac{1}{T}\int_0^T S(z, t)\, dt \\ &= \langle S \rangle \ \text{watt/m}^2 \end{aligned} \tag{2.32}$$

and for sound waves has the value given by Eq. 2.28. The human ear spans an incredibly large range of intensities. We can hear sounds as faint as 10^{-16} watt/cm^2, and pain sets in when the noise exceeds

10^{-3} watt/cm^2. Wilska [Skand. Arch. f. Physio. 72, 161 (1935)] actually measured the amplitude displacement s_o of the tympanic membrane itself, and his results show to what incredibly small amplitudes of oscillation the ear will respond. It is most sensitive at a frequency of ~3000 Hz at which frequency Wilska measured an amplitude displacement of the tympanic membrane of ~5×10^{-10} cm (0.05Å!). If we use this value in Eq. 2.28, remembering that $k = \omega/v$ and $\kappa = \gamma p_o$, we find the threshold value of intensity to be $I \approx 2 \times 10^{-16}$ watt/cm^2 which is about equal to the value stated earlier. When we compute the displacements of the tympanic membrane caused by random thermal fluctuations of air molecules hitting it, we get an average value of approximately 10^{-9} cm, which is seen to be the threshold computed above. Nature did well — the ear is an optimally designed instrument.

Because the ear responds to such tremendous ranges of intensity, it proves inconvenient to keep referring continuously to these high multiples of ten. Therefore we compress everything by taking the logarithm and thereby define new units, the bel and the decibel. If the intensity changes by a factor of 10 (up or down) we say that it has changed by 1 bel or 10 decibels. If it changes by one thousand, we say that it has changed by 3 bels or 30 decibels, and so on. In short, if I_o is the original intensity, and I the new intensity so that I/I_o is the intensity ratio, then

$$10 \log_{10} \frac{I}{I_o} = \text{decibel increase if } I > I_o$$

and (2.33)

$$10 \log_{10} \frac{I_o}{I} = \text{decibel decrease if } I < I_o.$$

Thus, for example, a change of 0.1 dB corresponds to a ratio I/I_o = 1.023, or a 2.3 percent change in intensity, which is a variation that the ear can just about detect. What are loud and what are normal kinds of sounds expressed on our new decibel scale? Let us take I_o = 10^{-16} watt/cm^2 as the standard which in fact is close to the threshold of hearing a sound signal of approximately 1000 Hz. Relative to I_o average conversation occurs at around 60 dB above I_o; the noisiest location under Niagara Falls is approximately 90 dB above I_o, the racket 10 ft away from a pneumatic drill is approximately 100 dB above I_o and a Boeing 707 at 50 m is 120 dB.*

Note that the decibel is not an absolute unit of anything, but is a measure of an intensity change. It is used not only in sound but also in electronics, microwaves and light — in short, whenever large intensity changes occur in oscillatory systems.

Example

The noise intensity level from a Boeing 707 at a distance of a few hundred feet is approximately 120 dB above the reference level I_o of 10^{-12} watts/m^2. Compute the root-mean-square pressure variation associated with the 707 noise, assuming a frequency ω of 200 sec^{-1}. Compute also the root-mean-square amplitude of the displacement.

Solution

The intensity of 120 dB above I_o is given by Eq. 2.33 as

$$10 \log_{10} \frac{I}{I_o} = 120$$

$$I = 10^{12} I_o = 1 \text{ watt/m}^2.$$

The intensity is related to the amplitude of the air displacement by Eqs. 2.28 and 2.32 as

$$I = \frac{1}{2} \kappa \omega k \overline{s_o^2}$$

where we have indicated an average value of s_o^2 because the noise is neither steady nor really confined to a single frequency. We insert numerical values,

$$\kappa = 1.4 \times 10^5 \text{ N/m}^2 \quad \text{(see Eq. 2.23)}$$

$$\omega = 2 \times 10^2 \text{ sec}^{-1}$$

*An interesting article on noise will be found in "Noise Pollution — What Can be Done?" by A. G. Shaw, Physics Today 28, 47 (1975).

$$k = \frac{\omega}{v}$$

$$v = 3.31 \times 10^2 \text{ m/sec}$$

and get the result

$$\sqrt{\overline{s_o^2}} = 3.43 \times 10^{-6} \text{ meters.}$$

The pressure variation can be related to the intensity and the displacement by Eqs. 2.18, 2.26, 2.28, and 2.32. The result is

$$I = \frac{\omega \overline{p^2}}{2\kappa k} = \frac{v \overline{p^2}}{2\kappa}$$

$$\sqrt{\overline{p^2}} = \sqrt{\frac{2\kappa I}{v}} = 0.029 \text{ N/m}^2.$$

This can be compared with the normal atmospheric pressure of 10^5 N/m^2.

2.3 OSCILLATING MODES OF A CLOSED SYSTEM

On an endless string, or in an endless pipe, the wave launched at the transmitter flows ever onward with a perpetual succession of crests and troughs following one another as if in a dance. A very different state of events occurs when the system is closed off by definite boundaries, causing the wave energy to remain within these confines. Now the wave is characterized by certain "standing" configurations known as modes of oscillation and they are the subject of this section.

The standing wave

We know that when a one-dimensional system (string, fluid in a long pipe, etc.) is excited at the origin by a sinusoidal signal $\psi_o \cos \omega t$ of frequency ω a wave $\psi_o \cos(\omega t - kz + \phi)$ can propagate to the right, say, that is, in the positive z direction. The temporal and spatial behavior of this disturbance is a solution of the wave equation $\partial^2\psi/\partial t^2 = v^2\partial^2\psi/\partial z^2$, as is readily verified by substituting $\psi_o \cos(\omega t - kz + \phi)$ in the differential equation, and checking that the left-hand side balances the right-hand side. We can imagine a similar situation in which the transmitter is stationed to the far right, thus causing a wave $\psi_o\cos(\omega t + kz + \phi)$ to propagate in the opposite direction. With both transmitters working simultaneously a more general type of disturbance

$$\psi = \psi_{o1} \cos(\omega t - kz + \phi_1) + \psi_{o2} \cos(\omega t + kz + \phi_2) \tag{2.34}$$

will travel along the system. The principle of superposition allows us to write the equation as shown, that is, as a sum of the two disturbances; substitution of Eq. 2.34 in the wave equation confirms that ψ is indeed a solution. The special case where the two traveling waves have equal amplitudes ($\psi_{o1} = \psi_{o2} = \psi_o$) and the same (zero) phase ($\phi_1 = \phi_2 = 0$) is of particular interest. Equation 2.34 then yields

$$\psi = 2\psi_o \cos(kz) \cos(\omega t). \tag{2.35}$$

Figure 2.8 illustrates how this new form of oscillation arises. The dashed and dotted lines represent the oppositely traveling waves. At the instant $t = 0$ shown at the top of the diagram the crests and troughs of the two waves are assumed to be coincident. The solid continuous line shows the resultant displacement ψ. The remaining three diagrams give the subsequent positions of the two waves and the resultant displacement ψ, as observed at time $t = T/2$, T and $3T/4$, with $T = 2\pi/\omega$ as the period. It will be observed that at certain points in space labeled N the displacement of the medium is <u>always</u> zero either because the displacement due to each wave is zero or because the displacements due to the two waves are equal and opposite. At certain other points labeled A the displacement is first double upwards, then zero, then double downwards, zero, and double upwards again to its original position. The points of zero motion are called nodes and points of maximum motion antinodes. The wave is known as a <u>standing wave.</u> It has an entirely different

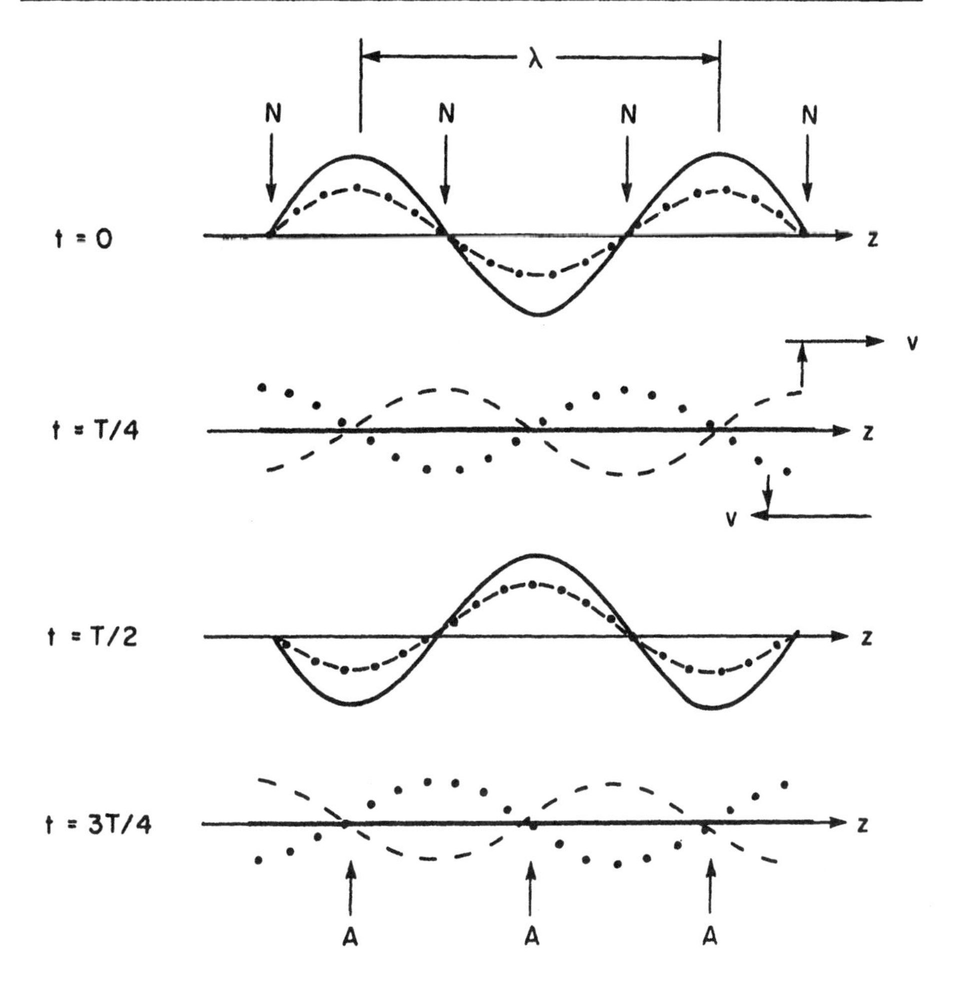

Fig. 2.8 A standing wave (solid line) generated by two identical waves (one shown dotted, the other dashed) traveling in opposite directions. N denotes a node and A an antinode. Neighboring nodes are separated by one half of a wavelength of the traveling waves.

character than the traveling wave $\psi_0 \cos(\omega t \pm kz + \phi)$ which, for example, does not have the property that certain points in space have zero motion at all times. Moreover, there is no net transport of energy in a standing wave as there is in a traveling wave. To verify this we substitute Eq. 2.35 in the equation for the flux S (cf. Eq. 2.25)

$$S(z, t) = \text{constant} \times \left(\frac{\partial\psi}{\partial z}\right)\left(\frac{\partial\psi}{\partial t}\right)$$
$$= \text{new constant} \times \sin(2kz)\sin(2\omega t). \tag{2.36}$$

When we time-average this flux over one period of oscillation we obtain $\langle S \rangle = 0$, which follows from the fact that $\langle \sin(2\omega t) \rangle$ is zero. This does not mean that there is no energy in a standing wave; there is, but it remains confined to the immediate vicinity of a given antinode. Observe that the antinodes occur when (kz) takes on the values 0, π, 2π, 3π, etc.; thus they are separated by a distance $(z_2 - z_1) = \pi/k$, or half a wavelength of the oppositely propagating waves. The nodes lie midway between the antinodes and are likewise separated by one half wavelength from one another.

Measuring the distance between nodes or antinodes provides a very powerful way of determining the wavelength of the radiation one is dealing with. Hertz used this idea to measure the wavelength of radio waves. And a few years later, in the year 1890, O. Wiener

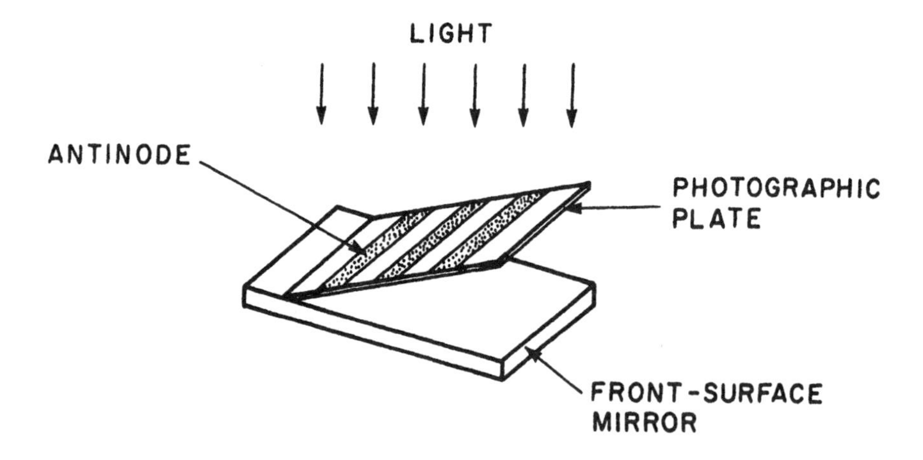

Fig. 2.9 Wiener's method of observing light waves. The dark lines observed after developing the film correspond to the spatial positions of the antinodal planes of the electric field of the standing wave [Ann. der Physik 40, 203 (1890); for a more modern version of the experiment see H. E. Ives and T. C. Fry, J. Opt. Soc. Am. 23, 73 (1933)].

was the first to demonstrate directly the existence of standing waves of light. His arrangement is shown schematically in Fig. 2.9. A parallel beam of light of a single wavelength (or rather, of a very narrow band of wavelengths) was allowed to strike a front-silvered mirror at normal incidence. The incident and reflected waves combined as we have discussed to form a standing wave in front of the mirror. Then a very thin transparent photographic film, less than 1/20 of a wavelength thick deposited on a glass flat was positioned in the standing wave as shown in the figure. The film was inclined to the mirror making an angle of approximately 1/100 of a degree with its surface, and in this way the film cut across the standing wave pattern. On developing the emulsion a series of black stripes equidistant from one another was found. They represent regions where the photographic film intersects the antinodes of the standing wave (light consists of electric and magnetic fluctuations and therefore we should specify whether we are talking of antinodes of electric fluctuations or magnetic fluctuations; it turns out that it is the $\vec{E}$ field to which film is sensitive, but this is a subject for a later chapter).

Modes of oscillation on a string

The standing wave of Eq. 2.35 has characteristics reminiscent of the oscillators we studied in Chapter 1. The disturbance is harmonic in time and is frozen in space — nothing propagates from here to there. It differs from the oscillators discussed earlier only insofar as the disturbance is spread out spatially and the amplitudes $2\psi_0 \cos(kz)$ at adjacent points are interrelated. The form of Eq. 2.35 suggests to us that we look for solutions of the wave equation of the type

$$\psi(z, t) = f(z) \cos(\omega t), \tag{2.37}$$

where f(z), a function of position only, is yet to be determined. Inserting Eq. 2.37 in Eq. 2.1 gives

$$\frac{\partial^2 f}{\partial z^2} + \frac{\omega^2}{v^2} f = 0. \tag{2.38}$$

The form of this relation is familiar to us; it is the equation of simple harmonic oscillation with time t replaced by position z. Its solution (cf. Eq. 1.10) is

$$f(z) = A \cos\left(\frac{\omega}{v} z\right) + B \sin\left(\frac{\omega}{v} z\right) \tag{2.39}$$

where A and B are arbitrary constants. The full result then follows on inserting Eq. 2.39 in Eq. 2.37:

$$\boxed{\begin{aligned}\psi(z, t) &= \left[A \cos\left(\frac{\omega}{v} z\right) + B \sin\left(\frac{\omega}{v} z\right)\right]\cos(\omega t)\\ &= [A \cos(kz) + B \sin(kz)] \cos(\omega t)\end{aligned}} \tag{2.40}$$

where the second, slightly more convenient formula follows from our definition (2.9) of the propagation constant $k \equiv \omega/v$.

This is as far as we can go in our attempt to describe the standing waves, that is the normal modes, of a more or less general system. To go beyond this we must become very specific.

To begin with, let us look at the mode structure of transverse displacements s on a stretched, perfectly flexible string. Assume that the string is fixed to rigid supports at the origin $z = 0$ and $z = L$ as shown in Fig. 2.10. These two physical statements provide the so-called boundary conditions necessary in order that the problem be uniquely determined. Let us take them one by one. When $z = 0$, the displacement $\psi(0, t) = s(0, t)$ of Eq. 2.40 is required to be zero at all times. This is possible only if the coefficient A is zero. Therefore,

$$\psi(z, t) = s(z, t) = B \sin(kz) \cos(\omega t). \tag{2.41}$$

Now, at the other end, $z = L$, $s(L, t)$ is again zero at all times. This can be achieved by setting $B = 0$, but if we take this as our solution, we are left with nothing. The only other way to satisfy this boundary condition is to restrict the possible values which the propagation constant $k = \omega/v$ (and hence the frequency ω) can have; that is, to allow it to have only those values for which

$$kL = \frac{\omega}{v} L = m\pi \tag{2.42}$$

where m is any positive integer. Each value of m represents a distinct mode of the string and for that particular mode, its "shape" is given by inserting Eq. 2.42 in 2.41, with the result that

$$s_m(z, t) = B_m \sin\left(\frac{m\pi z}{L}\right) \cos(\omega_m t). \tag{2.43}$$

Here B_m is the amplitude of mode m (its size depends on how hard the string is displaced) and ω_m its radian frequency

$$\begin{aligned} \omega_m &= \frac{m\pi}{L} v \\ &= \frac{m\pi}{L}\sqrt{\frac{\mathscr{T}}{\mu}} \qquad (m = 1, 2, 3 \ldots) \end{aligned} \tag{2.44}$$

Here $v = \sqrt{\mathscr{T}/\mu}$ is the phase speed with which a traveling wave would propagate, $\mathscr{T}$ is the tension and μ the mass per unit length of string (see Eq. 2.15). From this we see that the frequency of a given mode m is inversely proportional to the length L, directly proportional to the square root of the stretching force and inversely proportional to the diameter, for wires of the same material. These three facts were discovered by Galileo (1564-1642) and at about the same time by a friar, Marin Mersenne of Paris, who published his findings in the year 1636. The "shapes" of the fundamental $m = 1$ and of the next two harmonics $m = 2$ and 3 are shown in Fig. 2.10.

The easiest way to evoke separately the different harmonics of a stretched string is by resonance. A wire is stretched on a sounding board and the tension adjusted to give a frequency of, say, 128 Hz. When small paper riders are placed on the string, all will be thrown off when a tuning fork of frequency 128 Hz is sounded and its shaft rested on the bridge over which the wire passes. If riders are placed at L/4, L/2, 3L/4 from one end, and a fork of frequency 256 Hz is sounded, the first and third riders are thrown off but the middle one stays in place, and so on. But why all these precautions? Why not just grab the wire at some arbitrary place and pluck it? The reason is, of course, that by doing so many modes are likely to be excited simultaneously and the "shape" of the string will be some complicated combination of the different individual modes. Indeed, the most

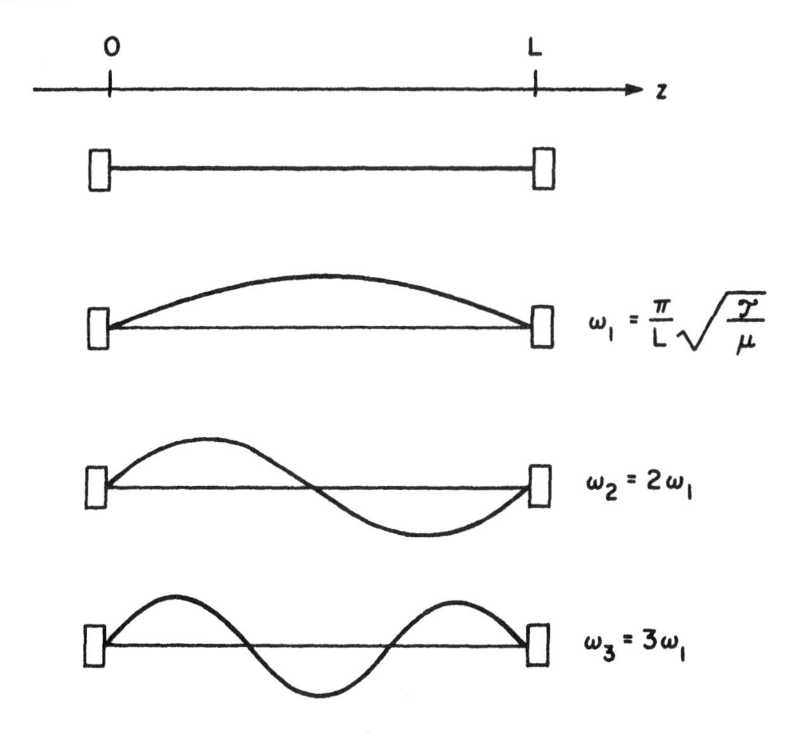

Fig. 2.10 The three lowest modes of oscillation of a string with fixed ends. The top picture is the equilibrium position of the stretched string. The three remaining pictures represent instantaneous snapshots of the modes at times of maximum displacement, say $t = 0$.

general type of vibration must be a superposition of all the infinite possible modes like that given by Eq. 2.43, that is,

$$
\begin{aligned}
s(z, t) &= \sum_m a_m \sin\left(\frac{m\pi z}{L}\right) \cos(m\omega_o t) \\
&\quad + \sum_m b_m \sin\left(\frac{m\pi z}{L}\right) \sin(m\omega_o t)
\end{aligned}
\tag{2.45}
$$

where $\omega_o = (\pi v/L)$ is the fundamental frequency (the second series comes from the fact that, depending on initial conditions, both sine and cosine variations in time may be required to satisfy them). The coefficients a_m and b_m represent the amplitudes of the fundamental and its higher harmonics. How big their relative values are will depend on how and where the string is plucked. The mathematical apparatus required to determine the coefficients is referred to as Fourier analysis and will be the subject of section 2.5.

Example

The lowest frequency obtainable from a vibrating wire under tension T, fixed at both ends, is 500 hertz (cycles per second). The mass of the wire is 20 gm and its length is 2 meters.

a) What is the tension in the wire?
b) If the wire is set into vibration and then clamped at a point one third the distance from one end, so that its displacement is zero at that point, what frequencies are possible?

Solution

For the lowest mode of oscillation of a wire fixed at both ends, the wavelength must be twice the length of the wire. Thus

$$\lambda_o = 2L = 4 \text{ meters.}$$

The phase velocity v is given by

$$v = \sqrt{\frac{T}{\mu}} = \lambda_o \nu_o = (4)(500) = 2000 \text{ meters/sec.}$$

The mass per unit length μ is

$$\mu = \frac{M}{L} = \frac{0.02}{2} = 10^{-2} \text{ kg/m.}$$

The tension is therefore

a) $$T = \mu v^2 = 10^{-2} \times 4 \times 10^6 = 4 \times 10^4 \text{ newtons.}$$

With the wire clamped at a point one-third the distance from one end only those modes which have zero amplitude at that point are possible.

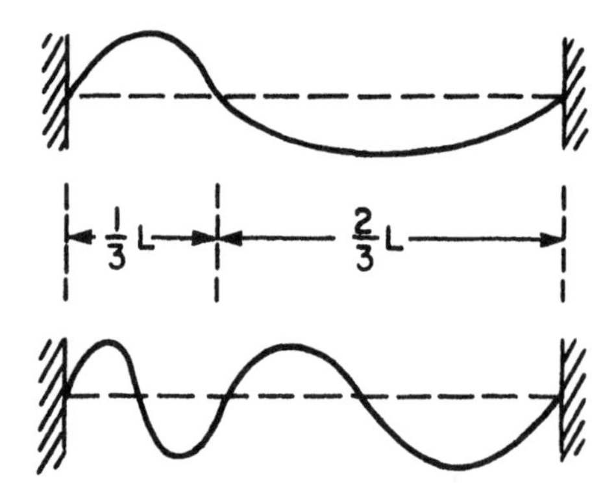

Thus we must have

$$\frac{n}{2}\lambda = \frac{1}{3}L \qquad n = 1, 2, 3....$$

and

$$\frac{n'}{2}\lambda' = \frac{2}{3}L \qquad n' = 1, 2, 3....$$

The possible wavelengths are

$$\lambda = \frac{2L}{3n} \qquad \lambda' = \frac{4L}{3n'}$$

but

$$\nu = \frac{v}{\lambda} \quad \text{and} \quad \nu_o = \frac{v}{2L} = 500 \text{ hertz},$$

so we find

$$\nu = 3n\nu_o = 1500\ n \text{ hertz}$$

$$\nu' = \frac{3n'}{2}\nu_o = 750\ n' \text{ hertz}.$$

b) Therefore the possible frequencies are all the harmonics of 750 hertz.

Vibrating air columns

The organ pipe and most wind instruments rely for their operation on the modes of vibration of air columns. There are two types of pipes that are of practical interest: the pipe that is open at both ends and the pipe that has one end closed. To tackle these problems, the first task is to determine what the boundary conditions open and closed imply. At an open end it is quite unlikely that a sizeable pressure differential can be maintained between a point just in the pipe and one just outside the pipe; if a finite pressure differential could be maintained, an element of air would experience a sudden, discontinuous change in acceleration over zero distance. The pressure differential p is given by the general expression (2.18), and since it must be zero (or close to zero) it follows that

$$\frac{\partial s}{\partial z} = 0 \quad \text{at open end.} \tag{2.46}$$

At a closed end, on the other hand, where a perfectly rigid plug is assumed to exist, an element of air cannot move longitudinally and its velocity must therefore be zero; thus,*

$$\frac{\partial s}{\partial t} = 0 \quad \text{at closed end.} \tag{2.47}$$

With these boundary conditions, let us then proceed to solve for the mode structure in a pipe having one end open at $z = 0$ and the other end closed at $z = L$ (we leave to the reader to work out the problem of the pipe with both ends open). We differentiate Eq. 2.40 with respect to z and then apply condition (2.46), $\partial s/\partial z = \partial \psi/\partial z = 0$ at $z = 0$. For this to be so at all times t, it is necessary that the coefficient B be zero, with the result that

$$s(z, t) = A \cos\left(\frac{\omega}{v} z\right) \cos(\omega t). \tag{2.48}$$

When we now differentiate s with respect to time we obtain $\partial s/\partial t = -A\omega \cos(\omega z/v) \sin(\omega t)$, and for this to be zero at $z = L$ requires that

*Prove to yourself that at a rigid wall, not only is the normal component of the displacement and the normal component of the velocity equal to zero, but so is the normal component of the acceleration $\partial^2 s/\partial t^2$. As a result of this and of Eq. 2.20, it follows that $\partial p/\partial z = 0$. This is an alternate form for the boundary condition at a rigid wall.

$\omega L/v$ be equal to an odd multiple of $\pi/2$. Hence

$$\omega_m = \frac{(2m-1)\pi}{2} \frac{v}{L} \tag{2.49}$$

where $v = \sqrt{\gamma p_o/\rho_o}$ is the sound speed and m is any integer. The equation for the displacement s_m of the mth mode is therefore

$$s_m(z, t) = A \cos\left[\frac{(2m-1)\pi}{2} \frac{z}{L}\right] \cos(\omega_m t). \tag{2.50}$$

From these equations we see that the frequency $\nu = \omega/2\pi$ of the fundamental equals v/4L hertz and that the pipe can resonate only at the odd harmonics of the fundamental, that is, 3(v/4L), 5(v/4L), etc.

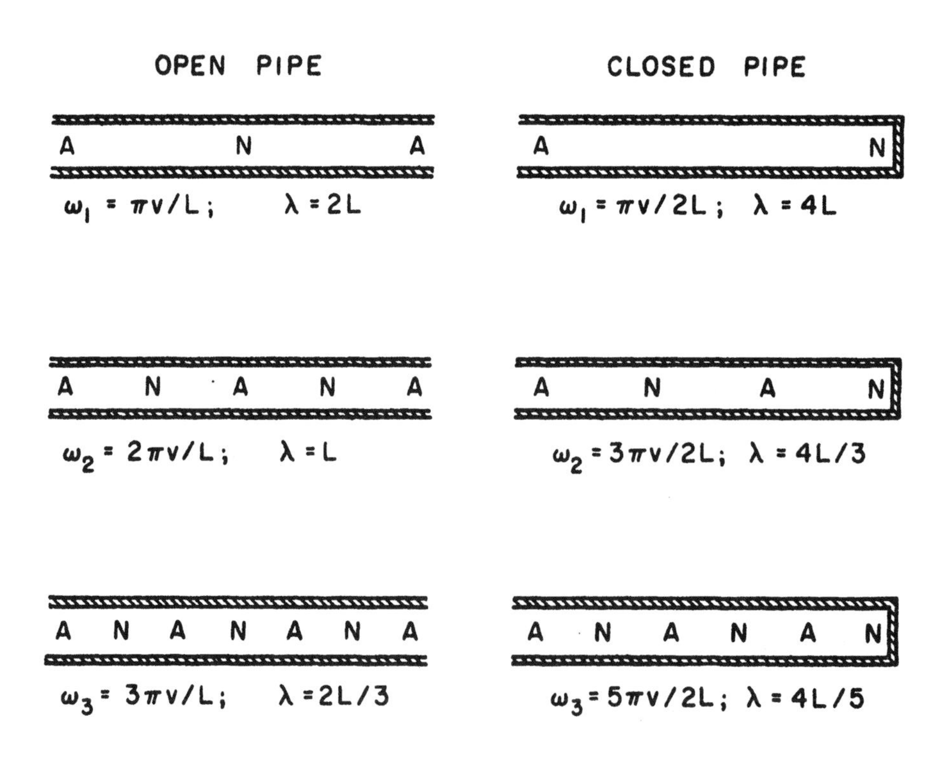

Fig. 2.11 The nodes N and antinodes A in open and closed pipes, for the fundamental frequency ω_1 and the next two harmonics ω_2 and ω_3. The corresponding wavelengths are derived from $\lambda = 2\pi v/\omega$ (or $k = \omega/v$). The nodes and antinodes refer to the displacement s of the air.

This differs from the pipe open at both ends, where all harmonics can be supported. Figure 2.11 compares the two systems.

In actual practice the antinode does not coincide precisely with the end of the pipe even if the latter has a clean-cut end. The effective length of a pipe is always a little greater than its physical length L. The "end correction" is approximately equal to 0.6 R where R is the pipe radius. Therefore, with two pipes of the same physical length but different radii, the wider pipe gives the lower note. Recall that we discussed the end-correction also in our discussion of the Helmholtz resonator (section 1.2).

A normal organ pipe has an antinode at its bottom end (where it is open) and may be closed or open at its top end. The flute and the piccolo act on the organ pipe principle and are essentially open organ pipes. The lips project a jet of air which impinges on the edge of the mouth hole. This gives rise to a fairly complicated nonlinear phenomenon known as the jet tone which then initiates the normal mode of vibration of the pipe, appropriate to the frequency of the jet tone. By varying the blowing pressure and the height of the mouth the performer can change the frequency of the jet tone over a wide range. We have presented here the barest of outlines. For a very interesting detailed account of the physics of musical instruments, see *Science and Music* by J. Jeans (Cambridge University Press, New York,. 1961); also A. Benade, *Horns, Strings and Harmony* (Doubleday Doran and Company, New York, 1960).

2.4 WAVES IN SPACE

All the waves we have discussed so far have been largely one-dimensional; in each case the nature of the problem was such that one axis (we chose the z axis) sufficed to describe the propagation of the wave. But, for example, take sound waves in a pipe whose diameter is about the same or even greater than its length. One may well imagine that the physics of the wave in such a pillbox will be more complex than in the slender flute. In this section we shall discuss some of the implications of opening up space and letting waves propagate in three dimensions.

The plane wave

The wave front of a wave can be defined as the locus of points in space which are all in identical phase of their vibration. The plane wave is the simplest example of a three-dimensional wave in that its wave front is a flat plane of infinite extent. All the wave fronts (in the case of sound, the succession of compressions and rarefactions) are parallel to one another as is illustrated in Fig. 2.12. The wave propagates in a direction perpendicular to the wave fronts. Its equation is already familiar to us from previous sections, that is,

$$\frac{\partial^2 \psi}{\partial t^2} = v^2 \frac{\partial^2 \psi}{\partial z'^2}. \tag{2.51}$$

For a harmonic disturbance and in the absence of boundaries it has a solution $\psi = \psi_0 \cos(\omega t - kz' + \phi)$. The wave propagates along the $+z'$ direction with a phase velocity v equal to ω/k. The wave front is a geometrical surface lying in the x'-y' plane over which the phase

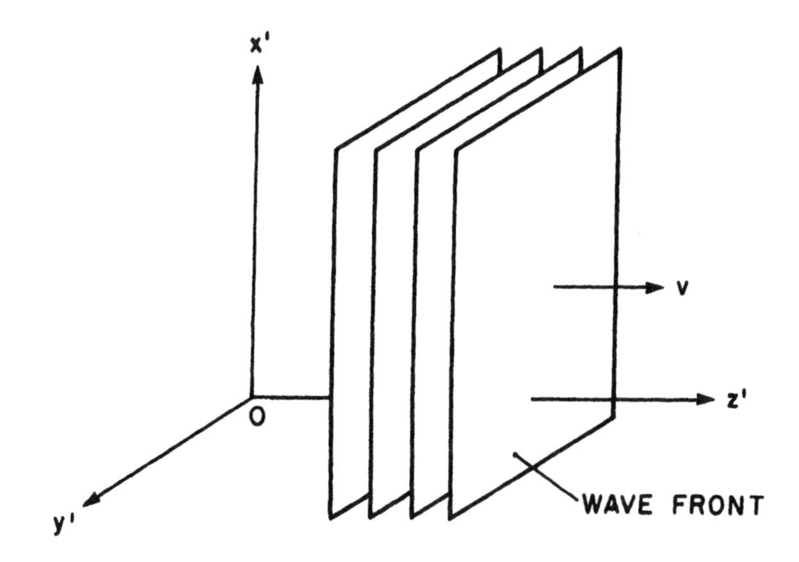

Fig. 2.12 A section of a plane wave traveling along the z' axis. The sheets are successive wave fronts corresponding to ever decreasing values of the phase $(\omega t - kz' + \phi)$.

$(\omega t - kz' + \phi)$ is everywhere a constant equal to, say, C. The successive parallel wave fronts are surfaces over which ever decreasing numerical values of the constant C are to be taken.

The propagation vector $\vec{k}$

Choosing a specific coordinate axis (z') as the propagation direction was a sensible way of going about things. However, in three-dimensional space we may not always have the luxury of this simplicity; suppose we have two plane waves propagating at an angle to one another and we want to combine them. How do we go about choosing the right coordinates? For these and similar reasons we shall now allow the plane wave to travel at an arbitrary direction relative to a

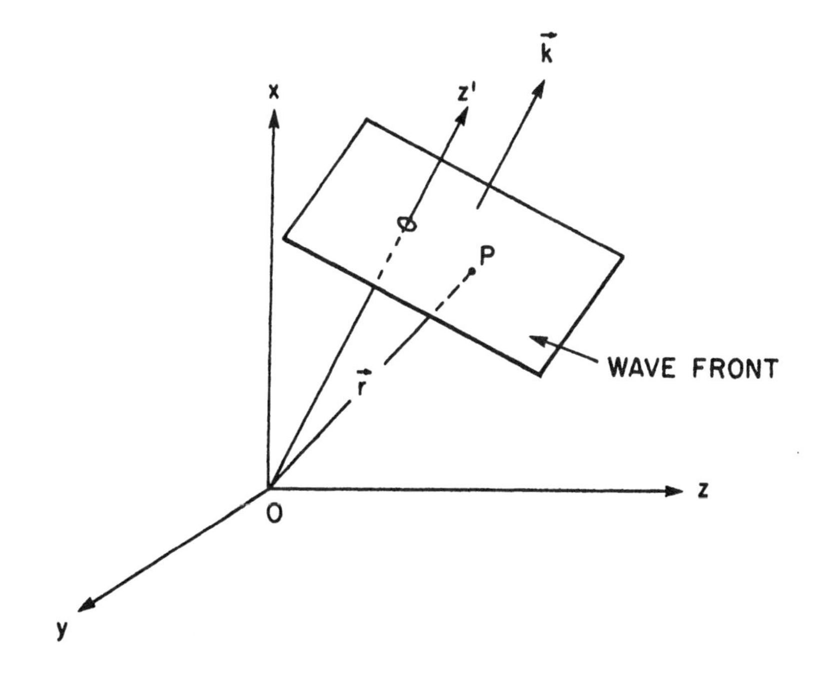

Fig. 2.13 Section of plane wave front of a wave traveling along the z' axis. The coordinates (x', y', z') of Fig. 2.12 are at an arbitrary angle relative to the new coordinates (x, y, z). The radius vector $\vec{r}$ joins the origin 0 with a point P on the wave front. The propagation vector $\vec{k}$ lies along $\vec{z}'$.

new coordinate system (x, y, z) illustrated in Fig. 2.13, and we shall attempt to express the disturbance $\psi = \psi_o \cos(\omega t - kz' + \phi)$ in terms of it. Let the radius vector $\vec{r} = \hat{x}x + \hat{y}y + \hat{z}z$ represent a point on a given wave front as measured from the origin 0 ($\hat{x}$, $\hat{y}$ and $\hat{z}$ are unit vectors along x, y and z, respectively). Then the wave front $(\omega t - kz' + \phi)$ = constant C is described in the new (x, y, z) system by the plane $[\omega t - k(\vec{r} \cdot \hat{z}') + \phi] = C$ where $\hat{z}'$ is a unit vector along the old coordinate system (x', y', z'). This follows from the fact that $z' = r \cos\theta = \vec{r} \cdot \hat{z}'$ where θ is the angle between $\vec{r}$ and $\vec{z}'$. Let us now do a little vector manipulation on the term $k(\vec{r} \cdot \hat{z}')$ and write it as

$$k(\vec{r} \cdot \hat{z}') = (k\hat{z}') \cdot \vec{r}, \tag{2.52}$$

from which we see that the product $k\hat{z}'$ is a vector lying along the propagation direction $\vec{z}'$ and having a magnitude equal to $k \equiv \omega/v$. We call this vector the propagation vector and we denote it by the symbol $\vec{k}$. Hence the traveling wave disturbance $\psi = \psi_o \cos(\omega t - kz' + \phi)$ can be expressed in vectorial notation

$$\begin{aligned} \psi &= \psi_o \cos[\omega t - \vec{k} \cdot \vec{r} + \phi] \\ &= \psi_o \cos[\omega t - (k_x x + k_y y + k_z z) + \phi] \end{aligned} \tag{2.53}$$

where the second form is the consequence of the fact that if $\vec{k}$ is a vector it must have components which in the Cartesian system are k_x, k_y, k_z such that

$$\vec{k} = \hat{x}k_x + \hat{y}k_y + \hat{z}k_z$$

and (2.54)

$$k = \pm\sqrt{k_x^2 + k_y^2 + k_z^2}.$$

The meaning of the magnitude of k is clear enough; it equals (2π/wavelength λ) where λ is the distance between, say, two crests of the wave as measured along the propagation direction $\vec{z}'$. On moving through a distance equal to one wavelength the phase of the wave C = $(\omega t - kz' + \phi)$ changes by 2π radians. How about the meaning of

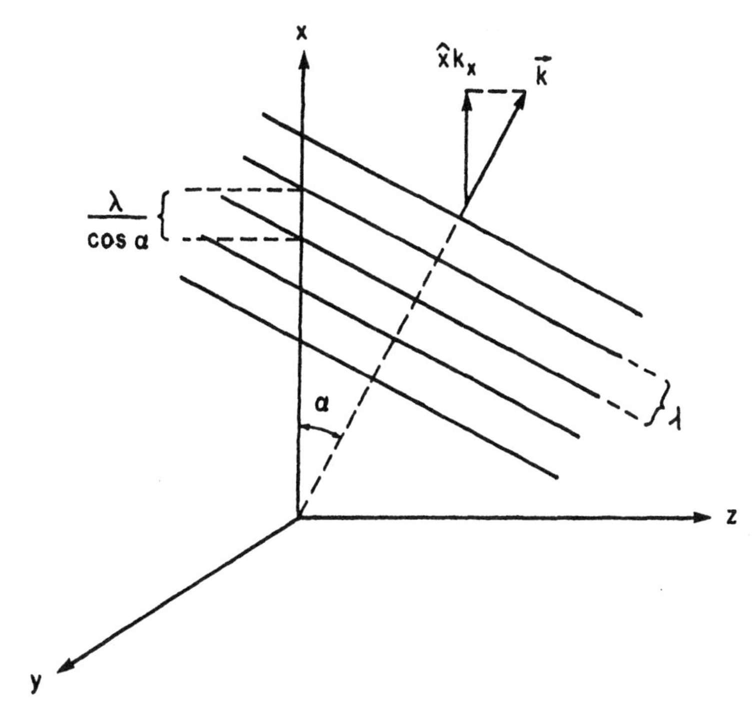

Fig. 2.14 A wave whose wave fronts are assumed to be planes lying perpendicular to the page. The propagation vector $\vec{k}$ is inclined at an angle a to the axis. The wave fronts are separated by a distance of one wavelength λ when measured along $\vec{k}$ and by a distance of $\lambda/\cos a$ when measured along x.

k_x, k_y and k_z? As an example, let the axis z' make an angle a with the new coordinate x. Then an advance of one wavelength along z' is equivalent to a displacement of $(\lambda/\cos a)$ along x (see Fig. 2.14). Therefore k_x equals the number of radians of phase change per unit displacement along the x axis, just as k equals the number of radians of phase change per unit displacement along z'. Similar meanings apply to the components k_y and k_z.

Example

A traveling wave disturbance is given by the expression

$$\psi = \psi_o \cos\left[6\pi\times 10^{10} t + 60\pi(\sqrt{5}x + 2y - 4z) + \phi\right]$$

a. What is the frequency of the wave?
b. What is the direction of propagation?
c. What is the wavelength?
d. What is the phase velocity of the wave?

Solution

The expression above should be compared with Eq. 2.53:

$$\psi = \psi_o \cos\left[\omega t - \vec{k}\cdot\vec{r} + \phi\right]$$

where

$$\vec{r} = x\hat{x} + y\hat{y} + z\hat{z}$$

and

$$\vec{k} = k_x\hat{x} + k_y\hat{y} + k_z\hat{z}.$$

Thus we see immediately that

$$\omega = 6\pi \times 10^{10}\ \text{sec}^{-1}.$$

a) $$\nu = \frac{\omega}{2\pi} = 3\times 10^{10}\ \text{hertz}.$$

The propagation vector $\vec{k}$ must evidently be

$$\vec{k} = -60\pi(\sqrt{5}\,\hat{x} + 2\hat{y} - 4\hat{z})$$

which is a vector in the direction of propagation. As a vector normalized to unity, it is

b) $$\frac{\vec{k}}{k} = -\frac{\sqrt{5}\,\hat{x} + 2\hat{y} - 4\hat{z}}{5}.$$

Since $k = 2\pi/\lambda$ and $k = 300\pi$, we find

c) $$\lambda = \frac{2\pi}{k} = \frac{2\pi}{300\pi} = 0.667\times 10^{-2}\ \text{meters}.$$

Finally, the phase velocity of the wave must follow from

d) $$v = \lambda\nu = 2\times 10^{8}\ \text{meters/sec}.$$

The three-dimensional wave equation

A wave in three dimensions needs a three-dimensional differential equation to express its motion. We shall not derive it, but rather guess at it, using our experience gathered heretofore. To generalize the one-dimensional Eq. 2.1 we note that in Cartesian coordinates the position variables x, y, z must appear symmetrically in a three-dimensional version; there is no distinguishing feature for any one of the axes, and we should be able to change at will from z to x, say, without altering the shape of the equation. As a result of this we have in place of Eq. 2.1 the formula

$$\frac{\partial^2 \psi}{\partial t^2} = v^2 \left(\frac{\partial^2 \psi}{\partial x^2} + \frac{\partial^2 \psi}{\partial y^2} + \frac{\partial^2 \psi}{\partial z^2} \right). \tag{2.55}$$

Let us check that the plane wave representation (2.53) is indeed a solution of this differential equation. Differentiating the first-named twice with respect to time we obtain $\partial^2\psi/\partial t^2 = -\omega^2\psi$; and differentiating twice with respect to x, y and z respectively, we obtain $[\partial^2\psi/\partial x^2 + \partial^2\psi/\partial y^2 + \partial^2\psi/\partial z^2] = -\left(k_x^2 + k_y^2 + k_z^2\right)\psi$. Inserting these values in Eq. 2.55 it follows that

$$\omega^2 \psi = v^2 \left(k_x^2 + k_y^2 + k_z^2\right) \psi. \tag{2.56}$$

But, from Eq. 2.54 we have $\left(k_x^2 + k_y^2 + k_z^2\right) = k^2$ and since $(vk)^2 = \omega^2$, we see that the two sides of the equation balance correctly.

Modes of vibration in three dimensions

In section 2.3 we showed that when a one-dimensional system is "clamped" in two places, the region between the two boundaries is capable of oscillating in one of an infinite number of discrete modes each of which is characterized by a unique frequency ω_m; and we found that the disturbance corresponding to the mth mode obeyed an equation of the form $\psi(z, t) = \psi_o \cos(\omega t) \sin(kz)$ [or $\psi_o \cos(\omega t) \cos(kz)$]

where the parameter $k = \omega/v$ took on discrete values determined by boundary conditions at the "clamped" positions. A similar phenomenon occurs also in three dimensions, but now we find that the solution of the three-dimensional wave equation (2.55) is a somewhat more complicated expression:

$$\psi(x, y, z, t) = \psi_0 \cos(\omega t) \sin(k_x x) \sin(k_y y) \sin(k_z z)$$

or (2.57)

$$\psi(x, y, z, t) = \psi_0 \cos(\omega t) \cos(k_x x) \cos(k_y y) \cos(k_z z)$$

where k_x, k_y and k_z are the Cartesian components of $\vec{k}$ defined by Eqs. 2.54. Which of the two equations 2.57 is to be used depends on the boundary conditions. Consider a rectangular enclosure bounded by perfectly rigid surfaces whose linear dimensions are L_x, L_y and L_z. The enclosure is filled with air. We would like to know the structure of the natural modes of oscillations that can exist, and their frequencies. Take the origin of a Cartesian coordinate system at one corner of the box with the axes oriented as shown in Fig. 2.15. Since the walls are taken to be perfectly rigid, we require that the air acceleration normal to the surface of the wall be zero (see footnote following Eq. 2.47). In one dimension, it is clear from Eq. 2.20 that the

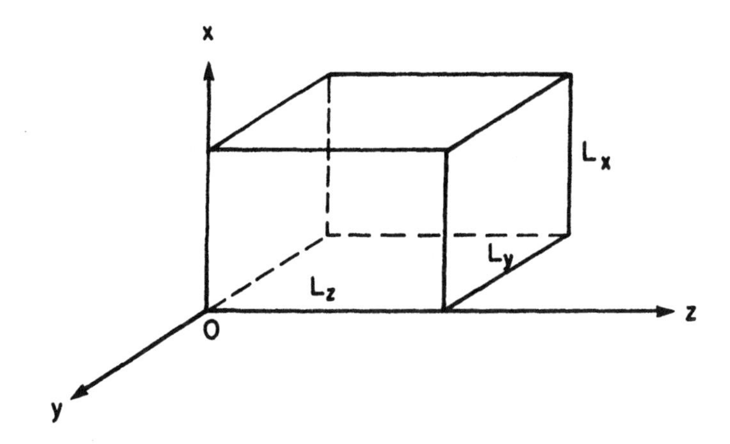

Fig. 2.15 Coordinate system used in calculating the modes of oscillation of waves enclosed in a rectangular box of sides L_x, L_y, L_z.

gas acceleration is proportional to $\partial p/\partial z$ where p is the (excess) gas pressure. Similarly, in three dimensions the components a_x, a_y and a_z of the vector acceleration are proportional to $\partial p/\partial x$, $\partial p/\partial y$ and $\partial p/\partial z$, respectively. Now, following the prescription given by Eqs. 2.57, we let the pressure $p = p_0 \cos \omega t \cos(k_x x) \cos(k_y y) \cos(k_z z)$. This choice automatically satisfies the boundary condition that the normal component of $\vec{a}$ vanishes on three of the walls, those for which x = 0, y = 0, and z = 0. To satisfy the boundary condition on the opposite walls $x = L_x$, $y = L_y$ and $z = L_z$, it is necessary that

$$k_x L_x = \ell\pi; \quad k_y L_y = m\pi; \quad k_z L_z = n\pi \quad (\ell, m, n = 0, 1, 2, 3 \ldots) \tag{2.58}$$

where now three integers ℓ, m, n (rather than one integer as in the one-dimensional problem) are required to determine fully the mode structure of this system. When we substitute these values of k_x, k_y and k_z in the second equation (2.54) and use the fact that the magnitude of k equals ω/v, we obtain the sought-after mode frequencies $\omega(\ell,m,n)$:

$$\omega(\ell, m, n) = \pi v \sqrt{\frac{\ell^2}{L_x^2} + \frac{m^2}{L_y^2} + \frac{n^2}{L_z^2}} \tag{2.59}$$

with v as the phase speed of sound.

Each set of three numbers (ℓ,m,n) determines one mode of oscillation of the rectangular box. The structure of a mode is quite complicated; whereas in the one-dimensional case (see Figs. 2.10 and 2.11) we have a line of nodes and antinodes, in the three-dimensional problem the nodes and antinodes lie on planes. In the one-dimensional case the frequencies of the higher harmonics are simple multiples of the frequency of the fundamental — they form a harmonic series. A quick perusal of Eq. 2.59 shows that this is not the case in three dimensions.

A very good approximation to the type of resonant box discussed here is an empty room with rigid (concrete) walls. Its lowest frequency mode is one for which $\ell = 1$ and $m = n = 0$ so that

$$\omega(1,0,0) = \frac{\pi v}{L_x}. \tag{2.60}$$

Knudsen [J. Acoust. Soc. Am. 4, 20 (1932)] explored the mode structure of a chamber 8 ft × 8 ft × 9.5 ft in size. Using Eq. 2.59 and the value of v equal to 1125 ft/sec he computed the first 34 possible values of ω and found these to lie between the frequency of $\omega/2\pi$ = 59.1 Hz and 231.4 Hz. By placing a variable frequency audio source in the room he was able to verify experimentally every one of the frequencies predicted from theory. Knudsen was not so much interested in the modes while the source was on; rather his interest centered on the way the different modes decayed as a result of absorption (by the walls, etc.). This time of decay, known as the reverberation time, is one of the parameters that must be considered in the design of buildings, lecture rooms, and music halls, but this fascinating subject lies outside the scope of this book.

Let us consider for a moment a cubical resonator with equal sides $L_x = L_y = L_z = L$. Then Eq. 2.59 for the mode frequencies becomes $\omega(\ell, m, n) = (\pi v/L)[\ell^2+m^2+n^2]^{1/2}$. From this we see that the same frequency can be obtained if the integers ℓ, m, and n are interchanged. In other words a single frequency may correspond to geometrically different patterns of vibration so that the number of normal modes is greater than the number of distinct frequencies. Such a system is said to be degenerate. Recall that in a one-dimensional system each mode is associated with a different frequency and degeneracy cannot occur.

The spherical wave

Imagine a transmitter so small that it can be considered as an idealized point source. Assume that it pulsates radially inward and outward and thus disturbs the medium surrounding it uniformly in all directions. It is then plausible to assume that wave fronts emanating from it are concentric spheres and that the disturbance $\psi(r)$ is a function only of the radial distance r measured from the position of the source. Such a transmitter is said to be isotropic. To determine the wave, we need to cast Eq. 2.55 into spherical coordinates. We note $r^2 = x^2 + y^2 + z^2$, and therefore in differentiating with respect to x we obtain $2r(\partial r/\partial x) = 2x$ and consequently

$$\frac{\partial r}{\partial x} = \frac{x}{r}. \tag{2.61}$$

With this result in mind, we then have the following transformations

$$\frac{\partial \psi}{\partial x} = \frac{\partial \psi}{\partial r}\frac{\partial r}{\partial x} = \frac{x}{r}\frac{\partial \psi}{\partial r} \tag{2.62}$$

and

$$\left.\begin{aligned} \frac{\partial^2 \psi}{\partial x^2} &= \frac{1}{r}\frac{\partial \psi}{\partial r} - \frac{x}{r^2}\frac{\partial \psi}{\partial r}\frac{\partial r}{\partial x} + \frac{x}{r}\frac{\partial^2 \psi}{\partial r^2}\frac{\partial r}{\partial x} \\ &= \frac{\partial^2 \psi}{\partial r^2}\frac{x^2}{r^2} + \frac{\partial \psi}{\partial r}\left(\frac{1}{r} - \frac{x^2}{r^3}\right) \\ &= \frac{\partial^2 \psi}{\partial r^2}\frac{x^2}{r^2} + \frac{\partial \psi}{\partial r}\left(\frac{y^2 + z^2}{r^3}\right) \end{aligned}\right\} \tag{2.63}$$

with similar expressions for $\partial^2\psi/\partial y^2$ and $\partial^2\psi/\partial z^2$. Substituting all of these in the wave equation (2.55) we get

$$\frac{\partial^2 \psi}{\partial t^2} = v^2\left[\frac{\partial^2 \psi}{\partial r^2} + \frac{2}{r}\frac{\partial \psi}{\partial r}\right]. \tag{2.64}$$

This result can be cast into

$$\frac{\partial^2 (r\psi)}{\partial t^2} = v^2\frac{\partial^2 (r\psi)}{\partial r^2}, \tag{2.65}$$

which we recognize immediately as having the same form as the one-dimensional wave equation, but containing the product $(r\psi)$ rather than ψ itself, as is the case in one dimension. Hence we know the solution: for a harmonically driven wave propagating radially outwards (cf. Eq. 2.8)

$$r\psi(r) = A\cos(\omega t - kr + \phi) \tag{2.66}$$

which yields the sought-after result,

$$\psi(r) = \frac{A}{r}\cos(\omega t - kr + \phi). \tag{2.67}$$

At any fixed value of time, this equation represents a cluster of concentric spheres filling all of space. Each wave front corresponds to a different value of the phase $(\omega t - kr + \phi)$.

Observe that the amplitude of a spherical wave falls off inversely as the distance r from the transmitter, unlike the plane wave whose amplitude remains constant, independent of distance. Far enough away from the source the radius of curvature of the wave fronts becomes very large. Thus, if we take a hunk of space far from the source whose size is small compared with r, the wave fronts there will approach closely those of a plane traveling wave discussed in earlier sections.

Since the amplitude falls off as $1/r$, the radiation flux, being proportional to amplitude squared falls off as $1/r^2$ and thus obeys the inverse square law. The total energy per second passing through a sphere of radius r equals the flux multiplied by the area $4\pi r^2$ and is thus independent of the distance r, as it must be if energy is to be conserved, and if there is to be no energy dissipation in the medium through which the wave travels.

At the beginning of this subsection we called the quantity $\psi(r)$ the "disturbance" caused by the spherical wave propagating through the medium. We were purposely a little vague in our definition but now is the time to firm up what we really mean. In the case of plane waves in pipes and in free space, we identified ψ with the longitudinal displacement s of an element of fluid from its equilibrium position. It turns out that on deriving the equation of motion for an element of fluid in a spherical wave, the "proper" variable that emerges is not the displacement s but the pressure variation p. Thus for sinusoidal sound waves

$$p = \frac{A}{r}\cos(\omega t - kr + \phi). \tag{2.68}$$

To deduce the radial diaplacement $s(r, t)$ we use the relationship

(cf. Eq. 2.20)

$$\frac{\partial^2 s}{\partial t^2} = -\frac{1}{\rho_o}\frac{\partial p}{\partial r} \tag{2.69}$$

and readily find that

$$s(r, t) = \frac{A}{\omega^2 \rho_o}\left[\frac{k}{r}\sin(\omega t - kr + \phi) - \frac{1}{r^2}\cos(\omega t - kr + \phi)\right]. \tag{2.70}$$

Equations (2.68) and (2.70) show that in a spherical sound wave the displacement s and the pressure p satisfy different functional relationships: p varies as $1/r$, whereas s contains terms that vary both as $1/r$ and $1/r^2$. Note that at sufficiently large values of r, the second term of Eq. 2.70 is negligible compared with the first and here p and s have the same spatial dependences. In a plane wave (i.e., when $r \to \infty$) these complications clearly do not arise.

2.5 ANHARMONIC WAVES

The sinusoidal wave with its periodic sine or cosine variation is a special case of a broad class of very general wave disturbances. Such disturbances need not be sinusoidal, nor do they need to be periodic. The repeated bursts of electromagnetic radiation reaching us from outer space, having been emitted by a pulsar, have a periodicity of marvelous precision; but they are not sinusoidal. The sound generated by a single clap of the hands is neither sinusoidal nor is it periodic. Nonetheless, the propagation characteristics of these and many other similar so-called "anharmonic" disturbances can be described by the wave equation (2.1). Our experience with sinusoidal waves suggests that the class of disturbances $\psi(z, t)$, which satisfy this wave equation, should perhaps have a space-time dependence of the form $(vt-z)$. This simply says that as time increases the wave retains its "shape" but in order to see the shape, we must shift our observation point by moving with the disturbance along z with a velocity v. Our intuition does not fail us. It turns out that <u>any</u> arbitrary function $\psi = f(vt-z)$ is a solution of the wave equation (2.1). One such is the familiar $\psi = \psi_0 \cos[a(vt-z)]$ where a is a constant; another is a pulse described by the equation $\psi = a[b+(vt-z)^2]^{-1/2}$; yet another is the Gaussian pulse $\psi = \psi_0 \exp[-a(vt-z)^2]$ illustrated in Fig. 2.16, and there are many others. To show this, let the variable $q \equiv (vt-z)$, so that $\psi(z,t) = f(q)$. Then, differentiating ψ twice with respect to z yields

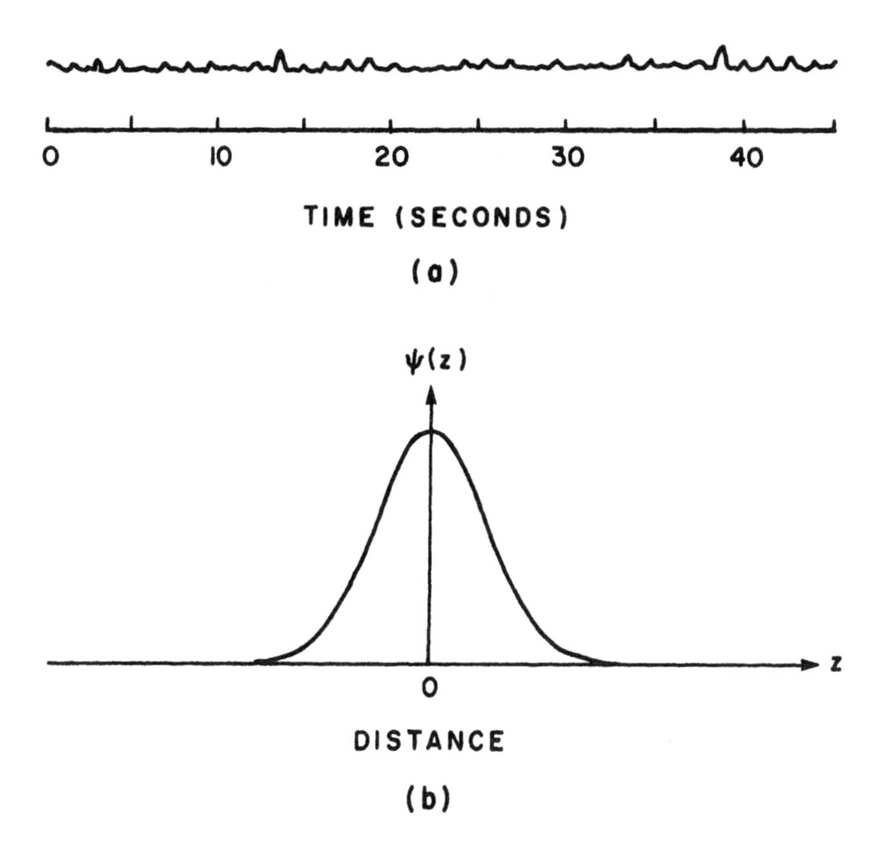

Fig. 2.16 Two examples of nonsinusoidal waves. (a) Recorder trace of the radio emission from CP 1919, the first pulsar ever observed. Twenty pulses can be distinguished in each 20-second period. Although the amplitudes of consecutive pulses show marked differences, their periodicity is of great precision, 1.33730113 seconds between pulses. (b) A single pulse. Its shape, known as a Gaussian $\psi(z,t) = \psi_0 \exp[-a(vt-z)^2]$, is much loved by theoretical physicists and mathematicians because of some of its remarkable properties, which will be discussed later. Here the pulse is shown at a fixed time $t = 0$.

$$\left.\begin{aligned} \frac{\partial\psi}{\partial z} &= \frac{\partial f}{\partial z} = \left(\frac{\partial f}{\partial q}\right)\left(\frac{\partial q}{\partial z}\right) \\ &= -\frac{\partial f}{\partial q} \\ \frac{\partial^2\psi}{\partial z^2} &= -\left(\frac{\partial^2 f}{\partial q^2}\right)\left(\frac{\partial q}{\partial z}\right) \\ &= \frac{\partial^2 f}{\partial q^2} \end{aligned}\right\} \tag{2.71}$$

Similarly, differentiating ψ twice with respect to time t yields

$$\begin{aligned} \frac{\partial\psi}{\partial t} &= \left(\frac{\partial f}{\partial q}\right)\left(\frac{\partial q}{\partial t}\right) = v\,\frac{\partial f}{\partial q} \\ \frac{\partial^2\psi}{\partial t^2} &= v\left(\frac{\partial^2 f}{\partial q^2}\right)\left(\frac{\partial q}{\partial t}\right) = v^2\,\frac{\partial^2 f}{\partial q^2}. \end{aligned} \tag{2.72}$$

Substituting these results in Eq. 2.1 we see that the two sides of the equation balance, which proves our point. In an identical fashion it can be shown that Eq. 2.1 is also a solution of another arbitrary function $g(vt+z)$ with the argument $(vt+z)$. And therefore, by the principle of superposition, the most general solution of the one-dimensional wave equation is a sum of the two "partial" waves

$$\psi(z, t) = f(vt-z) + g(vt+z). \tag{2.73}$$

We recognize that the first term, $f(vt-z)$ is a disturbance that propagates from left to right along the positive z axis, whereas $f(vt+z)$ is a disturbance moving from right to left, along the negative z axis. Despite the seemingly great generality of solution (2.73), there is one restriction implicit in the proof just given: the velocity of propagation v must be constant independent of time or position. In other words, using the language of previous sections, the medium in which the wave propagates must not be dispersive.

As an example, consider an infinite flexible string a short length of which is pulled aside and let go at time $t = 0$. The initial shape of the string is taken to be triangular as is illustrated at the top of Fig. 2.17. Such a shape can be "manufactured" from the two partial waves by requiring that f(0, z) and g(0, z) have identical triangular forms. As time increases, the triangular shape represented by f(vt−z) travels to the right and a similar shape represented by g(vt+z) travels to the left (Fig. 2.17). Perhaps a more interesting experiment is to pluck the string simultaneously at opposite ends. Two pulses travel toward one another, collide, seemingly annihilate one another, and emerge marvelously enough, unscathed and undistorted (Fig. 2.18). The principle of superposition of transverse displacements (and also of transverse

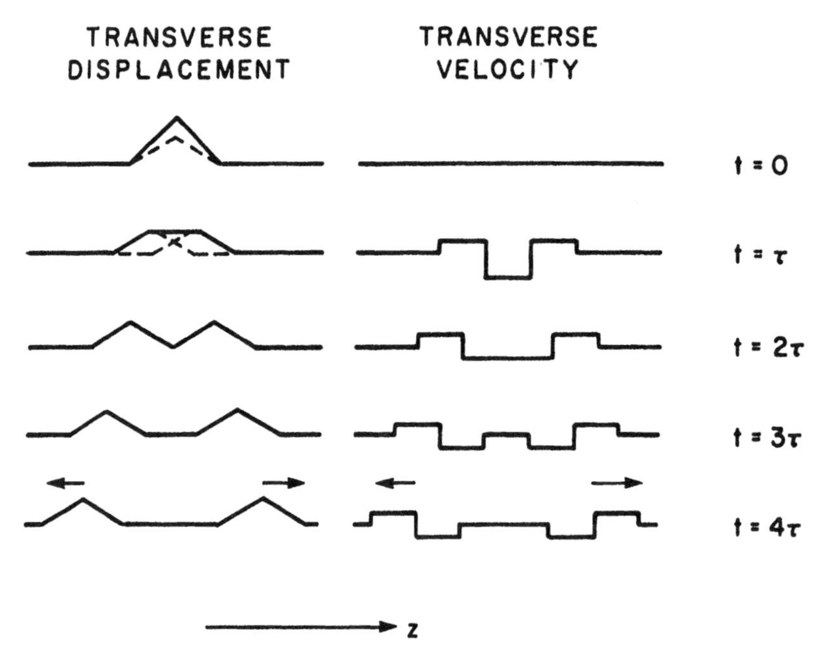

Fig. 2.17 A triangular wave pulse on a very long string giving rise to two pulses, one traveling to the left and the other traveling to the right. The left-hand sequence shows the string displacement s(z, t) at successive times τ, and the right-hand sequence shows the corresponding transverse string velocity $u(z,t) = \partial s/\partial t$. The dashed lines illustrate the two partial waves f(vt−z) and g(vt+z) which superpose to yield the resultant shape. Note that the perfectly sharp corners shown on the triangular pulse are a mathematically convenient abstraction which, of course, never occurs in nature, where all pulses have rounded corners.

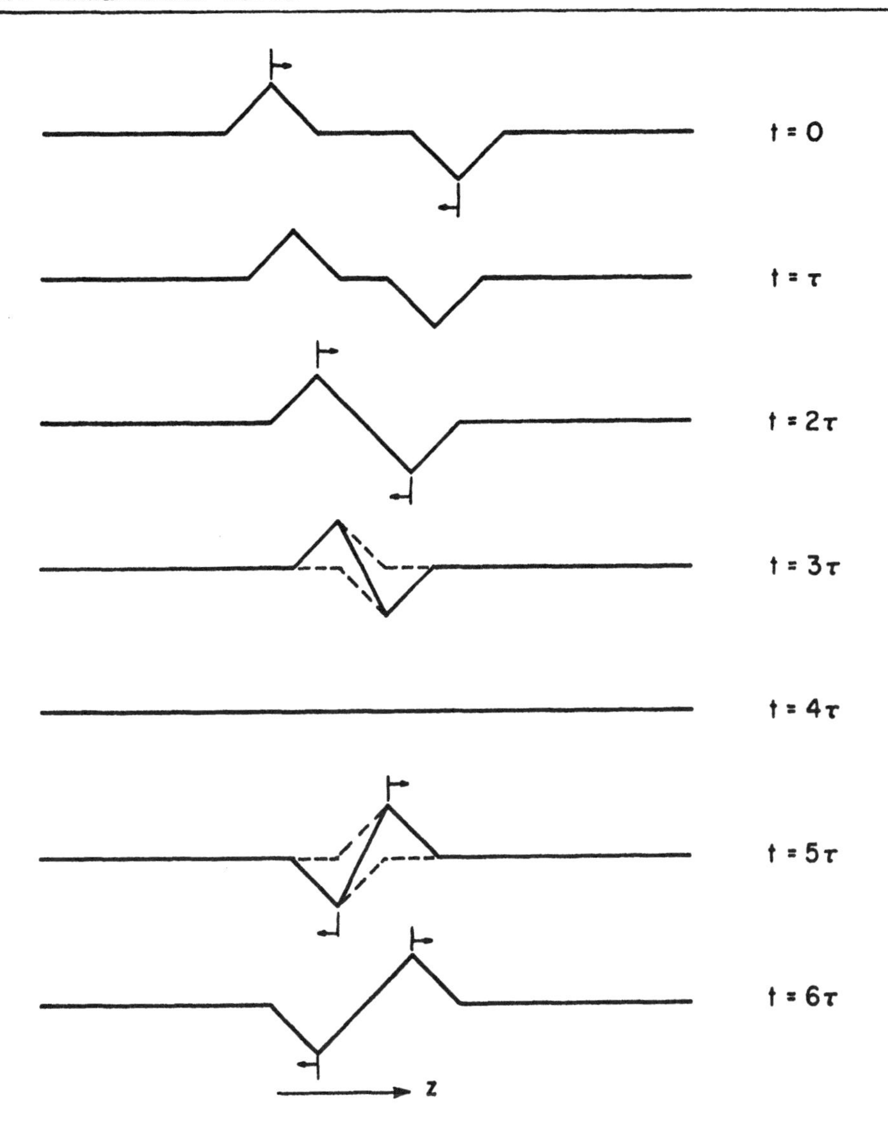

Fig. 2.18 Two equal but opposite transverse pulses on a string, traveling toward one another. In the fifth sequence from the top the net displacement of the string is zero. At this time all the energy in the pulse is in the form of kinetic energy of motion (the displacements s of the two partial waves cancel, but their velocities $u = \partial s/\partial t$ add).

string velocities) affords a full explanation of this curious behavior.

Next, let us consider a transverse disturbance on a very long string fastened to a rigid support at, say $z = 0$ (Fig. 2.19). We know that a pulse approaching the support will be reflected from it. We wish to study the details of the string's motion just before, during, and just after the reflection takes place. Because the support is rigid at $z = 0$, the displacement $\psi(vt+0)$ must be zero at all times. To

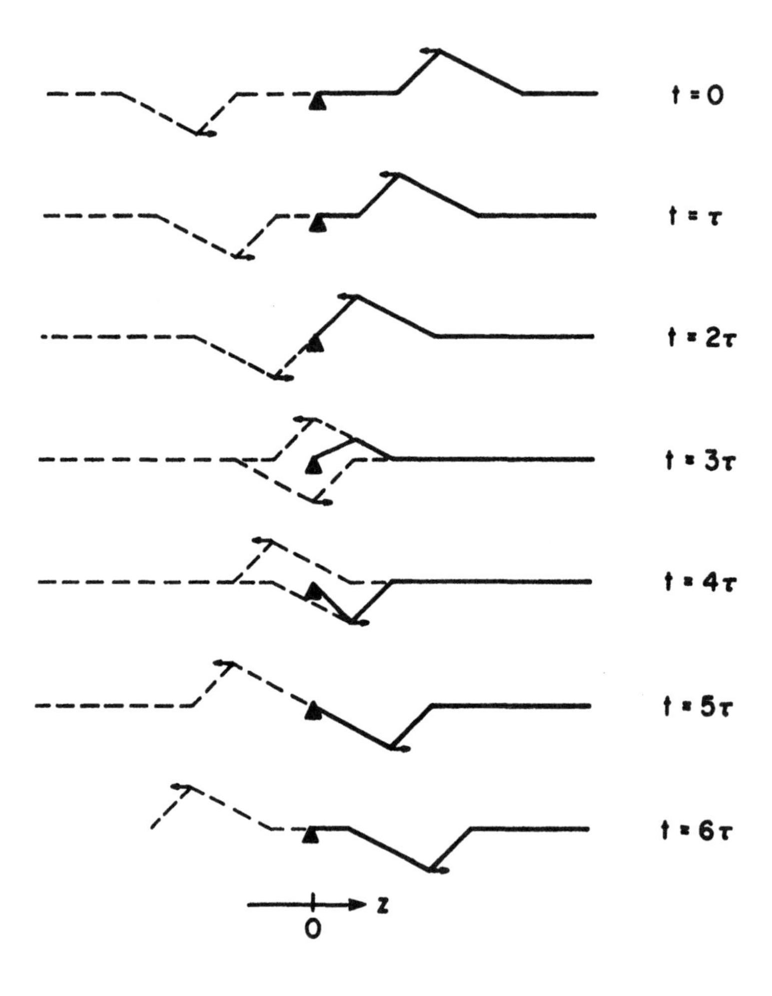

Fig. 2.19 Reflection of a pulse traveling toward the fixed end marked ▲. The dashed lines show the pulse shapes of the partial waves traveling on the imaginary portion of the string. The shapes are shown at successive intervals τ.

satisfy this condition, we must have, in addition to the wave $g(vt+z)$ approaching the support from the right, another imaginary disturbance $-g(vt-z)$ which approaches the boundary from the left. The combined solution $\psi(z, t) = g(vt+z) - g(vt-z)$ then correctly satisfies the boundary condition at $z = 0$. We can imagine that the disturbance $g(vt+z)$ continues to propagate to the left even into the imaginary nonexistent section of the string ($z < 0$), while the other pulse $-g(vt-z)$ emerges from $z < 0$, traverses the discontinuity at $z = 0$, and continues on along the physical part of the string. The net effect obtained by superposing the two disturbances is precisely the phenomenon of reflection. Observe that although the pulse suffers no distortion on reflection, it flips over from being positive (up) to negative (down). We leave to the reader to show that if the end of the string is perfectly free (rather than being rigidly fixed), the reflected pulse does not flip over.

When the string is anchored at two positions $x = 0$ and $x = L$, the character of the disturbance changes in a most important way. The presence of the second boundary requires that now the motion be periodic (but not necessarily sinusoidal). We can readily understand the reason for this: a pulse started at $x = 0$ travels to the other support $x = L$ in a time L/v, is reflected there, travels back to $x = 0$, and so on. To deal with this problem quantitatively, let us return to the method of partial waves. As before, we write for the displacement of the string that $\psi(z, t) = f(vt+z) - f(vt-z)$; it satisfies the boundary requirement that $\psi(0, t) = 0$ at $z = 0$. But we also require that $\psi(L, t) = 0$ at $z = L$, which means that $f(vt-L) = f(vt+L)$. To see what this means we define a length $p \equiv vt - L$, so $f(p) = f(p+2L)$. The implications of this mathematical statement are now obvious; the arbitrary function $f(p)$, which must be defined for all values of p, is periodic in p, and repeats itself at intervals of $2L$ along the entire length. Figure 2.20 shows the motion of a plucked string fixed at its two extremities. The dashed lines are the spatially periodic partial waves from which the combined displacement is constructed. Figure 2.21 plots the displacement of a point on the string as a function of time. It shows that while the motion is periodic in time, it is not sinusoidal.

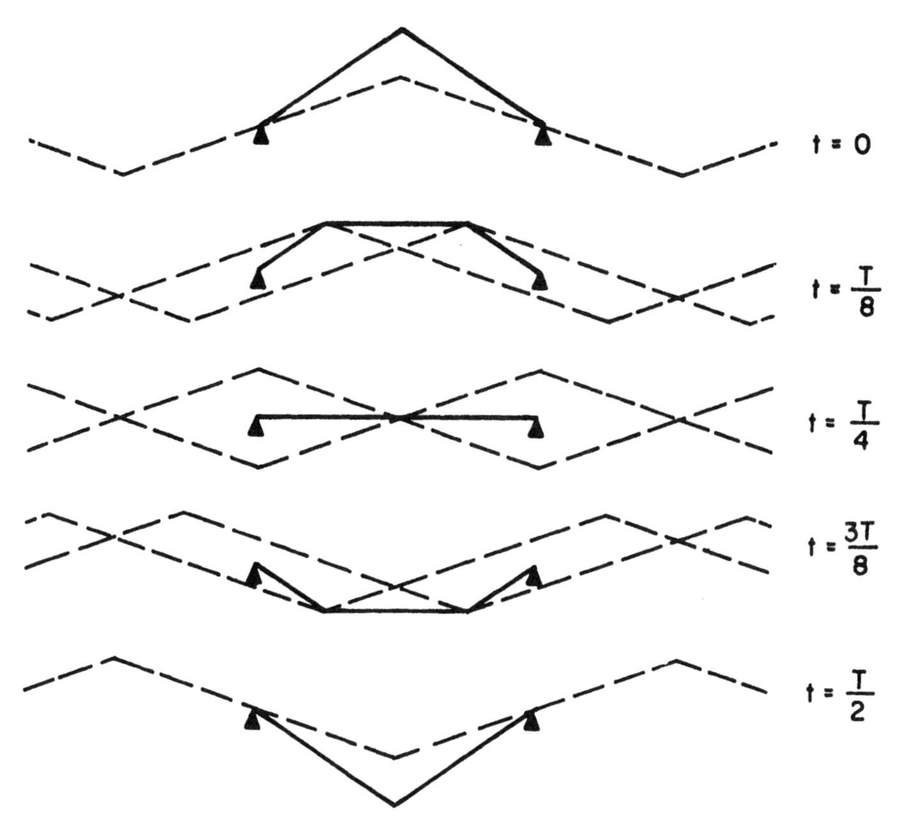

Fig. 2.20 Motion of a plucked string fixed at both ends. In contrast to the previous three figures the motion is now periodic. The two partial waves are seen to be identical periodic waves traveling in opposite directions.

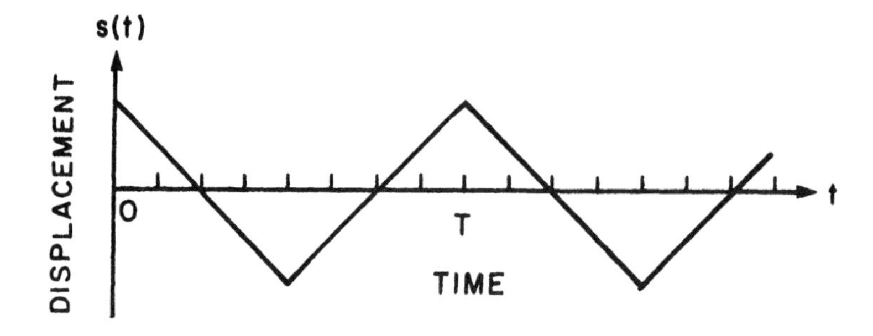

Fig. 2.21 Displacement s(t) of the center of the string of Fig. 2.20 plotted as a function of time, illustrating the fact that although the motion is periodic, it is not sinusoidal.

Periodic disturbances — Fourier series representation

The foregoing method of finding the shape of a disturbed elastic string, or for that matter of any other kind of wave motion, gives elegant results when applied to fairly simple problems. It becomes less useful in more intricate geometries or in the presence of more complicated boundary conditions. For that reason we shall now turn to a very powerful and exceptionally beautiful technique devised by Fourier (1768-1830). It is a mathematical technique; however, to a physicist and engineer it represents yet another vivid application of the principle of superposition. But, unlike the situation in the previous subsection where we superposed two quite general partial waves f and g, in the Fourier technique we superpose an infinite number of simple-harmonic waves, that is, sines and cosines. The Fourier theorem may, for our present purpose, be expressed as follows: Any periodic, single-valued function f(x) defined in the interval $-\pi < x < \pi$ and thus having a periodicity of 2π can be represented by the trigonometric series

$$\boxed{\begin{aligned} f(x) = \frac{A_o}{2} &+ \sum_{m=1}^{\infty} A_m \cos mx \\ &+ \sum_{m=1}^{\infty} B_m \sin mx \end{aligned}} \qquad (2.74)$$

where A_m and B_m are the amplitudes of the successive harmonics of the series. The trick of determining the coefficients A_m and B_m is simple enough. Integrate both sides of the equation with respect to x between the limits $x = -\pi$ and $x = \pi$.* Since the integrals of all the sine and cosine terms are equal to zero, we are left with only one term, $\int f(x)\, dx = \pi A_o$. Again, if we multiply both sides of Eq. 2.74 by cos nx [n = 1, 2, 3. . . .] and integrate between $-\pi$ and π, all terms on the right-hand side become equal to zero (see below) except the

*Observe that x is a dimensionless variable and not a length.

term for which $m = n$ and we then have

$$\begin{aligned} \int_{-\pi}^{\pi} f(x) \cos(mx)\, dx &= A_m \int_{-\pi}^{\pi} \cos^2(mx)\, dx \\ &= \pi A_m. \end{aligned} \tag{2.75}$$

In a similar manner, multiplication by sin(nx) yields the values of B_m. Thus, the two sets of coefficients of the Fourier series (2.74) become

$$\left. \begin{aligned} A_m &= \frac{1}{\pi} \int_{-\pi}^{\pi} f(x) \cos(mx)\, dx \qquad m = 0, 1, 2, 3\ldots \\ B_m &= \frac{1}{\pi} \int_{-\pi}^{\pi} f(x) \sin(mx)\, dx \qquad m = 1, 2, 3\ldots \end{aligned} \right\} \tag{2.76}$$

and they can be evaluated once the function f(x) is specified explicitly. This is known as Fourier analysis of the function f(x). We observe that the marvelously simple trick we used is predicated on what is called the "orthogonality" of certain mathematical functions. For sines and cosines this orthogonality takes the form

$$\left. \begin{aligned} &\int_{-\pi}^{\pi} \sin(mx) \cos(nx)\, dx = 0 \\ &\int_{-\pi}^{\pi} \sin(mx) \sin(nx)\, dx = \begin{cases} \pi & \text{when } n = m \\ 0 & \text{when } n \neq m \end{cases} \\ &\int_{-\pi}^{\pi} \cos(mx) \cos(nx)\, dx = \begin{cases} \pi & \text{when } n = m \\ 0 & \text{when } n \neq m \end{cases} \end{aligned} \right\} \tag{2.77}$$

These relationships can easily be verified. One quick way of arriving at these results is to express the products of the trigonometric functions appearing in the integrands as sums, and then carrying out the appropriate integrations; for example, write $2 \sin(mx) \sin(nx) = \cos[(m-n)x] - \cos[(m+n)x]$ and follow this up by integration over x.

There are certain symmetry conditions which, if recognized early, can save a lot of computational labor in evaluating Eqs. 2.76. If f(x) is an even function of x, by which we mean that $f(-x) = f(x)$, all B_m coefficients are identically zero and the resulting Fourier series will contain only cosine terms. The reason is that an even function multiplied by sin(mx) becomes odd, and the integration of this product from $x = -\pi$ to $x = +\pi$ gives zero contribution. Similarly, when f(x) is odd so that $f(-x) = -f(x)$, all A_m coefficients vanish and the Fourier series will contain only sine terms.

Because the functions cos(mx) and sin(mx) have a periodicity of 2π, the function f(x) of Eq. 2.74 will likewise repeat every 2π. Indeed, this is just how we set up the problem in the first place. It may appear at first sight that because x is confined to the interval $-\pi < x < \pi$, the usefulness of the Fourier method is of necessity limited to a very restricted class of problems. But this is not so, since the range of the interval can easily be altered by a simple change of variable. The choice of the range is determined by what physical quantity x represents and what specific question is being answered. In this chapter and in Chapter 1 we have dealt with two major types of problems: temporal disturbances in which the period T (or its equivalent $2\pi/\omega$) is a measure of the time periodicity; and spatial disturbances in which the wavelength λ or its equivalent $2\pi/k$ are the proper parameters to describe their periodicities. Thus, it will come as no surprise when the argument x appearing on the right-hand side of Eq. 2.74 is replaced by other variables such as ωt or kz (and others). Indeed, by an appropriate change of variable, the origin of the function f(x) can be shifted from the range $-\pi < x < \pi$ to, say, $0 < x < 2\pi$. Table 2.2 gives a few useful examples of such manipulations.

As an example of the technique, let us Fourier-analyze the periodic square wave illustrated at the top of Fig. 2.22. With the origin set as shown, we see that

Table 2.2 Examples of changes of variable in the Fourier series. The substitutions for the variable x indicated below are to be made in all terms appearing on the right-hand side of Eqs. 2.74 and 2.76.

Function	Variable	Interval
f(x)	x	$-\pi < x < \pi$
f(t)	$x \to \omega t$	$-(T/2) < t < (T/2)$
f(z)	$x \to (\pi/L)z$	$-L < z < L$
f(z)	$x \to (\pi/L)z - \pi$	$0 < z < 2L$
f(z)	$x \to (2\pi/\lambda)z$	$-(\lambda/2) < z < (\lambda/2)$

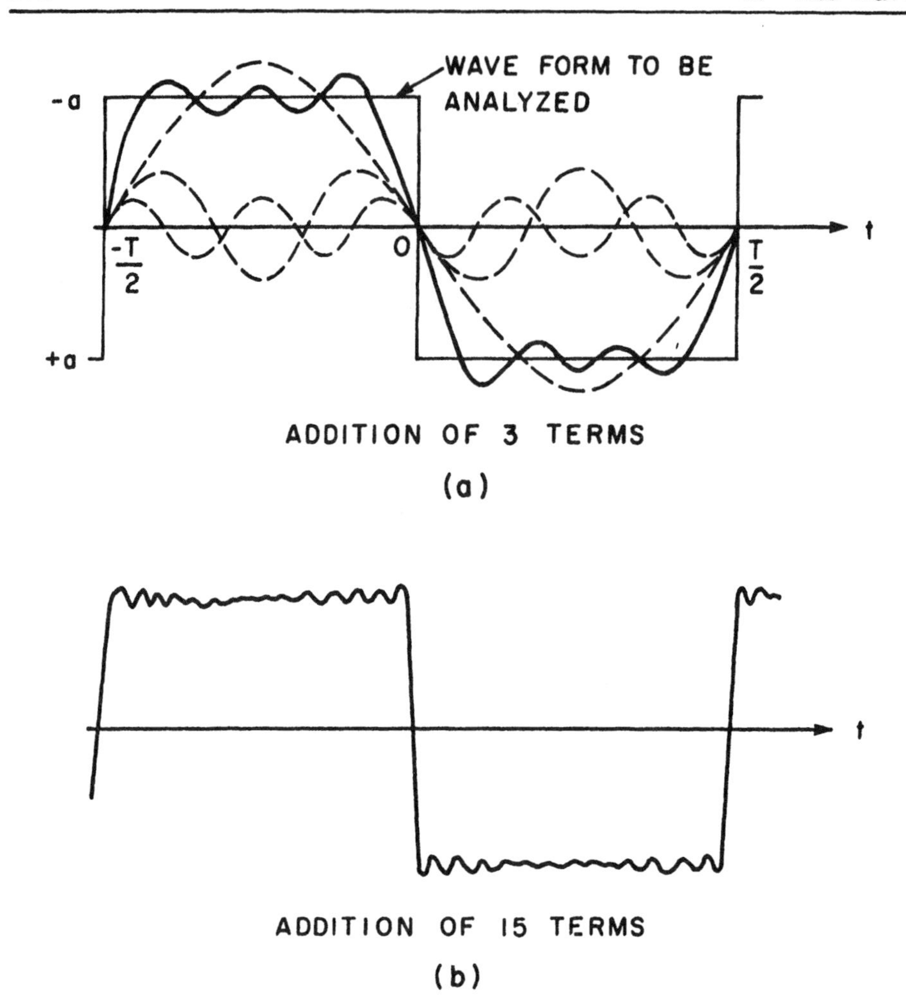

Fig. 2.22 Fourier analysis of a square wave. The dashed lines represent the partial waves as given by the successive terms of the series (2.80).

$$\begin{aligned} f(t) &= +a \quad \text{when} \quad 0 < t < (T/2) \\ &= -a \quad \text{when} \quad -(T/2) < t < 0. \end{aligned} \tag{2.78}$$

This is an odd function and therefore all coefficients A_m are zero. To find the coefficients B_m, make the change of variable prescribed by the second row of Table 2.2 and then substitute Eq. 2.78 in the second equation of (2.76). This yields

$$
\left.
\begin{aligned}
B_m &= \frac{\omega}{\pi}\int_{-T/2}^{T/2} f(t)\sin(m\omega t)\,dt \\
&= \frac{\omega}{\pi}\int_{0}^{T/2} a\sin(m\omega t)\,dt - \frac{\omega}{\pi}\int_{-T/2}^{0} a\sin(m\omega t)\,dt \\
&= \frac{2a}{m\pi}\left[1-\cos(m\pi)\right] \qquad [m = 1, 2, 3\ldots]
\end{aligned}
\right\} \tag{2.79}
$$

from which the Fourier series of the square wave follows directly:

$$
f(t) = \frac{4a}{\pi}\left[\sin(\omega t) + \frac{1}{3}\sin(3\omega t) + \frac{1}{5}\sin(5\omega t) + \ldots\right]. \tag{2.80}
$$

Such a series gives us a great deal of insight about the disturbance. It tells, for instance, that the second, fourth, sixth harmonics, etc., are absent from the motion. Also, since the intensity of the wave is proportional to the square of the amplitude, the equation tells that the fundamental frequency is 9 times stronger than the third harmonic and 25 times stronger than the fifth harmonic, etc. It is fairly clear that there can be exact correspondence between the square wave and the series expansion only when all of the infinite number of terms are added together. However, fair correspondence can be obtained even if only relatively few terms are taken, as is illustrated in Fig. 2.22. As a rule, smooth functions require fewer terms than highly discontinuous functions (like the square wave above with its sharp corners). The latter can be approximated with good accuracy by a series only if a large number of terms is used. In other words, highly irregular functions contain many high harmonics of the fundamental.

The quantity $f(t)$ of Eq. 2.80 can be looked upon as the temporal variations of a square wave viewed at a fixed position $z = 0$. If we wish to pass from the time to the space domain and obtain the spatial variations of the square wave at, say, $t = 0$, all we need to do is to replace the quantity ωt in Eq. 2.80 and Fig. 2.22 by the quantity kz.

As our second example, we go back to the problem of the plucked string discussed earlier. We know (section 2.3) that the displacement s_m of a string of length L held fixed at its two ends $z = 0$ and $z = L$, and oscillating in one of its normal modes, is

$$s_m(z, t) = \sin\left(\frac{m\pi z}{L}\right)[A_m \cos(\omega_m t) + B_m \sin(\omega_m t)] \tag{2.81}$$

where $\omega_m = (m\pi v/L)$ and v is the phase velocity $v = \sqrt{\mathcal{T}/\mu}$. The most general standing wave on the string is, by superposition, given by

$$s(z, t) = \sum_{m=1}^{\infty} s_m(z, t). \tag{2.82}$$

Suppose the initial displacement $s(z, t=0) \equiv s_o(z)$, and the initial velocity $(\partial s/\partial t)_{t=0} \equiv u_o(z)$ are given. For a string plucked in the middle (see Fig. 2.23) their value is

$$\left.\begin{aligned} s_o(z) &= \frac{2az}{L} && 0 \leqslant z \leqslant \frac{L}{2} \\ &= 2a\left[1 - \frac{z}{L}\right] && \frac{L}{2} \leqslant z \leqslant L \\ u_o(z) &= 0 \end{aligned}\right\} \tag{2.83}$$

What is the expression for the displacement s(z, t) at all subsequent times t > 0?

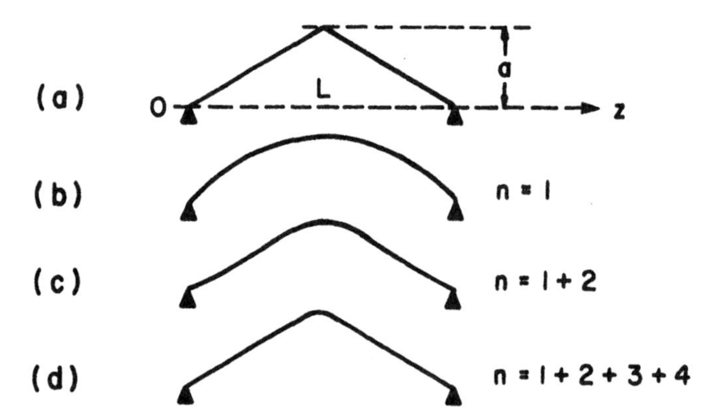

Fig. 2.23 A string of length L fastened at its two ends z = 0 and z = L. In (a) is shown its initial shape s(z, 0) at t = 0. Curves (b), (c), and (d) represent Fourier syntheses of the original shape obtained by taking the first term, the first two terms, and the first four terms, respectively, of the series given by Eq. 2.87. The case n = 1 + 2 + 3 is not shown.

It is clear that our task is to determine the coefficients A_m and B_m of Eq. 2.81, subject to the initial conditions just stated. At time $t = 0$ we have

$$s_o(z, 0) = \Sigma A_m \sin\left(\frac{m\pi z}{L}\right)$$

and (2.84)

$$u_o(z, 0) = \Sigma \omega_m B_m \sin\left(\frac{m\pi z}{L}\right),$$

which are just two Fourier series, one containing only sine terms and the other only cosine terms. Since in our problem the string is initially at rest $[u(z, 0) = 0]$, the B coefficients must be zero and they need not concern us further. To find the A coefficients, we multiply both sides of the first equation (2.84) by $\sin(n\pi z/L)$ and integrate from $z = 0$ to $z = L$; using the orthogonality relations (2.77) we then obtain

$$A_m = \frac{2}{L} \int_0^L s_o(z, 0) \sin\left(\frac{m\pi z}{L}\right) dz. \tag{2.85}$$

Alternately, we can, in Eqs. 2.76, make the change of variable $x \rightarrow [(\pi/L)z]$ (see Table 2.2) and obtain the same result. We insert the specific value for s_o as given by Eq. 2.83 and performing the called-for integrations, obtain

$$\left.\begin{aligned} A_m &= \frac{2}{L} \int_0^{L/2} \left(\frac{2a}{L}\right) z \sin\left(\frac{m\pi z}{L}\right) dz \\ &\quad + \frac{2}{L} \int_{L/2}^{L} (2a)\left[1 - \frac{z}{L}\right] \sin\left(\frac{m\pi z}{L}\right) dz \\ &= \frac{8a}{(m\pi)^2} \sin\left(\frac{m\pi}{2}\right) \\ &= 0 \qquad \text{when } m = 2, 4, 6 \ldots \\ &= \frac{8a}{\pi^2},\ -\frac{8a}{9\pi^2},\ +\frac{8a}{25\pi^2} \ldots\ldots \qquad \text{when } m = 1, 3, 5 \ldots \end{aligned}\right\} \tag{2.86}$$

Therefore, the displacement of the plucked string, at any position $0 \leq z \leq L$, and at any time $t \geq 0$, is

$$s(z,t) = \frac{8a}{\pi^2}\left[\sin\left(\frac{\pi z}{L}\right)\cos\left(\frac{\pi vt}{L}\right) - \frac{1}{9}\sin\left(\frac{3\pi z}{L}\right)\cos\left(\frac{3\pi vt}{L}\right)\right.$$

$$\left. + \frac{1}{25}\sin\left(\frac{5\pi z}{L}\right)\cos\left(\frac{5\pi vt}{L}\right)\ldots\right]. \tag{2.87}$$

Figure 2.23 illustrates the Fourier series representation of the initial form $[s(z,0)]$ of the string as obtained from the first few terms of the series. We see that as few as four terms give a reasonable approximation to the actual form. Observe that by plucking the middle of the string as was done in the above example, all even harmonics are missing. To the human ear this gives the feeling of a tone of poor quality. If, on the other hand, the string is plucked near the end, the full series of harmonics is present and the tone is richer and more brilliant. (On the piano the string is struck at a distance of about 1/8 of its length from one end.) Had we carried out our computations for the more general case in which the string is plucked at a distance d from one end, we would find instead of Eq. 2.86, Fourier coefficients given by the expression

$$A_m = \frac{2aL^2}{(m\pi)^2\, d[L-d]}\sin\left(\frac{m\pi d}{L}\right) \tag{2.88}$$

from which we obtain the general rule that in the motion of any plucked string all those harmonics are absent that have a node at the point pulled aside ($A_m = 0$ whenever $m = L/d, 2L/d, 3L/d\ldots$).

It is obvious from this discussion that the quality of the tone generated by a plucked string depends sensitively on the point of plucking. It also depends on the instrument used for plucking. A soft plucking object like the finger gives a fairly rounded form to the string at the point of displacement and this will give rise to a few high harmonics, and their amplitude will be weak. On the other hand, the hard plectrum (i.e., a pick) often used with guitars gives a sharp discontinuity to the string much like the perfectly sharp "mathematical" discontinuity assumed in the foregoing calculations, and the sound will be rich in harmonics. Based on these simple examples, we begin to appreciate the importance of Fourier analysis in the theory of music.

It is often advantageous to express the function f(x) not in a sine and cosine series, but in terms of a complex exponential series (cf. Eq. 1.16 where we use complex representation for a single harmonic oscillation).

$$f(x) = \sum_{m=-\infty}^{m=\infty} C_m \, e^{-jmx}; \qquad (-\pi \leqslant x \leqslant \pi). \tag{2.89}$$

The coefficients C_m are determined from

$$C_m = \frac{1}{2\pi} \int_{-\pi}^{\pi} f(x) \, e^{jmx} \, dx \qquad (m = 0, \pm 1, \pm 2, \ldots), \tag{2.90}$$

a result obtained by multiplying both sides of Eq. 2.89 by $\exp(+jnx)$ and integrating over x in the range $-\pi \leqslant x \leqslant \pi$. In view of the mathematical relationship $\exp[-jmx] = \cos(mx) - j\sin(mx)$, the complex coefficients C_m are connected with the real coefficients A_m and B_m of Eqs. 2.74 through

$$\left.\begin{aligned} C_m &= \tfrac{1}{2}[A_m + jB_m] && m > 0 \\ C_o &= \tfrac{1}{2} A_o && m = 0 \\ C_m &= \tfrac{1}{2}[A_{-m} - jB_{-m}] && m < 0 \end{aligned}\right\} \tag{2.91}$$

Traveling pulses and wave packets — the Fourier integral

It is shown in Table 2.2 that by an appropriate substitution for the variable x, the basic periodicity $-\pi \leqslant x \leqslant \pi$ of the function f(x) can be modified. Setting $x = (2\pi/\lambda)z$, where λ is the wavelength, Eqs. 2.89 and 2.90 become

$$\begin{aligned} f(z) &= \sum_{m=-\infty}^{m=\infty} C_m \, e^{-j(2\pi/\lambda)mz} \\ C_m &= \frac{1}{\lambda} \int_{-\lambda/2}^{\lambda/2} f(z) \, e^{j(2\pi/\lambda)mz} \, dz. \end{aligned} \tag{2.92}$$

The function f(z) has thus acquired a basic periodicity equal to λ, as is illustrated schematically in Fig. 2.24. This act of stretching the

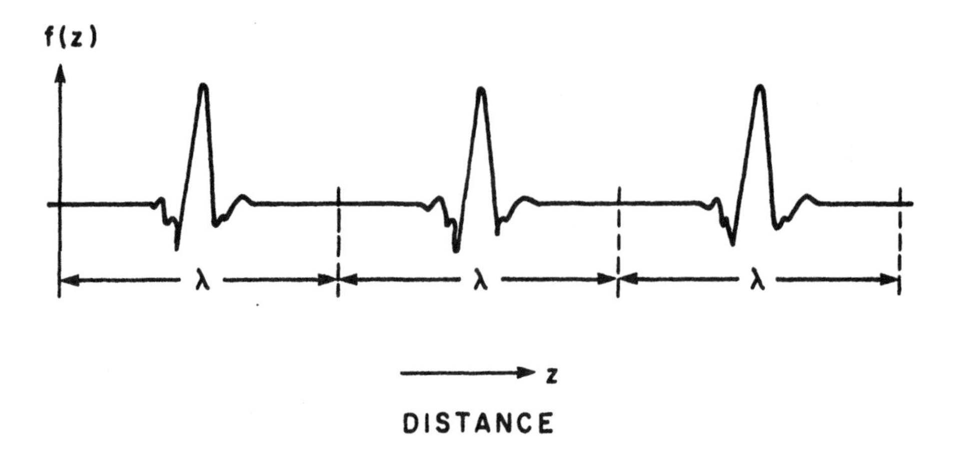

Fig. 2.24 A localized periodic wave train f(z) repeating at successive distances of λ.

basic interval from 2π to λ, where λ is arbitrarily large, suggests the possibility of passing to the limit $\lambda \rightarrow \infty$ and thereby obtaining a Fourier representation of nonperiodic functions. The realization of this is of utmost importance since it would allow us for the first time to analyze a new broad class of nonrepetitive wave disturbances, for example, a single pulse of radiation. The passage of Eqs. 2.92 to the limit $\lambda \rightarrow \infty$ is a mathematical exercise that we shall not carry out here. Let it just be said that the resulting nonperiodic function f(z) can no longer be represented by a fundamental and a discrete sequence of harmonics as was done above. It must now be represented by continuous superposition of waves of all wavelengths. This means that the sum must become an integral. The basic integral representations, corresponding to Eqs. 2.92 of the Fourier series are

$$f(z) = \frac{1}{2\pi}\int_{-\infty}^{\infty} F(k)\, e^{-jkz}\, dk$$

where (2.93)

$$F(k) = \int_{-\infty}^{\infty} f(z)\, e^{jkz}\, dz.$$

The function F(k) plays the role of the coefficients C_m of the series; it is generally spoken of as the "Fourier transform" of f(z); and f(z) is the "inverse Fourier transform" of $F(k)$. The first equation says that any (sufficiently well behaved) function of z can be expressed as a superposition of purely sinusoidal motions of all possible wavelengths $\lambda = 2\pi/k$, each having an amplitude given by $F(k)$. The equation therefore is no more than a mathematical statement of the superposition principle. The second equation (2.93) provides us with the means by which we can analyze, into its sinusoidal wave components, any disturbance whose mathematical form $f(z)$ is known.

Consider the following example. A source, say an atom, begins to radiate at some time, and then at a later time stops. In the process, a wave train of finite length L is created (Fig. 2.25). We wish to know the spectral distribution of waves of which this train is comprised. Let us assume that during the emission process an idealized sinusoidal train of the form $f(z) = f_o \cos(\omega_o t - k_o z)$ is generated where ω_o is the frequency and k_o the propagation constant $k_o = \omega_o/v$. Writing this disturbance in complex form and assuming, for convenience, that the wave analysis is carried out at time $t = 0$, we find that

$$f(z) = \frac{1}{2} f_o \left[e^{jk_o z} + e^{-jk_o z} \right] \qquad -L/2 \leq z \leq L/2$$

$$= 0 \qquad |z| > L/2. \qquad (2.94)$$

Substitution of Eq. 2.94 in the second equation (2.93) then gives

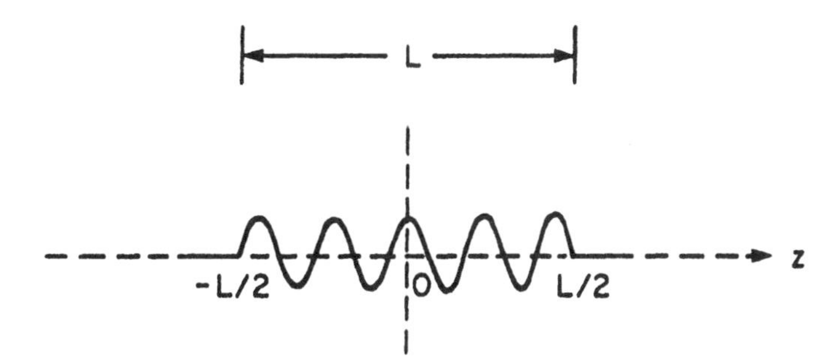

Fig. 2.25 A snapshot picture taken at time $t = 0$ of a sinusoidal wave train of finite length L. Note that as time increases the entire wave shape displaces along z. In a nondispersive medium this train travels along z undistorted with a velocity equal to the phase velocity v.

$$F(k) = \int_{-L/2}^{L/2} \left(\frac{f_o}{2}\right) e^{j(k+k_o)z}\, dz + \int_{-L/2}^{L/2} \left(\frac{f_o}{2}\right) e^{j(k-k_o)z}\, dz$$

$$= f_o \left[\frac{\sin[(k+k_o)L/2]}{(k+k_o)} + \frac{\sin[(k-k_o)L/2]}{(k-k_o)}\right]. \qquad (2.95)$$

For waves with wave numbers k near k_o, the second term dominates and, to good approximation,

$$F(k) \approx f_o \frac{\sin[(k-k_o)L/2]}{(k-k_o)} \qquad \text{(for } k \text{ near } k_o\text{)} \qquad (2.96)$$

which is plotted in Fig. 2.26. This then is the spectrum of wavelengths that must be superposed in order to create the wave train of finite length L. This assembly of a continuous distribution of amplitudes of pure sinusoidal oscillations is spoken of as a wave packet or wave group. As expected, the waves of largest amplitude are those

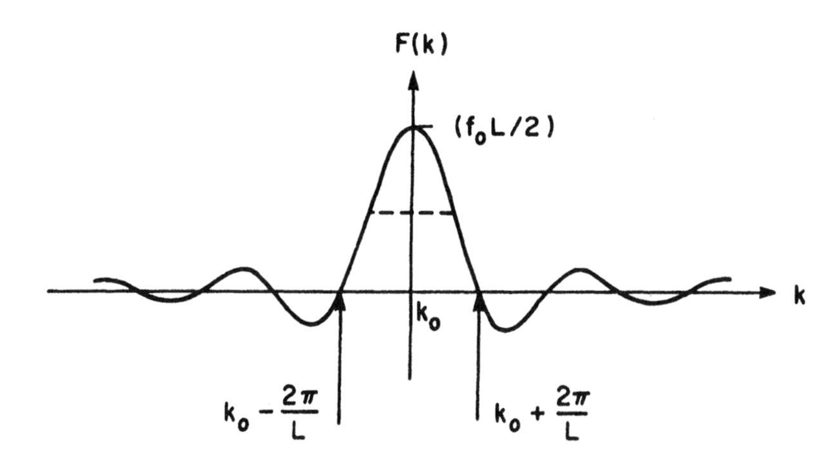

Fig. 2.26 The Fourier transform F(k) of the finite wave train of Fig. 2.25. The general shape of this wave packet is that of the function F(x) = sin x/x which has its principal maximum at x = 0 and goes through zero whenever $x = \pm\pi, \pm 2\pi, \pm 3\pi \ldots$. The secondary maxima on both sides of the main maximum occur when $x = \pm 1.43\,\pi,\ 2.46\,\pi,\ 3.47\,\pi \ldots$, that is, approximately midway between the zeros. Observe that F(k) falls to half its maximum value when $k \approx k_o \pm (\pi/L)$.

whose wave numbers lie closest to the fundamental wave number k_o. The fact that some amplitudes are negative should not be disturbing. This just says that waves with these values of k are out of phase. The intensity of the waves is equal to the square of the amplitude, and a plot of $|F(k)|^2$ as a function of k is referred to as the "power spectrum." It is shown plotted in Fig. 2.27.

We began with a seemingly ideal monochromatic wave train of single frequency ω_o and a single wavelength $\lambda_o = 2\pi/k_o$. However, our analysis which culminated in Eq. 2.96 and Fig. 2.26 tells us this is not correct. There is, in fact, a spread of wavelength. Is there something wrong? Not really. The finiteness of the wave train spoils matters. Since the oscillations do not continue forever, the wave train

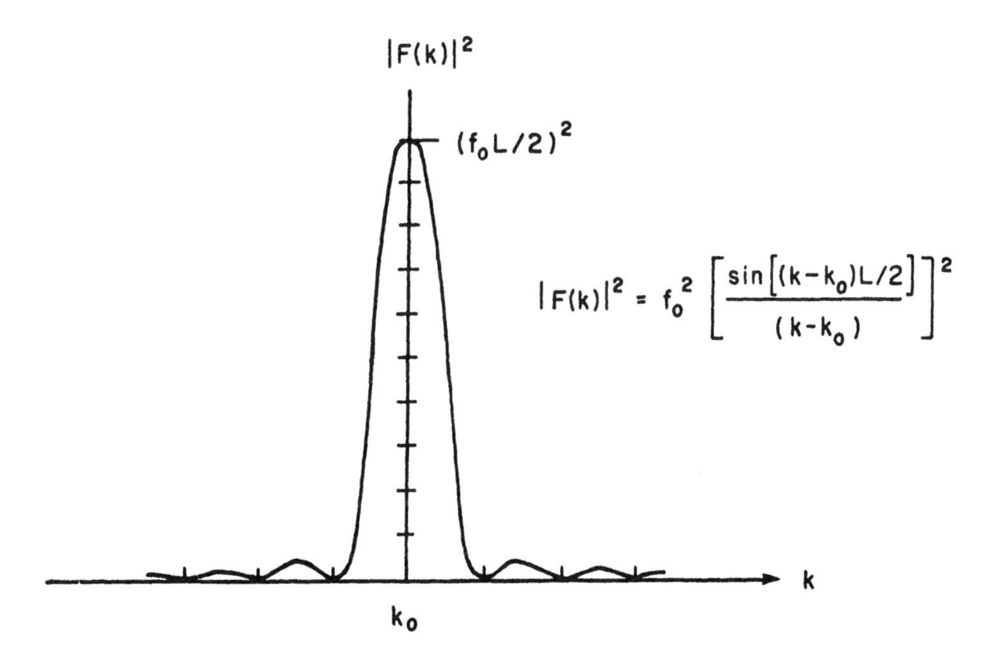

Fig. 2.27 The power spectrum of the wave packet shown in Fig. 2.26. Most of the energy is concentrated in waves whose wave numbers k lie close to k_o. The successive secondary maxima are down in intensity from the principal maximum in the ratios 1:0.047, 1:0.017, 1:0.0083, etc., a result which is found from solving the equation $[\sin x/x]^2$ for its maxima.

is not infinitely long, and the disturbance cannot be purely sinusoidal. The longer the L the more sinusoidal the oscillation will be. To show this, make L in Eq. 2.96 larger and larger. As a result the term $\sin[(k-k_o)L/2]$ will oscillate ever more rapidly with k, and the successive side lobes will crowd in more and more toward the center of the pattern. Ultimately as $L \to \infty$ the whole wave packet will have nonzero amplitude only at $k = k_o$. To define the width Δk of the wave packet quantitatively let us proceed as follows: read off the peak of F(k) in Fig. 2.26 and find the values k_1 and k_2 on either side of the peak where F(k) has fallen to half its maximum value. The width Δk then equals $k_1 - k_2$. Since $k_1 \approx k_o + (\pi/L)$ and $k_2 \approx k_o - (\pi/L)$, it follows that $\Delta k \approx 2\pi/L$. Calling the length L of the wave train Δx,

$$\boxed{\Delta k \Delta x \approx 2\pi.} \tag{2.97}$$

Although this result was derived for a specific type of wave packet, the result is of great generality and holds for almost any kind of reasonable pulse shape (the numerical value 2π may of course be a little different from one shape to another). The important result embodied in this statement is that a very long wave train or pulse has a small spread in Δk and thus is of great spectral purity. Since infinite wave trains do not exist in nature, where everything has a beginning and an end, there is no such thing as a purely monochromatic disturbance. On the other hand, if one attempts to localize the wave train spatially by making Δx small, then the wave is very impure and contains many, many wavelengths.

Thus far we have concerned ourselves with the spatial domain, that is, how a wave pulse develops in position (at a given time). To examine the temporal behavior of a pulse we must turn to Fourier integral equations written in the time domain. It is clear from previous discussions that the desired analogues of Eqs. 2.93 must have the form

$$f(t) = \frac{1}{2\pi} \int_{-\infty}^{\infty} F(\omega)\, e^{-j\omega t}\, d\omega$$

$$F(\omega) = \int_{-\infty}^{\infty} f(t)\, e^{j\omega t}\, dt. \tag{2.98}$$

Again, the first equation is an embodiment of the superposition of a continuous distribution of simple harmonic oscillations having frequencies ω and amplitudes $F(\omega)$. And the second equation is a statement of the fact that a known function of time $F(t)$ can be Fourier-analyzed into its frequency components. By analogy with Eq. 2.97 we also expect that

$$\boxed{\Delta\omega\Delta t \approx 2\pi} \tag{2.99}$$

which says that a pulse of duration Δt will, when Fourier-analyzed, contain frequencies extending from ω_o to approximately $\omega_o \pm (\pi/\Delta t)$ rad/sec.

Let us apply the foregoing to an analysis of the important case of a pulse with a Gaussian envelope (Fig. 2.28) whose equation is

$$f(t) = f_o\, e^{-at^2} \cos(\omega_o t). \tag{2.100}$$

The positive constant a determines the width of the pulse — the larger the a the shorter is its duration. Writing $\cos(\omega_o t)$ in complex form as $[\exp(j\omega_o t) + \exp(-j\omega_o t)]/2$ and inserting Eq. 2.100 in the second equation (2.98) gives

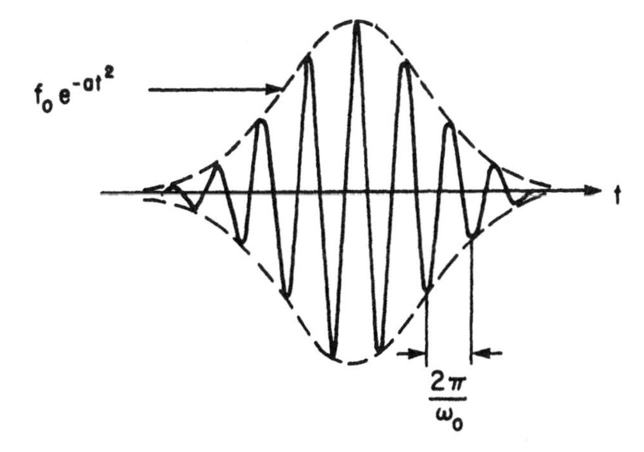

Fig. 2.28 A pulse with a Gaussian envelope represented by the equation $f(t) = f_o \{\exp(-at^2)\} \cos(\omega_o t)$.

$$F(\omega) = \int_{-\infty}^{\infty} \left(\frac{f_o}{2}\right) e^{j(\omega+\omega_o)\, t - at^2}\, dt$$

$$+ \int_{-\infty}^{\infty} \left(\frac{f_o}{2}\right) e^{j(\omega-\omega_o) t - at^2}\, dt. \qquad (2.101)$$

The integrals look formidable but in fact are easy to evaluate. The trick is to complete the square in the exponents by writing

$$\left.\begin{aligned} j(\omega+\omega_o)\, t - at^2 &= -\frac{(\omega+\omega_o)^2}{4a} - \left[\frac{j(\omega+\omega_o)}{2\sqrt{a}} - \sqrt{a}\, t\right]^2 \\ &\equiv -\frac{(\omega+\omega_o)^2}{4a} - x^2 \\ j(\omega-\omega_o)\, t - at^2 &= -\frac{(\omega-\omega_o)^2}{4a} - \left[\frac{j(\omega-\omega_o)}{2\sqrt{a}} - \sqrt{a}\, t\right]^2 \\ &\equiv -\frac{(\omega-\omega_o)^2}{4a} - y^2 \end{aligned}\right\} \qquad (2.102)$$

where the variables x and y are just shorthand ways of writing the full expressions within the square brackets $[x]^2$ and $[y]^2$. When the foregoing results are inserted into the exponents of Eq. 2.101 and variables of integration are changed from t to x and y, respectively, we find that

$$F(\omega) = \frac{f_o}{2\sqrt{a}}\, e^{-(\omega+\omega_o)^2/4a} \int_{-\infty}^{\infty} e^{-x^2}\, dx$$

$$+ \frac{f_o}{2\sqrt{a}}\, e^{-(\omega-\omega_o)^2/4a} \int_{-\infty}^{\infty} e^{-y^2}\, dy. \qquad (2.103)$$

The two integrals appearing above are well known and can be found in most standard tables; they have a magnitude equal to $\sqrt{\pi}$. Thus

$$F(\omega) = \sqrt{\frac{\pi}{4a}}\, f_o\, e^{-(\omega+\omega_o)^2/4a} + \sqrt{\frac{\pi}{4a}}\, f_o\, e^{-(\omega-\omega_o)^2/4a} \tag{2.104}$$

which is the sought-after result. This shows that a Gaussian pulse in time leads to Gaussians in the frequency domain. This interesting property that the Fourier transform of a Gaussian is yet another Gaussian is widely used to simplify analysis. When a in Eq. 2.104 is small enough and ω is near ω_o, the second term dominates and

$$F(\omega) \approx \sqrt{\frac{\pi}{4a}}\, f_o\, e^{-(\omega-\omega_o)^2/4a} \tag{2.105}$$

The form of $F(\omega)$ is illustrated in Fig. 2.29.

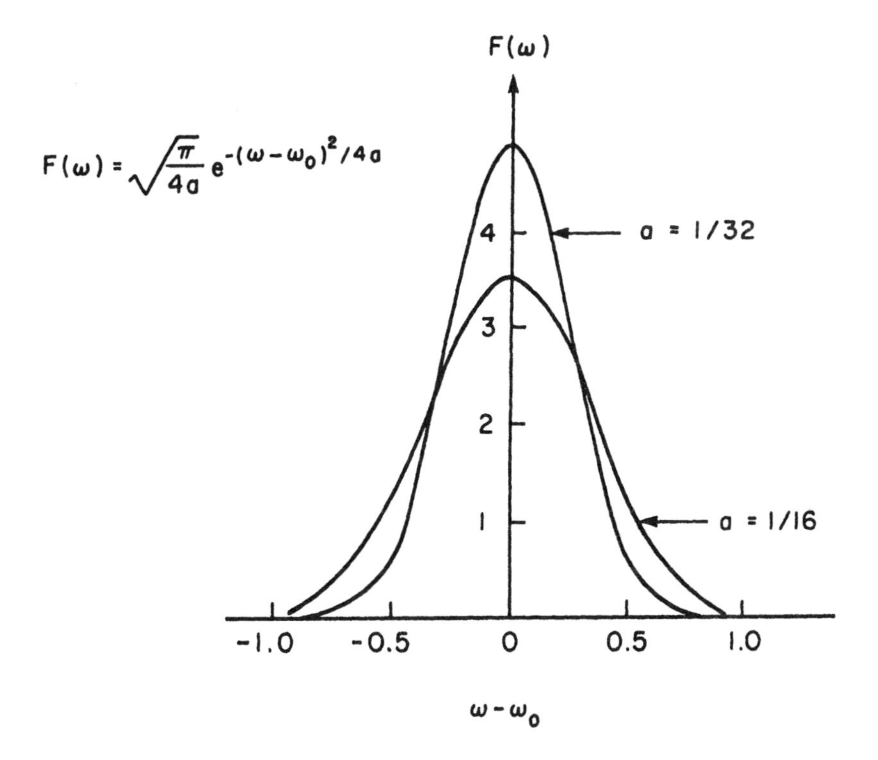

Fig. 2.29 Fourier transform $F(\omega)$ of the pulse f(t) illustrated in Fig. 2.28, plotted for two cases a = 1/16 and a = 1/32. The amplitude f_o is, for convenience, set equal to unity.

The temporal width Δt of the pulse given by Eq. 2.100 is obtained in the manner suggested earlier: that is, measure off the peak width at the place where f(t) has dropped off to half its value, and find that $\Delta t = 2\sqrt{\ln 2}/\sqrt{a}$; similarly, the frequency width of the packet given by Eq. 2.105 is $\Delta\omega = 4\sqrt{\ln 2}\sqrt{a}$. Multiplying $\Delta\omega$ by Δt then gives

$$\begin{aligned} \Delta\omega\Delta t &= 8 \ln 2 \\ &\approx 5.6 \end{aligned} \tag{2.106}$$

which is a value not too far removed from the factor 2π appearing on the right-hand side of Eq. 2.99.

Equation 2.99 has many applications. Some are mundane: for example, you wish to observe a pulse having a duration of 5 nsec. What must be the minimum frequency bandwidth of the oscilloscope so as to give a true, undistorted picture of the pulse? The answer is $\Delta\omega \approx 2\pi/\Delta t$ rad/sec, or $\Delta\omega/2\pi \approx 1/\Delta t = 200$ MHz, which means that you need a pretty good and pretty expensive oscilloscope.

Let us consider another example. The bound electrons of atoms make spontaneous jumps in energy, and on jumping from a higher state to a lower state, they emit light. The typical duration of the transition is 10^{-8} seconds, which is called the lifetime of the state. Because the oscillations are of finite duration, the emitted radiation will not be purely monochromatic but will have a spread in frequencies, $\Delta\omega/2\pi \approx 10^{8}$ Hz. We can express this result in wavelength: since $(\omega/2\pi)\lambda = c$, it follows that $|\Delta\lambda/\lambda| = |\Delta\omega/\omega|$. Suppose the light has a wavelength of 5000 Å [5×10^{-5} cm]; then $\Delta\lambda \approx 8 \times 10^{-12}$ cm or 8×10^{-4} Å. This is known as the natural linewidth of the spectral line. But, we note that as a result of several other broadening mechanisms, the typical spectral line observed in the laboratory is very much wider than the value just computed. Since the oscillations are of finite duration ($\Delta t \approx 10^{-8}$ sec), the emitted wave train $\Delta x = c\Delta t$ is also finite with a length of approximately 3 meters.

Finally, we wish to point out that these ideas carry over to quantum physics. In the wave mechanics of particles the frequency is identified as the particle energy ϵ divided by the Planck constant h, and the wavelength as the Planck constant divided by the momentum p.

That is,

$$\omega \to \frac{2\pi\epsilon}{h}$$

$$k = \frac{2\pi}{\lambda} \to \frac{2\pi p}{h}. \qquad (2.107)$$

When this reinterpretation is inserted in Eq. 2.97 we get

$$\Delta p \Delta x \approx h \qquad (2.108)$$

which is Heisenberg's uncertainty principle. It says that the more precisely one specifies the position of a particle the less precisely one knows its momentum.

CHAPTER 3

THE ELECTROMAGNETIC FIELD

Even in the remotest regions of our universe far-away from any star, where it may take a hundred years or more for a chance encounter between two atoms, there is, nonetheless, activity — activity of a very subtle nature which we call electric and magnetic fields. Indeed, there is no need to travel to such distant voids to find the fields; they permeate everything around us and make themselves felt on the atomic scale as well as on the cosmic scale of things. What do they look like, these fields which we customarily designate by $\vec{E}$ and $\vec{B}$, thereby indicating that they are vector fields possessing magnitude and direction? None of us has ever seen them or experienced them in any direct way. But that they do "exist" we know by inference. They have energy and we can measure that. And they exert a force and we can measure that. So, when we take a charge of magnitude q coulombs and move it around this electromagnetic field, exploring here and there, and measure the force on q, we find its magnitude and direction to be given by the formula

$$\boxed{\vec{F} = q[\vec{E} + (\vec{v} \times \vec{B})] \quad \text{newtons.}} \tag{3.1}$$

What this equation due to Lorentz tells us is that the net force at any given point in space and any instant of time is a combination of a velocity independent part $(q\vec{E})$ due to $\vec{E}$ alone, and a part $q(\vec{v} \times \vec{B})$ due to $\vec{B}$ alone which exists only by virtue of the fact that q has velocity $\vec{v}$. This, then, is one way in which electric and magnetic fields can be made to manifest themselves to us. How they are generated is a different question altogether; for its complete answer we must turn to the four fundamental equations of electromagnetism known as

Maxwell's equations. We shall review them in the next section assuming, however, that the reader has had some previous contact with them. In presenting this material we shall use the elegant and concise language of vector calculus where the ∇ operator plays a central role.

3.1 THE FUNDAMENTAL LAWS OF ELECTROMAGNETISM

Electric fields are produced by charges, and magnetic fields are produced by currents which are just charges in motion. Moreover, electric fields can also be "generated" by varying magnetic fields, and magnetic fields by varying electric fields. Therefore, it comes as no surprise that the equations describing the two kinds of fields are intimately coupled to one another.

Gauss' law

The most elementary of all electric fields is one produced by a point charge resting in vacuum. If q is such a charge, then Coulomb's law requires that the field at a point P, distance $\vec{r}$ away be

$$\vec{E} = \frac{q}{4\pi\epsilon_o r^2}\hat{r} \tag{3.2}$$

where the unit vector $\hat{r}$ points radially outward from the position of the charge, as is shown in Fig. 3.1. When many charges are present, experiment shows that the net electric field at P is obtained by

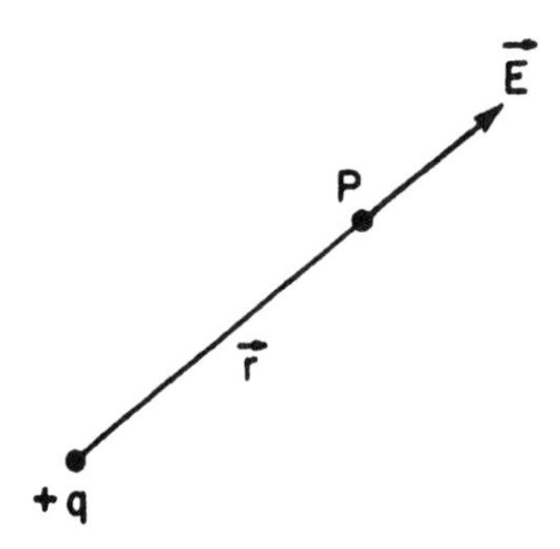

Fig. 3.1 The electric field $\vec{E}$ at a distance r from a positive point charge of magnitude q situated in vacuum.

addition of the fields of the individual charges:

$$\vec{E} = \sum_i \frac{q_i}{4\pi\epsilon_o r_i^2} \hat{r}_i. \tag{3.3}$$

If a small positive test charge q_t is placed at the point P, it experiences a force $\vec{F}_t = q_t\vec{E}$ directed along the resultant of the individual fields.

In the rationalized MKS system of units used throughout this book the electric field, being the force per unit charge, is given in newtons per coulomb. To make things come out right in regard to magnitude in Eqs. 3.2 and 3.3 the constant ϵ_o, referred to as the "permittivity of free space," must have the value

$$\epsilon_o = 8.854 \times 10^{-12} \text{ coulomb}^2 \text{ newton}^{-1} \text{ meter}^{-2}. \tag{3.4}$$

With this value, the electric field at a distance of one meter from a point charge of one coulomb equals $1/4\pi\epsilon_o$ newton per coulomb, as measured by placing a small test charge at the point in question. While newtons per coulomb is a perfectly appropriate unit for the electric field strength, the more common "electrical" unit used in practice is the equivalent – volt per meter (V/m). This makes ϵ_o of Eq. 3.4 take on the alternate units of coulombs per volt-meter ($CV^{-1}m^{-1}$).

Coulomb's law does not belong to the brotherhood of the four great laws of electromagnetism, but Gauss' law, which is traditionally derived from it, does. The reason is that the electric field $\vec{E}$ like that given by Eq. 3.2 is correct only if the charges producing it are at rest in the observer's frame of reference. But the elegant Gauss' law is always true despite the fact that it was derived from the more restricted statement of Coulomb! (This illustrates how important discoveries are made.) Gauss' law states that when the electric flux, defined as

$$\Phi_E = \int_{\text{surface}} \vec{E} \cdot d\vec{a}, \tag{3.5}$$

is summed over any <u>closed</u> but arbitrarily shaped surface S, its net value is proportional to the charge lying inside S (see Fig. 3.2):

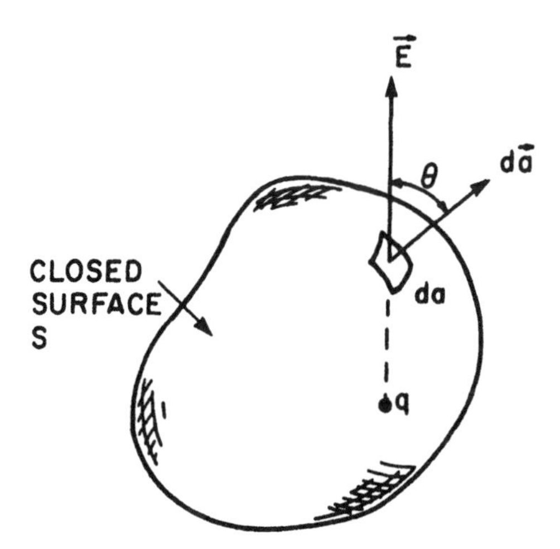

Fig. 3.2 An arbitrary closed surface S enclosing a single point charge q; da is an element of area of the surface. The vector $d\vec{a}$ has magnitude da and is directed along the outward normal to the surface. The element of flux $\vec{E} \cdot d\vec{a} = E\,da \cos\theta$ is positive when $\vec{E}$ points in the outward direction as shown. Any charges outside the closed surface S make zero contribution to the total flux $\int_S \vec{E} \cdot d\vec{a}$.

$$\int_{\text{closed S}} \vec{E} \cdot d\vec{a} = \frac{1}{\epsilon_o} \sum q(\text{inside S}). \tag{3.6}$$

The element of flux $\vec{E} \cdot d\vec{a}$ is taken to be positive if $\vec{E}$ points in the outward direction with respect to the closed surface. If we describe the location of charges in terms of a charge density ρ, we can consider that each infinitesimal element of volume dV contains the amount of charge ρdV. The sum over all such charges is then an integral over the volume V of which S is the surface, with the result that Eq. 3.6 becomes

$$\int_S \vec{E} \cdot d\vec{a} = \frac{1}{\epsilon_o} \int_V \rho\, dV. \tag{3.7}$$

This is known as the integral form of Gauss' law and is valid whether or not ρ is stationary or time-varying.

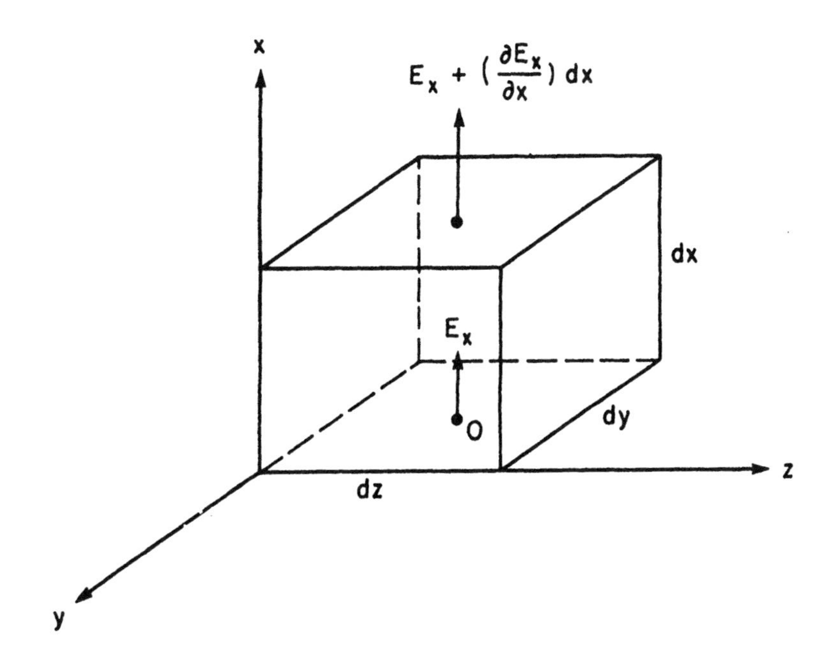

Fig. 3.3 An elementary cubical volume used in calculating the electrical flux flowing out of it. The field at the center of the bottom surface is, say, $E_x(0)$. At the top surface distance dx away, the field has a different value $E_x(0+dx)$. Expanding this function in a Taylor's series gives $E_x(0) + (\partial E_x/\partial x)_o\,(dx) + 1/2(\partial^2 E_x/\partial x^2)_o\,(dx)^2$ + etc. As $dx \to 0$, only the first two terms of the expansion need to be retained in the computation of the flux. The subscript 0 implies that the derivatives are to be evaluated at the lower point 0.

Since the volume V can be of any size (or shape) it is instructive to explore what happens as we shrink it to an infinitesimal size. To this purpose let us choose for simplicity a cubical volume element dxdydz as illustrated in Fig. 3.3. The electric field $\vec{E}$ can be in any arbitrary direction but let us now concentrate only on its x component. If E_x is its value at the bottom face, then the flux entering the bottom face is $E_x dydz$. Now suppose the field has a different value $[E_x + (\partial E_x/\partial x)dx]$ at the top surface. Therefore the flux leaving it equals $[E_x + (\partial E_x/\partial x)dx]$ dydz, and the net flux "flowing" vertically out of the box has the value $(\partial E_x/\partial x)$ dxdydz. Similar expressions give the outflow of flux in the y and z directions with the result that

$$\Delta\Phi_E = \left(\frac{\partial E_x}{\partial x} + \frac{\partial E_y}{\partial y} + \frac{\partial E_z}{\partial z}\right) dxdydz. \quad (3.8)$$

But this flux must result from charge within the elementary box. If ρ is the charge density, the amount of charge equals ρdxdydz. In accordance with Gauss' law the flux given by Eq. 3.8 must then be equated to the product of charge multiplied by $(\epsilon_o)^{-1}$ yielding

$$\left(\frac{\partial E_x}{\partial x} + \frac{\partial E_y}{\partial y} + \frac{\partial E_z}{\partial z}\right) = \frac{\rho}{\epsilon_o}. \quad (3.9)$$

This is the differential form of Gauss' law. It tells what happens in terms of derivatives evaluated at a point rather than integrals evaluated over a finite surface, as is the case of Eq. 3.7. Since derivatives are usually more easily manipulated than integrals, Eq. 3.9 is often a more useful form of the physical law.

A concise way of writing the left-hand side of Eq. 3.9 is through use of the mathematical operator ∇(del) which in Cartesian coordinates is defined as

$$\nabla = \hat{x}\frac{\partial}{\partial x} + \hat{y}\frac{\partial}{\partial y} + \hat{z}\frac{\partial}{\partial z}. \quad (3.10)$$

It is a vector operator and strictly speaking we should put an arrow above it ($\vec{\nabla}$) but for the sake of simplicity this will be omitted. Of course, the operator ∇ means nothing by itself – we must make it do something. Suppose we take the dot product ∇ with the vector $\vec{E}$ (or any other vector) and use the usual rules of scalar multiplication,

$$\begin{aligned} \nabla \cdot \vec{E} &= \left(\hat{x}\frac{\partial}{\partial x} + \hat{y}\frac{\partial}{\partial y} + \hat{z}\frac{\partial}{\partial z}\right) \cdot (\hat{x}E_x + \hat{y}E_y + \hat{z}E_z) \\ &= \frac{\partial E_x}{\partial x} + \frac{\partial E_y}{\partial y} + \frac{\partial E_z}{\partial z} \end{aligned} \quad (3.11)$$

which is just the left-hand side of Eq. 3.9. Thus we have the satisfactory result, Gauss' law, written as

$$\boxed{\nabla \cdot \vec{E} = \frac{\rho}{\epsilon_o}.} \qquad (3.12)$$

The quantity $\nabla \cdot \vec{E}$ is known as the "divergence of the vector $\vec{E}$," or in short, div $\vec{E}$.

Given the foregoing definition of div $\vec{E}$, a purely mathematical theorem (also due to Gauss) often comes in handy. It states that

$$\int_{\text{closed } S} \vec{E} \cdot d\vec{a} = \int_V (\nabla \cdot \vec{E})\, dV \qquad (3.13)$$

where V is the volume enclosed by the surface S. Equation 3.13 could be taken as a starting point and used to deduce the differential form of Gauss' law (3.12) from the integral form (3.7).

Faraday's law of induction

A second way of producing electric fields is by causing magnetic fields to change in time and (or) in space. This discovery by Faraday in 1831 is of great significance because this was the first time that a connection between electric fields and magnetism was shown to exist – until then the two subjects were considered as quite unrelated.

The "flux rule" states that the electromotive force (emf) in a circuit is equal to the rate at which the magnetic flux through the circuit is changing. The magnetic flux, in exact analogy with the electric flux of Eq. 3.5, is defined as

$$\Phi_B = \int_S \vec{B} \cdot d\vec{a} \qquad (3.14)$$

where the integration is to be carried out over any surface, plane or curved, which is enclosed by the circuit. The electromotive force is defined as the component of force per unit charge tangential to the circuit integrated once around the entire closed path. This quantity therefore equals the total work done on a unit charge traveling once around, and in mathematical language

$$\text{emf} = \oint_C \vec{E} \cdot d\vec{s} \qquad (3.15)$$

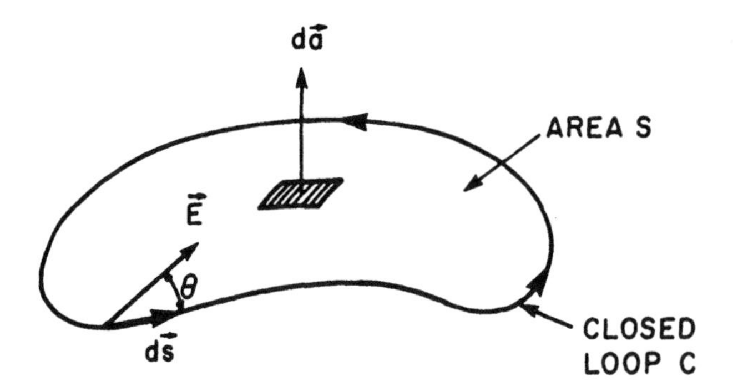

Fig. 3.4 The direction of the vector area $d\vec{a}$ (i.e., the normal to the surface) is defined to be positive when it is connected to the sense of rotation along the loop C through the right-hand screw rule. The electromotive force equals the line integral of $\vec{E} \cdot d\vec{s} = E\, ds \cos\theta$ taken around the loop.

where $d\vec{s}$ is an element of length of the circuit C. The circle on the integral $\oint$ is meant to denote that a line integration around a closed contour (circuit) is to be performed. Putting definitions (3.14) and (3.15) into our verbal statement of Faraday's law yields

$$\oint_C \vec{E} \cdot d\vec{s} = -\frac{d}{dt}\int_S \vec{B} \cdot d\vec{a} \tag{3.16}$$

where, by definition, the sense of the line integral and the direction in which the flux is called positive are given by the right-hand-screw rule (see Fig. 3.4). That is, go along the boundary in the clockwise direction when looking along the normal to the surface. Note that once we have accepted this definition, the negative sign of Eq. 3.16 describes an essential <u>physical</u> property of the system, its tendency to resist change. This property is called Lenz's law. More specifically, imagine that a given change in flux causes an emf in the circuit. If the circuit is a real physical conductor, current can flow in it. This current will give rise to a magnetic field and hence a magnetic flux directed in such a way as to counteract the primary change in flux. The negative sign of Eq. 3.16 does precisely this.

We have used the word "circuit" in a broad sense. The loop C could indeed be a real physical conductor or a portion of a conductor.

On the other hand, C may be merely a geometrical loop in empty space. The consequence of the last statement has far-reaching implications: it means that electric fields in vacuum can be created as if by magic wherever there are time-variant magnetic fields!

The electric field appearing on the left-hand side of Eq. 3.16 has units of volts per meter and therefore the emf = $\oint \vec{E} \cdot d\vec{s}$ is in volts. The magnetic flux must have the units of volt-seconds and B itself is in volt-seconds per meter squared. This same unit is also called one weber per square meter or one tesla and represents a lot of magnetic field. In terms of gauss, a unit for which we have a little more feeling,

$$\text{one} \rightarrow \left.\begin{matrix} \text{volt-sec-m}^{-2} \\ \text{or} \\ \text{weber m}^{-2} \\ \text{or} \\ \text{tesla} \end{matrix}\right\} = 10,000 \text{ gauss.} \tag{3.17}$$

In interstellar space there are magnetic fields of the order 10^{-6} gauss; the Earth's magnetic field is approximately 0.1 gauss. The most powerful air core solenoids can produce a steady magnetic field of approximately 300,000 gauss and the largest pulsed man-made fields may be as high as 5-10 million gauss. They are produced by passing a tremendously large pulsed current through a one-turn coil and, of course, blowing it up in the process.

We now want to do with Faraday's induction law what we did earlier with Gauss' law — transform it from the integral to the differential form. Therefore we take a rectangular element of surface area dxdy and compute the product $\vec{E} \cdot d\vec{s}$, moving in the sense shown in Fig. 3.5. The contributions from the four sides of the rectangle are

$$\oint \vec{E} \cdot d\vec{s} = E_x dx + \left[E_y + \frac{\partial E_y}{\partial x} dx\right] dy - \left[E_x + \frac{\partial E_x}{\partial y} dy\right] dx - E_y\, dy$$
$$= \left[\frac{\partial E_y}{\partial x} - \frac{\partial E_x}{\partial y}\right] dxdy \tag{3.18}$$

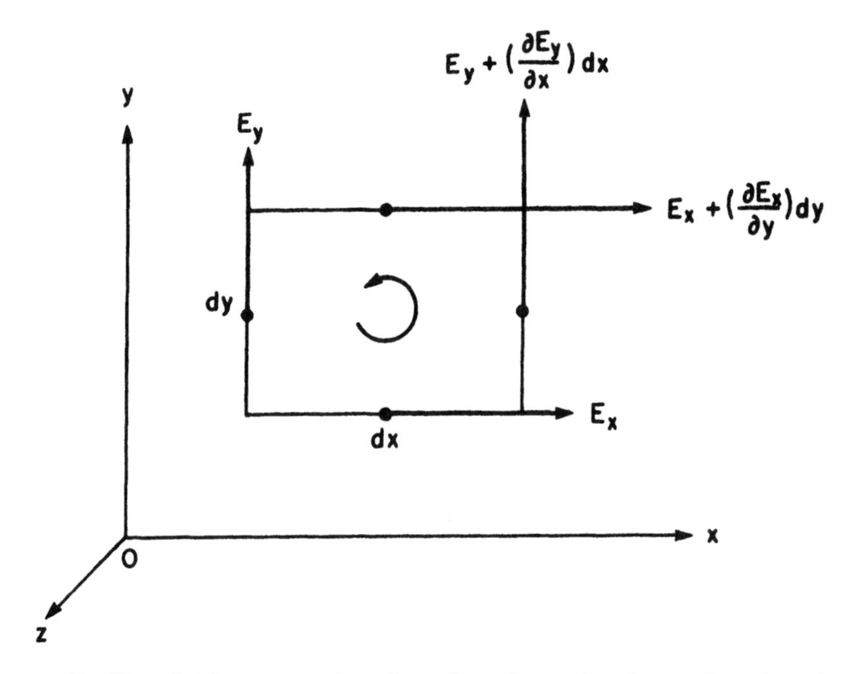

Fig. 3.5 Circulation around a closed rectangular loop of sides dx, dy lying entirely in the x-y plane. The increments in E_x and E_y at different positions of the loop are obtained by a Taylor's expansion as discussed in the caption of Fig. 3.3. When the circulation is in the direction shown, the z component of the curl $\vec{E}$ ($\equiv \nabla \times \vec{E}$) points out of the page, in the direction of the right-hand screw.

So much then for the left-hand side of Eq. 3.16. The magnetic flux on the right-hand side is

$$\Phi_B = B_z dxdy \tag{3.19}$$

and inserting this, together with Eq. 3.18, in Eq. 3.16 we get

$$\left[\frac{\partial E_y}{\partial x} - \frac{\partial E_x}{\partial y}\right] dxdy = - \frac{\partial B_z}{\partial t} dxdy \tag{3.20}$$

which relates the z component of the time-variant magnetic field with the space derivatives of the x and y components of the electric field. Because of our special choice of orientation of the element dxdy of Fig. 3.5 we have picked out but one of the three possible components of $\vec{B}$. If we repeat the entire process and orient the elemental area in the xz and yz planes, respectively, we obtain formulas similar to the above:

$$\left[\frac{\partial E_x}{\partial z} - \frac{\partial E_z}{\partial x}\right] dxdz = -\frac{\partial B_y}{\partial t} dxdz$$
$$\left[\frac{\partial E_z}{\partial y} - \frac{\partial E_y}{\partial z}\right] dydz = -\frac{\partial B_x}{\partial t} dydz. \tag{3.21}$$

The three equations (3.20) and (3.21) can be nicely packaged with the aid of the ∇ operator. Taking the cross product of ∇ and $\vec{E}$

$$\nabla \times \vec{E} = \left[\hat{x}\frac{\partial}{\partial x} + \hat{y}\frac{\partial}{\partial y} + \hat{z}\frac{\partial}{\partial z}\right] \times \left[\hat{x}E_x + \hat{y}E_y + \hat{z}E_z\right] \tag{3.22}$$

we note that the left-hand sides of Eqs. 3.20 and 3.21 are precisely the z, y, and x components respectively, of the quantity $\nabla \times \vec{E}$. Thus

$$\boxed{\nabla \times \vec{E} = -\frac{\partial \vec{B}}{\partial t}} \tag{3.23}$$

is the sought-after differential form of the induction law; the vector $\nabla \times \vec{E}$ is referred to as the curl of $\vec{E}$. The purely mathematical relationship

$$\oint_C \vec{E} \cdot d\vec{s} = \int_S (\nabla \times \vec{E}) \cdot d\vec{a}, \tag{3.24}$$

known as Stoke's theorem, applied to Eq. 3.16 yields Eq. 3.23 virtually on inspection. It provides a somewhat different and more direct route than the one adopted above. Observe that the curl of a vector tells us something about the "circulation" of the quantity, in contrast to the divergence which tells about its "flow."

Maxwell's generalization of Ampère's circuital law

Magnetic fields generated by steadily flowing currents obey Ampère's circuital law which states that the line integral of $\vec{B}$ around the closed path C is proportional to the total current I flowing through

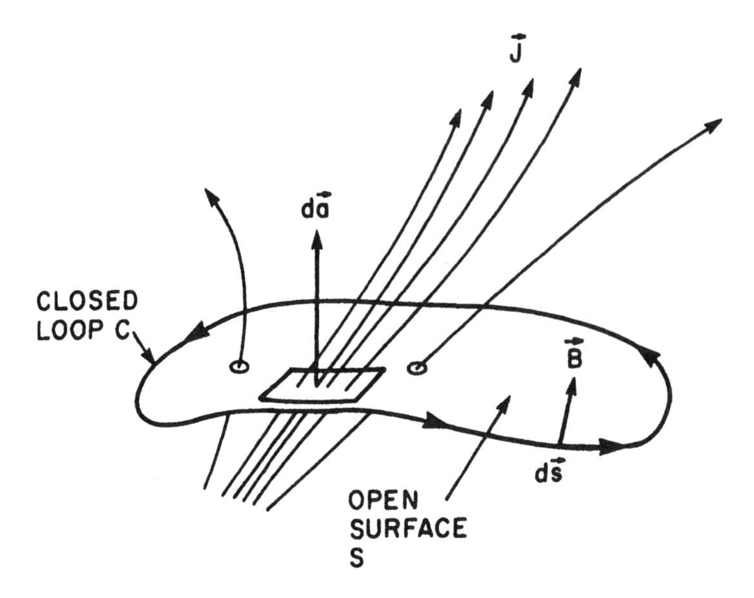

Fig. 3.6 The line integral of $\vec{B} \cdot d\vec{s}$ around a closed loop C depends only on the current $\int \vec{J} \cdot d\vec{a}$ enclosed by C.

the area of which C is the perimeter (see Fig. 3.6). That is,

$$\oint_C \vec{B} \cdot d\vec{s} = \mu_o I. \tag{3.25}$$

The constant μ_o known as the "permeability of free space" has the magnitude

$$\mu_o = 4\pi \times 10^{-7} \text{ webers m}^{-1} \text{ amp}^{-1} \tag{3.26}$$

so that when $\vec{B} \cdot d\vec{s}$ is measured in webers per meter, the current is in amperes. In terms of the current density $\vec{J}$ (amp per m^2) defined as

$$\int_S \vec{J} \cdot d\vec{a} = I \tag{3.27}$$

the circuital law, in its more general form, becomes

$$\oint_C \vec{B} \cdot d\vec{s} = \mu_o \int_S \vec{J} \cdot d\vec{a}. \tag{3.28}$$

Magnetic field effects caused by varying electric fields are not described by Ampère's law and were unknown at his time. It fell upon Maxwell to amend Eq. 3.28 in such a way as to accomodate the presence of time-varying $\vec{E}$ fields. Maxwell argued that a time-varying electric field is equivalent to an electric current whose density $\vec{J}_D$ is proportional to the time rate of change of $\vec{E}$. Specifically,

$$\vec{J}_D = \epsilon_o \frac{\partial \vec{E}}{\partial t} \text{ ampere-meter}^{-2} \tag{3.29}$$

This quantity is known as the displacement current density. In a way it is a fictitious quantity and is quite unlike the $\vec{J}$ of Eq. 3.28 which is due to the motion of real physical charges. The addition of $\vec{J}_D$ to the right-hand side of Eq. 3.28 yields the generalization of Ampère's circuital law which is essential whenever time-varying electric fields are present:

$$\oint_C \vec{B} \cdot d\vec{s} = \mu_o \int_S \left[\vec{J} + \epsilon_o \frac{\partial \vec{E}}{\partial t}\right] \cdot d\vec{a}. \tag{3.30}$$

Application of Stoke's theorem (3.24) allows us to write the foregoing equation as

$$\int_S (\nabla \times \vec{B}) \cdot d\vec{a} = \mu_o \int_S \left[\vec{J} + \epsilon_o \frac{\partial \vec{E}}{\partial t}\right] \cdot d\vec{a}. \tag{3.31}$$

But this must be true for any shape or size of area S, a fact which can be fulfilled only if

$$\boxed{\nabla \times \vec{B} = \epsilon_o \mu_o \frac{\partial \vec{E}}{\partial t} + \mu_o \vec{J}} \tag{3.32}$$

which is a differential representation of the integral form (3.30).

<u>The absence of the magnetic monopole and its consequences</u>

We are quite familiar with electric charges surrounded by an electric field directed radially outward or inward. The electron is the smallest elementary entity of this type. Is there an equivalent magnetic charge or magnetic monopole as it is called? Several searches have been made [see, for example, K. Ford, Scientific American <u>209</u>, 30 (1963); also, E. Goto, H. H. Kolm, and K. W. Ford,

Phys. Rev. 132, 387 (1963)] but all proved fruitless. We are therefore forced to conclude that the only sources of magnetic field are electric currents and time-varying electric fields as we have discussed.

In a region of space containing an isolated monopole, there would be a net flux of magnetic field. Therefore, an equation equivalent to Gauss' law (3. 6) or (3. 7) would exist for magnetism. The apparent total absence of such an object requires that the flux of $\vec{B}$ through any closed surface always be zero:

$$\int_{\text{closed S}} \vec{B} \cdot d\vec{a} = 0. \tag{3.33}$$

From this it follows that the fluxes of $\vec{B}$ through two different open surfaces having the same perimeter are equal. This has an important bearing on Eq. 3.16 where we recall that the open surface S remained arbitrary: that is, with the closed contour C fixed, the surface S could be just the plane surface bounded by C or, for that matter, any bulging surface as shown in Fig. 3.7. Now, as a result of Eq. 3.33 we know that it does not matter which surface is chosen since the fluxes $\int \vec{B} \cdot d\vec{a}$ through all surfaces having the same C are exactly the same.

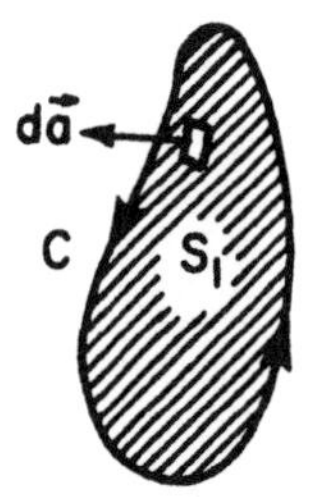

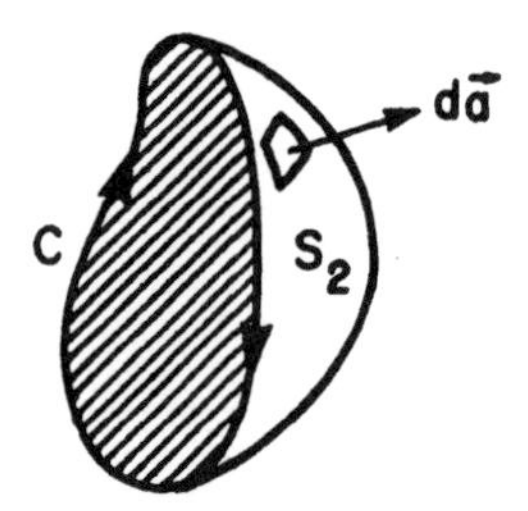

Fig. 3.7 The magnetic flux $\int_{S_1} \vec{B} \cdot d\vec{a}$ integrated over the plane surface S_1 is identical to the flux $\int_{S_2} \vec{B} \cdot d\vec{a}$ integrated over the bulging surface S_2 provided that S_1 and S_2 have the same perimeters C. This comes from the fact that when we use S_1 and S_2 to make a closed surface S, $0 = \int_S \vec{B} \cdot d\vec{a} = \int_{S_1} \vec{B} \cdot d\vec{a} + \int_{S_2} \vec{B} \cdot d\vec{a}$.

The differential form of Eq. 3.33 is obtained by use of Eq. 3.13, or by analogy with Gauss' law (3.12):

$$\boxed{\nabla \cdot \vec{B} = 0} \tag{3.34}$$

Charge conservation

Currents have to obey the law of charge conservation. In other words, no charge can flow away from some place without a corresponding decrease of charge in that region. Consider, then, a closed surface S surrounding the volume V. Suppose that at some instant of time an amount of current

$$I = \int_{\text{closed S}} \vec{J} \cdot d\vec{a} \tag{3.35}$$

is leaving the volume V which at that time contains an amount of charge $\int \rho dV$. Then,

$$\int_S \vec{J} \cdot d\vec{a} = -\frac{d}{dt} \int_V \rho dV \tag{3.36}$$

which is the integral representation of the law of conservation of charge. The differential form is

$$\boxed{\nabla \cdot \vec{J} = -\frac{\partial \rho}{\partial t}} \tag{3.37}$$

a result which the reader will have no trouble proving to himself.

Maxwell's equations

We have reached the point in our summary where there is little for us to do other than perhaps write down explicitly the very fundaments of classical electrodynamic theory. These are embodied in the four equations traditionally called Maxwell's equations, which we wrote out in bits and pieces on earlier pages. In differential form, they are:

$$\begin{aligned} &\nabla \cdot \vec{E} = \frac{\rho}{\epsilon_o} && \text{(a)} \\ &\nabla \times \vec{E} = -\frac{\partial \vec{B}}{\partial t} && \text{(b)} \\ &\nabla \times \vec{B} = \mu_o \epsilon_o \frac{\partial \vec{E}}{\partial t} + \mu_o \vec{J} && \text{(c)} \\ &\nabla \cdot \vec{B} = 0 && \text{(d)} \end{aligned} \tag{3.38}$$

To this we must add the Lorentz force equation

$$\vec{F} = q(\vec{E} + \vec{v} \times \vec{B}). \tag{3.39}$$

A useful supplemental equation is the equation of charge conservation

$$\nabla \cdot \vec{J} + \frac{\partial \rho}{\partial t} = 0 \tag{3.40}$$

but this equation is not essential because its content is implicit in Maxwell's equations.

The foregoing formulas contain all of our knowledge of classical electricity and magnetism. They cover a vast territory incredibly rich in physical phenomena. In general they are also very difficult to solve. And although they have been with us for a long time, they continue to yield important technological and scientific information even today, some hundred years after Maxwell's momentous discoveries. Our task henceforth will be to learn to use them, find out what they mean, and most important, to see what interesting things we can learn from them.

Summary of results from the algebra of vectors

We shall now list some results from the algebra of vectors useful in manipulating Maxwell's equations. They are purely mathematical relationships and we shall not prove their validity. To check that they are correct, the most straightforward though laborious way is to expand the vectors in Cartesian coordinates, perform the vectorial operation on each component of the vector, and confirm that the right-hand side balances the left-hand side. A pleasant text on this subject of vectors is by H. M. Schey, Div, Grad, Curl, and All That (Norton, New York, 1973).

Consider two vectors $\vec{A}$ and $\vec{B}$ whose components in Cartesian coordinates are

$$\begin{aligned} \vec{A} &= \hat{x}A_x + \hat{y}A_y + \hat{z}A_z \\ \vec{B} &= \hat{x}B_x + \hat{y}B_y + \hat{z}B_z. \end{aligned} \tag{3.41}$$

The rules of scalar and vector multiplication require that

$$\vec{A} \cdot \vec{B} = A_xB_x + A_yB_y + A_zB_z \quad \text{scalar} \tag{3.42}$$

$$\left.\begin{aligned} \vec{A} \times \vec{B} = \hat{x}&[A_yB_z - A_zB_y] \\ + \hat{y}&[A_zB_x - A_xB_z] \\ + \hat{z}&[A_xB_y - A_yB_x] \end{aligned}\right\} \quad \text{vector} \tag{3.43}$$

The last relation can also be written in a nice compact way as the determinant

$$\vec{A} \times \vec{B} = \begin{vmatrix} \hat{x} & \hat{y} & \hat{z} \\ A_x & A_y & A_z \\ B_x & B_y & B_z \end{vmatrix} \tag{3.44}$$

The next four results are helpful, particularly the triple products between three vectors $\vec{A}$, $\vec{B}$, and $\vec{C}$:

$$\vec{A} \times \vec{A} = 0 \tag{3.45}$$

$$\vec{A} \cdot (\vec{A} \times \vec{B}) = 0 \tag{3.46}$$

$$\begin{aligned} \vec{A} \cdot (\vec{B} \times \vec{C}) &= \vec{B} \cdot (\vec{C} \times \vec{A}) \\ &= \vec{C} \cdot (\vec{A} \times \vec{B}) \end{aligned} \tag{3.47}$$

$$\vec{A} \times (\vec{B} \times \vec{C}) = (\vec{A} \cdot \vec{C})\vec{B} - (\vec{A} \cdot \vec{B})\vec{C}. \tag{3.48}$$

The del operator ∇ defined in Cartesian coordinates as

$$\nabla = \left(\hat{x}\frac{\partial}{\partial x} + \hat{y}\frac{\partial}{\partial y} + \hat{z}\frac{\partial}{\partial z}\right) \tag{3.49}$$

can be treated as a regular vector, and vector manipulations like those given above can be applied to it directly. There are certain pitfalls which must be carefully avoided if other than Cartesian coordinates (x, y, z) are used, and for that reason the last named will be assumed implicitly henceforth. To begin with, let ∇ operate on the scalar quantity ϕ. Then the object

$$\nabla\phi = \hat{x}\frac{\partial\phi}{\partial x} + \hat{y}\frac{\partial\phi}{\partial y} + \hat{z}\frac{\partial\phi}{\partial z} \tag{3.50}$$

is known as the "gradient of phi" and represents a vector whose components are derivatives of ϕ. The scalar product of ∇ with the vector $\vec{A}$ defines the divergence of $\vec{A}$,

$$\begin{aligned}\nabla\cdot\vec{A} &= \left[\hat{x}\frac{\partial}{\partial x} + \hat{y}\frac{\partial}{\partial y} + \hat{z}\frac{\partial}{\partial z}\right]\cdot[\hat{x}A_x + \hat{y}A_y + \hat{z}A_z] \\ &= \frac{\partial A_x}{\partial x} + \frac{\partial A_y}{\partial y} + \frac{\partial A_z}{\partial z}\end{aligned} \tag{3.51}$$

which is seen to be a scalar quantity. The vector product between ∇ and $\vec{A}$ defines the curl of $\vec{A}$:

$$\begin{aligned}\nabla\times\vec{A} = \hat{x}&\left[\frac{\partial A_z}{\partial y} - \frac{\partial A_y}{\partial z}\right] \\ + \hat{y}&\left[\frac{\partial A_x}{\partial z} - \frac{\partial A_z}{\partial x}\right] \\ + \hat{z}&\left[\frac{\partial A_y}{\partial x} - \frac{\partial A_x}{\partial y}\right].\end{aligned} \tag{3.52}$$

In exact analogy with Eq. 3.44, this can also be written as

$$\nabla\times\vec{A} = \begin{vmatrix}\hat{x} & \hat{y} & \hat{z} \\ \frac{\partial}{\partial x} & \frac{\partial}{\partial y} & \frac{\partial}{\partial z} \\ A_x & A_y & A_z\end{vmatrix} \tag{3.53}$$

Triple products involving the ∇ operator come up quite regularly and the following set is helpful. On taking the divergence of both sides of Eq. 3.50 one obtains

$$\nabla \cdot (\nabla\phi) = (\nabla \cdot \nabla)\phi$$

$$= \frac{\partial^2 \phi}{\partial x^2} + \frac{\partial^2 \phi}{\partial y^2} + \frac{\partial^2 \phi}{\partial z^2}. \tag{3.54}$$

This scalar quantity is quite widely used in physics and is given a new symbol $\nabla^2\phi$ and a new name — the Laplacian. Recall that the right-hand side of this equation is precisely equal to the space-dependent part of the three-dimensional wave equation (2.55) which we can now write in the compact form.

$$\frac{\partial^2 \psi}{\partial t^2} = v^2 \nabla^2 \psi. \tag{3.55}$$

Additional relationships are:

$$\nabla \times (\nabla\phi) = 0 \tag{3.56}$$

$$\nabla \cdot (\nabla \times \vec{A}) = 0 \tag{3.57}$$

$$\nabla \cdot (\vec{A} \times \vec{B}) = B \cdot (\nabla \times \vec{A}) - \vec{A} \cdot (\nabla \times \vec{B}) \tag{3.58}$$

$$\nabla \times (\nabla \times \vec{A}) = \nabla(\nabla \cdot \vec{A}) - (\nabla \cdot \nabla)\vec{A}. \tag{3.59}$$

In the last equation we have a somewhat unfamiliar object, that is, the Laplacian $\nabla^2 = (\nabla \cdot \nabla)$ operating on a vector rather than on a scalar, as was the situation in Eq. 3.54 above. The outcome of a Laplacian operating on a vector is another vector whose components are

$$\nabla^2\vec{A} \equiv (\nabla \cdot \nabla)\vec{A}$$

$$= \hat{x}\left[\frac{\partial^2 A_x}{\partial x^2} + \frac{\partial^2 A_x}{\partial y^2} + \frac{\partial^2 A_x}{\partial z^2}\right] + \hat{y}\left[\frac{\partial^2 A_y}{\partial x^2} + \frac{\partial^2 A_y}{\partial y^2} + \frac{\partial^2 A_y}{\partial z^2}\right]$$

$$+ \hat{z}\left[\frac{\partial^2 A_z}{\partial x^2} + \frac{\partial^2 A_z}{\partial y^2} + \frac{\partial^2 A_z}{\partial z^2}\right] \tag{3.60}$$

showing that there are three "regular" Laplacians, one for each of the three components of the vector $\vec{A}$. The right-hand side is quite lengthy to write out in full, but now that we know precisely what it says we can abbreviate it as

$$\nabla^2\vec{A} = \frac{\partial^2\vec{A}}{\partial x^2} + \frac{\partial^2\vec{A}}{\partial y^2} + \frac{\partial^2\vec{A}}{\partial z^2}, \tag{3.61}$$

a form which is sometimes found in the literature.

Example

Differential vector operators are an extremely valuable "shorthand" which find their way into many branches of physics as they permit the writing of partial differential equations in a concise, compact form. Familiarity with these operators is strongly encouraged and the student is urged to prove Eqs. 3.56, 3.57, 3.58, and 3.59. The proof of the last is given below.

Solution

We seek to evaluate $\nabla\times(\nabla\times\vec{A})$ where $\vec{A}$ is any vector. Using Eq. 3.52 the x component is

$$[\nabla\times(\nabla\times\vec{A})]_x = \frac{\partial}{\partial y}(\nabla\times\vec{A})_z - \frac{\partial}{\partial z}(\nabla\times\vec{A})_y$$

$$= \frac{\partial}{\partial y}\left(\frac{\partial A_y}{\partial x} - \frac{\partial A_x}{\partial y}\right) - \frac{\partial}{\partial z}\left(\frac{\partial A_x}{\partial z} - \frac{\partial A_z}{\partial x}\right)$$

$$= \frac{\partial}{\partial x}\left(\frac{\partial A_y}{\partial y} + \frac{\partial A_z}{\partial z}\right) - \frac{\partial^2 A_x}{\partial y^2} - \frac{\partial^2 A_x}{\partial z^2}$$

$$= \frac{\partial}{\partial x}\left(\frac{\partial A_x}{\partial x} + \frac{\partial A_y}{\partial y} + \frac{\partial A_z}{\partial z}\right) - \frac{\partial^2 A_x}{\partial x^2} - \frac{\partial^2 A_x}{\partial y^2} \cdot - \frac{\partial^2 A_x}{\partial z^2}$$

$$= \frac{\partial}{\partial x}(\nabla\cdot\vec{A}) - \nabla\cdot\nabla A_x.$$

In this last line we have used Eqs. 3.51 and 3.54. Similar expressions can be developed for the y and z components. Thus the end result is

$$\nabla \times (\nabla \times \vec{A}) = \nabla(\nabla \cdot \vec{A}) - \nabla^2 \vec{A}$$

where we write $\nabla \cdot \nabla\vec{A} = \nabla^2\vec{A}$ (Eq. 3.60).

Example

It was stated in the text that the equation of charge conservation (3.40) is implicit in Maxwell's equations (3.38). Prove this statement.

Solution

Take the divergence of both sides of Eq. 3.38c:

$$\nabla \cdot (\nabla \times \vec{B}) = \mu_o \epsilon_o \nabla \cdot \left(\frac{\partial \vec{E}}{\partial t}\right) + \mu_o \nabla \cdot \vec{J}.$$

But, the divergence of the curl of any vector is identically zero (see Eq. 3.57). Thus $\nabla \cdot (\nabla \times \vec{B}) = 0$. Also, we can interchange time and space derivatives so that

$$\nabla \cdot \frac{\partial \vec{E}}{\partial t} = \frac{\partial}{\partial t}(\nabla \cdot \vec{E}).$$

It follows that

$$0 = \mu_o \epsilon_o \frac{\partial}{\partial t}(\nabla \cdot \vec{E}) + \mu_o \nabla \cdot \vec{J}.$$

Eliminating $\vec{E}$ with the help of Gauss' law (Eq. 3.38a) we get

$$0 = \frac{\partial \rho}{\partial t} + \nabla \cdot \vec{J}$$

which is the equation of charge conservation.

3.2 THE ELECTROMAGNETIC FIELD IN REGIONS FREE OF SOURCES

Although we said earlier that varying magnetic fields cause electric fields and varying electric fields cause magnetic fields, we realize that speaking precisely, the ultimate cause of all electromagnetic fields are the real charges and real currents somewhere or other in space (including of course charges and currents in matter). It is therefore the charge density ρ and the current density $\vec{J}$ which act as the sources of $\vec{E}$ and $\vec{B}$ in Maxwell's equations (3.38). If ρ and $\vec{J}$ are constant in time, the equations become $\nabla \cdot \vec{E} = \rho/\epsilon_0$, $\nabla \times \vec{E} = 0$; $\nabla \times \vec{B} = \mu_0 \vec{J}$ and $\nabla \cdot \vec{B} = 0$. From this we see that the four equations group themselves into two sets of two equations each. In the first set only the electric field appears and in the second set only the magnetic field. This means that electricity and magnetism are separate phenomena so long as charges and currents are static. Our present interests are with time-varying charges and currents in which case $\vec{E}$ and $\vec{B}$ become interdependent. The solution of Eqs. 3.38 for a given distribution of ρ and $\vec{J}$ becomes a particularly formidable task and we do not propose to tackle it just yet. Rather, we turn to a simpler question concerning the fields in a region of space far removed from all charges and currents. The solution will be interesting.

In the absence of sources, Eqs. 3.38 become

$$\left.\begin{aligned} &\nabla \cdot \vec{E} = 0 && \text{(a)} \\ &\nabla \times \vec{E} = -\frac{\partial \vec{B}}{\partial t} && \text{(b)} \\ &\nabla \times \vec{B} = \mu_0 \epsilon_0 \frac{\partial \vec{E}}{\partial t} && \text{(c)} \\ &\nabla \cdot \vec{B} = 0 && \text{(d)} \end{aligned}\right\} \qquad (3.62)$$

which are almost perfectly symmetrical in $\vec{E}$ and $\vec{B}$, respectively; the minus sign spoils somewhat the pleasing symmetry. The four vector equations in actuality represent a total of eight equations: the two (scalar) divergence equations, the three Cartesian components of the curl $\vec{E}$ equation and the three components of the curl $\vec{B}$ equation. For the time being let us set aside the two divergence equations.

The two remaining curl equations are coupled in that each contains both the $\vec{E}$ and $\vec{B}$ variables. To eliminate, say, $\vec{B}$ we take the curl of both sides of Eq. 3.62b with the result that

$$\nabla \times (\nabla \times \vec{E}) = -\frac{\partial}{\partial t} (\nabla \times \vec{B}) \qquad (3.63)$$

where we have interchanged the order of differentiation with respect to the time and space coordinates. We insert Eq. 3.62c into the right-hand side of (3.63) and thereby eliminate the magnetic field:

$$\nabla \times (\nabla \times \vec{E}) = -\mu_o \epsilon_o \frac{\partial^2 E}{\partial t^2}. \tag{3.64}$$

But, in accordance with Eq. 3.59, $\nabla \times (\nabla \times \vec{E}) = \nabla(\nabla \cdot \vec{E}) - (\nabla \cdot \nabla)\vec{E}$. The first term is identically zero because Gauss' law in the absence of charges demands that $\nabla \cdot \vec{E} = 0$ (see Eq. 3.62a); and the second term is the definition (3.60) of the Laplacian operating on the vector field $\vec{E}$. Hence it follows that

$$\nabla^2 \vec{E} \equiv \frac{\partial^2 \vec{E}}{\partial x^2} + \frac{\partial^2 \vec{E}}{\partial y^2} + \frac{\partial^2 \vec{E}}{\partial z^2} = \mu_o \epsilon_o \frac{\partial^2 \vec{E}}{\partial t^2} \tag{3.65}$$

and by similar manipulations, that

$$\nabla^2 \vec{B} \equiv \frac{\partial^2 \vec{B}}{\partial x^2} + \frac{\partial^2 \vec{B}}{\partial y^2} + \frac{\partial^2 \vec{B}}{\partial z^2} = \mu_o \epsilon_o \frac{\partial^2 \vec{B}}{\partial t^2}. \tag{3.66}$$

Thus we have arrived at wave equations, six in number, one for each of the three components of $\vec{E}$ and one for each of the three components of $\vec{B}$. We are thus led to the inescapable conclusion that the equations of electromagnetism, that is, the Gauss', Faraday's, Ampère's laws, etc. when considered in their totality allow the existence of electromagnetic waves — disturbances which to all appearances require no physical matter through which to propagate. They can exist in perfect vacuum, unlike the mechanical waves and oscillations which we studied in Chapters 1 and 2. And unlike mechanical waves, the electromagnetic wave is characterized by two interdependent oscillating quantities, the electric field and the magnetic field.

A comparison of Eqs. 3.65 and 3.66 with the three-dimensional wave equation (3.55) permits us to extract immediately the phase velocity v with which the electric and magnetic disturbances propagate. It is given by

$$v = \frac{1}{\sqrt{\mu_o \epsilon_o}} \tag{3.67}$$

Since $\mu_o = 4\pi \times 10^{-7}$ webers m^{-1} amp^{-1} and $\epsilon_o = 8.854 \times 10^{-12}$ coulomb $volt^{-1}$ m^{-1}, we obtain

$$\begin{aligned} v &= [4\pi \times 8.854 \times 10^{-19}]^{-1/2} \\ &= 2.9980 \times 10^{8} \text{ meters sec}^{-1} \end{aligned} \tag{3.68}$$

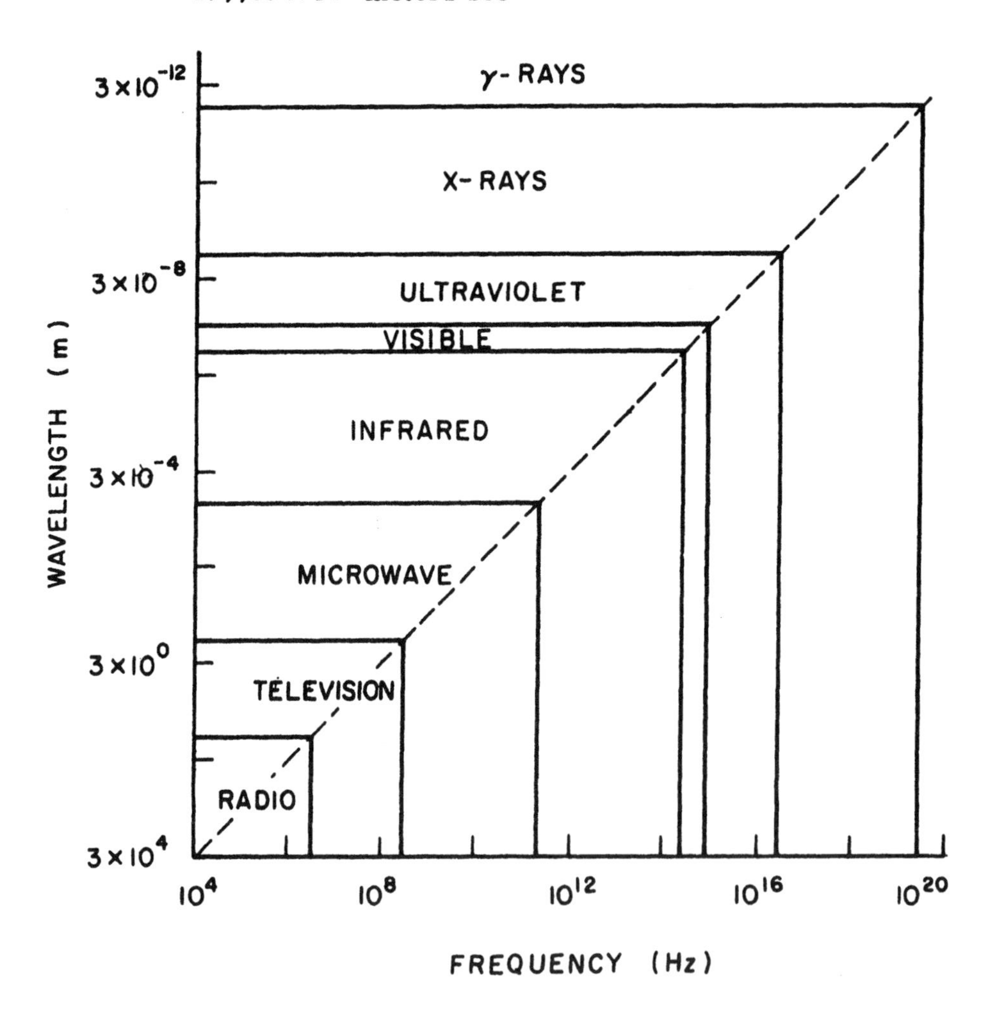

Fig. 3.8 The electromagnetic spectrum.

which is the speed of light in vacuum. On recognizing these facts Maxwell was led to postulate that light is an electromagnetic phenomenon and in 1865 with brilliant success he laid the foundations of the electromagnetic theory of light [see Philosophical Transactions R. Soc. London 155, 459 (1865)]. We now know that these considerations apply not only to visible light but to a vast range of other radiations of both lower and higher frequencies, extending from long radio waves at the low-frequency end to gamma rays at the other end. This is illustrated in Fig. 3.8. However, Clerk Maxwell's contribution to science transcends the specific discovery we have just described. As Einstein said in commemorating Maxwell's birthday:

> "We may say that, before Maxwell, Physical Reality, in so far as it was to represent the processes of nature, was thought of as consisting in material particles, whose variations consist only in movements governed by partial differential equations. Since Maxwell's time, Physical Reality has been thought of as represented by continuous fields, governed by partial differential equations, and not capable of any mechanical interpretation. This change in the conception of Reality is the most profound and the most fruitful that physics has experienced since the time of Newton...." [A. Einstein, in James Clerk Maxwell, A Commemoration Volume (The Macmillan Company, New York, 1931), p. 71].

3.3 THE TRANSVERSE UNIFORM PLANE WAVE

Equations 3.65 and 3.66 are just a little too complicated and a little too general to allow us to understand the fine details of electromagnetic propagation. Therefore, to discover the relative orientations of $\vec{E}$ and $\vec{B}$, their magnitudes, and so on, we shall return to the four basic equations (3.62) and apply them to the familiar standby – the plane wave.

Using our earlier experience with plane waves in mechanical systems, we make the plausible assumption that the oscillating quantities, that is, the three components of $\vec{E}$ and the three components of $\vec{B}$, depend only on time t and a single spatial coordinate, say z. Let us see what the consequences are when we apply this assumption to the electric field $\vec{E}$. We write out in full the first equation (3.62a), $\nabla \cdot \vec{E} = 0 = [\partial E_x/\partial x + \partial E_y/\partial y + \partial E_z/\partial z]$, and since we postulated that there are no variations with x and y, it follows that

$$\frac{\partial E_z}{\partial z} = 0 \tag{3.69}$$

which says that the component of the electric field in the z direction is constant in space. Now we examine the z component of the curl equation (3.62c) which written out in full is

$$\left[\frac{\partial B_y}{\partial x} - \frac{\partial B_x}{\partial y}\right] = \mu_o \epsilon_o \left(\frac{\partial E_z}{\partial t}\right). \tag{3.70}$$

Since $\vec{B}$ likewise has no variation with x and y it follows that

$$\frac{\partial E_z}{\partial t} = 0 \tag{3.71}$$

which says that E_z is constant in time. We now have the case of a field component that is independent both of space and time — it is a constant field like that which is encountered between the plates of a parallel plate capacitor charged with a (steady) dc battery. This field is of no interest to us since in our study of waves we are seeking dynamically varying fields. Thus, we are at liberty to set $E_z = 0$.* Using precisely the same arguments for the magnetic fields, we can convince ourselves that B_z is also zero. In summary, then: subject to the assumption that $\vec{E}$ and $\vec{B}$ vary only with z and t,

$$\begin{aligned} E_z &= 0 \\ B_z &= 0. \end{aligned} \tag{3.72}$$

We now write out in full the remainder of the two curl equations (3.62b and 3.62c) subject to our initial postulate:

*In other words: Maxwell's equations are linear differential equations. The principle of superposition holds. Thus, the steady electric field can be subtracted from the final solution.

$$\left.\begin{aligned} \frac{\partial E_x}{\partial z} &= -\frac{\partial B_y}{\partial t} && \text{(a)} \\ \frac{\partial B_y}{\partial z} &= -\mu_o \epsilon_o \frac{\partial E_x}{\partial t} && \text{(b)} \end{aligned}\right\} \tag{3.73}$$

$$\left.\begin{aligned} \frac{\partial E_y}{\partial z} &= \frac{\partial B_x}{\partial t} && \text{(a)} \\ \frac{\partial B_x}{\partial z} &= \mu_o \epsilon_o \frac{\partial E_y}{\partial t} && \text{(b)} \end{aligned}\right\} \tag{3.74}$$

These are our basic results from which many conclusions will be drawn. The equations group themselves naturally into two independent sets. The first set involves only variables E_x and B_y, and the second set, only variables E_y and B_x. We shall consider one set alone, say the first set, and see what conclusions can be obtained from it. Our findings will then apply to the second set also. The two equations (3.73) may be manipulated so as to eliminate one or the other variable E_x or B_y. To this purpose, we differentiate both sides of Eq. 3.73a with respect to z so that $(\partial^2 E_x/\partial z^2) = -\partial/\partial z(\partial B_y/\partial t)$. When we interchange the order of differentiation with respect to z and t and substitute for $\partial B_y/\partial z$ from Eq. 3.73b it follows that

$$\frac{\partial^2 E_x}{\partial z^2} = \mu_o \epsilon_o \frac{\partial^2 E_x}{\partial t^2} \tag{3.75a}$$

and from similar arguments that

$$\frac{\partial^2 B_y}{\partial z^2} = \mu_o \epsilon_o \frac{\partial^2 B_y}{\partial t^2}. \tag{3.75b}$$

These will be recognized as two one-dimensional wave equations, one for $\vec{E}$ and one for $\vec{B}$. They represent plane waves traveling either in the plus or minus z directions with a phase velocity $v = (\mu_o \epsilon_o)^{-1/2} = c$. Observe that this plane electromagnetic wave is a purely transverse vibration — it has no longitudinal components E_z or B_z. In this sense it is like the transverse wave on a stretched string. Also, the electric and magnetic fields are at right angles to one another: when $\vec{E}$ has a component along plus x only, $\vec{B}$ has a component along plus y only.

The plane sinusoidal traveling wave

Consider a sinusoidal wave of frequency ω traveling along the positive z direction. Equation 3.75a then demands that the solution be of the form

$$E_x = E_{ox} \cos(\omega t - kz + \phi_E) \tag{3.76}$$

where E_{ox} is the amplitude and ϕ_E the phase of the electric field of the wave; the quantities ω and k are connected via the relation $\omega/k = v = c$. The magnetic field oscillations determined by Eq. 3.75b will be of similar form,

$$B_y = B_{oy} \cos(\omega t - kz + \phi_B). \tag{3.77}$$

Since $\vec{E}$ and $\vec{B}$ fields are interrelated by Maxwell's equations, the amplitudes and phases of the two fields must bear definite ratios to one another. To determine them, substitute Eqs. 3.76 and 3.77 in 3.73a and 3.73b. This results in two equations

$$\begin{aligned} kE_{ox} \sin(\omega t - kz + \phi_E) &= \omega B_{oy} \sin(\omega t - kz + \phi_B) \\ kB_{oy} \sin(\omega t - kz + \phi_B) &= \mu_o \epsilon_o \omega E_{ox} \sin(\omega t - kz + \phi_E) \end{aligned} \tag{3.78}$$

which must be true at all times t and in all places z. The only way this can be fulfilled is if

$$\phi_E = \phi_B = \phi, \tag{3.79}$$

showing that the electric and magnetic fields are exactly in phase with one another. In addition, dividing the second equation (3.78) into the first, and using Eq. 3.79 gives $(B_{oy}/E_{ox}) = (\mu_o \epsilon_o)(E_{ox}/B_{oy})$, or equivalently

$$\begin{aligned} B_{oy} &= \sqrt{\mu_o \epsilon_o}\, E_{ox} \\ &= \frac{1}{c} E_{ox}, \end{aligned} \tag{3.80}$$

which is the sought-for relation giving the magnitude of the magnetic field in terms of the magnitude of the electric field. Combining the bits and pieces of information gathered heretofore we obtain for the plane sinusoidal wave traveling along +z, the results

$$\begin{aligned} E_x &= E_{ox} \cos(\omega t - kz + \phi) \\ B_y &= \frac{E_{ox}}{c} \cos(\omega t - kz + \phi) \end{aligned} \tag{3.81}$$

which are shown sketched in Fig. 3.9. We note that if the wave propagates in the negative z direction and if at any instant $\vec{E}$ points along plus x, then B_y points along minus y, a fact which is readily proved from Eqs. 3.73. Indeed, it is generally true that if $\vec{k}$ is the propagation vector (for the definition of $\vec{k}$ see section 2.4), then the vectors $\vec{E}$, $\vec{B}$, and $\vec{k}$ of a plane electromagnetic wave in free space are always orthogonal to one another and have directions specified by the right-hand-screw rule (the screw points along $\vec{k}$) as is illustrated in Fig. 3.10.

Equations 3.81 are a special case of a transverse, uniform, plane wave. It is special because the wave is propagating along the z-axis with $\vec{E}$ and $\vec{B}$ having only x and y components, respectively. A more general case of a transverse, uniform, plane wave is given in the following example. The wave is transverse because both $\vec{E}$ and $\vec{B}$ have no components along $\vec{k}$, the propagation vector. The wave is uniform because neither $\vec{E}$ nor $\vec{B}$ varies over a plane normal to the direction of propagation. The wave is a plane wave because the loci of points of common phase are planes normal to the direction of propagation. We have already encountered spherical waves in Chapter 2 (see Eq. 2.67) and we shall meet them again in Chapter 4 when we treat radiation. Finally, we shall encounter waves that are neither transverse nor uniform when we treat guided waves in Chapter 5.

The electric and magnetic fields are inseparable components of the electromagnetic wave and on equal footing in importance in determining the propagation. However, when it comes to detecting the

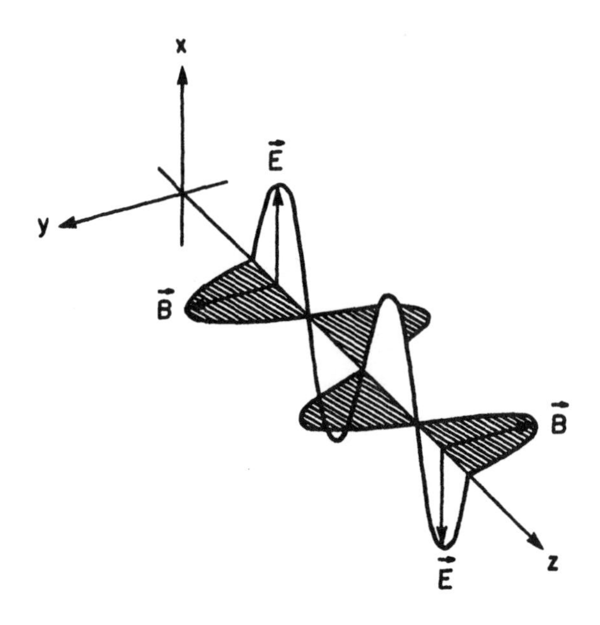

Fig. 3.9 Sketch of a plane sinusoidal traveling electromagnetic wave showing the instantaneous directions of the electric and magnetic field vectors. $\vec{E}$ and $\vec{B}$ are everywhere and at every instant in phase with one another.

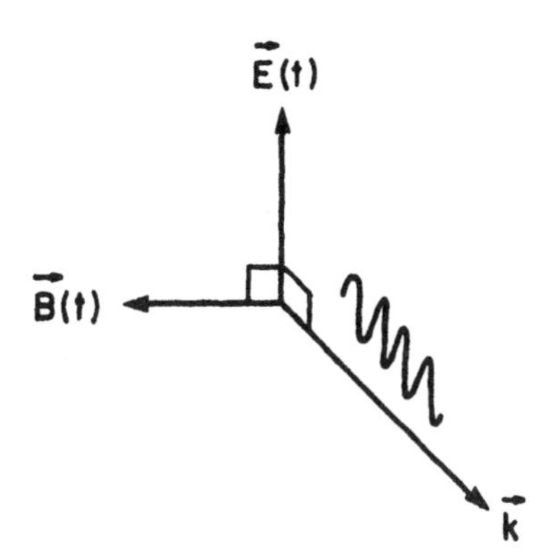

Fig. 3.10 The field vectors $\vec{E}$ and $\vec{B}$ and the propagation vector $\vec{k}$ of a plane wave in free space are at all times mutually orthogonal to one another.

presence of the wave, the electric vector is usually the more important of the two. The reason is that most detectors are sensitive to $\vec{E}$ rather than $\vec{B}$. Such is the case with photographic film and the eye; radio and microwave frequency antennas are often (but not always) designed to pick up the electric rather than the magnetic field of the wave. For this reason we shall generally stress the electric component to the exclusion of its magnetic counterpart. However, the magnetic field can always be obtained from the electric field by means of the relation

$$|\vec{B}| = \frac{|\vec{E}|}{v} \tag{3.82}$$

where v is the phase speed of the wave. Thus, for example, if the electric field at your radio antenna is 3×10^{-3} volts/m, then the associated magnetic field is $(3\times10^{-3})/(3\times10^{8})$ webers per m^2 or 10^{-7} gauss. On the other hand, when light from a high-power laser is focused with a lens, an electric field of 10^9 volts/m is fairly readily achieved at the focal spot. The corresponding oscillating magnetic field of the wave is 33,300 gauss!

Example

Section 3.3 treats the special case of a plane electromagnetic wave in charge-free space traveling along the z axis with electric field oriented along the x axis. We wish to generalize these results for a plane electromagnetic wave traveling in the direction $\vec{k}$ with an electric field specified by the constant vector $\vec{E}_o$, i.e., $\vec{E}_o$ is not a function of time or spatial coordinates. Specifically, take

$$\vec{E}(\vec{r}, t) = \vec{E}_o \, e^{j(\omega t - \vec{k}\cdot\vec{r})}$$

$$\vec{B}(\vec{r}, t) = \vec{B}_o \, e^{j(\omega t - \vec{k}\cdot\vec{r})}$$

and show that $\vec{E}_o$, $\vec{k}$, and $\vec{B}_o$ must be mutually orthogonal, and obtain the relation between $\vec{E}_o$ and $\vec{B}_o$.

Solution

If our expression for $\vec{E}(\vec{r}, t)$ is substituted in the wave equation, Eq. 3.65, and Eq. 3.67 is used, straightforward differentiation yields

$$\nabla^2 \vec{E}(\vec{r}, t) = -k^2 \vec{E}(\vec{r}, t)$$

and

$$\mu_o \epsilon_o \frac{\partial^2 \vec{E}(\vec{r}, t)}{\partial t^2} = -\frac{\omega^2}{v^2} \vec{E}(\vec{r}, t).$$

Thus

$$k^2 = k_x^2 + k_y^2 + k_z^2 = \frac{\omega^2}{v^2} = \left(\frac{2\pi}{\lambda}\right)^2,$$

a result obtained earlier in Eq. 2.56. In a vacuum, $v = c$ and $k = |\vec{k}| = \omega/c = 2\pi/\lambda$.

It follows, again by differentiation, from Eq. 3.62a and 3.62d that

$$\vec{k} \cdot \vec{E}_o = 0 \quad \text{and} \quad \vec{k} \cdot \vec{B}_o = 0.$$

Thus $\vec{k}$ is perpendicular to both $\vec{E}_o$ and $\vec{B}_o$. It remains to establish the vector relationship between $\vec{E}_o$ and $\vec{B}_o$. To do this, the remaining two Maxwell's equations (3.62b) and (3.62c) must be used. Substitution of our expressions for $\vec{E}_o$ and $\vec{B}_o$ in these equations gives

$$\vec{k} \times \vec{E}_o = +\omega \vec{B}_o$$

$$\vec{k} \times \vec{B}_o = \frac{-\omega}{v^2} \vec{E}_o.$$

Thus $\vec{B}_o$ is perpendicular to the plane of $\vec{k}$ and $\vec{E}_o$, but since $\vec{k}$ has been shown to be perpendicular to $\vec{E}_o$ it follows that $\vec{k}$, $\vec{E}_o$, and $\vec{B}_o$ are mutually orthogonal. Also, since $k = \omega/v$ we have

$$E_o = |\vec{E}_o| = vB_o = v|\vec{B}_o|.$$

Remember that the results of this example apply to transverse uniform plane waves. They do not apply in many cases to guided waves (Chapter 5); that is, to waves confined to limited regions of space.

3.4 POLARIZATION

We have already established that an electromagnetic wave in free space undergoes transverse oscillations, which means that the electric (and magnetic) vectors lie in a plane perpendicular to the propagation direction $\vec{k}$. The precise direction of the electric vector in this plane is said to determine the state of polarization of the wave. Indeed, one may have two waves of the same frequency and with the same propagation direction, but with their electric vectors at right angles to one another. Since the two waves have orthogonal field components, they are independent of one another and therefore their amplitudes and phases can be different. Exactly what form the resultant vibrations will take is a basic part of the study of polarization.

Two possible solutions have already been discussed. The one we have discussed in detail is the sinusoidal plane wave (3.81)

$$E_x = E_{ox} \cos(\omega t - kz + \phi_x) \tag{3.83}$$

which is a solution of the set of equations (3.73). The other solution

$$E_y = E_{oy} \cos(\omega t - kz + \phi_y) \tag{3.84}$$

comes from the analogous set (3.74). The electric vector of the combined waves is therefore

$$\vec{E} = \hat{x}E_{ox} \cos(\omega t - kz + \phi_x) + \hat{y}E_{oy} \cos(\omega t - kz + \phi_y) \tag{3.85}$$

where ϕ_x and ϕ_y are the phases of the two waves. A convenient way of picturing the state of polarization of the resultant is to choose a particular transverse plane x-y lying perpendicular to the propagation direction z, and watch the path traced by the tip of the electric vector as time progresses. To this purpose we set $z = 0$ and with little loss of generality obtain

$$\vec{E} = \hat{x}E_{ox} \cos(\omega t + \phi_x) + \hat{y}E_{oy} \cos(\omega t + \phi_y). \tag{3.86}$$

But, instead of plotting the vector $\vec{E}$ with its given components it is sufficient and simpler to give the location of its end point, that is, the coordinate values of $E_x(t)$ and $E_y(t)$. By eliminating the time t one can then construct the curve of E_x versus E_y. This is known as the vibration curve.

When E_{oy} is zero, the vibration curve is a horizontal straight line ending at $E_x = -E_{ox}$ and $E_x = +E_{ox}$. When E_{ox} is zero, one obtains a vertical straight line of length $2E_{oy}$. When $\phi_x = \phi_y$, $E_x/E_y = E_{ox}/E_{oy}$ and this leads to a straight line with the slope equal to $\tan \alpha = E_{oy}/E_{ox}$ and of length $2\left[E_{ox}^2 + E_{oy}^2\right]^{1/2}$. The only other case of a straight line

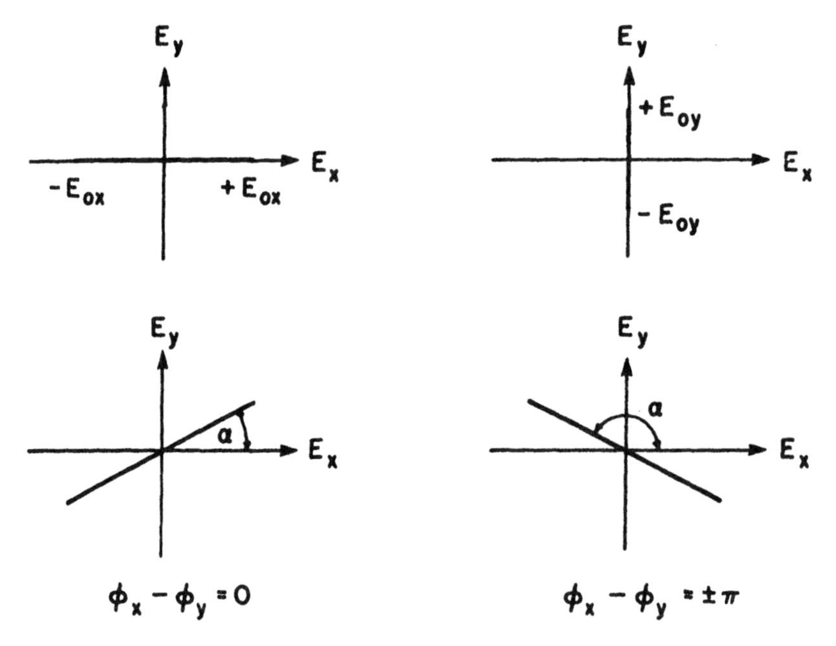

Fig. 3.11 Vibration curves of the different states of linear polarization of a plane electromagnetic wave. The heavy line is the trajectory of the tip of the electric vector as time progresses, viewed at a fixed position (say z = 0) along the propagation direction of the wave. The plane containing E_x, E_y is perpendicular to the propagation vector $\vec{k}$ which points out of the page along the +z axis.

is obtained when $\phi_y = \phi_x \pm \pi$. In this situation $E_x/E_y = -E_{ox}/E_{oy}$ and $\tan \alpha = -E_{oy}/E_{ox}$. All these cases are illustrated in Fig. 3.11. In these states, the wave is said to be plane or linearly polarized.

When $\phi_x - \phi_y = \pm(\pi/2)$, then $E_x = E_{ox} \cos(\omega t + \phi_x)$, $E_y = \pm E_{oy} \sin(\omega t + \phi_x)$, and elimination of t gives the equation of an ellipse

$$\left(\frac{E_x}{E_{ox}}\right)^2 + \left(\frac{E_y}{E_{oy}}\right)^2 = 1. \tag{3.87}$$

In the special case of equal amplitudes, $E_{ox} = E_{oy}$, the vibration curve is a circle, and the wave is said to be circularly polarized. Some of these cases are illustrated in Fig. 3.12.

In all other situations where the phase difference $\phi_x - \phi_y$ is any arbitrary constant, as is the ratio of amplitudes E_{ox}/E_{oy}, the vibration curve is once again an ellipse but with its major axis inclined to the chosen coordinate system. Now the appropriate equation replacing Eq. (3.87), from which the vibration curve is constructed, is

$$\left(\frac{E_x}{E_{ox}}\right)^2 + \left(\frac{E_y}{E_{oy}}\right)^2 - 2\left(\frac{E_x}{E_{ox}}\right)\left(\frac{E_y}{E_{oy}}\right)\cos\epsilon = \sin^2\epsilon \tag{3.88}$$

where $\epsilon \equiv (\phi_x - \phi_y)$. Figure 3.13 shows a series of vibration curves for a fairly general case of polarization.

If $\phi_x - \phi_y = +\pi/2$, then $E_x = E_{ox}$ and $E_y = 0$ when $(\omega t + \phi_x) = 0$. At a slightly later time, E_x becomes less than E_{ox} but E_y increases; when $(\omega t + \phi_x) = \pi/2$, $E_x = 0$ and E_y reaches its maximum value of E_{oy}. This signifies that the tip of the electric vector moves counterclockwise. On the other hand, when $\phi_x - \phi_y = -\pi/2$, the rotation is clockwise. In modern usage the wave is said to be right-hand polarized when as it propagates toward the observer the tip of the electric vector rotates counterclockwise, and vice versa for a left-hand polarized wave. (One way to remember this is to imagine a magnetic field applied along the direction of propagation $\vec{k}$ of the wave. The right-hand polarized wave corresponds to the case in which the tip of the electric vector rotates in the direction of an electron orbiting in the magnetic field.) Note, however, that in many books on optics precisely

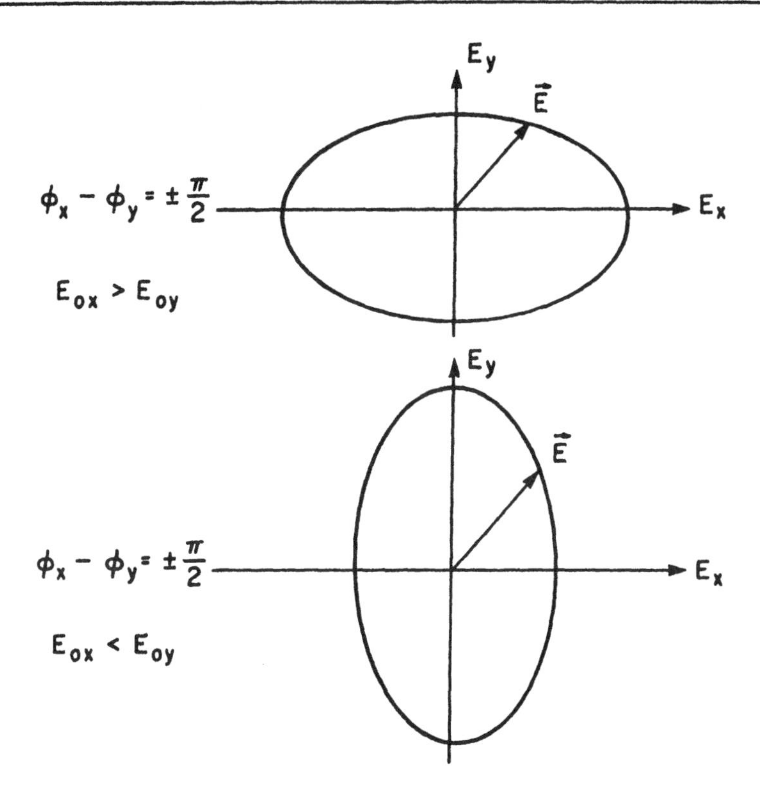

Fig. 3.12 Horizontal and vertical states of elliptic polarization of a plane electromagnetic wave propagating out of the page. When an observer, looking toward the source, sees the tip of $\vec{E}$ rotating counterclockwise ($\phi_x - \phi_y = +\pi/2$), the wave is said to be right-hand polarized. This is the "natural" definition that uses the concept of the right-hand screw; it is opposite to the traditional definition.

the opposite definitions are used. Figure 3.14 illustrates the motion of the $\vec{E}$ vector for several different states of polarization.

To see what happens as a function of space, we can freeze time and observe the wave along its propagation direction z. We take the special case of a vertically, right-elliptically polarized wave, that is, $\phi_x - \phi_y = +\pi/2$ and $E_{ox} < E_{oy}$. We let the time of observation be such that $(\omega t + \phi_x) = \pi$. Then Eq. 3.85 yields

$$\vec{E} = -\hat{x}E_{ox}\cos(kz) + \hat{y}E_{oy}\sin(kz). \tag{3.89}$$

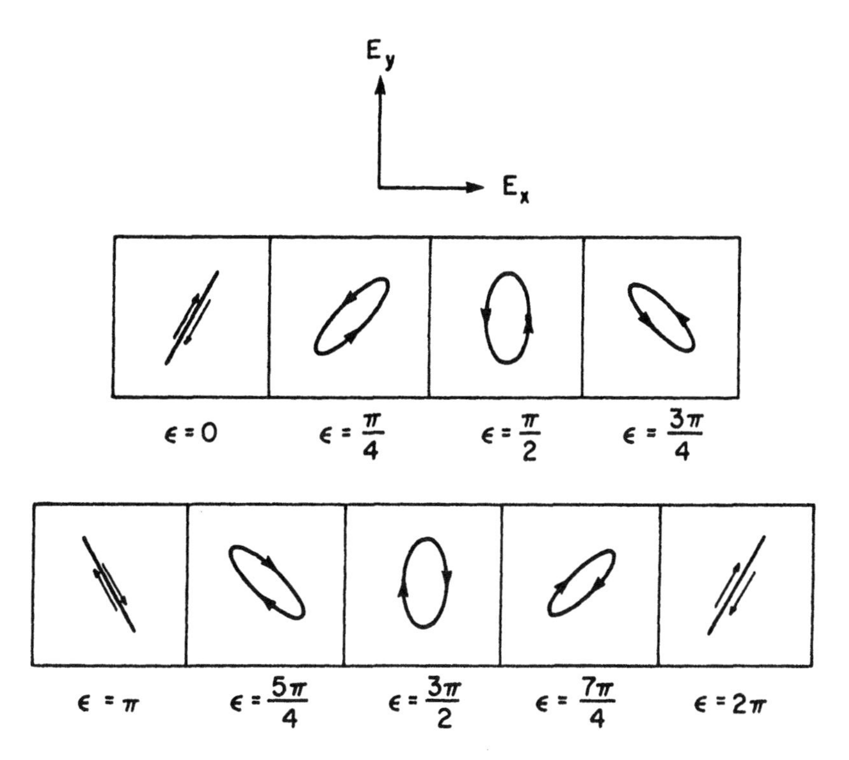

Fig. 3.13 Various states of polarization prescribed by the phase difference $\epsilon = \phi_x - \phi_y$. The amplitude ratio E_{ox}/E_{oy} is constant throughout. The wave propagates out of the page. Note the change in the sense of the polarization for $\epsilon < \pi$ and $\epsilon > \pi$.

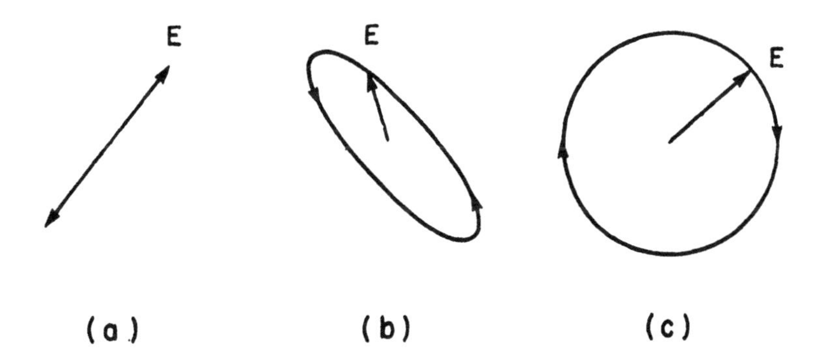

Fig. 3.14 Motion of the tip of the electric vector for (a) a linearly polarized wave, (b) a right-hand elliptically polarized wave, and (c) a left-hand circularly polarized wave. The propagation vector $\vec{k}$ is taken to point out of the page.

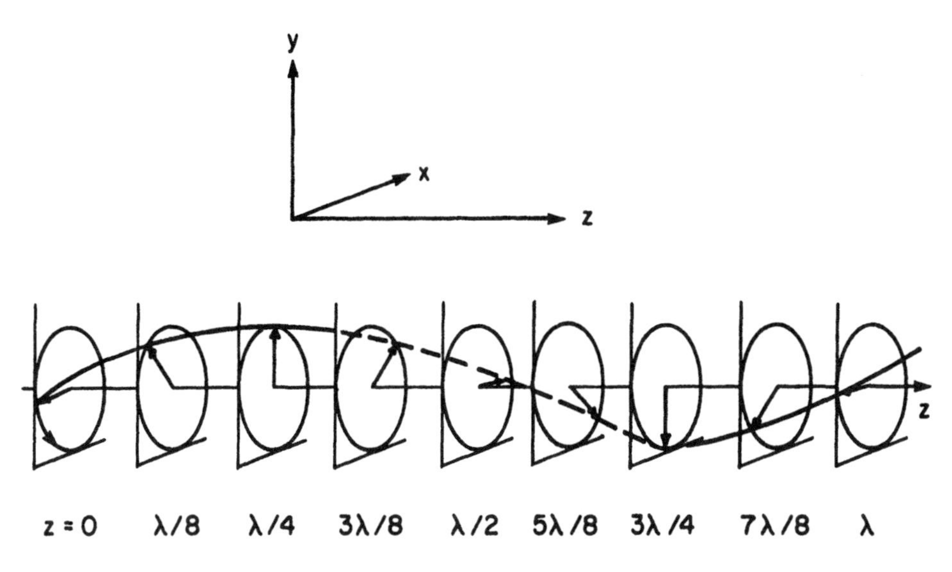

Fig. 3.15 The spatial vibration curve showing the helix of a right-hand circularly polarized wave. The arrows give the directions of the electric vector.

This equation allows us to plot a "spatial" vibration curve. As shown in Fig. 3.15, it is a helix on a cylinder of elliptical cross section. The pitch of the helix is the wavelength. At a time, say, Δt later than that considered in Eq. 3.89, the helix is the same except that it is displaced along +z by a distance $c\Delta t$. Note that as time progresses, the helix moves with a velocity (ω/k) without twisting or rotating. For the special case of a circular wave $(E_{ox} = E_{oy})$ the helix is in the shape of the groove on a screw. For linearly polarized waves (say, $E_{oy} = 0$) it is a cosine curve.

One often hears it said that light from the Sun or from a lamp is unpolarized. This statement is very imprecise and needs to be qualified. There is no solution of Maxwell's equations corresponding to some state of "unpolarized" light. Every atom or oscillating charge produces a wave train of a precisely specified state of polarization. At any given instant of time the emissions from all the individual atoms radiating at a given frequency ω can be summed to form a single resultant electric vector which has a definite direction and thus is in a definite state of polarization. Where then is the problem? It lies in the words "at any given instant of time." An atom will normally radiate a coherent wave train of definite phase for time τ of the order of 10^{-8} seconds; over longer periods than this its radiation is constantly being interrupted by collisions with other atoms and a variety of other effects. Hence the emitting atoms cannot maintain a definite phase for times much longer than the "coherence time" τ and the phases of the successive waves bear no fixed relation to the phase of the earlier waves. Therefore, if measurements of the polarization are made with instruments that average over times much greater than τ, no single state of polarization will be discernible. In this sense then the light from a flash lamp is unpolarized, or as it is sometime called, randomly polarized.

In general, however, electromagnetic radiations whether they are natural or artificially generated are neither completely unpolarized nor are they absolutely perfectly polarized, but the state of polarization lies somewhere between. The production and detection of polarized radiation are subjects that will be treated in later chapters.

Example

Describe the polarization, with a sketch, of the following electric fields and derive the associated magnetic fields:

a) $$\vec{E} = E_o\{-\hat{y} + \hat{z}\}\cos(\omega t - kx)$$

b) $$\vec{E} = E_o\left\{\hat{y}\cos(\omega t - kx) + \hat{z}\cos\left(\omega t - kx - \frac{\pi}{6}\right)\right\}$$

Solution

The temporal behavior of the electric vector can be seen by plotting the locus of the tip of the vector at any constant value of x, say, x = 0. Thus we have

$$\vec{E} = E_o(-\hat{y} + \hat{z})\cos\omega t.$$

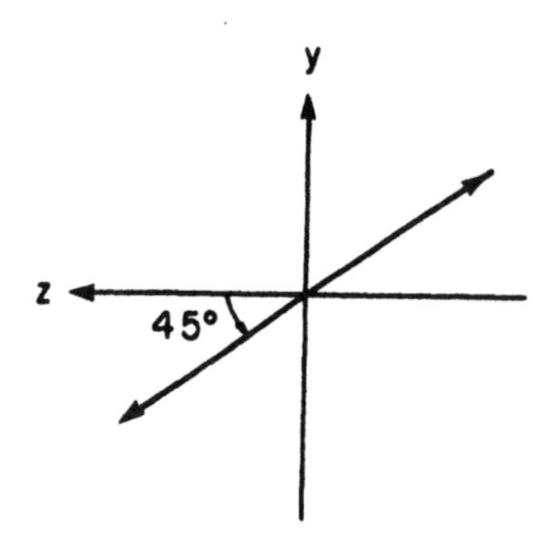

Wave is linearly polarized at an angle of 45° to the z axis and 135° to the y axis.

The associated magnetic field can be found from Eq. 3.62b or from the relations in the Example that concludes section 3.3. Since

$$\vec{k} = k\hat{x} = (\omega/v)\,\hat{x}$$

$$\vec{B} = \frac{1}{\omega} k \times \vec{E} = \frac{E_o}{v} \hat{x} \times (-\hat{y} + \hat{z}) \cos(\omega t - kx)$$

a) $$\vec{B} = \frac{-E_o}{v} (\hat{y} + \hat{z}) \cos(\omega t - kx).$$

Turning now to the second expression and setting x = 0:

$$\vec{E} = E_o \left\{ \hat{y} \cos \omega t + \hat{z} \cos\left(\omega t - \frac{\pi}{6}\right) \right\}.$$

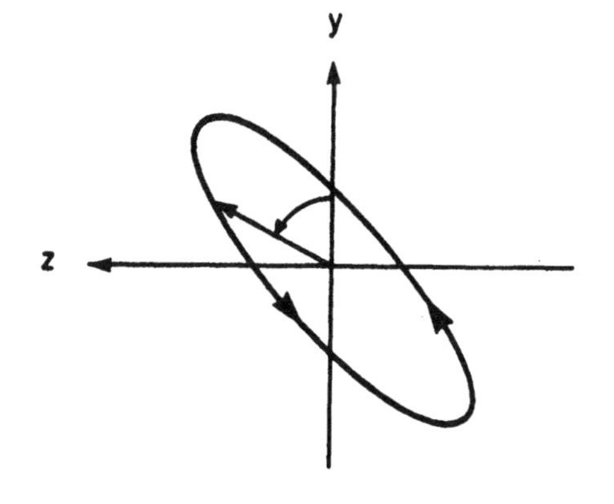

Wave is right-hand elliptically polarized.

This time we use Eq. 3.62b to find the magnetic field:

$$\nabla \times \vec{E} = -\frac{\partial \vec{B}}{\partial t}$$

which gives

$$\frac{\partial B_y}{\partial t} = \frac{\partial E_z}{\partial x} = kE_o \sin\left(\omega t - kx - \frac{\pi}{6}\right)$$

$$-\frac{\partial B_z}{\partial t} = \frac{\partial E_y}{\partial x} = kE_o \sin(\omega t - kx).$$

These can now be integrated to give

b) $$\vec{B} = \frac{E_o}{v}\left\{-\hat{y}\cos\left(\omega t - kx - \frac{\pi}{6}\right) + \hat{z}\cos(\omega t - kx)\right\}.$$

3.5 THE STANDING ELECTROMAGNETIC WAVE

Just as in the case of mechanical waves, Eq. 3.75a has, in addition to traveling wave solutions, solutions of the form

$$E_x = E_{ox} \sin(kz) \sin(\omega t) \tag{3.90}$$

which represent a standing oscillation of the type discussed in section 2.3 (recall that combinations like [sin(kz) cos(ωt)], [cos(kz) cos(ωt)], and [cos(kz) sin(ωt)] and sums thereof are also permissible but for the sake of brevity we just stay with the form (3.90)).

With the electric field E_x given, the magnetic field is fully determined by the pair of equations (3.73). Inserting Eq. 3.90 in Eq. 3.73a and performing the differentiation with respect to z yields

$$\frac{\partial B_y}{\partial t} = -kE_{ox} \cos(kz) \sin(\omega t), \tag{3.91}$$

a result which when integrated with respect to time gives

$$B_y = \frac{E_{ox}}{c} \cos(kz) \cos(\omega t). \tag{3.92}$$

Equations 3.90 and 3.92 tell us that, just as in the traveling wave, $\vec{E}$ and $\vec{B}$ are always at right angles to one another and that, just as in the traveling wave, the amplitude of B equals E_{ox}/c. However, unlike the traveling wave, the standing wave has its electric and magnetic fields out of phase with one another. That is, when E_x is proportional

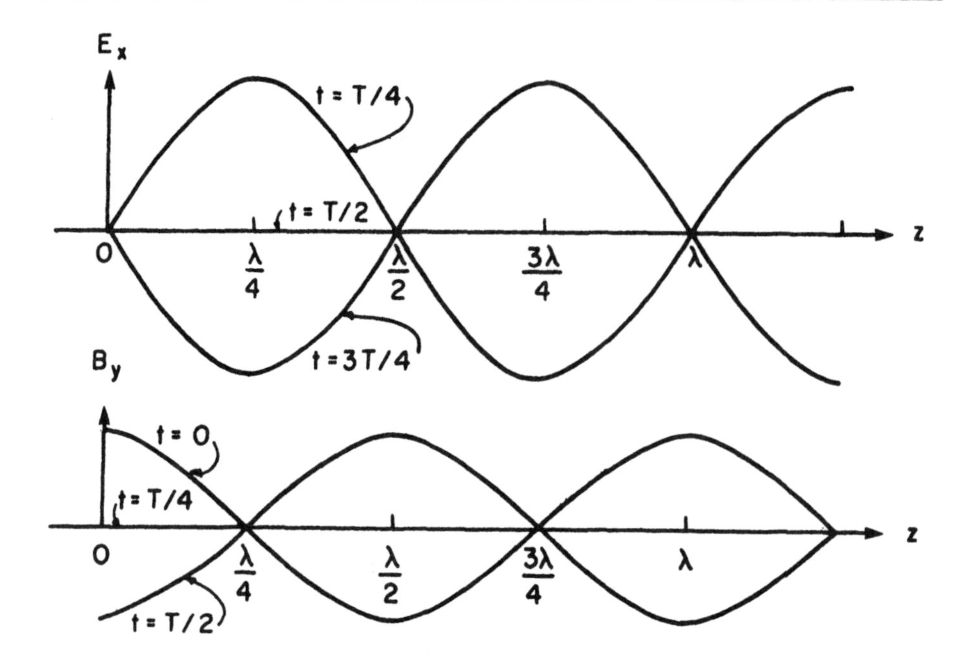

Fig. 3.16 Plane linearly polarized standing electromagnetic wave. E_x and B_y oscillate out of phase with one another, both in time and in space. The nodes of E_x and the nodes of B_y are displaced relative to one another by $\lambda/4$.

to $\sin(\omega t)$, B_y is proportional to $\cos(\omega t)$ and when E_x is proportional to $\sin(kz)$, B_y is proportional to $\cos(kz)$. Therefore the two field components are out of phase both in time and in space. This is illustrated in Fig. 3.16 which shows that the electric and magnetic nodes are displaced relative to one another by a quarter of a wavelength.

At the time $t = 0$, $E_x = 0$ everywhere. Hence, at this time the electrical energy density is zero and there is no storage of electrical energy. However, the magnetic field has its largest value, B_{oy}. At the antinodes, the average amount of magnetic energy stored is then $\langle U_M \rangle = B_{oy}^2/4\mu_o$ joules/m^3. At a quarter period later, the magnetic field has vanished and the stored energy is all electrical. The amount stored at the electrical antinodes is $\langle U_E \rangle = (1/4)\,\epsilon_o E_{ox}^2$ joules/m^3. But, since $E_{ox} = cB_{oy}$ and $\sqrt{\epsilon_o \mu_o} = 1/c$ it follows that $\langle U_E \rangle = \langle U_M \rangle$. Therefore, the energy stored is always the same, it only shifts its location from magnetic to electrical and back again. A similar situation was discovered in section 1.2 in connection with the LC circuit.

3.6 THE POYNTING'S VECTOR AND THE FLOW OF ENERGY

As electromagnetic waves propagate through space from their source to the distant receiving points, there is a transfer of energy from the source to the receiver. We shall show that there exists a simple relationship between the rate of energy flow and the amplitudes of the electric and magnetic fields. To discuss this relationship we must write an equation that expresses quantitatively the conservation of energy for electromagnetism, much as we wrote in section 2.2 conservation equations for mechanical waves. The form taken by all such relations is a statement that the rate at which energy changes in some volume V is made up by the flux flowing out of the surface which encloses V. A typical equation is relation (3.36) for charge conservation. By analogy we can write a similar expression for energy conservation

$$\frac{d}{dt}\int_V U\,dV = -\int_{\substack{\text{closed}\\ \text{surface}}} \vec{S}\cdot d\vec{a} \tag{3.93}$$

either in integral form as above, or in differential form (cf. Eq. 3.37) as

$$\frac{\partial U}{\partial t} = -\nabla\cdot\vec{S}. \tag{3.94}$$

The expression for the energy density U is known to us already. If $\vec{E}$ is the instantaneous value of the electric field of the wave, then the energy density stored in the electric field is

$$U_E = \frac{1}{2}\epsilon_o E^2 \quad \text{joules/m}^3 \tag{3.95}$$

where $E^2 \equiv \vec{E}\cdot\vec{E}$ denotes the magnitude squared of the field vector. Similarly, the energy stored in the magnetic field is

$$U_M = \frac{1}{2\mu_o} B^2 \quad \text{joules/m}^3 \tag{3.96}$$

so that the total energy density U in the wave becomes

$$\boxed{U = \frac{1}{2}\epsilon_o E^2 + \frac{1}{2\mu_o} B^2} \qquad (3.97)$$

We now understand quantitatively the left-hand side of Eqs. 3.93 and 3.94; our task is to determine the form of the radiation flux $\vec{S}$ on the right-hand side, consistent with the result (3.97). The problem is mainly a mathematical one, and to help us on our way we make the simplifying assumption that the wave in question is a plane linearly polarized wave propagating along the z axis and having components E_x and B_y. Therefore

$$U = \frac{1}{2}\epsilon_o E_x^2 + \frac{1}{2\mu_o} B_y^2. \qquad (3.98)$$

We construct a cubical volume element dxdydz as illustrated in Fig. 3.17 with faces parallel to our chosen coordinate system and consider the energy flow through this box. Since the wave travels along z, the only nonzero component of the flux $\vec{S}$ is the z component. From

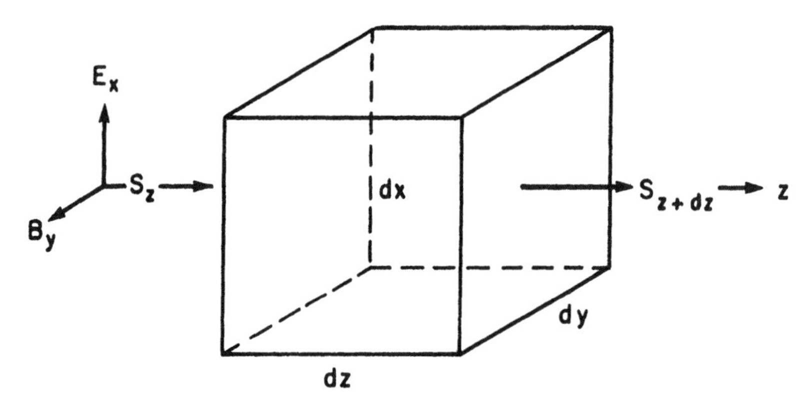

Fig. 3.17 The energy conservation equation (3.93) applied to the case of a plane linearly polarized wave passing through the element of volume dxdydz. If U is the energy density, then for this element (3.93) says that

$$\frac{dU}{dt}\,dxdydz = -[S_{z+dz} - S_z]\,dxdy \qquad \text{or} \qquad \frac{\partial U}{\partial t} = -\frac{\partial S_z}{\partial z}$$

which is an alternate way of arriving at Eq. 3.99.

Eqs. 3.94 and 3.98 it then follows that

$$
\begin{aligned}
-\frac{\partial S_z}{\partial z} &= \frac{\partial}{\partial t}\left[\frac{1}{2}\epsilon_o E_x^2 + \frac{1}{2\mu_o} B_y^2\right] \\
&= \epsilon_o E_x \frac{\partial E_x}{\partial t} + \frac{1}{\mu_o} B_y \frac{\partial B_y}{\partial t}.
\end{aligned}
\tag{3.99}
$$

But Eqs. 3.73 permit us to replace $(\partial E_x/\partial t)$ by $(-1/\mu_o\epsilon_o)[\partial B_y/\partial z]$ and $(\partial B_y/\partial t)$ by $(-\partial E_x/\partial z)$, with the result that

$$
\begin{aligned}
-\frac{\partial S_z}{\partial z} &= -\left[\frac{1}{\mu_o} E_x \frac{\partial B_y}{\partial z} + \frac{1}{\mu_o} B_y \frac{\partial E_x}{\partial z}\right] \\
&= -\frac{1}{\mu_o}\frac{\partial}{\partial z}(E_x B_y)
\end{aligned}
\tag{3.100}
$$

from which it follows that

$$S_z = \frac{1}{\mu_o} E_x B_y. \tag{3.101}$$

In a similar manner, and for the case of a wave propagating along z but with $\vec{E}$ along +y and $\vec{B}$ along −x, we find that

$$S_z = -\frac{1}{\mu_o} E_y B_x \tag{3.102}$$

so that the net flux of a wave having electric field components along both x and y is the sum of Eqs. 3.101 and 3.102:

$$S_z = \frac{1}{\mu_o}[E_x B_y - E_y B_x] \quad \text{watt/m}^2. \tag{3.103}$$

The results we have just presented are special cases of a more general theorem known as Poynting's theorem, which gives the flux per unit area of any electromagnetic disturbance:

$$\boxed{\vec{S} = \frac{1}{\mu_o}(\vec{E} \times \vec{B}) \ \text{watt/m}^2.} \tag{3.104}$$

Note that the direction of $\vec{S}$ is given by the right-hand rule and is seen to be collinear with the wave's propagation vector $\vec{k}$ (we mention in passing that in certain cases of wave propagation in crystalline media and in anisotropic plasmas, $\vec{S}$ and $\vec{k}$ need not be collinear).

Examples of energy flow

For the case of a plane linearly polarized electromagnetic wave, substitution of Eqs. 3.81 in Eq. 3.101 or 3.104 gives

$$S_z = \epsilon_o c E_{ox}^2 \cos^2(\omega t - kz + \phi) \tag{3.105}$$

as the value for the instantaneous flux. From this and Fig. 3.18 we see that S_z varies between zero and $\epsilon_o c E_{ox}^2$ and that it oscillates periodically about a finite average value with a frequency of 2ω. The more interesting quantity is the time-average flux usually referred to as the wave intensity (or irradiance; see section 2.2). Its magnitude is

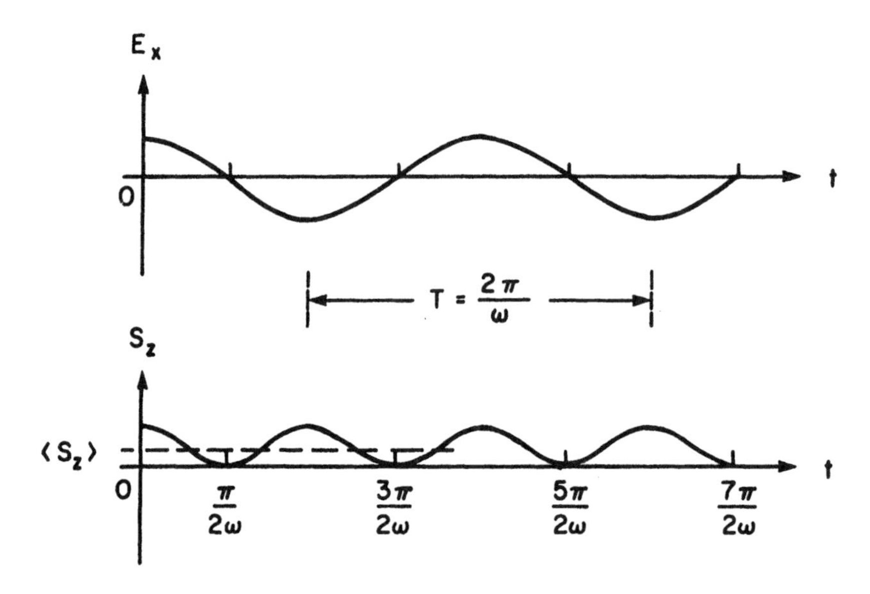

Fig. 3.18 The electric vector E_x and the associated Poynting flux S_z plotted as a function of time. The plane linearly polarized wave travels along the z axis. The time-average value $\langle S_z \rangle$ is shown as a dashed line. Note that $\langle E_x \rangle = 0$.

$$\langle S_z \rangle = \frac{1}{2}\epsilon_o c E_{ox}^2 \ \text{watt/m}^2 \tag{3.106}$$

showing, just as in the case of mechanical waves, that the flow of energy is proportional to the square of the amplitude.

The instantaneous energy density in the wave is in accordance with Eq. 3.98 given by the expression

$$\begin{aligned} U &= \frac{1}{2}\epsilon_o E_{ox}^2 \cos^2(\omega t - kz + \phi) + \frac{1}{2\mu_o} B_{oy}^2 \cos^2(\omega t - kz + \phi) \\ &= \epsilon_o E_{ox}^2 \cos^2(\omega t - kz + \phi) \end{aligned} \tag{3.107}$$

where in obtaining the second result we made use of the relationships $B_y = E_x/c$, $c = (\mu_o\epsilon_o)^{-1/2}$. Therefore, the time-average energy density is

$$\langle U \rangle = \frac{1}{2}\epsilon_o E_{ox}^2 \ \text{joules/m}^3. \tag{3.108}$$

Comparison of this equation with the flux formula (3.106) leads us to the familiar result

$$\langle S_z \rangle = c\langle U \rangle \tag{3.109}$$

proved earlier for the special case of mechanical waves (cf. Eq. 2.31). This pleasant result gives us greater physical understanding of the somewhat unfamiliar Poynting flux.

Consider, for example, a high-power pulsed CO_2 laser emitting 100 joules of energy at a wavelength of 10^{-5} meters. The pulse duration τ is 10 nanoseconds and the laser beam has a circular cross section of radius r = 1 cm. The average energy density $\langle U \rangle$ in the beam is therefore $100/\pi r^2 c\tau$ or 1.1×10^5 J/m^3. The time-average flux $\langle S \rangle = c\langle U \rangle$ has the value of 3.3×10^{13} watt/m^2. The electric field amplitude calculated from Eq. 3.106 and the known value of $\langle S \rangle$ equals 1.6×10^8 V/m and the magnetic field amplitude $B_{oy} = E_{ox}/c$ equals 5300 gauss. These are very large numbers, but (fortunately?) everything is over in a very short time. Compared to this, how does our Sun fare as it pours its rays continously upon the Earth's surface? The "solar constant" specifies the flux of radiation flowing toward the

Earth. At the top of the Earth's atmosphere it has a value of 1.94 calories per cm^2 per minute which is equivalent to 1352 watts per square meter. Use of Eq. 3.106 then gives the electric field amplitude as 1009 V/m.

The energy conservation equation (3.94) when written out in full is

$$\frac{\partial}{\partial t}\left[\frac{1}{2}\epsilon_o\vec{E}\cdot\vec{E}+\frac{1}{2\mu_o}\vec{B}\cdot\vec{B}\right]+\nabla\cdot\left(\frac{1}{\mu_o}\vec{E}\times\vec{B}\right)=0. \qquad (3.110)$$

It is a correct statement in regions of space free of charges. In the presence of matter the equation cannot really be correct because we know that electromagnetic fields do work on charges and this work must represent a drain of energy as far as Eq. 3.110 is concerned. Can we account for this energy loss? The force on a charge q is given by the Lorentz formula (3.1) and if v is the velocity of the charge, the rate of doing work dW/dt (that is the power) is $\vec{F}\cdot\vec{v}$, or

$$\begin{aligned}\frac{dW}{dt}&=q[\vec{E}+\vec{v}\times\vec{B}]\cdot\vec{v}\\&=q\vec{v}\cdot\vec{E}\end{aligned} \qquad (3.111)$$

where the contribution from the magnetic field drops out as a result of the vector identity $\vec{A}\cdot(\vec{A}\times\vec{B})=\vec{B}\cdot(\vec{A}\times\vec{A})=0$. If we have N charges per unit volume, the work done per second per unit volume equals $(Nq\vec{v})\cdot\vec{E}$. But, the object $Nq\vec{v}$ is the current density $\vec{J}$, and thus the scalar product $\vec{J}\cdot\vec{E}$ is the power per unit volume expended by the fields. As a result, Eq. 3.110 must be modified to read,

$$\frac{\partial}{\partial t}\left[\frac{1}{2}\epsilon_o\vec{E}\cdot\vec{E}+\frac{1}{2\mu_o}\vec{B}\cdot\vec{B}\right]+\nabla\cdot\left(\frac{1}{\mu_o}\vec{E}\times\vec{B}\right)=-\vec{J}\cdot\vec{E} \qquad (3.112)$$

where the minus sign signifies a leakage of energy out of the electromagnetic field. The energy appears as heat and represents joule heating of the system.

Funny things can happen when we play with Poynting's vector

Consider a straight piece of conductor of length L having resistance and carrying a steady dc current I. Because there is a potential drop V along the wire there must be an electric field $E = V/L$ just outside it and parallel to its surface. There is, in addition, an azimuthal magnetic field caused by the current. The magnetic field goes around the wire in circles and its value, found from Ampère's law, is $B_\phi = \mu_0 I/2\pi r$ where r is the distance of the field point measured from the center of the wire. The Poynting flux $(\vec{E}\times\vec{B})/\mu_0$ is therefore directed radially into the wire. Its magnitude at the surface of the wire, $r = a$, is

$$\left.\begin{aligned} |\vec{S}_r| &= \frac{1}{\mu_0}|\vec{E}\times\vec{B}| \\ &= \frac{1}{\mu_0}\frac{V}{L}\frac{\mu_0 I}{2\pi a} \\ &= \frac{VI}{2\pi a L}\ \text{watt/m}^2. \end{aligned}\right\} \tag{3.113}$$

From this we see that the total power flowing into the wire is VI which is just the expected joule rate of energy loss due to the ohmic resistance in the wire. And although the magnitude of the energy loss is precisely correct, the derivation implies a very strange happening: the energy of the moving electrons (eventually dissipated in heat) apparently does not flow down the wire as intuition would tell us, but it enters the wire radially through its surface! This is just one of many similar examples encountered when one plays with Poynting's theorem. It suggests that although it is possible to obtain the correct result for the magnitude of the energy flow (indeed no discrepancies with experiment are known to exist) it is often not possible to say just where the energy is.

We wish to make an additional remark. Whereas the total power flow through any closed surface (obtained by integrating $(\vec{E}\times\vec{B})/\mu_0$ over the surface) always, to the best of our knowledge, gives the correct and unique answer, it does not necessarily follow that the flux $(\vec{E}\times\vec{B})/\mu_0$ correctly represents the power flow at each point. This is because we can add to the Poynting vector any other vector having zero divergence without changing the total energy flowing across the closed surface. To show this, let us manufacture a new formula for the flux at a point S(new) having the form

$$\vec{S}(\text{new}) = \left(\frac{1}{\mu_o}\vec{E}\times\vec{B}\right) + (\nabla\times\vec{G}) \text{ watt/m}^2 \tag{3.114}$$

where $\vec{G}$ is some arbitrary vector. Then the net power flow through a closed surface is

$$P = \int_S \frac{1}{\mu_o}(\vec{E}\times\vec{B}) \cdot d\vec{a} + \int_S (\nabla\times\vec{G}) \cdot d\vec{a}. \tag{3.115}$$

But, there is the purely mathematical relation (3.13) which allows us to transform the second surface integral into an integral over the volume V of which S is the surface. The result is that

$$P = \int_S \frac{1}{\mu_o}(\vec{E}\times\vec{B}) \cdot d\vec{a} + \int_V \nabla \cdot (\nabla\times\vec{G})\, dV. \tag{3.116}$$

Since the divergence of the curl of any vector is identically zero (see Eq. 3.57) it follows that

$$P = \int_S \frac{1}{\mu_o}(\vec{E}\times\vec{B}) \cdot d\vec{a} \tag{3.117}$$

which is the customary relation for the outflow of electromagnetic energy and is independent of the presence or absence of the artificially introduced vector $\vec{G}$.

We see that our interpretation of $(\vec{E}\times\vec{B})/\mu_o$ as being the electromagnetic flux crossing an elementary area is not necessarily the unique answer to a given problem. From time to time various alternative forms to Poynting's theorem have been published (see, for example, Mason and Weaver, The Electromagnetic Field (University of Chicago Press, Chicago, 1929)). The various theories are quite complex and add little to our understanding of the problem. For that reason, the simple, straightforward Poynting's theorem is generally accepted as the best available statement of energy flow.

Example

The magnetic field of a uniform plane wave in vacuum is given by

$$\vec{B}(\vec{r}, t) = 10^{-6} [\hat{x} + 2\hat{y} + B_z\hat{z}] \cos(\omega t + 3x - y - z)$$

in mks units. Determine the following:

a) The direction of propagation.

b) The wavelength λ.

c) The angular frequency ω.

d) The z component of the $\vec{B}$ field.

e) The associated electric field.

f) The energy flux (energy flow per unit area).

g) The magnetic and electric energy densities.

Solution

The argument of the cosine factor is $\omega t - \vec{k} \cdot \vec{r}$, therefore

$$\vec{k} = -3\hat{x} + \hat{y} + \hat{z} \qquad |\vec{k}| = k = \sqrt{11}\ \text{m}^{-1}$$

so the direction of propagation is

a) $$\vec{n} = \frac{\vec{k}}{k} = \frac{1}{\sqrt{11}} [-3\hat{x} + \hat{y} + \hat{z}].$$

The wavelength λ is

b) $$\lambda = \frac{2\pi}{k} = \frac{2\pi}{\sqrt{11}} = 1.89 \text{ meters.}$$

The angular frequency is

c) $$\omega = kc = \sqrt{11} \times 3 \times 10^8 = 9.93 \times 10^8 \text{ sec}^{-1}.$$

The z component of the magnetic field is found from the requirement that

$$\nabla \cdot \vec{B} = \frac{\partial B_x}{\partial x} + \frac{\partial B_y}{\partial y} + \frac{\partial B_z}{\partial z} = 0,$$

which gives

$$-3 + 2 + B_z = 0 \qquad B_z = 1.$$

d) Thus the z component of $\vec{B}$ is 10^{-6} webers/m^2 or 10^{-2} gauss.

The electric field is found from the condition (see solutions to example in section 3. 3)

$$\vec{k} \times \vec{B} = \frac{-\omega}{v^2}\vec{E} = -\frac{k}{v}\vec{E}.$$

Since $\vec{k}$ and ω have just been given and $v = c$ in vacuum, we find

e) $$\vec{E}(\vec{r}, t) = 3\sqrt{6} \times 10^{+2}\left[\frac{+\hat{x} - 4\hat{y} + 7\hat{z}}{\sqrt{66}}\right]\cos(\omega t + 3x - y - z)\ \mathrm{V/m}.$$

The energy flux is found from Poynting's vector

$$\vec{S} = \frac{1}{\mu_o}\vec{E} \times \vec{B}\ \ \mathrm{watts/m^2}$$

Substituting our expressions for $\vec{E}$ and $\vec{B}$ and using

$$\mu_o = 4\pi \times 10^{-7}\ \mathrm{webers\ m^{-1}\ amp^{-1}}$$

gives

f) $$\vec{S} = 1.43 \times 10^3\ \cos(\omega t + 3x - y - z)\ \vec{n}\ \ \mathrm{watts/m^2}$$

where $\vec{n}$ is a unit vector along $\vec{k}$. Note that the magnitude of $\vec{S}$ is approximately 140 $\mathrm{mW/cm^2}$. This can be compared with the maximum recommended safe power density for long-term human exposure to microwave radiation of 10 $\mathrm{mW/cm^2}$.

The electric and magnetic energy densities are given by Eqs. 3. 95 and 3. 96 with $\epsilon_o = 8.85 \times 10^{-12}$ coulomb volt^{-1} meter^{-1}.

$$U_E = \frac{1}{2}\epsilon_o E^2 = 2.39 \times 10^{-14}\cos^2(\omega t + 3x - y - z)\ \ \mathrm{joules/m^3}$$

$$U_M = \frac{1}{2\mu_o}B^2 = 2.39 \times 10^{-14}\cos^2(\omega t + 3x - y - z)\ \ \mathrm{joules/m^3}.$$

It is hardly an accident that the electric and magnetic energy densities come out to be the same, as can be seen from the discussion of Eq. 3. 107.

3.7 RADIATION PRESSURE

In the previous section we showed that electromagnetic waves carry energy which, when incident on a material object, can be partly or totally absorbed. It is plausible to assume that this transfer of energy from wave to matter is accompanied by a transfer of momentum. We would like to know the magnitude of this momentum.

Let us abandon, for the moment, the wave picture of radiation and turn to the quantum theory which says that in many respects light acts like a bunch of particles. These particles, the so-called photons, are relativistic and therefore their momentum-energy relation is given by the familiar result

$$W^2 = c^2p^2 + \left(m_oc^2\right)^2. \tag{3.118}$$

Since a photon has zero rest mass m_o, it follows that

$$\boxed{p = \frac{W}{c}} \tag{3.119}$$

showing that the momentum p equals the energy W divided by the speed of light c. This is true for every individual photon and so it must be true for the entire wave.

The simple result (3.119) just derived must also be obtainable, as we shall now show, from the classical electromagnetic theory of light. Let us consider a perfectly absorbing sheet of material placed in the path of a plane linearly polarized electromagnetic wave propagating along the positive z axis (Fig. 3.19). We pick out one electron of this sheet situated at $z = 0$. Initially, the electron is at rest, but as the wave passes over it the following occurs: The electric vector E_x of the wave sets the electron in motion and imparts to it a velocity v_x. Because the charge is now moving, the magnetic field B_y can exert a force on it of magnitude qv_xB_y directed along the positive z direction, that is, in the direction of the propagation velocity of the wave. This driving force is called the radiation pressure of light. At the end of section 3.3 it was mentioned that ordinarily the effects of the

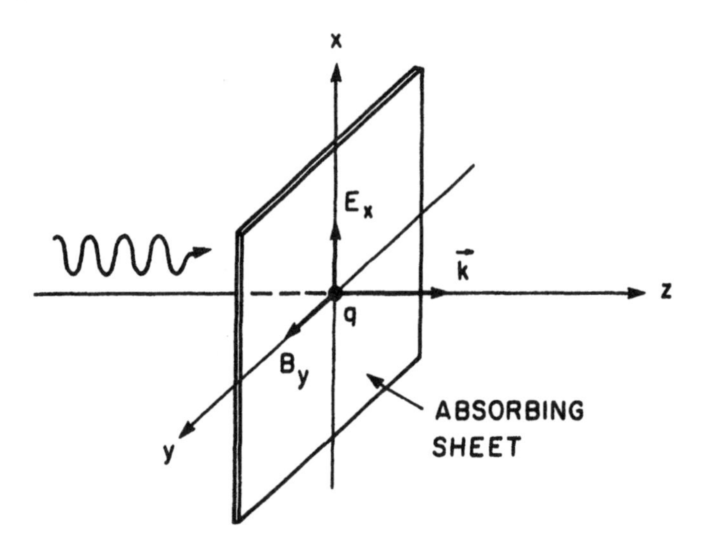

Fig. 3.19 A perfectly absorbing sheet placed at the origin of coordinates at right angles to an incident electromagnetic wave. A charge q of this sheet experiences a force qE_x due to $\vec{E}$; it acquires a velocity v_x. The magnetic force qv_xB_y directed along +z is responsible for the radiation pressure.

magnetic field of a wave are small in comparison with the effects caused by the electric field. In the case of radiation pressure, however, we have a situation where the phenomenon is directly the consequence of the $\vec{B}$ field.

The motion of the electron being driven by the wave at position $z = 0$ is given by the equation (cf. Eq. 1.121)

$$\frac{d^2x}{dt^2} + \beta \frac{dx}{dt} + \omega_o^2 x = \frac{q}{m} E_{ox} \cos \omega t \tag{3.120}$$

where β is the damping coefficient and ω_o the natural frequency of simple harmonic oscillations. The inclusion of damping is mandatory here, otherwise the electron has no means of removing energy from the wave and our initial postulate of an absorbing sheet would be violated. The velocity acquired by the electron is computed by the methods of sections 1.4 and 1.5 with the result that

$$v_x = -\frac{(q/m)\omega}{\sqrt{\left(\omega_o^2 - \omega^2\right)^2 + (\beta\omega)^2}} E_{ox} \sin(\omega t + \delta) \quad (3.121)$$

where δ is the phase factor $\tan \delta = (\beta\omega)/\left(\omega^2 - \omega_o^2\right)$. In view of the fact that the magnetic field of the wave is given by $B_y = (E_{ox}/c) \cos(\omega t)$, the force of radiation $F_z = qv_xB_y$ becomes

$$F_z = -\frac{(q^2/mc)\omega}{\sqrt{\left(\omega_o^2 - \omega^2\right)^2 + (\beta\omega)^2}} E_{ox}^2 \cos(\omega t) \sin(\omega t + \delta). \quad (3.122)$$

The absorption of energy by the electron is the highest at resonance $(\omega = \omega_o)$ and here therefore

$$F_z = -\frac{q^2}{mc\beta} E_{ox}^2 \cos(\omega t) \sin(\omega t + \delta). \quad (3.123)$$

Now, the rate of work done by the wave on the charge q equals $\vec{F} \cdot \vec{v} = q[\vec{E} + (\vec{v} \times \vec{B})] \cdot \vec{v}$ and since the second term is identically zero, $\vec{F} \cdot \vec{v} = q\vec{E} \cdot \vec{v}$, with the result that at resonance

$$\frac{dW}{dt} = -\frac{q^2}{m\beta} E_{ox}^2 \cos(\omega t) \sin(\omega t + \delta). \quad (3.124)$$

On comparing the two foregoing equations, it follows that $F_z = (1/c)(dW/dt)$. Since force equals the rate of change of momentum p of the charge, $dp/dt = (1/c)(dW/dt)$ from which it follows that

$$p = \frac{W}{c} \text{ kg-meter sec}^{-1}. \quad (3.125)$$

We interpret this by saying that this momentum is being carried by the wave, a result precisely equal to that obtained on the photon model. Since energy density U rather than energy is a more appropriate quantity associated with a wave, division of both sides of Eq. 3.125 by the volume yields

$$g = \frac{U}{c} \text{ kg-meter}^{-2} \text{ sec}^{-1} \quad (3.126)$$

where g is the momentum per unit volume associated with the wave, and is directed along the wave's propagation vector $\vec{k}$. From the relation $|\vec{S}| = cU$ between U and the Poynting flux $\vec{S}$, the momentum per unit volume can be written as a vector

$$\vec{g} = \frac{\vec{S}}{c^2} = \frac{\vec{E} \times \vec{B}}{\mu_o c^2}. \tag{3.127}$$

In the foregoing analysis we have assumed that the wave is perfectly absorbed when interacting with the electrons of the absorbing sheet. In the presence of reflection, we must not neglect the momentum carried by the reflected wave. One finds that in the presence of reflection, the momentum transferred to the sheet is given by

$$p = \frac{W_A}{c} + 2\frac{W_R}{c} \tag{3.128}$$

where W_A and W_R are the absorbed and reflected wave energies, respectively. From this we see that for a perfect mirror the momentum transferred is twice the wave energy divided by c. This doubling of momentum is readily understood when light is thought of as made up of discrete particles, photons.

Let us now compute the pressure, that is, the force per unit area exerted by the wave incident on a fully absorbing sheet. Take a cylindrical section of wave of cross sectional area A and length $c\Delta t$. The amount of momentum $gc\Delta tA$ contained in the cylinder must leave through its front surface (where the absorbing sheet is located) in a time Δt seconds. Therefore the pressure P which equals the change of momentum per second per unit area is gc, and by Eq. 3.127 is given by

$$P = \frac{1}{c}|\vec{S}| = \frac{1}{\mu_o c}|\vec{E} \times \vec{B}| \quad \text{newton/m}^2. \tag{3.129}$$

Let us see how large this pressure is. In section 3.6 we calculated that the flux of sunlight at the top of the Earth's atmosphere equals $\langle S \rangle$ = 1352 watts/m^2 and therefore the time-average pressure $\langle P \rangle = 1352/(3\times10^8) = 4.51\times10^{-6}$ newton/m^2 which is an exceedingly small pressure indeed*; it corresponds to a total force of 7×10^4 metric tons pushing on the Earth, to be compared with the force of 3×10^{18} tons with which the Sun pulls on the Earth as a result of the mutual gravitational attraction! The existence of such weak forces is quite difficult to demonstrate in the laboratory. The first to succeed was P. N. Lebedev in 1899 who suspended a vane in a highly evacuated vessel from a fine quartz torsion fiber and observed the vane's rotation when strong light was allowed to impinge on it. If the vessel is not highly evacuated, bombardment of the vane by the gas atoms can cause extraneous effects. Let us point out, however, that since the discovery in the early sixties of very powerful lasers, an accurate measurement of radiation pressure becomes a rather simple matter. In section 3.5 we found that a value of $\langle S \rangle = 3.3\times10^{13}$ watt/m^2 is readily achieved with what is now a fairly conventional high-power laser. Equation 3.129 therefore gives a radiation pressure equal to 1.1×10^5 newtons/m^2 or 11 newtons/cm^2. The laser beam with a 1-cm radius therefore pushes an object in its path with the force of about a 3 kg weight! †

Although the pressure of solar or other stellar radiation is negligible in the case of large celestial bodies like the planets it can be very important when it acts on smaller bodies. The reason is that at a given distance from the Sun, the radiation force is proportional to area, that is, on the square of the linear dimensions of the body, whereas gravity depends on mass and the force is therefore proportional to the cube of the linear dimension. Hence, as the radius of the body decreases, the radiation force becomes ever more important; and since the radiant intensity falls off as the square of the distance, just like the gravitational force, the ratio of radiation force F_R to gravitational force F_G

---

*Atmospheric pressure is approximately 10^5 newton/m^2.

†See the article "Pressure of Laser Light" by A. Ashkin, Scientific American, February 1972, page 63.

remains constant independent of the distance from the Sun. We have seen that at a distance of 1 A. U. (one astronomical unit A. U. equals the mean Earth-Sun distance, or 1.4964×10^{11} m) the solar pressure $\langle P \rangle$ equals 4.51×10^{-6} newton/m^2, and the force on an object of radius a equals

$$F_R = 4.51 \times 10^{-6} \pi a^2 \text{ newton.} \tag{3.130}$$

The gravitational force on the same object and at the same distance from the Sun is readily shown to be

$$F_G = 5.93 \times 10^{-3} \left[\frac{4}{3}\pi a^3 \rho\right] \text{ newton} \tag{3.131}$$

where $(4/3)\pi a^3 \rho$ is the mass of the body in kilograms and ρ its mass density in kg/m^3. Therefore the ratio of radiation to gravitational force equals

$$\begin{aligned} R &= \frac{F_R}{F_G} \\ &= \frac{5.70 \times 10^{-4}}{a\rho}. \end{aligned} \tag{3.132}$$

Suppose that the object under discussion has a mass density of 5.7×10^3 kg/m^3 (which is 5.7 times the density of water and is about the mean density of the Earth and of meteorites); then $R = 1$ when $a = 10^{-7}$ meters, or 1000 Å.

Some important conclusions can be drawn from that simple calculation concerning the behavior of small cosmic dust particles. When the particles have a critical size (~1000 Å) such that $R = 1$ their trajectories in the solar system will not be affected by gravity, which has just been balanced out by the radiation pressure. For larger particles than this (i. e., $R < 1$) the effect of the Sun's gravity will be somewhat reduced but the particles will continue to pursue elliptical orbits around the sun. When $R > 1$, the particles will be repelled away from the Sun and they will execute hyperbolic orbits around it. The actual size of the grains is not known with any degree of precision. However, it is interesting to note that certain types of cosmic clouds known as Barnard's clouds are believed to be composed of dust particles of approximately the critical size of 1000 Å. An important consequence of this coincidence is that when the Sun or some other star passes through such clouds, it would not be capable of attracting the dust particles. In this case stellar buildup by slow accretion of the interstellar dust, a hypothesis that is sometimes mentioned, would apparently not take place. Figure 3.20 shows one of the most spectacular examples of light obstruction produced by interstellar grains concentrated in a cloud known as the Horsehead Nebula. It is approximately 1000 light years distant and is located in the constellation Orion. For an interesting article on the subject of grains, see "Interstellar Grains" by J. M. Greenberg, Scientific American, October 1967.

One of the most striking phenomena which have in the past been ascribed to radiation pressure is in connection with comets. The tails of these bodies are directed away from the Sun and they are curved in such a way as to suggest that a repulsive force varying inversely as the square of the distance exists in competition with the gravitational attractive force. It turns out that radiation pressure for many comets is much too small to explain the observations. Other mechanisms* have been proposed in recent years in an attempt to explain the physical origin of the repulsion experienced by the matter constituting comets. This subject is discussed in a most readable text Cosmic Dust by A. Dauvillier (Philosophic Library, New York 1963).

In concluding this section, we want to mention that in some cases the transfer of energy from the wave to the body is also accompanied by a transfer of angular momentum. This happens when the wave is circularly or elliptically polarized. If an absorbing sheet is placed in the path of this wave, it will acquire an angular momentum about an axis parallel to the propagation direction of the wave. With respect to an observer looking toward the oncoming wave, the rotation is counterclockwise for a right-hand polarized wave, and clockwise for

*It is now thought that matter ejected from the Sun is the primary cause of comet-tail dust deflection. This matter is ejected continuously and is the so-called "solar wind"; see E. N. Parker, "Solar Wind," Scientific American, April 1964.

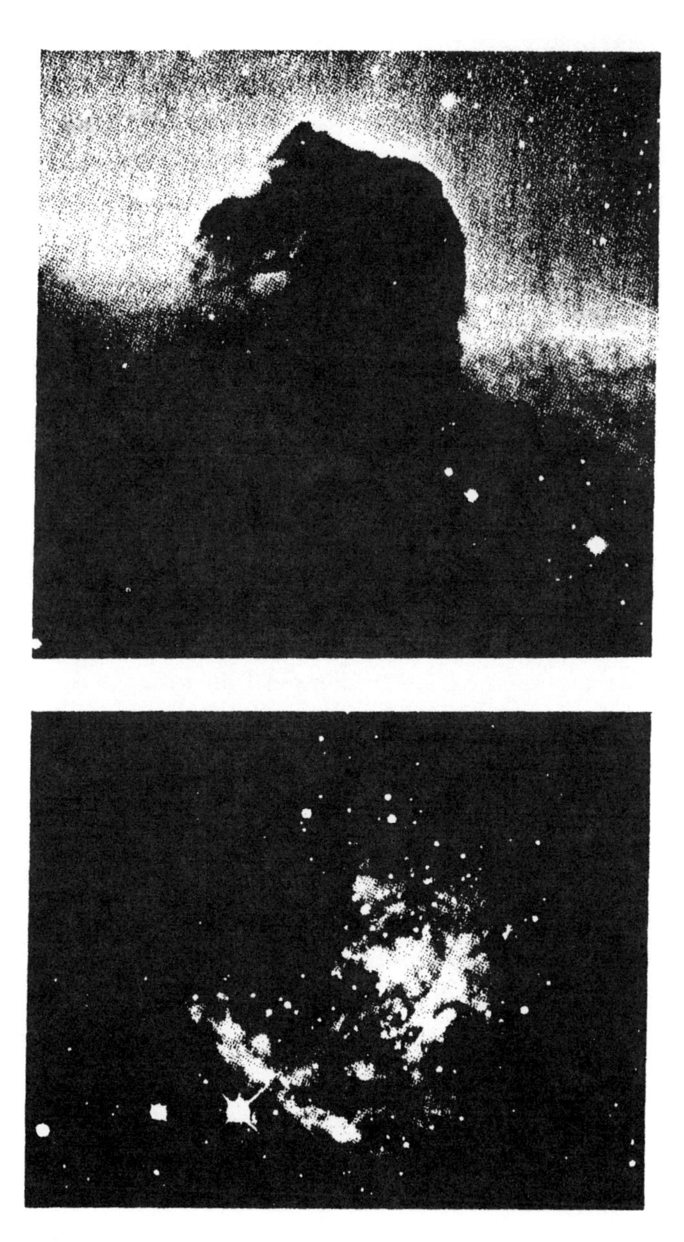

Fig. 3.20 The Horsehead Nebula (above), and the Orion Nebula (below). [From "Interstellar Grains" by J. M. Greenberg, Scientific American, October 1967.]

a left-hand polarized wave (see section 3.4). The magnitude of the angular momentum can be computed with the aid of the same model as that used in obtaining Eq. 3.125 in which the motion of one electron of the absorbing sheet was singled out and analyzed.

Consider a right-hand circularly polarized wave propagating along the positive z axis. At the origin of coordinates where the electron is assumed to be situated, the electric vector of the wave has the form (see section 3.4),

$$\vec{E} = \hat{x}E_o \cos(\omega t) + \hat{y}E_o \sin(\omega t). \tag{3.133}$$

In exact analogy with Eq. 3.121 this wave induces an electron velocity which at resonance, $\omega = \omega_o$, is given by

$$\vec{v} = -\hat{x}\left(\frac{q}{m\beta}\right) E_o \sin(\omega t + \delta) - \hat{y}\left(\frac{q}{m\beta}\right) E_o \cos(\omega t + \delta). \tag{3.134}$$

Here β is the coefficient of friction and δ is the phase factor given by $\tan \delta = \beta\omega/\left(\omega^2 - \omega_o^2\right)$ which at resonance equals $\pi/2$. The displacement $\vec{r}$ from the origin is obtained by integrating Eq. 3.134 with respect to time:

$$\vec{r} = \hat{x}\left(\frac{q}{m\beta\omega}\right) E_o \cos(\omega t + \delta) + \hat{y}\left(\frac{q}{m\beta\omega}\right) E_o \sin(\omega t + \delta). \tag{3.135}$$

The torque $\vec{\tau}$ exerted by the electrical force* is given by

$$\begin{aligned} \vec{\tau} &= \vec{r} \times \vec{F} \\ &= q\vec{r} \times \vec{E} \end{aligned} \tag{3.136}$$

and substituting for $\vec{r}$ and $\vec{E}$ from Eqs. 3.133 and 3.135 gives

$$\vec{\tau} = -\hat{z}\,\frac{q}{m\beta\omega}\,E_o^2. \tag{3.137}$$

Now, the rate of doing work by the wave on the charge q equals $\vec{F} \cdot \vec{v} = q\vec{E} \cdot \vec{v}$. Substituting for $\vec{E}$ and $\vec{v}$ from the above equations gives

*The torque produced by the magnetic force is negligible for nonrelativistic speeds. Moreover, it is zero when averaged over a cycle.

$$\frac{dW}{dt} = -\frac{q}{m\beta} E_o^2 \cos^2(\omega t) - \frac{q}{m\beta} E_o^2 \sin^2(\omega t)$$
$$= -\frac{q}{m\beta} E_o^2. \tag{3.138}$$

Comparing Eqs. 3.137 and 3.138 we see that $|\vec{\tau}| = (dW/dt\omega)$. But, $\vec{\tau} = d\vec{L}/dt$ where $\vec{L}$ is angular momentum. Thus $|\vec{\tau}| = d|\vec{L}|/dt = \omega^{-1} dW/dt$ with the result that angular momentum is

$$|\vec{L}| = \frac{W}{\omega} \text{ kg m}^2\text{-sec}^{-1} \tag{3.139}$$

where ω is the angular frequency of the wave $\omega = 2\pi\nu$ in radians per second.

Note, this result is independent of β, the damping coefficient, as, of course, it must be, because L is a property of the electromagnetic wave, not the material upon which it impinges.

The first precise measurement of the angular momentum imparted by an electromagnetic wave is due to R. A. Beth (Phys. Rev. 50, p. 115 (1936)). He suspended a very sensitive torsional pendulum in a chamber evacuated to a pressure of 10^{-6} Torr. The fine quartz fiber supported a series of circular plates illuminated from the bottom by a vertical beam of circularly polarized light. Changes in torque caused by the beam as small as 4×10^{-13} N-meter could thus be detected. Using a resonance technique together with a clever optical arrangement which greatly enhanced the effect, Beth was able to obtain a good check on Eq. 3.139 (or its equivalent). Recall that on the basis of quantum theory, each of the photons of which the wave is comprised carries an energy $h\nu$ where h is Planck's constant. Therefore, from Eq. 3.139 we deduce that a photon of a circularly or elliptically polarized wave carries angular momentum equal to $h/2\pi$.

CHAPTER 4

SOURCES OF RADIATION

Maxwell's equations tell us how electric and magnetic fields are generated by charges and by charges in motion, that is, currents. But what special properties must these charges have for them to give rise to that fascinating interplay of $\vec{E}$ and $\vec{B}$ we call the electromagnetic wave, and which we studied in such detail in Chapter 3? The complete answer to this question in the context of classical, that is, nonquantum mechanical physics is contained in the four equations of Maxwell. However, to extract it requires fairly elaborate mathematical procedures developed during the last years of the nineteenth century by men like Hertz, Larmor, Liénard, and Wiechert. Accordingly, a full treatment of radiation lies outside the scope of this book and we shall content ourselves with a less exhaustive (and exhausting) description of the problem. Fortunately, some of the main features of the radiation can be deduced by fairly elementary and physically vivid methods.

When we think of radiation from a charge or system of charges, we can exploit with advantage the somewhat analogous problem of radiation of sound waves discussed in section 2.4. Here, just as in that section, we expect to see at a very large distance r from the source a spherical or pseudospherical wave propagating radially outward and transporting energy. In a nonabsorbing medium (like vacuum) the total energy carried across a sphere of radius r should be independent of the size of r, since in this case the wave can neither gain nor lose energy. Because the area of the sphere surrounding the source is proportional to r^2, the expected time-averaged radiation flux $\langle|\vec{S}|\rangle = \langle|\vec{E}\times\vec{B}|\rangle/\mu_0$ must vary as $1/r^2$ so as to maintain the product of $\langle|\vec{S}|\rangle$ with $4\pi r^2$ invariant. And, just as in sound propagation, the wave at large r when viewed over a restricted volume, should look very much like a plane wave. This means that $\vec{E}$, $\vec{B}$, and $\vec{S}$ are orthogonal to one another and the magnitudes of $\vec{E}$ and $\vec{B}$ are simply related through $|\vec{B}| = |\vec{E}|/c$. Thus, for a sinusoidal spherical electromagnetic wave we expect that as $r \to \infty$,

$$|\vec{E}| = \frac{\text{"strength of source"}}{r} \cos(\omega t - kr)$$

$$|\vec{B}| = \frac{\text{"strength of source"}}{rc} \cos(\omega t - kr). \tag{4.1}$$

In the words "strength of source" lie buried all the physical processes by which the charge sheds its electromagnetic wave. Before calculating the details of what these processes must be, we can make some very general statements about what they cannot be. For one, a stationary charge clearly cannot emit. To be sure, electric field lines emanate radially from such a charge like the spokes of a wheel, but this is not emission; no energy passes off to infinity. Not only is there no magnetic field, but $\vec{E}$ is everywhere radial, so $\vec{S}$ has no component along the radius vector $\vec{r}$. In addition, the coulombic field of the charge falls off as $1/r^2$ rather than as $1/r$ in contradiction to the requirements of the first equation (4.1).

If a stationary charge cannot radiate, can a charge in uniform motion? A uniformly moving charge has associated with it an electric field and since there is now a current, there is also a magnetic field. We shall not compute these fields but just state their magnitudes and directions.* They are

$$\vec{E} = \frac{q}{4\pi\epsilon_0 r^2} \frac{1-\beta^2}{(1-\beta^2 \sin^2\theta)^{3/2}} \hat{r}$$

$$\vec{B} = (\vec{u} \times \vec{E})/c^2. \tag{4.2}$$

*See, for example, Electricity and Magnetism, E. M. Purcell (McGraw Hill Book Company, New York, 1965), Chapter 5.

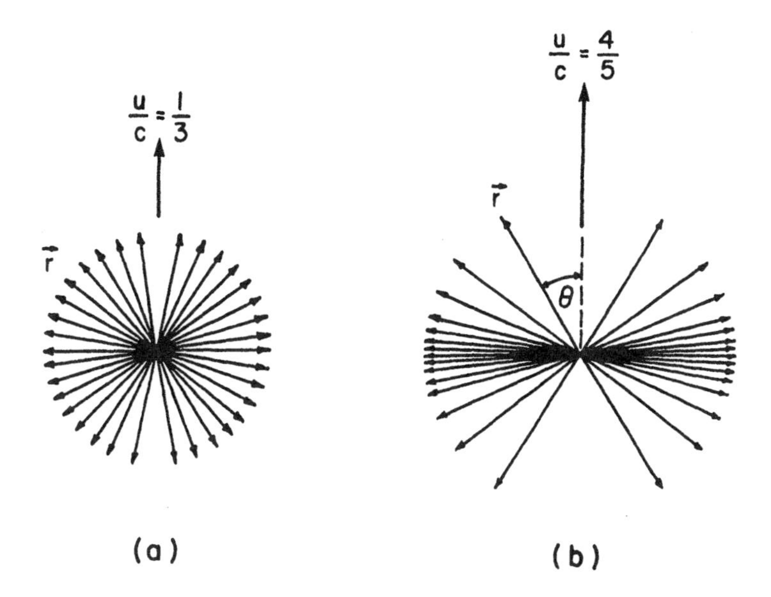

Fig. 4.1 The electric field lines around a positive charge moving at constant velocity. In (a) the speed is low, in (b) it is relativistic. In either case the field lines point radially from the instantaneous position of the charge. The density of lines is proportional to the magnitude of $\vec{E}$.

Here r is the distance from the instantaneous position of the charge q moving with a velocity $\vec{u}$, and $\beta = u/c$ is the ratio of the speed of the charge to the speed of light c; θ is the angle subtended between the unit vector directed along $\vec{r}$ and the velocity vector $\vec{u}$, as is shown in Fig. 4.1. From the foregoing equations and the figure several important observations can be made: (a) The electric field always points precisely radially outward along a line drawn from the instantaneous position of the charge q. This sounds like instantaneous transmission of information from the charge to some observer who might be hundreds of miles away from q, and observing its field $\vec{E}$. But this is not so, because the charge has been moving with constant velocity for a long time before reaching its present position and has been, so to speak, announcing its position all along its path. (b) At large velocity, $u \to c$, the electric field is much stronger at right angles to $\vec{u}$ than along $\vec{u}$, a fact which manifests itself in the drawing of Fig. 4.1 as a crowding of the lines of force around $\theta \approx \pi/2$. (c) At low, nonrelativistic speeds, $(u/c)^2 \ll 1$, the electric field is virtually the same as that of a stationary charge. (d) The magnitudes of $\vec{E}$ and $\vec{B}$ fall off with distance as $1/r^2$ and hence the magnitude of the Poynting flux $\vec{S}$ falls off as $1/r^4$. Therefore, when we compute the power flowing across a sphere of radius r, we find that it vanishes as $1/r^2$, contrary to the foregoing discussion in which we demanded that the power be independent of r. We must conclude that a charge moving at constant velocity cannot radiate energy. This is consistent with our understanding of special relativity theory, for, if $\vec{u}$ is the relative velocity between q and the observer, a frame of reference can always be found in which q is at rest and the observer in uniform motion. But we said already that a static charge cannot radiate energy. And, if it is possible to find an inertial frame with respect to which q is at rest, then radiation cannot occur in any other inertial frame. You may object to this argument on the grounds that Maxwell's equations were discovered long before relativity and that they therefore need to be "fixed up" to agree with Einstein's precepts, before arguments like ours can be accepted. The answer is that unlike Newton's equations, Maxwell's equations do not require doctoring up. Indeed, they possess that wonderful property of being in full agreement with special theory of relativity, a fact that was not perceived until Einstein.

And hence, we have reached the very general and profound result of classical electrodynamic theory, that a charge in empty space cannot radiate electromagnetic waves unless it is undergoing accelerated motions.* Since charges can execute many different types of accelerations, they can give rise to a rich variety of radiations. The sinusoidal linear oscillations of electrons in a metal wire lead to the typical antenna emissions in radio and television broadcasting. The impinging and stopping of a beam of electrons as it smashes into a metal target gives rise to x rays or so-called bremsstrahlung (stopping radiation).

*There is an interesting exception in which a charge moving at a constant velocity can in fact emit electromagnetic radiation. But it must be moving through a medium like a dielectric. This is called Čerenkov radiation.

And the centripetal acceleration suffered by electrons as they move in circular orbits in betatrons and synchrotrons and in interstellar magnetic fields results in the familiar synchrotron emissions. These will be the subjects discussed in this chapter. First, however, we must address ourselves to finding the proper relations between $\vec{E}$, $\vec{B}$, and the "strength of the source" as this appears in Eq. 4.1. We now know that this source strength is a function of the charge's acceleration.

4.1 RADIATION FROM AN ACCELERATED CHARGE

We pointed out earlier that a complete calculation of the radiation with Maxwell's equations as a starting point is fraught with mathematical difficulties. For this reason we shall follow here a more indirect method suggested originally by J. J. Thomson. In this model we investigate the following special situation.* A point charge +q is initially at rest at the origin of coordinates $x = y = z = 0$. Its field lines are of course pointing radially outwards from 0. Suddenly, at time $t = 0$, the charge is given an acceleration a for a very short time Δt, during which time it travels from 0 to 0' as is shown in Fig. 4.2. On completing the acceleration, the charge acquires a speed $u = a\Delta t$ and proceeds to move with this speed along the positive x axis. After an additional time t (in the post-acceleration period) much greater than Δt, the charge will have reached the new position 0" such that $0'0'' = ut$. Let us find the field at the end of this additional time t when the charge is at 0".

Let the line 0A represent any field line drawn from the charge q while it was stationary at 0, and draw a sphere of radius $c(t + \Delta t)$ with 0 as the origin and c as the speed of light. The field emitted at the beginning of the acceleration period Δt has just had time to reach the surface of the sphere. Therefore, an observer situated outside this sphere will not have received news yet concerning the fact that the particle experienced a change, that is, an acceleration. In other

*For another discussion see J. R. Tessman and J. T. Finnell, Am. J. Physics 35, 523 (1967).

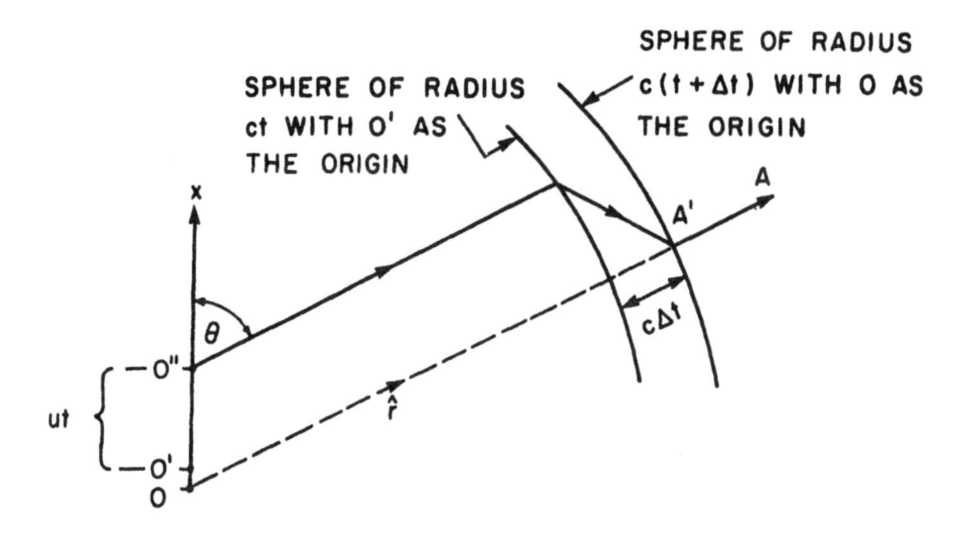

Fig. 4.2 Field line from an accelerated point charge. The charge is initially at rest at 0; it accelerates for a short time Δt, reaching point 0' and then it coasts at constant velocity $u = a\Delta t$. It reaches a point 0" in a time t. The kink in the field line propagates outward with velocity c as time progresses. It is assumed that $u \ll c$ and $\Delta t \ll t$. The piece of field line A'A comes from the initial field of the stationary charge when it was originally at 0.

words, the field just outside the sphere is precisely the field of the stationary charge at 0. Now, draw a second sphere of radius ct having its center at 0'. At any point inside this sphere an observer will see the field of the uniformly moving charge traveling with the post-acceleration velocity u. We know from Fig. 4.1 and Eqs. 4.2 that this field points radially outward from the instantaneous position of the charge; and if u is a nonrelativistic speed assumed here, the structure of the field is virtually identical with that of the stationary charge in the preacceleration period. Between the two spheres is a shell of thickness $c\Delta t$ within which the transition from one type of field to the other occurs. But fields cannot stop or start in empty space and we must join the ends of the field lines. This of necessity creates a kink. As time progresses the shell with its kink propagates outward with the speed of light. It is our task to show that the kink has associated with it a component of electric (and magnetic) field

Fig. 4.3 The electric field lines around a charge moving at constant velocity along the x axis (i.e., up the page) until, at a certain time, it is abruptly decelerated and brought to rest. The kinks in the field lines and formation of the spherical wave front are clearly visible. [Snapshot from the computer-generated film strip "Electric Fields of Moving Charges" by the Education Development Center, Inc., Newton, Mass.]

which carries a certain amount of energy into space. By drawing the entire picture as in Fig. 4.3 we see that we have here the makings of a spherical, outwardly propagating wave.

Because the velocity u is assumed to be very small compared with the speed of light c the distance ut traveled by the charge is tiny compared with the radius $r = ct$ of the spherical cap drawn in Fig. 4.2. Therefore, the directions of the two field lines shown in the figure are virtually the same, and it is safe to assume that they are exactly parallel to one another and thus parallel to the radius vector we call $\vec{r}$. The electric field in the kink can now be resolved into components $E_{\perp}$ and $E_{\parallel}$ relative to the direction $\vec{r}$, as is shown in Fig. 4.4. By similar triangles, the ratio of these two fields is seen to be

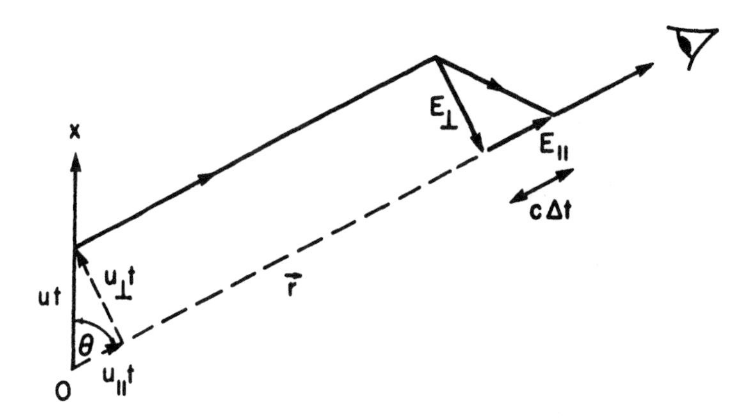

Fig. 4.4 The electric field in the kink resolved into components parallel and perpendicular to the direction of observation $\vec{r}$. By similar triangles $(E_\perp/E_\parallel) = (u_\perp t/c\Delta t)$. The distance r = ct is very large compared with the displacement ut of the charge from its origin 0.

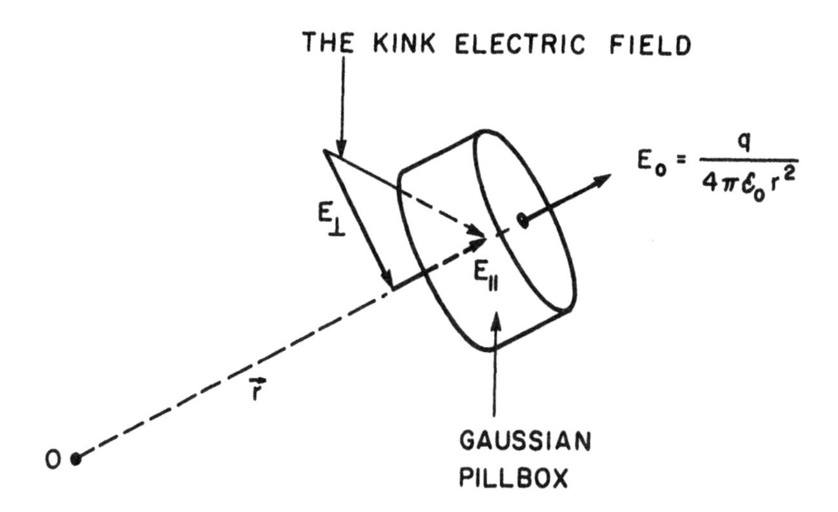

Fig. 4.5 A Gaussian surface drawn in the region of the kink. Since there are no charges within the surface, $E_\parallel$ must equal E_o, the original field of the charge before acceleration.

$$\frac{E_{\perp}}{E_{\parallel}} = \frac{u_{\perp}t}{c\Delta t} \tag{4.3}$$

where $u_{\perp}$ is the component of the charge's velocity resolved perpendicular to $\vec{r}$. Since $u = a\,\Delta t$ it follows that $u_{\perp} = a_{\perp}\,\Delta t$ with $a_{\perp}$ as the component of $\vec{a}$ normal to $\vec{r}$.

Hence, Eq. 4.3 can be rewritten as

$$\begin{aligned} E_{\perp} &= E_{\parallel}\left[\frac{a_{\perp}t}{c}\right] \\ &= E_{\parallel}\left[\frac{a_{\perp}r}{c^2}\right] \end{aligned} \tag{4.4}$$

where in obtaining the second form we used the fact that $r = ct$.

It remains for us to determine the magnitude of the parallel component $E_{\parallel}$ of the field in the kink. To do this we draw a pillbox whose axis points along the radius vector $\vec{r}$ (Fig. 4.5) and invoke Gauss' law. The only contribution to the electric flux comes from the flat top and bottom surfaces. At the top surface the flux is just the radial electric field $(q/4\pi\epsilon_o r^2)$ multiplied by the pillbox area A. At the bottom surface the flux is $E_{\parallel}A$. Since there are no charges present in the box, the two fluxes are equal and

$$E_{\parallel} = \frac{q}{4\pi\epsilon_o r^2}. \tag{4.5}$$

Inserting this result in formula (4.4) yields the sought-for result,

$$E_{\perp} = \frac{qa_{\perp}}{4\pi\epsilon_o rc^2}. \tag{4.6}$$

Observe that this field has precisely the properties we have been looking for in our attempt to manufacture the wave: The field is transverse to the direction of propagation $\vec{r}$ and it varies as $1/r$. It depends on the acceleration of the charge in a simple and direct way, and there is no $E_{\perp}$ when there is no acceleration. Moreover, $E_{\perp}$ at position $\vec{r}$ and time t is due to the acceleration a which took place at an earlier time t' given by

$$t' = t - \frac{r}{c}, \tag{4.7}$$

thus showing that time elapsed before the observer at $\vec{r}$ became aware of things happening to the charge at $|\vec{r}| = 0$. And, while we have not derived the value of the magnetic field $\vec{B}$, we are fairly confident that at large distances r to which the above analysis refers the accompanying magnetic vector is related to $\vec{E}$ in the same way as for plane waves; that is, it is orthogonal to $\vec{E}$ and has magnitude $|\vec{B}_\perp| = |\vec{E}_\perp|/c$. For the sake of simplicity we now drop the subscript $\perp$ on $\vec{E}$ and $\vec{B}$ and summarize all our findings by the following three vector equations:

$$\begin{aligned} \vec{E}(\vec{r},t) &= -\frac{q\vec{a}_\perp(t')}{4\pi\epsilon_0 rc^2} \text{ volt m}^{-1} \\ \vec{B}(\vec{r},t) &= \hat{r} \times \vec{E}(\vec{r},t)/c \text{ webers m}^{-2} \\ \vec{S} &= \frac{\vec{E} \times \vec{B}}{\mu_0} \text{ watt m}^{-2}. \end{aligned} \tag{4.8}$$

The minus sign in front of the first equation denotes the fact that $\vec{E}$ is antiparallel to $\vec{a}_\perp$ as can be seen readily from Figs. 4.2 and 4.3. Observe that the electric field $\vec{E}$ lies in a plane defined by the vectors $\vec{r}$ and $\vec{a}$ so that a linearly moving charge generates a linearly polarized wave.* The geometry of the situation expressed by Eqs. 4.8 is illustrated in Fig. 4.6.

The foregoing results were derived for a very special and fairly artificial motion of the charge q, and one therefore wonders about their degree of generality. An exact calculation for the radiation from a charge having arbitrary acceleration shows that Eqs. 4.8 are in fact

*In vector notation, the direction of $\vec{E}$ is given by the vector $\hat{r} \times (\hat{r} \times \vec{a})$, where $\hat{r}$ is a unit vector pointing from the charge to the observer.

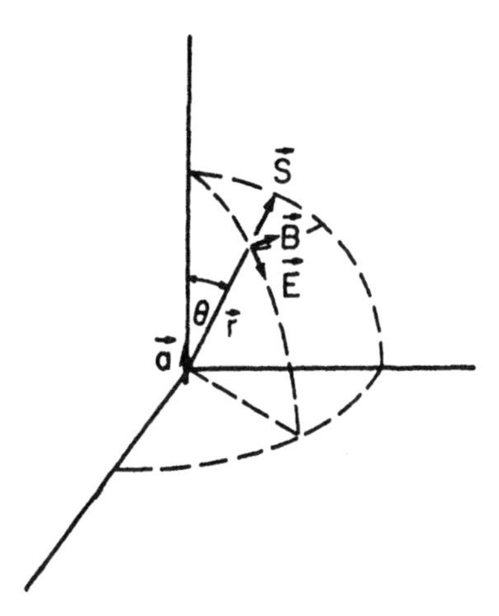

Fig. 4.6 The distant fields ($r \gg \lambda$) of an accelerated point charge situated at the origin of the coordinates.

excellent approximations to the truth, provided that the motions are so slow that relativistic corrections of order $(u/c)^2$ can be neglected, and provided that the observation distance r is large compared with the wavelength of the radiation, $\lambda = 2\pi c/\omega$. Moreover, the results are not only valid for a single point charge but for more extended charge distributions or a system of discrete charges as long as one remembers to sum the products $q\vec{a}_\perp$ over all the elementary charges participating in the motion. However, the size of this charge distribution cannot be arbitrarily large; if d is the characteristic dimension over which the accelerated charges are distributed, it is required that $d \ll \lambda$. In summary then, the three conditions that must be imposed on Eqs. 4.8 are

$$
\begin{aligned}
&(u/c)^2 \ll 1 \\
&r \gg \lambda \\
&d \ll \lambda.
\end{aligned}
\tag{4.9}
$$

Formulas (4.8) together with the imposed conditions (4.9) are often referred to as the "dipole approximation" to the radiation field.

The radiation pattern

Suppose θ is the angle subtended between the instantaneous direction of the acceleration vector $\vec{a}$ and the direction of wave propagation $\vec{r}$. Then $a_\perp$ of Eq. 4.8 is just a sin θ and

$$E(r,t) = -\frac{qa(t') \sin\theta}{4\pi\epsilon_0 rc^2}, \tag{4.10}$$

showing that at a fixed radius r the electric field varies as sin θ. Thus $\vec{E}$ (and $\vec{B}$) have maximum values at 90 degrees to the direction of acceleration and zero values along the direction of acceleration. Unlike, say, a tiny acoustic radiator which emits equally in all directions (see section 2.4), the elementary electromagnetic radiator emits anisotropically, the emission being predominant at right angles ($\theta = \pi/2$) to $\vec{a}$. This fact is, of course, reflected in the angular distribution of the Poynting flux whose value is readily deduced by inserting Eq. 4.10 in the second and third equations (4.8):

$$|\vec{S}(r,t)| = \frac{q^2 a^2(t') \sin^2\theta}{16\pi^2\epsilon_0 r^2 c^3}. \tag{4.11}$$

The angular dependence of this and other radiation fields can be vividly illustrated graphically in what is known as the radiation pattern. This is a polar diagram of $|\vec{S}|$ as function of θ obtained by drawing vectors from the position of the dipole of length proportional to $\sin^2\theta$ (or whatever other angular dependence the system obeys). Figure 4.7 shows the radiation pattern predicted by Eq. 4.11. Of course a complete radiation pattern gives the emission at all angles θ and ϕ and actually requires three-dimensional presentation. This difficulty is usually overcome (as in Fig. 4.7) by showing a cross section of the pattern in the plane of interest. To obtain the entire three-dimensional pattern one must rotate our figure eight curve about an axis which passes along the acceleration vector $\vec{a}$. This results in the doughnut-shaped object sketched in Fig. 4.8. Observe that even this most elementary of radiators exhibits quite pronounced directional character-

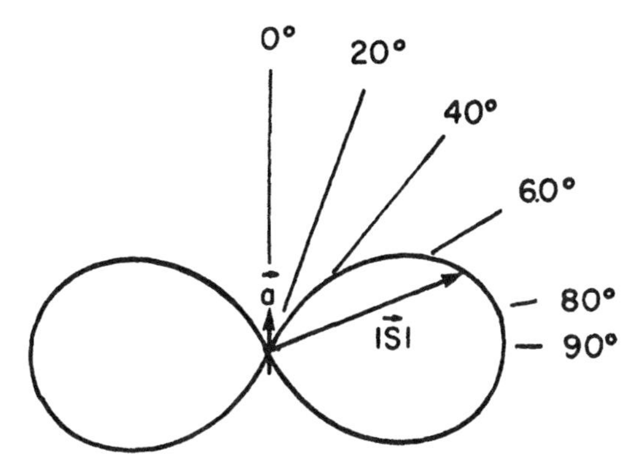

Fig. 4.7 Radiation pattern of an accelerated point charge whose acceleration vector points up the page. The figure represents a cross section of the three-dimensional pattern taken in the plane containing $\vec{a}$. In the plane perpendicular to $\vec{a}$, the radiation pattern is a circle.

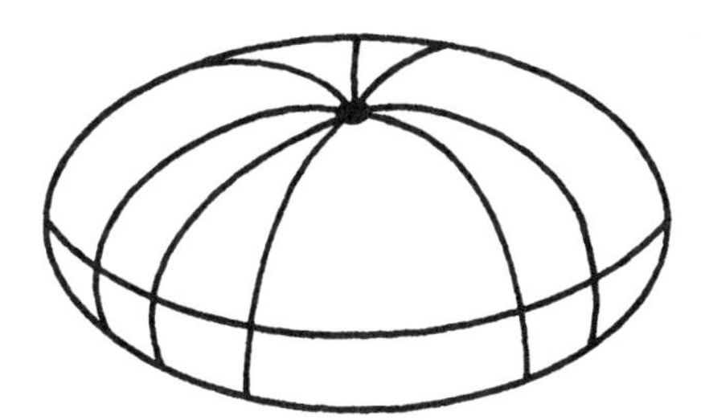

Fig. 4.8 Sketch of the three-dimensional radiation pattern of a nonrelativistic accelerated point charge.

istics, thus permitting appreciable "beaming" of the radiation into space. In fact, there is in electromagnetism no such thing as a perfectly isotropic radiator and it cannot be manufactured, however complex a charge distribution one may want to think up.

The Larmor formula for the radiated power

The total power radiated by a nonrelativistic accelerated charged particle is obtained by integrating the Poynting flux over the area of a large sphere whose origin coincides with the instantaneous position of

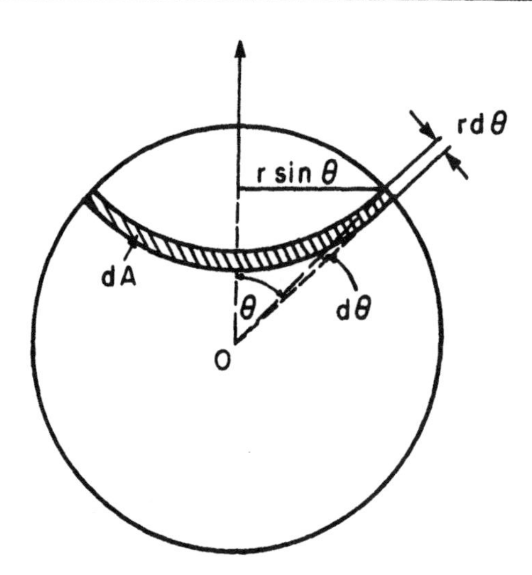

Fig. 4.9 An element of area $dA = 2\pi(r \sin \theta)\, r d\theta$ of a spherical surface of radius r used in integrating the Poynting flux from an accelerated charge.

q. In order to carry out the integration the spherical surface is subdivided into elementary circular strips whose area is $2\pi r^2 \sin \theta \, d\theta$ (see Fig. 4.9). The total power is then obtained by summing over all such strips,

$$P(t) = \int_0^{\pi} |\vec{S}(r,t)| \, 2\pi r^2 \sin \theta \, d\theta$$

$$= \frac{q^2 a^2(t')}{8\pi\epsilon_0 c^3} \int_0^{\pi} \sin^3\theta \, d\theta, \tag{4.12}$$

where the second equation results from substituting for $|\vec{S}|$ from Eq. 4.11. The integral over $\sin^3\theta$ equals 4/3 so that

$$\boxed{P(t) = \frac{q^2 a^2(t')}{6\pi\epsilon_0 c^3} \text{ watt.}} \tag{4.13}$$

This is the well-known Larmor result. Note that P(t) is independent of the radius of the sphere as it must be if energy is to be conserved in the ever-expanding spherical wave.

We are now in a position to apply our results to various practical cases, and we begin with a discussion of antennas.

Example

A proton is uniformly accelerated in a van de Graaff accelerator through a potential difference of 700 kilovolts. The length of the linear accelerating region is 3 meters.

a) Compute the ratio of the radiated energy to the final kinetic energy.

b) Show that for a particle moving in a linear accelerator the rate of radiation of energy is

$$\frac{dU}{dt} = \frac{q^2}{6\pi\epsilon_0 M^2 c^3}\left(\frac{dU_k}{dx}\right)^2$$

where U_k is the kinetic energy.

Solution

The energy radiated in time t is given by Eq. 4.13 as

$$U_{rad} = [P(t)]t = \left(\frac{dU}{dt}\right) t = \frac{q^2 a^2 t}{6\pi\epsilon_0 c^3}.$$

But $v = at$ and $s = \frac{1}{2} at^2 = \frac{1}{2} vt$, where s is the length of the accelerating region. Thus $t = 2s/v$ and $a = v/t = v^2/2s$ hence

$$U_{rad} = \frac{q^2 v^3}{12\,\pi\epsilon_0 c^3 s}.$$

The final kinetic energy U_k is given by

$$U_k = \frac{1}{2} Mv^2 = qV$$

$$v = (2qV/M)^{1/2}$$

where M is the proton mass and V is the accelerating voltage. The ratio of radiated energy to kinetic energy is

$$\frac{U_{rad}}{U_k} = \frac{q^2 v}{6\pi\epsilon_o Mc^3 s} = \frac{q^2}{6\pi\epsilon_o Mc^3 s}\left(\frac{2qV}{M}\right)^{1/2}.$$

Introducing numerical values:

$q = 1.60 \times 10^{-19}$ coulomb $\qquad V = 7 \times 10^5$ volts

$c = 3 \times 10^8$ m/sec $\qquad M = 1.67 \times 10^{-27}$ kg $\qquad s = 3$ m

$\epsilon_o = 8.85 \times 10^{-12}$ coulomb/volt-m

gives the result:

a) $$\frac{U_{rad}}{U_k} = 1.31 \times 10^{-20}.$$

Obviously radiation losses are negligible in linear accelerators.

The rate of radiation of energy in a linear accelerator can be easily related to the gain in kinetic energy per unit distance as follows:

$$U_k = \frac{1}{2} Mv^2$$

$$\frac{dU_k}{dx} = Mv\frac{dv}{dx} = Mv\frac{dv}{dt}\frac{dt}{dx} = Ma.$$

Hence

$$a = \frac{1}{M}\left(\frac{dU_k}{dx}\right)$$

and, using Eq. 4.13

b) $$\frac{dU}{dt} = \frac{q^2}{6\pi\epsilon_o M^2 c^3}\left(\frac{dU_k}{dx}\right)^2.$$

4.2 RADIATION FROM ANTENNAS

The oscillating current element is the simplest of all antennas. A uniform filamentary current I is pictured as flowing along an elementary length $\Delta\ell$. One can imagine this as occurring (in principle, at least) in a wire of such extremely short length that I remains virtually constant along the entire wire. While this may appear to be a fairly unrealistic concept, it is of great theoretical importance because any real circuit or antenna carrying an oscillatory current may be considered as made up of a large number of these elementary units, joined end to end. The principle of superposition then enables us to derive the total electromagnetic field at a point in space as a sum of fields of these individual units.

The oscillating current element

The quantity that determines the radiation field is the product $[qa(t')]$ which appears time and again in Eqs. 4.8 through 4.13. It is this quantity which we shall now examine. Instead of q being a point charge, let it be spread out spatially over a short length $\Delta\ell$. If μ is the charge per unit length, then $qa(t')$ becomes $\mu a(t')\,\Delta\ell$ and since the acceleration equals the rate of change of velocity u, we can write $qa(t')$ as $\Delta\ell[d(\mu u)/dt']$. But the product μv is just the current at any instant t' so that

$$qa(t') = \left[\frac{dI(t')}{dt'}\right]\Delta\ell. \tag{4.14}$$

The current is oscillatory and we assume that it undergoes sinusoidal variations of frequency ω such that $I(t') = I_o \cos(\omega t')$, where I_o is the amplitude. Equation 4.14 therefore becomes

$$\left.\begin{aligned} qa(t') &= -I_o \Delta\ell\omega \sin(\omega t') \\ &= -I_o \Delta\ell\omega \sin\left[\omega\left(t - \frac{r}{c}\right)\right] \\ &= -I_o \Delta\ell\omega \sin\left[\omega t - kr\right] \end{aligned}\right\} \qquad (4.15)$$

The second relation reflects the fact that the time t at which we observe the field at some distance r is retarded relative to the time t' at which the charge acceleration took place at $r = 0$. The connection between t and t' is as given by Eq. 4.7. The third relation merely presents the result in terms of the propagation constant k which is related to ω and the phase velocity c through $k = \omega/c$.

Inserting the last equation (4.15) in Eq. 4.10 yields the electric field

$$E(r,t) = \frac{I_o \Delta\ell \sin\theta}{4\pi\epsilon_o r c^2}\, \omega \sin(\omega t - kr) \qquad (4.16)$$

and since $B = E/c$, also the corresponding magnetic field

$$B(r,t) = \frac{I_o \Delta\ell \sin\theta}{4\pi\epsilon_o r c^3}\, \omega \sin(\omega t - kr). \qquad (4.17)$$

The directions of $\vec{E}$ and $\vec{B}$ are exactly as is shown in Fig. 4.6, with the understanding that the current element $I\Delta\ell$ is oriented so that the vector $\Delta\vec{\ell}$ is coincident with the acceleration vector $\vec{a}$. Likewise, the radiation pattern of this current element is just like the patterns shown in Figs. 4.7 and 4.8.

To obtain the radiated power, we substitute Eq. 4.15 in Eq. 4.13 and find that

$$P(t) = \left[\frac{(\Delta\ell\omega)^2}{6\pi\epsilon_o c^3}\right] I_o^2 \sin^2(\omega t - kr) \text{ watt}. \qquad (4.18)$$

This formula looks like the formula $P = I^2R$ for the power dissipation in a resistor R, with the term in square brackets having dimensions of resistance. However, unlike R of the joule heating formula, the quantity

$$R_{rad} = \frac{(\Delta\ell\omega)^2}{6\pi\epsilon_o c^3} = 80\pi^2\left(\frac{\Delta\ell}{\lambda}\right)^2 \text{ ohm} \tag{4.19}$$

is a measure of the efficiency with which the antenna radiates for a given input current I. It is known as the radiation resistance. The second form of the equation comes from replacing ω by ($2\pi c$/wavelength λ) and by noting that $\epsilon_o c^3 = c/\mu_o = c/(4\pi \times 10^{-7})$.

We wish to remark that instead of using the notion of a current element $Id\ell$ one often finds a different representation of essentially the same ideas, based on the fact that an oscillating charge also looks very much like a time-varying dipole. Conservation of charge requires that there be an accumulation of charge at the end of the element $\Delta\ell$ given by $dq/dt' = I_o \cos(\omega t')$. That is, the charge at one end is increasing and at the other it is decreasing by the amount of the current flow. Thus, at an instant of time there is an excess of positive charge q at the end toward which the current flows and a reduction of charge q at the opposite end. The magnitude of the charge accumulations is given by $q(t') = I_o \sin(\omega t')/\omega$. Hence the product $I_o\Delta\ell \sin(\omega t')$ becomes $q(t')\,\Delta\ell\omega$, or $p(t')\,\omega$ where $p(t') = q(t')\,\Delta\ell$ is the value of the time-varying dipole moment. It is for this reason that the oscillating current element is also frequently referred to as the oscillating electric dipole. The field equations governing its radiation are identical with those given by (4.16) and (4.17) with $I_o\Delta\ell\omega \sin(\omega t - kr)$ replaced with $p_o\omega^2 \sin(\omega t - kr)$.

The short "practical" antenna

The hypothetical current element discussed above has useful theoretical attributes but it is not a practical antenna. The reason is that in a real physical wire the current I cannot be the same

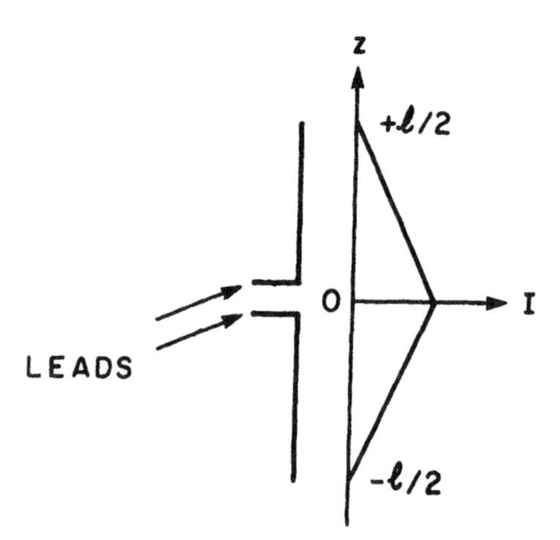

Fig. 4.10 The current I as a function of position z along a short "practical" antenna. The current drops off to zero at the two ends. The total length of the antenna is ℓ.

everywhere however short the wire may be, but is maximum at the point where the antenna is fed, and falls off to zero at the two ends. This is shown in Fig. 4.10 from which we see that the drop-off is pretty much linear. The radiation pattern of this antenna is virtually identical with that of the elementary current element dicussed earlier. However, for the same current I at the terminals the short practical antenna of length ℓ radiates only one quarter as much power as the current element of the same length which has the current I distributed uniformly over its entire length. This is because now the product $I\ell$ is but one half that of the current element, so that the field strengths at every point are reduced to one half and the Poynting flux is reduced to one quarter. Therefore, the radiation resistance of the practical short antenna is one quarter the value given by Eq. 4.19, or

$$R_{rad} \approx 20\,\pi^2\left(\frac{\ell}{\lambda}\right)^2 \text{ ohm.} \tag{4.20}$$

Although this formula holds strictly for very short antennas only, it is a good approximation even when ℓ is as large as a quarter of a wavelength. Taking this length then, R_{rad} becomes equal to approximately 12.5 ohm. Thus, when this antenna is fed with 100 amp of radio-frequency current I_o, it will radiate a total of 125 kW peak power at the wavelength λ in question.

Longer Antennas

In order to compute the electric and magnetic fields around longer antennas it is necessary to know the current distribution I(z) along the wire. This information can be obtained by an exact solution of Maxwell's equations but the problem is a difficult one and it is only fairly recently that a full solution has been obtained [for the details see R. W. P. King, Theory of Linear Antennas (Harvard University Press, Cambridge, Mass. 1956).] On the basis of such solutions, however, a fairly simple picture emerges. When the wire is very thin compared with its length, the current I(z) is very nearly sinusoidal and is zero at the ends of the wire (see Fig. 4.11). That is,

$$I(z) = I_o \sin\left[m\pi\left(\frac{|z|}{\ell}+\frac{1}{2}\right)\right]. \tag{4.21}$$

Here m is an integer which equals the number of half wavelengths contained in the length ℓ of the wire, and I_o is the peak current. Thus for a half-wave antenna (m = 1) the current peaks at the center of the wire z = 0 and falls off sinusoidally to zero at the two ends, $z = \pm\ell/2$.

To find the electric field at a distant observation point the procedure is as follows. One subdivides the antenna into tiny elements of length Δz with currents I(z) flowing through it. The field is given by Eq. 4.16. The total field at a point is then deduced by summing (that is, integrating) the conbributions of all such elements. Hence

$$E(r,t) = \int_{-\ell/2}^{\ell/2} \frac{I(z)\sin\theta}{4\pi\epsilon_o rc^2}\,\omega\sin(\omega t - kr)\,dz. \tag{4.22}$$

Now if r is very large compared with ℓ one may consider the quantity $\sin\theta/r$ to be a slowly varying function of z that can be removed from under the sign of integration. Note, however, that the phase

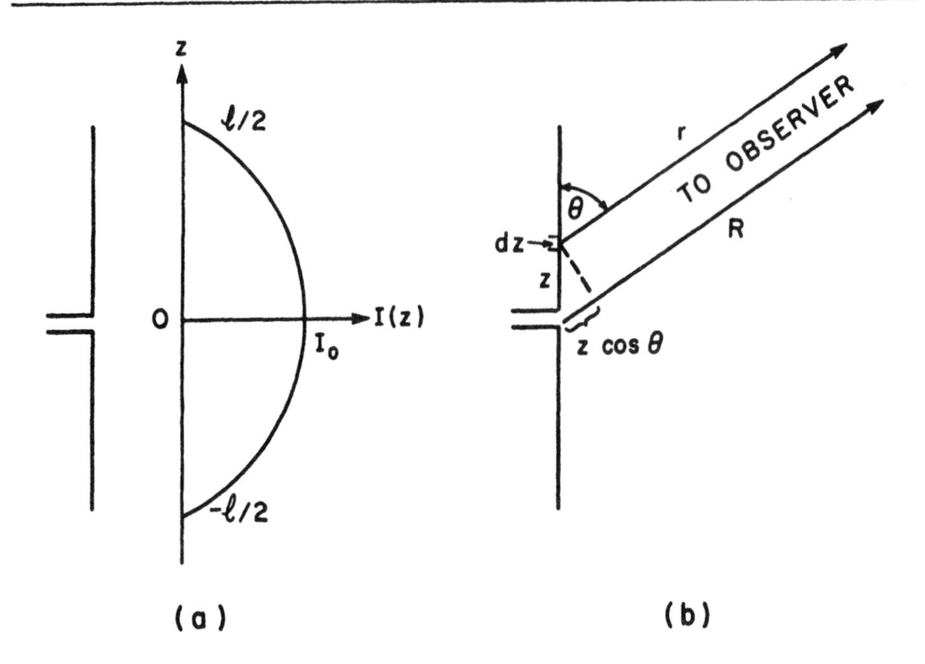

Fig. 4.11 (a) Approximate current distribution on a half-wave antenna $\ell = \lambda/2$. The distribution is given by $I = I_o \cos(\pi z/\ell) = I_o \cos(2\pi z/\lambda)$. (b) The path length difference at a distant observer between a wave emanating from a current element at $z = 0$ and another current element located at a distance z from 0. The path length difference of $z \cos \theta$ leads to a difference in arrival time of $z \cos \theta/c$. This causes an interference between the waves.

$(\omega t - kr)$ must be treated with more care. From Fig. 4.11 it is clear that the time required for a wave emitted from the element dz to reach the observer is less than that required for the wave originating from 0. The time lag is $(z \cos \theta/c)$ and the corresponding lag in phase is $(kz \cos \theta)$ where $k = 2\pi/\lambda$. Such a lag causes a reduction of the field at the observation point and if the lag becomes as much as half a wavelength in distance it cancels out the wave originating from 0. Hence it is imperative that the lags of waves from all the elements of the wire be kept track of in the phase term of Eq. 4.22. Since $r \approx R - z \cos \theta$, $\sin(\omega t - kr)$ becomes $\sin(\omega t - kR + kz \cos \theta)$ and Eq. 4.22 takes on the form

$$E(R,t) \approx \frac{\omega \sin\theta}{4\pi\epsilon_0 Rc^2}\int_{-\ell/2}^{\ell/2} I(z)\sin(\omega t - kR + kz\cos\theta)\,dz. \qquad (4.23)$$

For a half-wave antenna, say, Eq. 4.21 gives the current distribution as $I(z) = I_0 \sin[\pi z/\ell + \pi/2]$ a result which must now be inserted in Eq. 4.23, and the integration over z performed. The integration is straightforward but a little tedious and we shall leave it to the reader to verify that

$$E(R,t) = \frac{I_0 \ell}{4\pi\epsilon_0 Rc^2}\left[\frac{2}{\pi}\,\frac{\cos\left(\frac{\pi}{2}\cos\theta\right)}{\sin\theta}\right]\omega\sin(\omega t - kR); \qquad [\ell = \lambda/2] \qquad (4.24)$$

which differs from the formula (4.16) for the current element mainly in the form of the angular distribution of the electric field. The radiation pattern represented by this angular distribution of $\vec{E}$ gives a Poynting flux $|\vec{S}(\theta)|$ which has the form

$$|\vec{S}(\theta)| \propto \frac{\cos^2\left(\frac{\pi}{2}\cos\theta\right)}{\sin^2\theta} \qquad (4.25)$$

and is shown plotted in Fig. 4.12. We see that it is somewhat more directional than the pattern of the elementary current element $I\Delta\ell$. This represents a desirable feature of this antenna. The total power emitted by the antenna is obtained by integrating the flux over a complete sphere of radius R, in the manner outlined in an earlier subsection in connection with the radiating current element. Now, however, the integral over θ has no simple analytic solution and must be evaluated numerically. The result is

$$P(t) = 73\, I_0^2 \sin^2(\omega t - kR) \text{ watt} \qquad (4.26)$$

and thus the corresponding radiation resistance equals

$$R_{rad} = 73 \text{ ohm}. \qquad (4.27)$$

This is to be compared with the value of 12.5 ohm calculated earlier for the case of the quarter-wavelength long antenna. Therefore, for

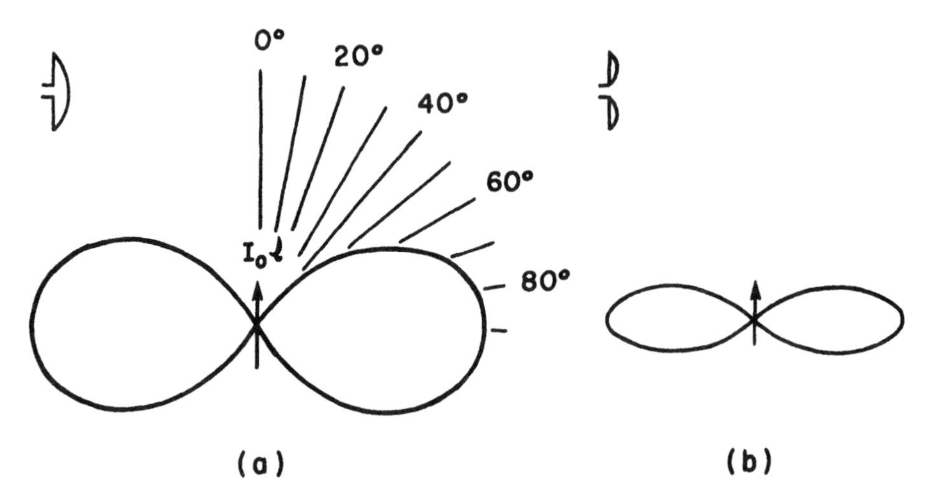

Fig. 4.12 (a) Radiation pattern of a half-wave antenna $\ell = \lambda/2$. As the antenna is made longer, constructive and destructive interference of waves emanating from different portions of the antenna becomes more and more pronounced. (b) Radiation pattern of a full-wave antenna $\ell = \lambda$, m = 2. Both antennas are center-driven. The small inserts illustrate the current distributions.

the same current input I_0, the half-wave antenna emits about six times more power than the quarter-wave antenna.

The near field of an oscillating current element

Heretofore we have concentrated our attention on the distant field of a radiating charge distribution. The reason is two-fold: first these fields represent the more interesting and important aspects of the problem; and second, we were able to derive them from simple physical arguments. This cannot be said about the near fields, that is, fields that are observed at distances r comparable to the wavelength of the radiation. They are quite complex and their derivation is not easy. We merely state the results:*

*E_θ of Eq. 4.28a is defined to be positive when it points in the direction of increasing θ, in conformity with Fig. 4.13. This differs from our earlier convention where $\vec{E}$ was defined to be positive when it points along the instantaneous acceleration vector. Hence the sign in front of Eq. 4.28a is opposite to that in front of Eq. 4.16.

$$\left.\begin{aligned}
E_\theta &= \frac{I_o \Delta\ell \sin\theta}{4\pi\epsilon_o}\left[-\frac{\omega \sin(\omega t')}{rc^2} + \frac{\cos(\omega t')}{r^2 c} + \frac{\sin(\omega t')}{\omega r^3}\right] && \text{(a)}\\
E_r &= \frac{2I_o \Delta\ell \cos\theta}{4\pi\epsilon_o}\left[\frac{\cos(\omega t')}{r^2 c} + \frac{\sin(\omega t')}{\omega r^3}\right] && \text{(b)}\\
B_\phi &= \frac{I_o \Delta\ell \sin\theta}{4\pi\epsilon_o}\left[-\frac{\omega \sin(\omega t')}{rc^3} + \frac{\cos(\omega t')}{r^2 c^2}\right] && \text{(c)}
\end{aligned}\right\} \qquad (4.28)$$

$$E_\phi = B_r = B_\theta = 0.$$

Here the subscripts r, θ, and ϕ denote the fact that we are now dealing in spherical coordinates where the three components of $\vec{E}$

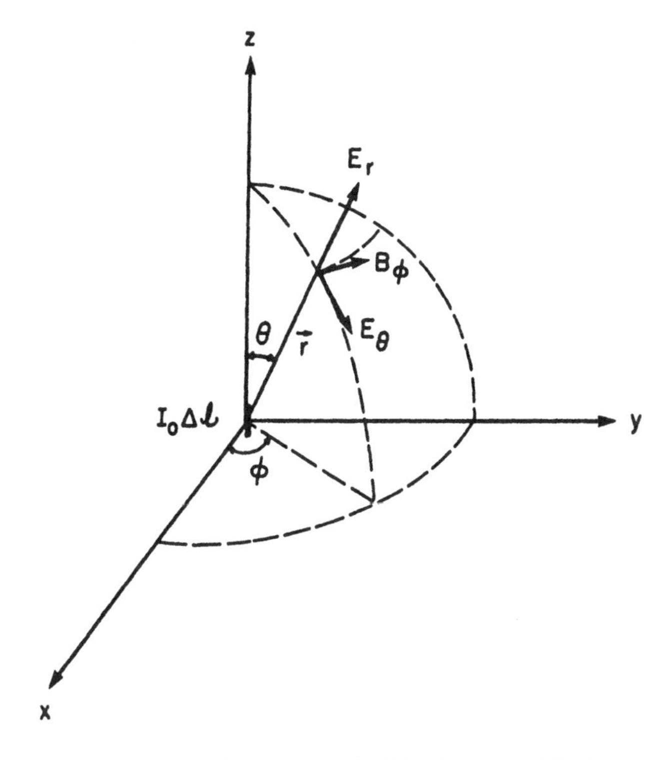

Fig. 4.13 Components of the near field of an oscillating current element $I_o \Delta\ell$ oriented along the z axis. The field components are expressed in spherical coordinates.

and the three components of $\vec{B}$ are resolved along the radial, polar, and azimuthal directions as is shown in Fig. 4.13.

At very large distances r, terms containing $(1/r^2)$ and $(1/r^3)$ become vanishingly small. Here the radial electric field approaches zero and the only terms that survive are the first term of Eq. 4.28a and the first term of Eq. 4.28c. These are precisely the electric and magnetic components of the radiation field we discussed extensively earlier in this and the previous sections.

When r is very small the dominant terms are those that vary as $(1/r^3)$, namely

$$E_\theta = \frac{I_o \Delta\ell \sin\theta \sin\omega t'}{4\pi\epsilon_o \omega r^3} = \frac{(p_o \sin\omega t') \sin\theta}{4\pi\epsilon_o r^3}$$

$$E_r = \frac{2I_o \Delta\ell \cos\theta \sin\omega t'}{4\pi\epsilon_o \omega r^3} = \frac{2(p_o \sin\omega t') \cos\theta}{4\pi\epsilon_o r^3}. \qquad (4.29)$$

These will be recognized as the fields that would exist around an electrostatic dipole of moment $p = p_o \sin\omega t'$. The fields of course are not static but their structure at any instant of time is precisely that of an electrostatic dipole.

We see that at very large and very small distances r, Eqs. 4.28 can readily be understood. At intermediate distances new terms that vary as $(1/r^2)$ make their appearance and now the field structure becomes quite complex. Fields that vary as $1/r^2$ are often referred to as the induction fields of the radiating antenna.

One may well inquire how the radially directed Poynting flux $S_r = (E_\theta B_\phi)/\mu_o$ fares when the appropriate values for E_θ and B_ϕ are inserted from Eqs. 4.28. The expression looks awfully messy and contains terms that vary as $1/r^5$, $1/r^4$, $1/r^3$ and $1/r^2$. However, the time average value $\langle S_r \rangle$ undergoes dramatic simplification and only one term survives:

$$\langle S_r \rangle = \frac{(I_o \Delta\ell)^2 \omega^2 \sin^2\theta}{32\pi^2 \epsilon_o r^2 c^3}. \qquad (4.30)$$

It arises from those components of $\vec{E}$ and $\vec{B}$ which vary as $1/r$. At large distance r the radiation terms are the only ones that have appreciable value, but now we know that even close to the current element where the electrostatic and induction fields dominate, only $(1/r)$ terms contribute to the average outward power flow of energy. It is probably unnecessary to point out that Eq. 4.30 is nothing more than the time-average value of Eq. 4.11 with $\langle [qa(t')]^2 \rangle = \langle [I_0 \Delta\ell\omega \sin(\omega t - kr)]^2 \rangle$.

4.3 BREMSSTRAHLUNG

When electrons pass through matter, they suffer collisions with other particles; they are accelerated and decelerated and as a result emit electromagnetic radiation. The emission suffered during these collisional processes is known as bremsstrahlung which, translated from German, means breaking or stopping radiation. The name stems from the time when it was discovered that radiation is emitted whenever a beam of high-energy electrons smashes into a thick metal target. The resulting x-ray emission which was then observed was interpreted as being caused by electrons that penetrate an atom, come into close proximity of the nucleus and are strongly accelerated by it. Today we know that bremsstrahlung is not confined to the x-ray region alone, but has been observed under a variety of situations over the

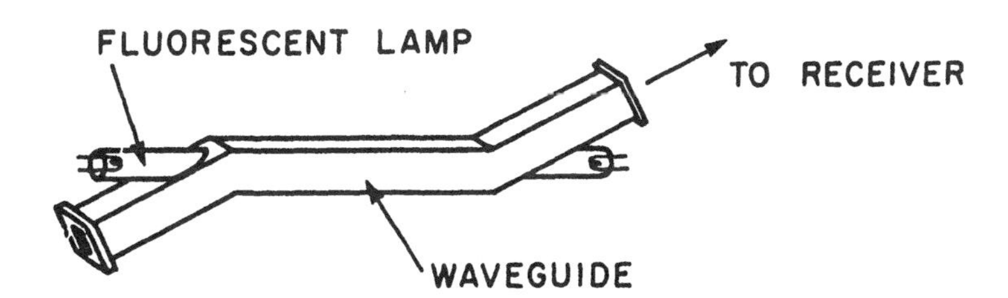

Fig. 4.14 Viewing the bremsstrahlung from electron-atom collisions in a fluorescent lamp observed at microwave frequencies. The frequency of measurement is 3000 MHz. An antenna in place of a waveguide could be used equally well to detect the radiation. [For details see G. Bekefi, J. L. Hirshfield, and S. C. Brown, Phys. Rev. 116, 1051 (1959.)]

entire electromagnetic spectrum. For example, drill a couple of holes in a microwave waveguide (see Fig. 4.14), insert a fluorescent lamp into the guide and connect one end of the guide to a very sensitive receiver. When the fluorescent lamp is turned on, the receiver picks up bremsstrahlung at microwave frequencies generated as a result of collisions of the free electrons with neutral argon atoms contained within the discharge tube. To be sure, the radiation is very weak, it may be as small as 10^{-15} watts within a receiving bandwidth of 1-MHz frequency, but it is there, nonetheless, and has been clearly identified. A more interesting source of bremsstrahlung is interstellar gas. Huge clouds of ionized matter are known to exist in outer space in an almost fully ionized state, by which we mean that virtually all the atoms have been stripped of at least one electron and thus exist as free ions only. Collisions of the free electrons with the free ions of this so-called plasma medium give emissions of radio and microwave signals which are readily picked up by Earth stations equipped with radio telescopes. The Orion nebula shown in Fig. 3.20 is one of many such known extraterrestrial sources. The Sun's chromosphere and corona furnish additional examples of sources from which this "breaking" radiation is known to emanate.

Power radiated in an electron-ion collision

To compute the power radiated by a colliding electron one needs to know its acceleration, and to find the acceleration one must understand the particle's motion. The problem can become quite complex in the case of an electron boring its way into a neutral atom. Not only is the electron moving in a complicated force field, but the entire calculation may require quantum mechanics for its solution. An equally interesting but much simpler problem is that of an electron interacting with an ion in ionized matter like that discussed earlier. Now the electron interacts with the familiar coulomb field of the ion, which for all practical purposes may be considered as a point charge. To be sure, some electrons will penetrate into the ion itself as was the case for the neutral atoms, but the majority of the electrons will, so to speak, feel out the long-range coulomb field of the ion at relatively large distances from the nucleus where the force is simply given by

$$|F(r)| = \frac{qQ}{4\pi\epsilon_o r^2}. \tag{4.31}$$

Here q is the charge on the electron, Q the charge on the ion and r the distance between the two. The motion of the electron in this central ($1/r^2$) field is well known and is illustrated in Fig. 4.15. The electron approaches the heavy, virtually immobile ion with a velocity $\vec{v}'$; its perpendicular distance to 0 is called the impact parameter b. The electron leaves with velocity $\vec{v}$ having suffered a deflection; its path is a hyperbola (to be quite correct, it is a hyperbola only when $|\vec{v}'| = |\vec{v}|$, but see below).

To find the radiation we need to know the acceleration at every point of the hyperbolic trajectory. It can be found, but it is time consuming; instead, we restrict our attention to only those electrons which pass the ion at appreciable distance. Most collisions will be of this type and therefore the error of neglecting all close collisions turns out to be not serious. At large distance, then, the hyperbolic orbit becomes almost a straight line (see Fig. 4.16) and this is the situation we shall now examine. The approximation becomes better and better as v' increases because a fast electron spends little time near the ion and will not be deflected appreciably. But, there is one more approximation we would like to make first. We would like to say that the incoming and outgoing speeds $|\vec{v}'|$ and $|\vec{v}|$ of the electron are virtually identical, which implies that the initial and final kinetic energies of q are almost unchanged. This can be true only if the energy lost by the electron due to all causes is negligible compared with its initial kinetic energy. One source of energy loss is just the elastic recoil caused by the fact that the approaching electron induces motion in the ion. A simple billiard-ball type of a calculation shows that the fraction of energy lost by the electron in this process is about twice the ratio of the electron to ion mass. For a hydrogen ion this means that only about 1 part in 900 of the energy is lost in recoil. The second source of energy loss is the possibility that the

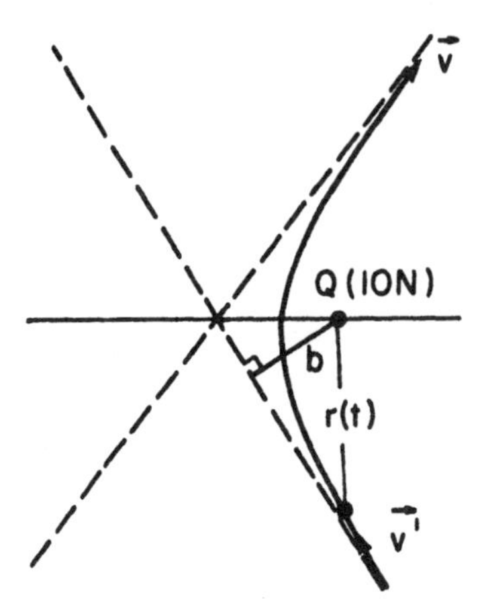

Fig. 4.15 Hyperbolic trajectory of an electron in the coulomb field of a positive ion. The length b is called the impact parameter; r(t) is the instantaneous distance between the ion and electron.

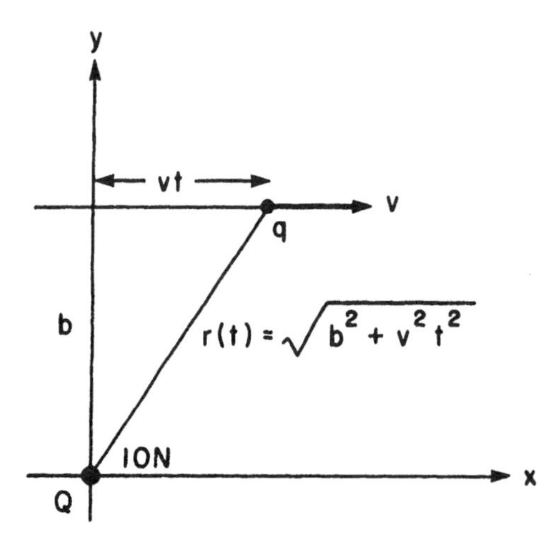

Fig. 4.16 Straight-line approximation to the electron trajectory in the field of an ion of charge Q.

electron can cause excitations of the bound electrons in the ion. In such a process, the free electron could, in fact, lose virtually all its initial energy in one fell swoop. Fortunately, the probability of this happening is small so that when one views the average motion of many electrons, only an occasional one will suffer such a dramatic fate. And third, the electron will lose speed by the very act of emitting bremsstrahlung. To see how important this is, one can calculate the radiated energy assuming $|\vec{v}'| = |\vec{v}|$ and then compare the radiated energy with $(mv'^2/2)$. One finds that for the distant collisions on which we are focusing our attention, the bremsstrahlung energy is indeed small compared with the particle's kinetic energy.

And so if m is the mass of the electron, its acceleration a obtained from Eq. 4.31 equals $(qQ/4\pi\epsilon_0 mr^2)$ and the power radiated is

$$
\begin{aligned}
P &= \frac{dW}{dt} \\
&= \frac{q^4Q^2}{96\,\pi^3\epsilon_0^3 m^2 c^3}\,\frac{1}{[x^2+b^2]^2} \text{ watt.}
\end{aligned}
\tag{4.32}
$$

In deriving this formula we substituted a in Eq. 4.13 and used the fact that for the straight-line trajectory shown in Fig. 4.17, $r^2 = x^2 + b^2$. Observe that the instantaneous radiated power varies with time because $x = vt$. To find the total energy W radiated, merely bring dt from the denominator of dW/dt into the numerator of the right-hand side, and integrate over the entire duration of the collision from $t = -\infty$ to $t = +\infty$:

$$
\begin{aligned}
W &= \frac{q^4Q^2}{96\,\pi^3\epsilon_0^3 m^2 c^3}\int_{-\infty}^{\infty}\frac{dt}{[v^2t^2+b^2]^2} \\
&= \frac{q^4Q^2}{192\,\pi^2\epsilon_0^3 m^2 c^3}\,\frac{1}{vb^3} \text{ joules.}
\end{aligned}
\tag{4.33}
$$

The second equation comes from evaluating the fairly standard integral appearing in the equation. The integral has the value $\pi/(2vb^3)$.

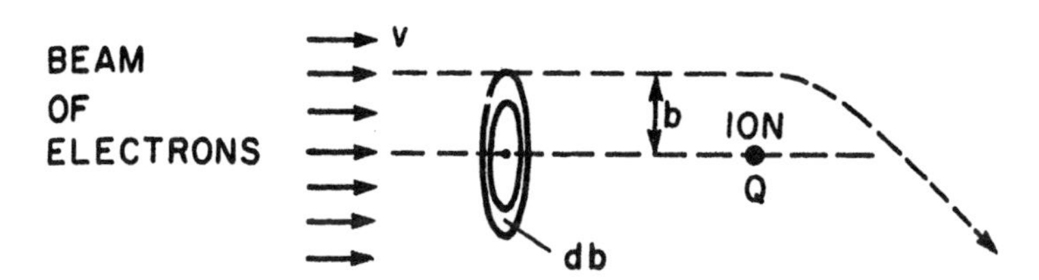

Fig. 4.17 A parallel, monoenergetic beam of electrons approaching an ion. Each electron approaches Q with a different impact parameter b.

W is the energy radiated by a single electron in a single collision. Now let us picture a parallel beam of electrons traveling towards the ion, each having a velocity v. Different electrons of the beam will approach the ion with different impact parameters b. If the electron density is N_e, the flux of electrons equals $N_e v$ and thus the number of electrons per second crossing the annular area $2\pi b db$ shown in Fig. 4.17 equals $N_e v\ 2\pi b db$. But this is just the rate at which the beam electrons in the annular ring "collide" with the ion, emitting W watts per "collision." Therefore the total energy radiated per second in the form of bremsstrahlung equals $WN_e v\ 2\pi b db$ integrated over all possible impact parameters b:

$$\begin{aligned} P_{brems.} &= \int WN_e v\ 2\pi b db \\ &= \frac{N_e q^4 Q^2}{96\ \pi\epsilon_o^3 m^2 c^3} \int_{b_o}^{\infty} \frac{db}{b^2} \end{aligned} \tag{4.34}$$

where in obtaining the second equation, we merely substitute for W from Eq. 4.33. Observe that we cut off the integration at a value $b = b_o$ rather than extending it all the way to zero. We do this from necessity because otherwise the integral diverges. Is our whole theory wrong then? Not really, it is only wrong at small distances b where our initial assumption of a straight-line trajectory is not valid, because here the electron is close to the ion and is likely to suffer large deflections. With these reservations the integral is readily evaluated with the result that

$$P_{brems.} = \frac{N_e q^4 Q^2}{96\, \pi \epsilon_o^3 m^2 c^3} \frac{1}{b_o}. \tag{4.35}$$

Here b_o is the minimum impact parameter which needs to be deduced from some other physical considerations. One such consideration comes from quantum mechanical modifications arising from the wave nature of particles. The uncertainty principle sets certain limits on the validity of classical orbit calculations like the above. More specifically, it says that a classical definition of a particle trajectory must lose its meaning over distances smaller than the particle's wavelength h/mv, where h is Planck's constant. Thus, it is fairly customary* to set b_o approximately equal to h/mv. More accurate calculations show that $b_o \approx h/(2\pi mv)$ with the result that

$$P_{brems.} = \frac{N_e q^4 Q^2 v}{48\, \epsilon_o^3 m c^3 h} \text{ watts per ion.} \tag{4.36}$$

We see that the bremsstrahlung power increases linearly with the particle speed, and the hotter the electrons, the larger the radiation. Note also that the power increases as the square of the ionic charge. For example, a completely stripped oxygen ion plasma will radiate 64 times more power than a proton plasma, all other quantities ($N_e v$) being equal.

To have some feeling for magnitude of the radiated power, let us take the case of hydrogen ions being bombarded by electrons. Then $q = Q = 1.602 \times 10^{-19}$ coulomb. We insert numerical values for ϵ_o, m, c, and h in Eq. 4.36 and obtain the result that

$$P_{brems.} = 1.85 \times 10^{-38} N_e \sqrt{U} \text{ watts per ion.} \tag{4.37}$$

*For low energy particles, one takes b_o to equal that value of b for which the electron makes a 90° deflection relative to its original direction. From the kinematics of the motion one finds that $b(90°) = qQ/4\pi\epsilon_o mv^2$. In a given problem choose that b_o which gives the larger of the two values.

Here N_e is the electron density in units of meter^{-3} and U is the kinetic energy of the electrons expressed in electron volts (recall that $U = mv^2/2$ when expressed in joules, and $mv^2/2e$ when expressed in eV; $1\ \text{eV} = 1.602 \times 10^{-19}$ joules). Now suppose that there are N_p protons per cubic meter with which the electrons interact. It follows from Eq. 4.37 that the power radiated by the medium equals

$$P_{\text{brems.}} = 1.85 \times 10^{-38}\, N_e N_p \sqrt{U} \ \text{watt per m}^3. \qquad (4.38)$$

Ionized interstellar clouds are electrically neutral so that $N_e = N_i = N$ with N typically equal to $10^7\ \text{m}^{-3}$; the electron energy is of the order of 2 eV. Substituting these values in Eq. 4.38 yields an emission of $\sim 3 \times 10^{-24}\ \text{watt/m}^3$. This is a small number, but the clouds are large so that the total radiated power is not negligible. Note that the bremsstrahlung intensity in such clouds is proportional to N^2. Therefore, a measurement of the power and knowledge of the size of the cloud can be used in estimating the charged particle density N.

Bremsstrahlung imposes rather important restrictions on the operation of future thermonuclear fusion reactors. In these devices one hopes to raise the energy of low atomic weight nuclear fuel to such heights that two nuclei will come sufficiently close to one another to fuse, giving up energy. The probability of a fusion event increases with energy as does the bremsstrahlung calculated above. It turns out that a low energy $P_{\text{brems.}}$ exceeds the production rate by nuclear fusion but is less at high energy. Thus, there is a break-even energy when the production rate is just balanced by the loss rate due to bremsstrahlung. As is shown in Fig. 4.18, the break-even energy for a D-D reactor is 36 keV and for a D-T reactor it is 4 keV. When the fuel temperature is less than these values, a self-sustained nuclear reaction cannot be maintained. Take a D-T reactor operating at the break-even energy of 4 keV, and having a particle density $N_e = N_p = 10^{21}\ \text{m}^{-3}$. Formula (4.38) tells us that it will radiate away approximately 1.2 megawatts of power per cubic meter of reactor fuel as bremsstrahlung — an amount which just balances the production rate from the nuclear fuel.

We recall that the power radiated is proportional to Q^2, the

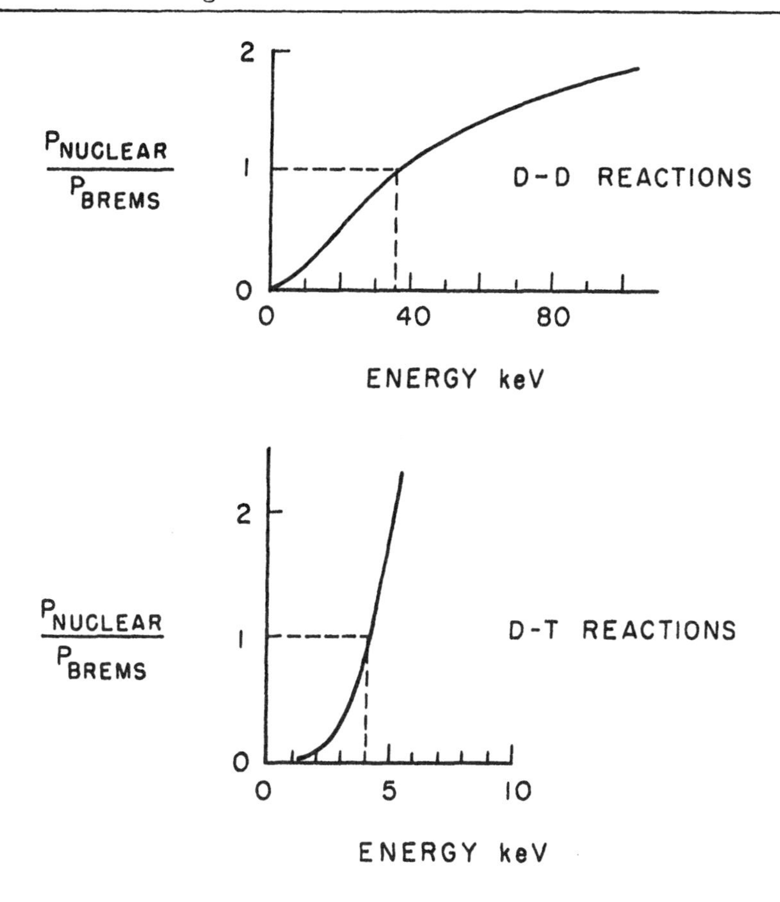

Fig. 4.18 Ratio of nuclear power generated, to bremsstrahlung power lost from reactor as a function of the electron energy.

square of the ionic charge. Hence, any high atomic weight impurities in the reactor can have disastrous effects on the energy balance of the machine. For this reason, purity control of the reactor plasma is of utmost importance, a problem that has as yet not been solved satisfactorily.

Polarization

Equation 4.8 and Fig. 4.6 tell us that the electric vector $\vec{E}$ of the emitted electromagnetic wave is polarized in a definite direction relative to the acceleration vector $\vec{a}$ of the radiating charge: if $\hat{n}$ is the unit vector directed from the charge to the position of the distant observer, then $\vec{E}$ points in the direction of $\hat{n} \times (\hat{n} \times \vec{a})$. At any one instant of time the radiation pattern is like that shown in Fig. 4.7. Therefore, the radiation emitted by a collimated beam of electrons is polarized and the radiation intensity is peaked in certain preferred directions. Matters are different in ionized clouds or in the fusion reactor discussed above. Here the electrons are incident on the ions from all possible angles and the electric field directions of the waves from the individual electrons are therefore randomly distributed with respect to one another. The electric field direction averaged over the time of observation is random and the bremsstrahlung is unpolarized.

Frequency spectrum

In section 4.2 we found that when the electrons of an antenna are set into sinusoidal oscillation with frequency ω, the antenna radiates electromagnetic waves precisely at the frequency of excitation. In the collision of an electron with an ion (or atom) the acceleration suffered by the light particle is far from sinusoidal – rather the acceleration is much like a sharp impulse lasting only for a short time while the electron is in the immediate vicinity of the ion. What then is the frequency distribution of the radiated power? To answer this problem let us refer back to section 2.5 in which Fourier analysis of such impulses was discussed. There we found a general result of great utility. It is this. When an impulse (having a quite arbitrary time history) lasts for a time τ seconds, it will generate all frequencies from zero to a maximum value ω_{max} in conformity with the relation (cf. Eq. 2.99)

$$\omega_{max}\tau \approx 2\pi. \qquad (4.39)$$

Moreover, the intensity of the signal at each frequency up to approximately ω_{max} will be roughly constant and then drop off to zero for frequencies $\omega \gtrsim \omega_{max}$. Accepting this result, we can sketch the frequency spectrum of the bremsstrahlung and this is shown in Fig. 4.19. Observe how different this is from our experience with more common

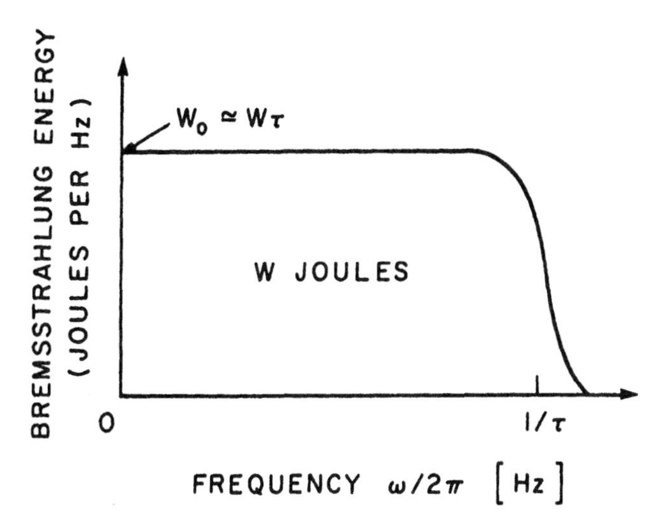

Fig. 4.19 Sketch of the energy spectrum of bremsstrahlung by a single nonrelativistic electron. The "plateau" magnitude W_o joules per Hz is approximately equal to the total energy W of Eq. 4.33 multiplied by τ. This follows from the fact that $W = \int W_o \frac{d\omega}{2\pi} \approx W_o \frac{\omega_{max}}{2\pi} = W_o \frac{1}{\tau}$.

type of radiators such as a radio signal at 89.7 MHz (Public Radio WGBH, Boston), the red light of a neon sign, or the bright yellow light of rock salt burned in the flame of a bunsen burner. The last three are essentially monochromatic sources, whereas bremsstrahlung is a broadband emitter of electromagnetic waves.

To find out how broadband a source we are dealing with, an estimate of the "collision time" τ needs to be made. If v is the speed of the electron and d the distance over which the accelerating force is large, then $\tau \approx d/v$. Inserting this in Eq. 4.39 gives

$$\omega_{max} \approx 2\pi \frac{v}{d}. \tag{4.40}$$

Suppose an electron comes within 10 Å (10^{-9} m) of a hydrogen ion. Assume it has a velocity of 5.93×10^6 m/sec, which corresponds to a kinetic energy of 100 eV. Then the maximum frequency of radiation $\omega_{max} \approx 3.7 \times 10^{16}$ rad/sec and the minimum wavelength $\lambda_{min} = 2\pi c/\omega_{max}$ or ~500 Å. From this we conclude that the bremsstrahlung from this electron will be approximately constant in intensity from zero frequency up to ~500 Å, a wavelength which lies in the soft x rays. It is a broadband signal indeed.

Equation 4.40 suggests that for a given speed v, ω_{max} can be made arbitrarily large by making d small enough. This is, however, not true. The photon picture of radiation throws light on this question. Suppose an electron of speed v' approaches an ion. Its initial kinetic energy before the collision is $mv'^2/2$. By energy conservation, this is the maximum that can go into creating a bremsstrahlung photon. Hence equating $mv'^2/2$ to $h\nu_{max}$ we get

$$\begin{aligned} \omega_{max} &= 2\pi\nu_{max} \\ &= \frac{2\pi}{h}\left(\frac{1}{2}mv'^2\right) \end{aligned} \tag{4.41}$$

where h is Planck's constant. This is known as the limit of Duane and Hunt, so named after its discoverers. It is clear that our classical orbit model must fail long before this limit is reached, if for no other reason than the fact that we assumed throughout that the bremsstrahlung loss in a collision is small compared with $mv'^2/2$; an assumption which was necessary in order that the electron's trajectory be given by a simple hyperbola.

4.4 CYCLOTRON AND SYNCHROTRON RADIATION

When an electron is injected at right angles to a steady uniform magnetic field B_0, it executes circular motion about the magnetic field lines. And while its orbital speed around the circle may be constant, it experiences a centripetal acceleration and as a result emits electromagnetic waves. This is known as cyclotron radiation when low-energy electrons are involved, and synchrotron radiation when relativistic electrons participate in the motion — names given to these phenomena because they were first observed in cyclotrons and synchrotrons.

Let us begin with the cyclotron emission by nonrelativistic electrons. The emitted power is readily calculated. Suppose v is the speed of the electron in its circular orbit. The magnetic force $\vec{F}_M = q(\vec{v} \times \vec{B}_o)$ gives the electron a centripetal acceleration whose magnitude equals v^2/R with R as the radius of the orbit. Equating qvB_o to mv^2/R gives the radius

$$R = \frac{mv}{qB_o} \tag{4.42}$$

and hence the acceleration

$$a = v\,\frac{qB_o}{m}. \tag{4.43}$$

The frequency of orbiting ω_c, also known as the cyclotron or gyrofrequency, is connected to v and R through $v = \omega_c R$ and use of Eq. 4.42 then gives the familiar result

$$\omega_c = \frac{qB_o}{m}\ \text{rad/sec}. \tag{4.44}$$

The acceleration is therefore $a = v\omega_c$ and substitution of this value in Larmor's formula (4.13) yields for the radiated power

$$P_{cycl} = \frac{q^2\omega_c^2 v^2}{6\pi\epsilon_o c^3}\ \text{watts per electron}. \tag{4.45}$$

If the radiating medium contains N_e electrons per cubic meter, then the power emitted per unit volume equals N_e times the value given by Eq. 4.45, or

$$\begin{aligned} P_{cycl} &= \frac{N_e q^2\omega_c^2 v^2}{6\pi\epsilon_o c^3} \\ &= 6.21 \times 10^{-20}\ N_e B_o^2 U\ \ \text{watt/m}^3 \end{aligned} \tag{4.46}$$

where the second version of the equation follows by substituting numerical values for q, m, ϵ_o, and c into the first version; the electron energy U appearing in the formula is expressed in electron volts.

Suppose we take the experiment of Fig. 4.14 and apply a magnetic field B_o along the axis of the discharge tube. The free electrons will

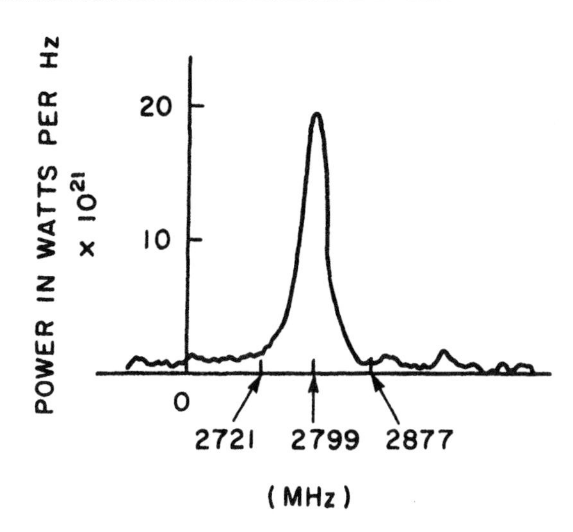

Fig. 4.20 Emission spectrum of cyclotron radiation from the non-relativistic charges of a fluorescent lamp subjected to a uniform magnetic field of 1000 gauss. The Q of this resonance line is approximately 60 and much of this damping comes from collisions of the electrons with the gas atoms. Part of the broadening comes from the Doppler effect. The electrons have random motions along the field lines; when they approach the detector the observed frequency is $\omega_c[1 + (v/c)]$ and when they recede, the frequency is $\omega_c[1 - (v/c)]$, with a different v for each electron. [For details see G. Bekefi and S. C. Brown, Am. J. Phys. **29**, 404 (1961).]

orbit around the field lines and emit cyclotron radiation. How much? Well, in this lamp typical values of N_e and U are 10^{17} m^{-3} and 1 eV, respectively. Let B_o equal 1000 gauss or 0.1 webers/m^2. With these values, Eq. 4.46 yields $P_{cycl} \approx 6 \times 10^{-5}$ watt/m^3. This quite appreciable power is readily observed. The result of such a measurement is illustrated in Fig. 4.20 which shows a plot of the power flowing down the waveguide as a function of the frequency of observation. The magnetic field was almost exactly 1000 gauss. We see that the emission is strongest at a frequency of $\nu = \omega/2\pi \approx 2800$ MHz and drops off rapidly on both sides of this value. But, this frequency is nothing other than the cyclotron frequency $\omega_c/2\pi$ given by Eq. 4.44 and computed for 1000 gauss. From this we infer that cyclotron emission occurs at a single frequency $\omega = \omega_c$, a result which is not really surprising. The motion of an electron in a perfectly circular orbit is equivalent to the superposition of two linear sinusoidal oscillators of frequency ω_c, vibrating at right angles to one another. The outcome of this must be an electromagnetic wave at $\omega = \omega_c$. The cyclotron emission seen in the discharge tube is in addition to the bremsstrahlung that occurs as a result of the electrons colliding every so often with atoms and ions. How do these two phenomena compare in regard to power? In the experiment just described the bremsstrahlung and cyclotron powers are roughly equal. However, the former is spread out over a huge frequency range (see section 4.3), whereas the latter is all concentrated around $\omega \approx \omega_c$. For this reason the cyclotron emission stands out prominently, being superposed on the weak background of bremsstrahlung.

At any one instant of time, the spatial distribution of the radiated intensity (that is, the radiation pattern) is like that given in Fig. 4.21. It is the typical pattern of an accelerated, nonrelativistic charge discussed in section 4.1. The emission is greatest along the tangent to the circular orbit, that is, at right angles to the acceleration vector; there is as much emission in the backward direction as in the forward direction. The radiation is polarized and it is noteworthy that the state of polarization depends on the direction of observation. Suppose one looks along the magnetic field lines towards the circulating electron. One sees the acceleration vector $\vec{a}$ as a function of time sweeping around a circle. The electric vector $\vec{E}$ of the wave at a large distance from the orbit is always in a direction given by $\hat{n} \times (\hat{n} \times \vec{a})$ where $\hat{n}$ is a unit vector pointing from the charge to the observer (see sections 4.1 and 4.3). Hence, the tip of the electric vector also moves on a circular path and the wave is circularly polarized. If, however, the orbit (and the wave) are viewed in a direction perpendicular to $\vec{B}_0$ only the linear projection of the electron's motion is seen, and the wave propagating toward the observer is therefore linearly polarized. In between these two positions, elliptical polarization occurs. Figure 4.22 illustrates the geometry of these situations.

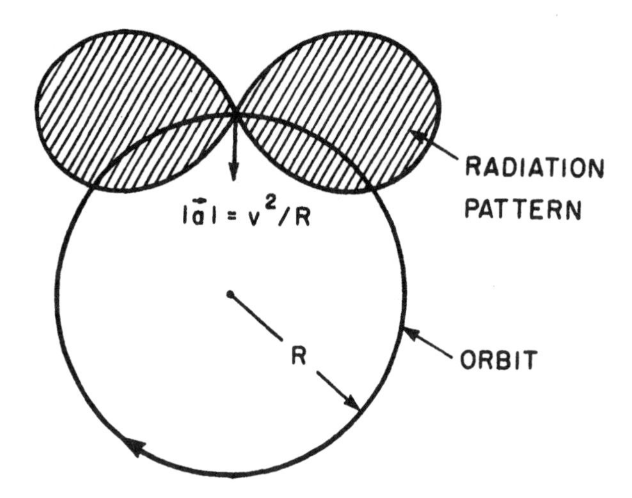

Fig. 4.21 Instantaneous radiation pattern of a nonrelativistic electron in a circular orbit in a uniform magnetic field $\vec{B}_o$ pointing into the page.

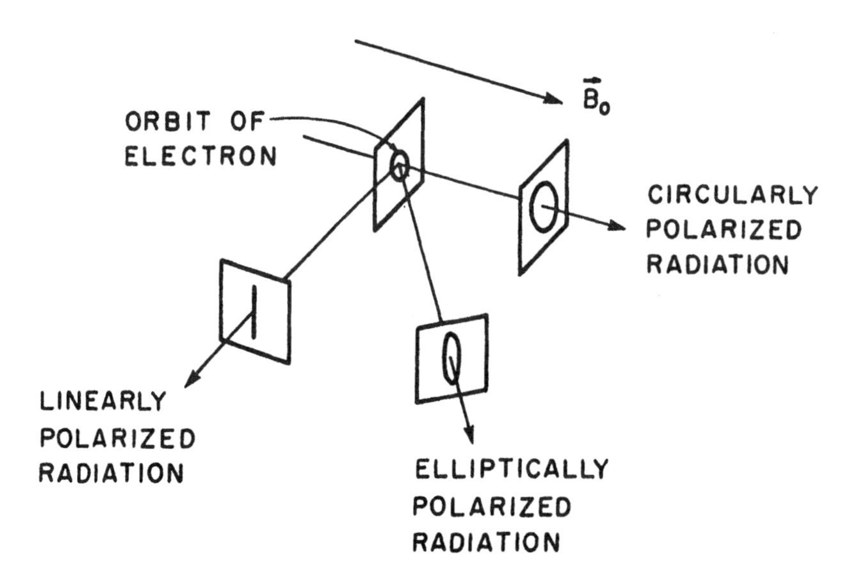

Fig. 4.22 Polarization of cyclotron radiation as a function of the angle of observation.

Example

A nonrelativistic electron of charge q and mass m rotates in a circular orbit with speed v in a uniform, steady, magnetic field of strength B_o. The power radiated by the electron is given by Eq. 4.45. Unless energy is continually supplied to the electron it will slow down. Show that the time taken for the initial energy to fall to (1/e) of its initial value is given by

$$\tau = \frac{6\pi\epsilon_o m^3 c^3}{q^4 B_o^2}.$$

Calculate this time for the electrons of a thermonuclear fusion reactor in which the electrons are confined by a magnetic field of 5 weber/m^2 (50 kG).

Solution

The kinetic energy ($1/2\ mv^2$) and the potential energy ($-\vec{\mu}\cdot\vec{B}$, where $\vec{\mu}$ is the magnetic moment of the one-electron current moving in a circle) are equal so the total energy U' is mv^2. From Eq. 4.45

$$P(t) = -\frac{dU'}{dt} = \frac{q^2\omega_c^2 v^2}{6\pi\epsilon_o c^3} = \frac{q^2\omega_c^2 U'}{6\pi\epsilon_o mc^3}.$$

But from Eq. 4.44

$$\omega_c = \frac{qB_o}{m};$$

hence we have

$$\frac{dU'}{U'} = -\frac{q^4 B_o^2}{6\pi\epsilon_o m^3 c^3}\,dt = -\frac{dt}{\tau}.$$

Thus

$$U'(t) = U_o e^{-t/\tau}$$

$$\tau = \frac{6\pi\epsilon_o m^3 c^3}{q^4 B_o^2}.$$

Inserting numerical values

$B_o = 5$ weber/m^2 $\qquad$ $m = 9.11 \times 10^{-31}$ kg

$c = 3 \times 10^8$ m/sec $\qquad$ $q = 1.60 \times 10^{-19}$ coulomb

$\epsilon_o = 8.85 \times 10^{-12}$ coulomb^2/newton-m^2

gives the result

$$\tau = 0.208 \text{ seconds.}$$

Synchrotron radiation*

We now turn to the radiation from an orbiting electron whose velocity v approaches that of light. In its own frame of reference the accelerating electron has the characteristic dipolar radiation pattern like that shown in Figs. 4.7, 4.8, and 4.21. But as v approaches c the radiation pattern becomes strongly peaked in the forward direction when viewed by an observer in the laboratory frame, as illustrated in Fig. 4.23. Therefore, when the observer looks in the orbital plane at the oncoming highly relativistic electron, he sees a bright point of light which flashes rapidly by him as the electron rushes headlong along its circular path. Calculations show that the angular width θ of this "searchlight" beam is approximately

$$\theta \approx \sqrt{1 - \frac{v^2}{c^2}} \text{ radians.} \tag{4.47}$$

*Formulas for the angular distribution of synchrotron radiation and its spectrum are derived in detail in Classical Electrodynamics by J. D. Jackson (John Wiley & Sons, Inc., New York, 1962), Chapter 14. We restrict ourselves to stating results.

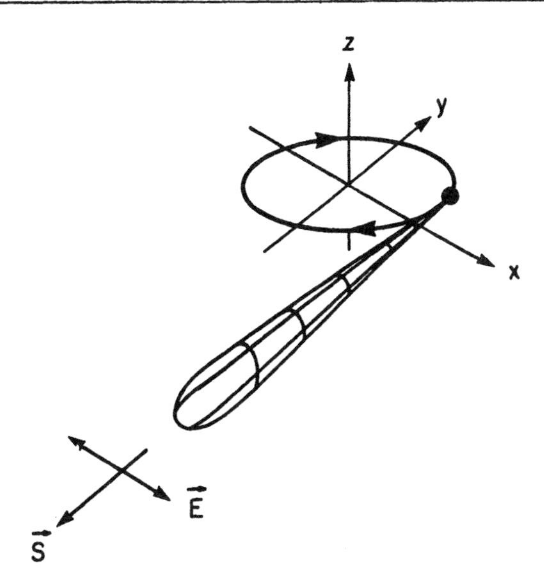

Fig. 4.23 The radiation pattern of synchrotron radiation by a highly relativistic electron. The pattern has the shape of a searchlight beam oriented along the instantaneous direction of the velocity vector. Observe that the light is linearly polarized with $\vec{E}$ vibrating in the plane of the orbit (i.e., the x-y plane). The radiant flux $\vec{S}$ is confined to a narrow cone of width $\approx \sqrt{1 - (v^2/c^2)}$ radians.

Now, the total relativistic energy (kinetic energy plus rest energy $m_o c^2$, with m_o as the rest mass) is given by

$$\begin{aligned} \epsilon &= mc^2 \\ &= \frac{m_o c^2}{\sqrt{1 - (v^2/c^2)}} \end{aligned} \qquad (4.48)$$

a result which states the well-known fact that the mass of a moving object is greater than its rest mass by the amount $[1 - (v^2/c^2)]^{-1/2}$ Using Eq. 4.48 and inserting it into Eq. 4.47 allows us to give the width of the searchlight beam in the following more convenient form:

$$\theta \approx \frac{m_o c^2}{\epsilon}. \tag{4.49}$$

Since $m_o c^2$ for an electron equals approximately 5×10^5 eV, we have, in the case of an electron of energy $\epsilon = 10^9$ eV (one BeV) the incredibly small angle $\theta \approx 5 \times 10^{-4}$ radians, or ~1.7 minutes of arc. The emission is therefore all concentrated in the direction of the particle's instantaneous velocity – virtually no emission takes place in the backward direction.

The searchlight effect has a profound influence on the emission spectrum of the radiation. No longer is the frequency of emission ω centered around the orbiting frequency ω_c as was the case for the non-relativistic electron. We are now dealing with an impulsive type of an event: Recall from section 4.3 that in such a situation the emission intensity is spread fairly uniformly over all frequencies ω ranging from zero to a maximum frequency ω_{max} given by $\omega_{max} \approx 1/\tau$ where τ is the duration of the impulse (see Eq. 4.39). Let us then compute τ. Since the angular width of the beam is θ (Eq. 4.47), the electron illuminates the observer only for a time interval $\tau' = R\theta/v$ where R is the orbit radius. But the particle speed v is almost c so that

$$\tau' \approx \frac{R\theta}{c}. \tag{4.50}$$

By writing τ' rather than τ we wish to stress the fact that the time interval in question is the interval measured in the electron's moving frame of reference. A transformation into the laboratory frame of the observer yields the result

$$\tau = \frac{R\theta}{c}\left(\frac{m_o c^2}{\epsilon}\right)^2. \tag{4.51}$$

The radius of curvature R and the orbital frequency ω_c are related via $R = v/\omega_c \approx c/\omega_c$ where the orbital frequency

$$\omega_c = \frac{qB_o}{m_o}\sqrt{1 - \frac{v^2}{c^2}} = \frac{qB_o}{m_o}\left(\frac{m_o c^2}{\epsilon}\right) \tag{4.52}$$

has been properly corrected for the relativistic mass change of the electron. When we eliminate R and θ from Eq. 4.51 using the results of Eqs. 4.49 and 4.52 we obtain

$$\tau = \left(\frac{m_o}{qB_o}\right)\left(\frac{m_o c^2}{\epsilon}\right)^2. \tag{4.53}$$

This pulse length determines for us the maximum frequency of synchrotron radiation, and since $\omega_{max} \approx 1/\tau$, it follows that

$$\begin{aligned} \omega_{max} &\approx \frac{qB_o}{m_o}\left(\frac{\epsilon}{m_o c^2}\right)^2 \\ &= 1.76 \times 10^{11} \; B_o \left(\frac{\epsilon}{m_o c^2}\right)^2 \text{ rad/sec.} \end{aligned} \tag{4.54}$$

Take as an example the high-energy machine SPEAR at Palo Alto, California, in which electrons moving in an orbit of radius R = 13 meters are accelerated to an energy $\epsilon = 3 \times 10^9$ eV. They are maintained in circular orbit by a magnetic field B_o = 8000 gauss or 0.8 webers/m^2. With these values we find from Eq. 4.54 that the maximum emission frequency $\omega_{max} \approx 4.9 \times 10^{18}$ rad/sec. This corresponds to a wavelength $\lambda = 2\pi c/\omega_{max}$ or about 3.9 Å! Thus, the machine emits virtually continuous radiation from the longest radio waves to quite hard x-rays. Of course, the visible light of synchrotron emission can also be seen clearly emanating from these high-energy beam facilities. Note that unlike cyclotron radiation where $\omega \approx \omega_c$ (see Fig. 4.20) ω_{max} greatly exceeds the orbital frequency of the gyrating electron. Substituting the above numbers in Eq. 4.52 gives $\omega_c \approx 2.3 \times 10^7$ rad/sec, to be compared with $\omega_{max} \approx 4.9 \times 10^{18}$ rad/sec.

Detailed calculations show that the intensity of radiation as a function of frequency is not entirely flat over the whole range, but that it rises slowly from low values to a maximum value reached at approximately $\omega = \omega_{max}$. The spectrum is sketched in Fig. 4.24a and a detailed plot for the SPEAR facility is shown in Fig. 4.24b.

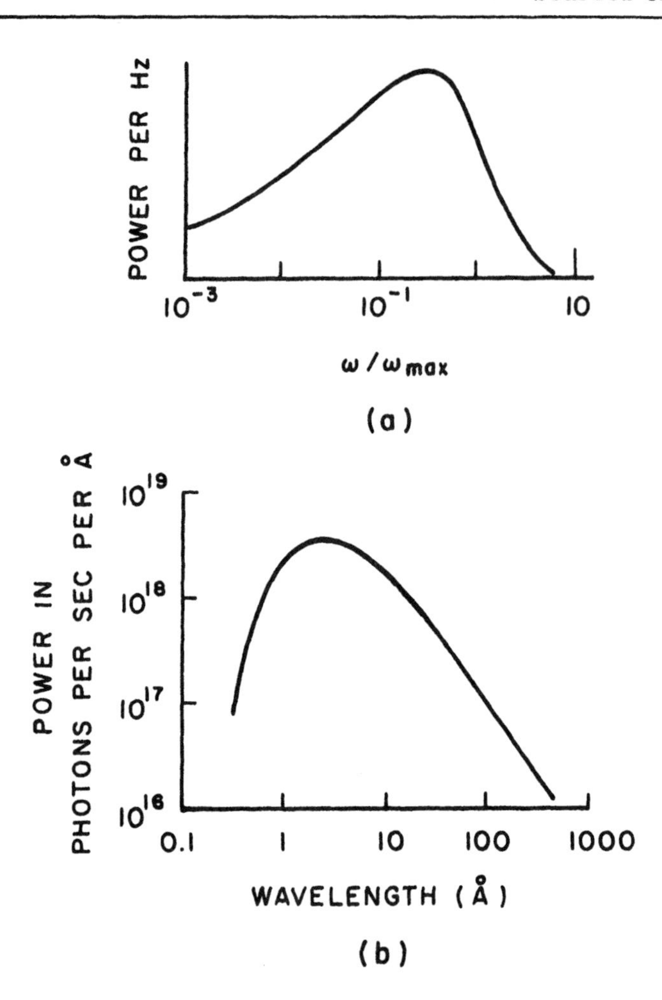

Fig. 4.24 (a) Sketch of the synchrotron power emitted as a function of frequency (ω_{max} is given by Eq. 4.54). (b) Power in a bandwidth of 1 Å emitted as a function of the wavelength from the accelerator SPEAR. [See M. L. Perlman, E. M. Rowe, and R. E. Watson, Physics Today 27, 30 (1974).]

Observe how rapidly the intensity falls off once ω exceeds ω_{max}. The power emitted in the synchrotron process is very substantial. The formula for it is given by

$$P_{synch.} = \frac{q^2 a^2}{6\pi\epsilon_o c^3}\left(\frac{\epsilon}{m_o c^2}\right)^4 \tag{4.55}$$

which is just the Larmor equation (4.13) modified by the relativistic correction $(\epsilon/m_o c^2)^4$. The vertical scale in Fig. 4.24b gives the emission expressed in photons per second emitted within a bandwidth of 1 Å. Thus we see that we have here an intense, broadband source of electromagnetic waves. The radiation is well collimated spatially and one can "tune in" for different wavelengths. It is therefore not surprising that synchrotron radiation is being used more and more as a tool in a number of disciplines including spectroscopy, photochemistry, material studies and biology [for further details see "Synchrotron Radiation — Light Fantastic" by M. L. Perlman, E. M. Rowe, and R. E. Watson, Physics Today 27, 30 (1974)].

One of the most spectacular sources of synchrotron radiation are certain extraterrestrial objects of which the Crab nebula is the most famous. In the year 1054 Chinese and Japanese chroniclers described a momentous event in which apparently a star exploded and the remnants of this explosion are with us still. Much of the light emanating from this object has quite unusual properties — it is broadband and it is polarized, quite unlike the emission from other stellar bodies where the light is unpolarized and narrow-band, originating primarily from atomic transitions (see Fig. 4.25). In 1953 these facts led the Soviet astronomer I. S. Shklovsky to suggest that here we are, in fact, seeing synchrotron radiation of highly relativistic electrons orbiting in weak interstellar magnetic fields. To explain the magnitude and spectrum of the emitted radiation relatively few high energy particles are required. Typical values are: electron density $N_e \approx 10^{-2} m^{-3}$ of highly relativistic electrons with energies between 10^{10} eV and 10^{12} eV, orbiting in a magnetic field of approximately 10^{-4} gauss. Since the emission is broadband, low-frequency waves, in addition to light emission, are also expected. Such waves have been documented by radio astronomers who observed polarized radio and microwaves over a range from ~10 MHz to ~10,000 MHz. A review of this fascinating field will be found in "Cosmic Magnetobremsstrahlung (Synchrotron Radiation)" by V. L. Ginzburg and S. I. Syrovatskii in _Annual Review of Astronomy and Astrophysics_ (Annual Review Inc., Palo Alto, California, 1965, Volume 3, page 297); a less mathematical discourse is

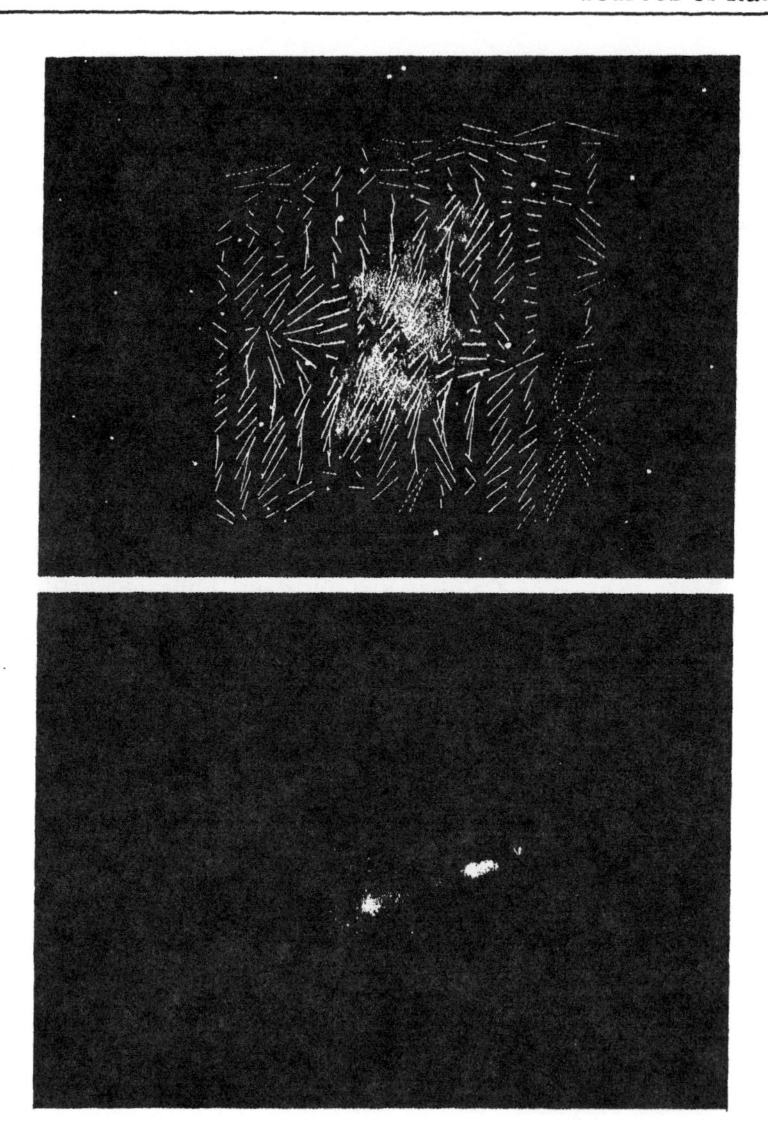

Fig. 4.25 The top picture shows the Crab Nebula with superimposed lines giving the polarization direction of the synchrotron radiation emanating from that section of the Nebula. The length of the line indicates the fraction of light polarized. The broken lines represent uncertain observations. The bottom picture shows a large galaxy in the constellation Virgo. The wisp of light near the center emits polarized synchrotron radiation. This source is 100 times larger than the Crab. [From "The Crab Nebula" by J. H. Oort, Scientific American, March 1957.]

given in Cosmic Radio Waves, I. S. Shklovsky (Harvard University Press, Cambridge, Massachusetts, 1960), and no mathematics is used in "The Crab Nebula," by J. H. Oort, Scientific American, March 1957.

4.5 BLACK-BODY RADIATION

We have found on previous pages that the emission by accelerated charges is determined by their motions. Details of the radiation, such as the power, the polarization, and the spectrum were found to be sensitive functions of the time history of the charges' accelerations. There also exists a very remarkable emission process called black-body radiation which is entirely independent of the minute of the motions, being a function of but one parameter – the temperature T of the emitting body. To be sure, the temperature is a measure of the average energy of jiggling of the electrons and atoms of the emitter, but it is quite insensitive to the way this energy is acquired. To make this so-called black-body radiation one can proceed as follows. Take a container impenetrable to any radiation, lined on the inside with highly reflecting walls. Introduce into it a tenuous gas of radiating atoms and (or) accelerated charges and wait a little while for things to come to equilibrium. If there are enough atoms, the cavity will quickly become filled with black-body radiation. To see it, drill a tiny hole in the container wall as shown in Fig. 4.26. Be sure to make the hole so tiny that the energy escaping is negligibly small compared with the radiant energy trapped in the cavity, otherwise you will disturb the equilibrium and the emission will change its character.

To understand qualitatively what is happening let us take one oscillating electron, radiating electromagnetic waves. If there were no container, these waves would escape out to infinity, the electron would keep losing energy and eventually stop radiating (see the worked-out example given in section 4.4). When the container is in place, this demise of the electron will not occur. The container walls reflect the radiation which, after many bounces, strikes the electron. The electric field of the wave reaccelerates the charge and rejuvenates

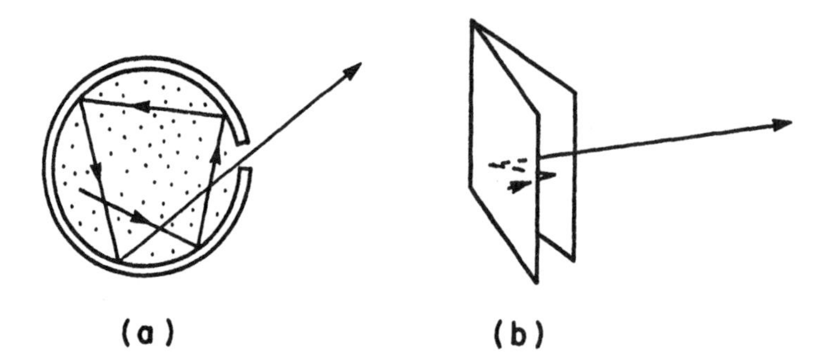

Fig. 4.26 (a) A theorist's black-body emitter. The cavity is filled with enough radiating atoms so that the photons have a good chance of being absorbed and reemitted a few times before escaping through the hole. (b) A heated strip of metal folded into a wedge of small angle forms a cavity whose inside approximates closely a black body. Here the wall atoms are the emitters. [See Edgerton and Milford, Proc. Roy. Soc. A 130, 111 (1930).]

its emission capability. Stated differently, the electron is bathed in its own radiation, absorbs it and reemits it in different directions, a process known as scattering. In addition, the electron is illuminated by the radiation from other electrons (and atoms) whose waves it likewise absorbs and then scatters. One may worry that some radiators are much better emitters than others so that the region where they are located will lose energy fast and cool down relative to neighboring regions of the cavity. But this does not happen because it turns out that good emitters are also good absorbers (this fact is known as Kirchhoff's law), and the emission and the absorption are so finely balanced that the entire cavity attains a constant, uniform temperature. The energy density of radiation fills the container uniformly.

To calculate the radiant energy in the container, taking into account all the innumerable emissions and absorptions, appears to be a superhuman undertaking. Indeed, even if one attempted such a program of calculations, Maxwell's equations (3.38) alone could not give the answer; the very parameter T, being a statistical quantity, has no place in these equations. But thermodynamics comes to our aid. One of its important conclusions is the following: in a system that has attained thermal equilibrium, be it composed of matter or of radiation or a mixture thereof, the energy "per degree of freedom" as it is called, or more loosely speaking, per mode of oscillation, is given by

$$\boxed{\begin{aligned} u &= \frac{1}{2}\kappa T \\ &= \frac{1}{2}\times[1.381\times 10^{-23}]\, T \text{ joules} \end{aligned}} \tag{4.56}$$

where T is the absolute temperature in degree Kelvin, and $\kappa = 1.381 \times 10^{-23}$ joule/°K is Boltzmann's constant. Now, our black-body cavity is literally crammed full of such modes of oscillation. The modes are just the standing waves in a box, and in section 2.4 (see also section 3.5) we showed how to go about calculating their properties (we did it for sound waves but the same ideas apply to electromagnetic waves). Therefore, all we need is to find the total number of modes in our container and use Eq. 4.56 to compute the total electromagnetic energy contained within it. Note that an electromagnetic mode actually has energy κT, of which $\kappa T/2$ resides in the magnetic field and $\kappa T/2$ in the electric field (see, for example, section 5.4, page 387).

Before embarking on this calculation let us examine first the simpler one-dimensional problem in which waves can travel only along one direction, say the z direction. An example of this is a long lossless transmission line of length L terminated at its two ends $z = 0$ and $z = L$ by reflecting metal plugs. Waves bounce back and forth between the two terminations and a standing wave pattern is set up. The system has many modes of oscillation each corresponding to a standing wave like that given by Eq. 2.43:

$$E_m(z,t) = A_m \sin\left(\frac{m\pi z}{L}\right)\cos(\omega_m t). \tag{4.57}$$

Here E_m is the electric field of the mth mode whose frequency ω_m is given by

$$\omega_m = \frac{m\pi}{L} c \qquad m = 1, 2, 3, \ldots \tag{4.58}$$

When this is written in terms of the wavelength, it yields the equation

$$\lambda_m = \frac{2L}{m} \tag{4.59}$$

which merely says that in order to fit a standing wave, L must equal an integral number of half wavelengths.

To obtain the spectrum of the radiation we must find the number of modes dm of oscillation in a narrow band of frequencies $d\omega$. This is readily obtained by differentiating* Eq. 4.58 with respect to ω:

$$dm = \frac{L}{\pi c}\, d\omega. \tag{4.60}$$

Each mode of the wave has on the average $\kappa T/2$ of energy in the electric field and $\kappa T/2$ of energy in the magnetic field so that the mean energy of radiation in the frequency interval $d\omega$ is given by

$$\kappa T \frac{L}{\pi c}\, d\omega \text{ joules} \tag{4.61}$$

showing that, in fact, the energy of radiation is frequency-independent.

A standing wave can be decomposed into two traveling waves, one going each way (see Fig. 2.8). The power in the two waves can now be easily calculated. Each wave is associated with half the energy density given by Eq. 4.61. This energy must cross the length L of transmission line in a time equal to L/c. Hence, the power flowing in one direction is

$$P = \frac{\kappa T}{2\pi}\, \Delta\omega \text{ watt} \tag{4.62}$$

and is likewise independent of frequency. A demonstration of this energy flow is carried out fairly readily. We take a coaxial transmission line and insert into it a hot radiator, as for example a piece of carbon which can be heated externally by passing current through it. We connect the other end of the transmission line to a broadband receiver (the bigger the bandwidth $\Delta\omega$ the larger the power P) of high sensitivity and low intrinsic noise. A power meter attached to the receiver will measure P in accordance with Eq. 4.62. Suppose the

*We assume that the modes are so closely spaced that m and ω are almost continuous variables, and thus can be differentiated.

carbon slug is heated to 1000°K and the bandwidth of the receiver is 100 MHz. Then, from Eq. 4.62 we obtain that $P \approx 1.4 \times 10^{-12}$ watts, a small but measurable power.*

We now turn from the lossless transmission line to our three-dimensional black box. For simplicity we assume it has rectangular sides L_x, L_y, and L_z with perfectly reflecting walls (see Fig. 2.15). Each mode of oscillation (i.e., standing wave) has a frequency ω determined by three integers ℓ, m, and n as was discussed in section 2.4. The frequency of a given mode in accordance with Eq. 2.59 is given by

$$\frac{\ell^2}{L_x^2} + \frac{m^2}{L_y^2} + \frac{n^2}{L_z^2} = \frac{\omega^2}{\pi^2 c^2}. \tag{4.63}$$

We must now count the total number of standing waves N, that is, sets of whole numbers (ℓ, m, n), with frequencies less than a given frequency ω. This is most easily done by plotting ℓ, m, and n along the three Cartesian axes, as is illustrated in two dimensions in Fig. 4.27. Each integer value (ℓ, m, n) then supplies a point on the graph, the sum total being a lattice of points with unity spacing. From Eq. 4.63 it follows that the surfaces of constant frequency ω are ellipsoids having semiaxes of lengths $(L_x\omega/\pi c)$, $(L_y\omega/\pi c)$ and $(L_z\omega/\pi c)$, respectively. All frequencies less than ω are represented by points inside the given ellipsoid. Since only positive integers are of interest, only one octant corresponding to positive integers (ℓ, m, n) needs to be considered. If the number of points in the octant is large (that is, a large container with many modes) the number of points N approaches closely the volume of the octant and

$$N \approx \frac{1}{8}\,\frac{4\pi}{3}\left(\frac{L_x\omega}{\pi c}\right)\left(\frac{L_y\omega}{\pi c}\right)\left(\frac{L_z\omega}{\pi c}\right) = \frac{\omega^3}{6\pi^2 c^3}\,V, \tag{4.64}$$

*For a very clever way of measuring powers even a thousand times weaker than this see "The Primeval Fireball" by P. J. E. Peebles and D. T. Wilkinson, Scientific American, June 1967; also R. H. Dicke, Rev. Sci. Instrum. 17, 268 (1946).

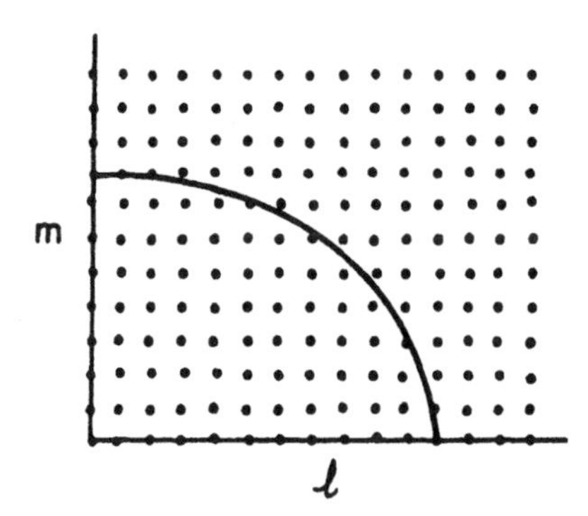

Fig. 4.27 Plot of the wave numbers ℓ, m for the purpose of counting the standing waves in a closed box. Observe that some error is made by taking the area under the smooth curve rather than by counting individual lattice points because the curve misses some points. But the error is small if the number of lattice points is large.

with V as the volume of the container. Suppose V is one cubic centimeter ($=10^{-6}$ m^3) and the wavelength $\lambda = 2\pi c/\omega$ is 5000 Å; then $N \approx 3.4 \times 10^{13}$. With this large number of lattice points we are assured that even if, by our method of counting, we missed some modes, the error in Eq. 4.64 must be small. The number of standing waves is actually twice the value given by this equation because for every direction of propagation, there are two possible polarizations of the wave. Hence the total N equals $(\omega^3/3\pi^2c^3)V$ and the number dN in a frequency interval $d\omega$ is

$$dN = \frac{\omega^2 V}{\pi^2 c^3}\, d\omega, \tag{4.65}$$

a result obtained by differentiation of N with respect to ω. By the theorem stated earlier each of these modes carries an energy κT and thus the energy density of black-body radiation in the frequency interval between ω and $\omega + d\omega$ is

$$U(\omega)\, d\omega = \frac{\kappa T \omega^2}{\pi^2 c^3}\, d\omega \quad \text{joules/m}^3 \tag{4.66}$$

which is an equation originally derived by Rayleigh and Jeans. Note that it differs by the factor ω^2 from the one-dimensional situation given by Eq. 4.61. Thus the energy density increases as the square of the frequency of observation.

Careful measurements of the radiation spectrum show that whereas Eq. 4.66 works well at low frequencies it fails completely at high frequencies. That there is something wrong with the equation is already obvious from the fact that when it is integrated over all frequencies it gives an infinite energy density in the medium. If this were correct, there could never be any thermal equilibrium, all energy being eventually absorbed into electromagnetic waves. Thus, a high-frequency cutoff of Eq. 4.66 is needed. The solution to this dilemma was supplied by Planck who, in finding it, ushered into being the era of quantum mechanics. He showed that the energy per mode instead of its classical value of κT has the value

$$\boxed{\frac{(h\omega/2\pi)}{e^{h\omega/2\pi\kappa T} - 1}} \tag{4.67}$$

with $h = 6.626 \times 10^{-34}$ joule-sec, as the familiar Planck's constant. Introducing this formula in Eq. 4.66 in place of κT yields the famous result

$$\boxed{U(\omega)\, d\omega = \frac{h\omega^3}{2\pi^3 c^3} \frac{1}{[e^{h\omega/2\pi\kappa T} - 1]}\, d\omega} \tag{4.68}$$

which is shown plotted in Fig. 4.28. We now see clearly the high-frequency cutoff which prevents the occurrence of the so-called "ultraviolet catastrophe" mentioned earlier. At low frequencies such that $h\omega/(2\pi\kappa T) \ll 1$, a series expansion of the exponential appearing in Eq. 4.68 and retention of the first two terms correctly reduces the Planck formula to the Rayleigh-Jeans result given by Eq. 4.66. The dashed line of Fig. 4.28 delineates approximately the region where classical theory is valid. If the product wavelength λ in meters and tempera-

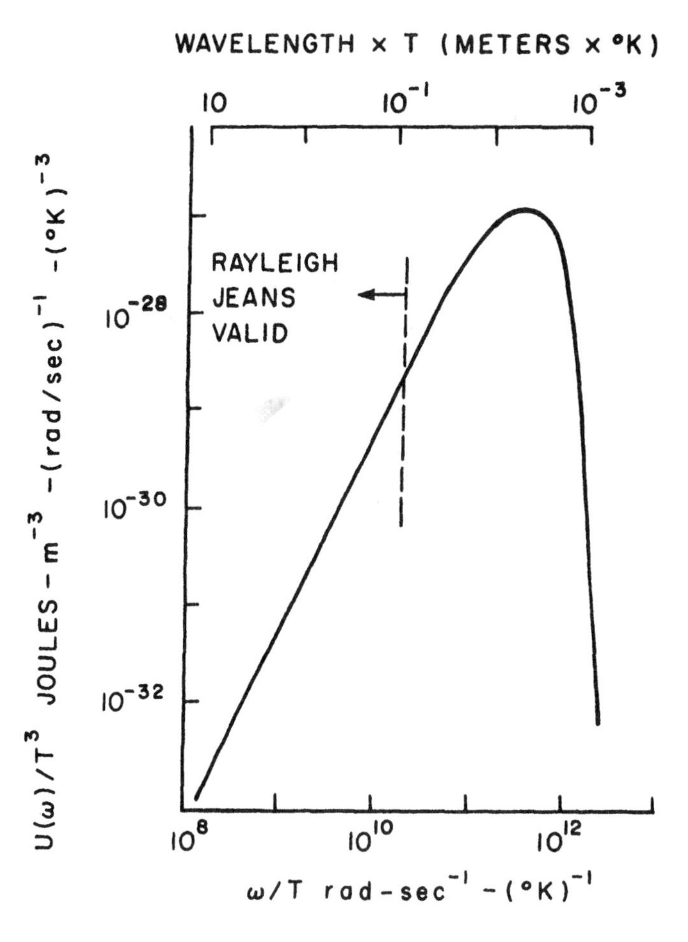

Fig. 4.28 Frequency spectrum of the energy density U of black-body radiation. Plotting U/T^3 vs ω/T gives the universal curve shown. If U were plotted as a function of ω, a different curve would need to be drawn for each temperature T, similar to the one shown. With increasing T, the maximum value of U increases and shifts to higher frequencies. U(max) occurs when $\omega = 3.69 \times 10^{11}$ T rad/sec. Its value then equals U(max) $\approx 10^{-27}$ T^3 $d\omega$ J $-$ m^{-3}; $d\omega$(rad/sec) is the narrow bandwidth over which the energy is measured. The total black-body energy density summed over all frequencies is obtained by an integration of Eq. 4.68. The result is U(total) $= 7.56 \times 10^{-16}$ T^4 J/m^3. Thus, the total black-body radiation varies as the fourth power of the body's temperature. This is known as the Stefan-Boltzmann law.

ture T in °K is larger than $\sim 10^{-1}$ the foregoing inequality is well satisfied and the classical result is applicable. For example, this means that if T = 1000°K, λ must be larger than $\sim 10^{-4}$ meters, but when T = 1,000,000°K, λ must only be larger than $\sim 10^{-7}$ meters, or 1000 Å.

The radiation from most bodies maintained at an elevated temperature approximates that of the ideal black body, at least over a restricted range of frequencies: the inside of a smelting furnace, the hot tungsten filament of a lamp, the surface of the Sun (whose effective black-body temperature is ~6000°K), and so on. Indeed, the entire universe appears to be bathed in a weak background of this radiation and we all appear to be sitting within a gigantic black-body enclosure having an energy density $U(\omega)$ corresponding to a temperature of ~3°K. The discovery of this weak emission [A. A. Penzias and R. W. Wilson, Astrophys. J. 142, 41 (1965)] is one of the most revolutionary developments of cosmology. It suggests that we are here witnessing the remnants of a "primeval fireball" created during the earliest days of the universe when the universe was extremely hot and contracted. Presumably the energy density of black-body radiation was enormously high when the universe began with the "big bang," but as a result of the expansion, a cooling took place and is taking place still. The spectrum of the radiation as it appears today is shown in Fig. 4.29. The experimental points fit well the black-body curve for a temperature of 2.8°K. Unfortunately, there are as yet no good points on the falling part of the curve and until such are obtained, a positive identification of this radiation as black-body emission cannot be made. Nonetheless, based on these findings, cosmologists have begun making quite detailed models of the evolution of the universe over its entire existence — believed to be approximately 9×10^{9} years or $\sim 3 \times 10^{17}$ seconds. Figure 4.30 shows the results of one such calculation made by R. A. Alphen, G. Gamow, and R. Herman [Proc. Nat. Acad. Sci. U.S.A. 58, 2179 (1967)]. From this model we see that when the universe was 100 seconds old its temperature was around 10^{9} °K. When it became approximately 10^{6} years old, it cooled down to some 200°K at which time an interesting transition occurred — the relative density

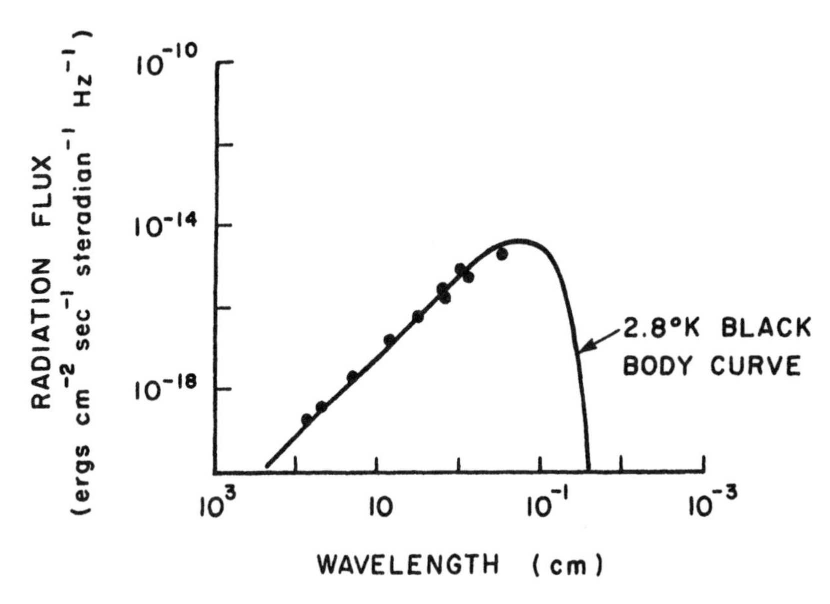

Fig. 4.29 The measured universal microwave background (dots) compared with the theoretical black-body curve for an assumed temperature T of 2.8°K. The radiation flux plotted here is derived from the energy density of Fig. 4.28 by the relation Flux = $(U \times c)/4\pi$ (cf. Eq. 3.109). Division by the factor 4π gives the flux per unit solid angle (i.e., per steradian), rather than the total flux in all directions. [For more details see P. Thaddeus, Ann. Rev. Astron. Astrophys. 10, 305 (1972).]

of matter and radiation reversed. (Here the definition of density of radiation is its mass equivalent obtained by dividing the total energy density by c^2.) Before the transition, radiation prevailed over matter and during this epoch it is believed that the radiation was so strong that it kept the matter spread out uniformly in the form of a tenuous gas. After the transition, when the mass density of matter exceeded that of radiation, gravitational forces came to the fore, condensed matter began forming and the universe began to take on an appearance similar to what it has today. It is noteworthy that some of the main features of this model were proposed long before the black-body background was discovered experimentally. This subject is discussed by G. Gamow in quite nonmathematical terms in "The Evolutionary Universe," Scientific American, September 1956. For recent theories see the difficult but fascinating book Gravitation and Cosmology by

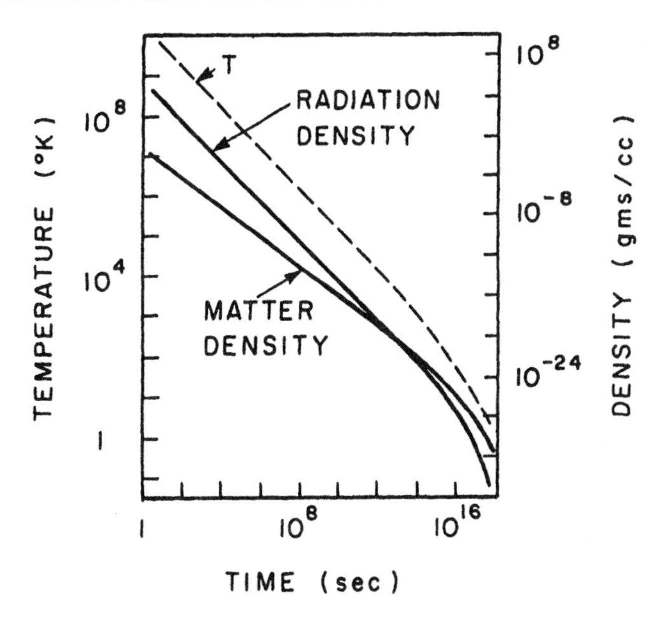

Fig. 4.30 The temperature T, matter density, and black-body radiation density (in gms/cc equivalent) from the time the universe was 1 second old, to the present day. [From R. A. Alphen, G. Gamow, and R. Herman, Proc. Nat. Acad. Sci. U.S.A. 58, 2179 (1967).]

S. Weinberg (John Wiley & Sons, New York, 1972).

Example

Justify the following statements of the caption of Fig. 4.28:

a) The frequency ω at which the maximum energy density occurs is given by $\omega = 3.69 \times 10^{11}$ T radians/sec.

b) The total energy density, integrated over all frequencies, is given by $U(T) = 7.56 \times 10^{-16}\ T^4\ J/m^3$.

Solution

Eq. 4.68 can be written as

$$U(x, T) = \frac{4k^3T^3}{h^2c^3}\left(\frac{x^3}{e^x - 1}\right) \qquad x = h\omega/2\pi kT.$$

This expression is differentiated with respect to ω to get

$$\frac{dU}{d\omega} = \frac{dU}{dx}\frac{dx}{d\omega} = \frac{2k^2T^2}{\pi hc^3}\left[\frac{3x^2}{e^x - 1} - \frac{x^3 e^x}{(e^x - 1)^2}\right].$$

For a maximum, $dU/d\omega = 0$. Setting the quantity in brackets equal to zero gives

$$(3 - x)\, e^x = 3$$

which can be solved by "trial and error" to yield

$$x = 2.82 = h\omega/2\pi kT$$

Thus

a) $\omega = (2.82)\left(\frac{2\pi k}{h}\right) T = 3.69 \times 10^{11}$ T rad/sec.

The total energy density is given by

$$U(T) = \int_0^\infty U(\omega, T)\, d\omega = \frac{8\pi k^4 T^4}{h^3 c^3}\int_0^\infty \frac{x^3 dx}{e^x - 1}$$

where x is as given above. The integral can be evaluated by standard techniques and equals $\pi^4/15$. Thus we find

b) $U(T) = aT^4$

where

$$a = \frac{8\pi^5 k^4}{15\, h^3 c^3} = 7.56 \times 10^{-16} \text{ J/m}^3\text{-deg}^4.$$

CHAPTER 5

GUIDED WAVES

An important problem in modern communication is the guidance of electromagnetic power from one point to another. The two points may be separated by only a few meters, just enough to bring the power from the receiving antenna to the living room; or they may be separated by thousands of miles as is the case for intercontinental links. At frequencies below a few gigahertz (GHz) the most commonly used means is the two-wire transmission line of which the coaxial cable is the best known example. However, two conductors are not always necessary to transfer power between points in space. A single conductor, usually in the form of a metal pipe called a waveguide is sufficient if the frequency is high enough; in the microwave region from, say, 3 GHz to 150 GHz this mode of guidance is, in fact, greatly preferred over the two-wire transmission line. Even a single conductor is not essential for the purposes of energy transfer. A dielectric rod or a dielectric pipe can do the job and much research is being carried out today with a view to guiding laser radiation at optical and infrared wavelengths on such structures.

In this chapter we wish to discuss the properties of guided radiation in two-wire transmission lines as well as in waveguides. But before we can begin tackling these problems, we must address ourselves first to the question of how fields behave in the immediate vicinity of conductors. Once we know the so-called "boundary conditions" (see section 2.3) we can then proceed with the topics central to this chapter.

5.1 ELECTRIC AND MAGNETIC FIELDS NEAR THE SURFACE OF A CONDUCTOR

Let us consider a piece of conductor and focus our attention on the region around the surface separating the bulk of metal on one side from vacuum on the other side. There are time-varying electric and magnetic fields above, at, and below the surface in question as is shown in Fig. 5.1. These fields are produced somewhere, somehow – say light is shone on the metal, but we need not concern ourselves at this point with the details by which the electromagnetic field is generated. We wish to find out all we can about these fields and, in particular, how they change as we cross the surface. The answers, of course, must lie in Maxwell's equations which we shall now apply one at a time.

Metals like silver, copper, and brass are such good conductors that we shall not err very greatly by assuming that they are, in fact, perfect conductors. Within this idealized medium the electric field must be zero. This is because the charges are assumed to be so mobile that they move instantly in response to whatever change occurs, and short out any imposed electric field. Let us then apply Gauss' law given by Eq. 3.7 to the elementary "pillbox" shown in Fig. 5.2 which encloses a portion of the boundary surface. We resolve the electric field $\vec{E}$ above the surface into a component E_n normal to the surface and a component E_t tangential to the surface. Use of Eq. 3.7 then yields for the electric flux the value

$$\epsilon_o E_n da + \Phi_E(\text{curved}) = \rho \, d\ell da \qquad (5.1)$$

where da is the cross-sectional area of the pillbox, $d\ell$ its height, and ρ is any time-varying charge density residing within it. Φ_E(curved) is the electric flux flowing through the curved surface of the pillbox resulting from the presence of the tangential field E_t. Now, E_t is assumed to be finite everywhere, so that, as the height of the box $d\ell$ is allowed to go to zero, Φ_E also vanishes. Hence we are left with the result that $E_n = \rho \, d\ell/\epsilon_o$. But, $\rho \, d\ell$ as $d\ell \to 0$ is nothing more than the surface charge density on the conductor; denoting it by ρ_S we

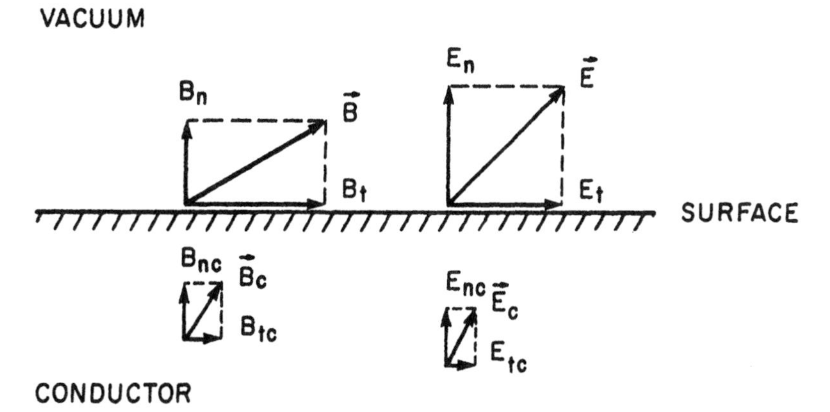

Fig. 5.1 Region around a surface separating a conductor from vacuum. The surface need be neither plane nor smooth. What is shown above is just a tiny section, small enough so that it is smooth and flat. All fields are taken to be time-varying, and the length and direction of the vectors represent the situation at some instant of time. Because of finite conductivity, some field penetration into the metal occurs; these fields are labeled with a subscript c for conductor.

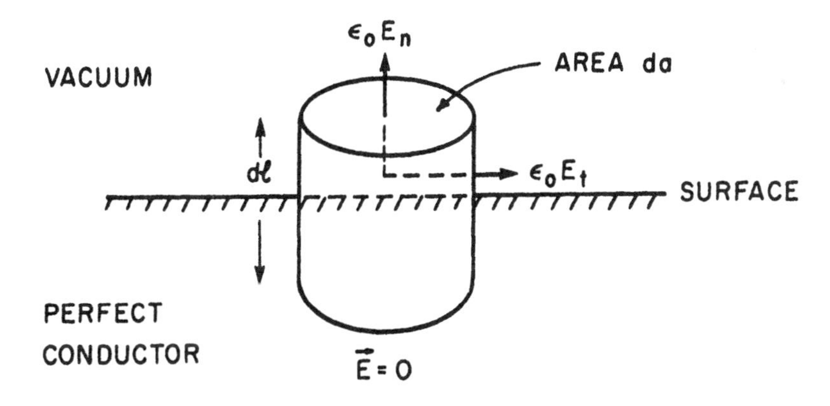

Fig. 5.2 A Gaussian "pillbox" enclosing a portion of the boundary surface. As will be found in Chapter 6, the flux of the quantity $\kappa_e \epsilon_o \vec{E}$ rather than $\vec{E}$ itself is appropriate when dealing with fields in matter. Inside the perfect conductor $\kappa_e \epsilon_o \vec{E}$, just like $\vec{E}$ itself, is zero. Above the surface, in vacuum, $\kappa_e = 1$.

obtain the familiar result of electrostatics which says that the normal component of electric field above a conductor is given by

$$E_n = \frac{\rho_S}{\epsilon_0} \text{ volt/m} \tag{5.2}$$

with ρ_S given in coulomb/m^2. Note, however, that now E_n is the time-varying field of some electromagnetic oscillation and ρ_S is the associated time-varying surface charge density on which the field lines end. One may well ask where ρ_S came from in the first place. The answer is that free conduction electrons are drawn from the bulk of the metal by the tiny but nonetheless finite electric field which can penetrate part way below the metal surface (the conductivity is, after all, not really infinite and some field penetration occurs).

When we use Eq. 3.33 instead of Eq. 3.7 and draw a pillbox similar to the one shown in Fig. 5.2, we obtain by analogy with the preceding calculation a condition on the normal components of the magnetic field,

$$B_n\, da + \Phi_B(\text{curved}) + \Phi_B(\text{bottom}) = 0\,. \tag{5.3}$$

Here Φ_B(curved) is the magnetic flux contributed by the curved surface, which again vanishes as the pillbox is squinched to zero thickness ($d\ell \to 0$). Φ_B(bottom) is the magnetic flux from the flat bottom surface immersed within the conductor. However, because the time-varying electric field within the perfect conductor is zero, the time-varying magnetic field must also be zero, a fact which follows directly from Eq. 3.38b. Note that a steady magnetic field can exist within the conductor but we are not interested in steady fields. Hence Φ_E(bottom) = 0 and Eq. 5.3 reduces to

$$B_n = 0\,. \tag{5.4}$$

It says that the normal component of the magnetic field just above the surface of a perfect conductor must be identically equal to zero.

In obtaining information about E_n and B_n at the surface of the conductor we have used up two of the four Maxwell equations. The remaining two equations will tell us how the tangential components E_t and B_t behave. Let us begin with E_t. Draw a small rectangle of length db and width $d\ell$ enclosing the surface, as is illustrated in Fig. 5.3, and apply to it Faraday's law of induction given by Eq. 3.16. In going around the loop and remembering that $\vec{E} = 0$ within the metal we obtain

$$E_n(1)\,\frac{d\ell}{2} - E_t\,db - E_n(2)\,\frac{d\ell}{2} = -\left(\frac{d}{dt}\,B_z\right)\,d\ell db \tag{5.5}$$

where B_z is the component of the time-varying magnetic field pointing out of the page, spatially averaged over the rectangle $d\ell db$. Now, squinch the width $d\ell$ of the rectangle to zero keeping at all times the surface of discontinuity positioned in between the two sides of the rectangle. Since B_z is assumed to be finite, the right-hand side of Eq. 5.5 approaches zero. Similarly, if E is everywhere finite, the

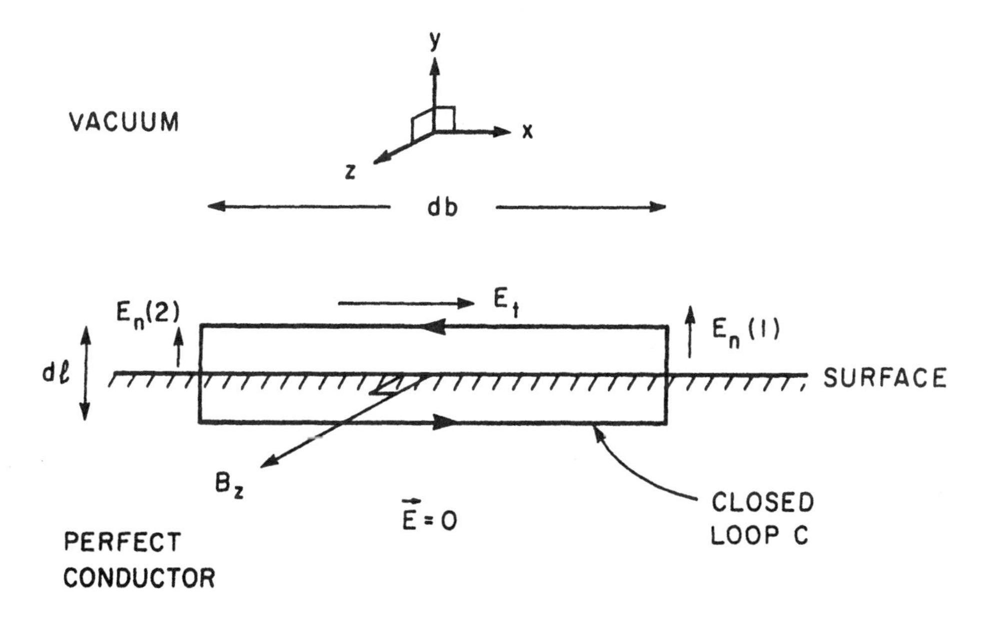

Fig. 5.3 Closed rectangular loop, enclosing a section of the boundary surface, used in evaluating the integrals appearing in Faraday's induction law. $E_n(1)$ and $E_n(2)$ is the normal field component whose value may vary with position along the surface.

terms multiplied by $\Delta\ell/2$ vanish, with the result that

$$E_t = 0. \tag{5.6}$$

The conclusion is that any electric field parallel to the conductor surface is zero, a result familiar to us already from electrostatics.

In a manner entirely similar to the foregoing, draw the rectangle shown in Fig. 5.4 and apply to it the last of Maxwell's equations (3.30); then

$$\frac{B_n(1)}{\mu_o}\frac{d\ell}{2} - \frac{B_t}{\mu_o}\,db - \frac{B_n(2)}{\mu_o}\frac{d\ell}{2} = \epsilon_o\frac{dE_z}{dt}\,d\ell db + J_z\,d\ell db. \tag{5.7}$$

Again letting $d\ell$ approach zero and employing arguments similar to

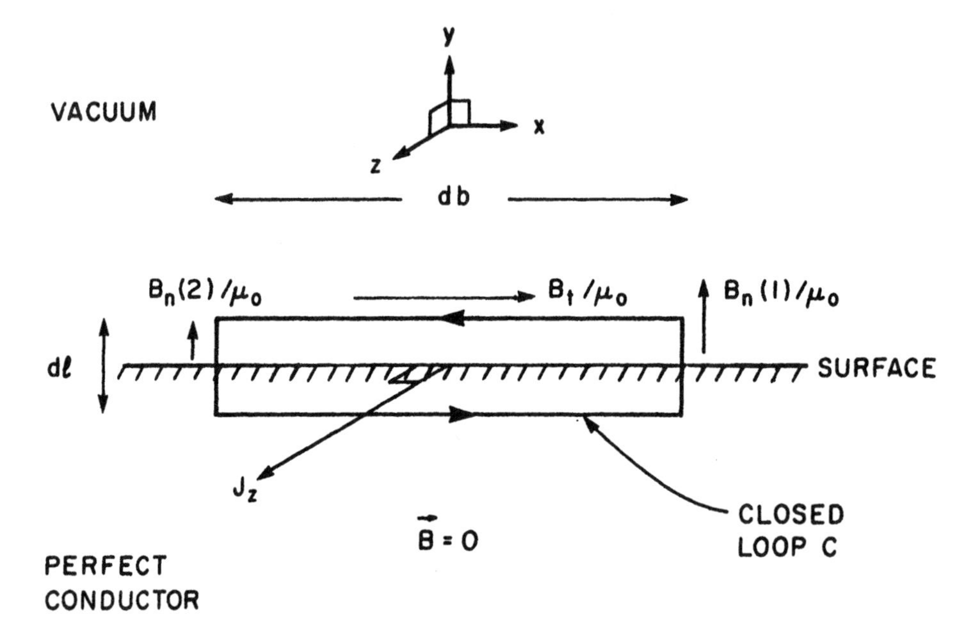

Fig. 5.4 Closed rectangular loop, enclosing section of boundary surface, used in evaluating the integrals appearing in Maxwell's extension of Ampère's law (3.30). As will be found in Chapter 6, the circulation of the quantity $\vec{B}/\kappa_m\mu_o$ rather than $\vec{B}$ itself is appropriate when dealing with fields in matter. Inside the perfect conductor, $\vec{B}/\kappa_m\mu_o$ just like $\vec{B}$ itself, is zero. Above the surface, in vacuum, $\kappa_m = 1$.

those just used in deriving Eq. 5.6, we obtain

$$-\frac{B_t}{\mu_o} \doteq J_z\, d\ell. \tag{5.8}$$

But observe that in going to the limit $d\ell \to 0$, we treated the two terms on the right-hand side of Eq. 5.7 differently. We allowed the first term to vanish in accord with the assumption that E_z is finite, so that the product $E_z\, d\ell db \to 0$ as $d\ell \to 0$. The second term $J_z\, d\ell db$, is the time-varying current through the rectangular strip flowing in a direction pointing out of the page. In a conductor of large but finite conductivity one finds that the oscillating current (just like the oscillating electric field) is confined to a thin layer situated at the metal-vacuum boundary. The thickness of the layer, that is, the depth of penetration, decreases as the conductivity and frequency increase. Thus, in a good conductor, a high frequency current will flow in a thin sheet near the surface, the depth of the sheet approaching zero as the conductivity goes to infinity. This gives rise to the useful but fictitious concept of the <u>current sheet</u> having a surface current density J_S defined as

$$\vec{J}_S \doteq \lim_{d\ell \to 0} \vec{J}\, d\ell \quad \text{amp/m}. \tag{5.9}$$

For this limit to be finite the current density itself must go to infinity. In terms of J_S, then, Eq. 5.8 becomes

$$B_t = -\mu_o J_{Sz}\,. \tag{5.10}$$

Equation 5.10 states that the current per unit width along the surface of the metal is proportional to the tangential component of the magnetic field just outside the surface. To the question, what drives this current, one answers that for large but finite conductivity the tangential electric field in the metal is small but not identically zero. And it is this tangential electric field in a thin sheet near the surface that furnishes the force on the conduction electrons, causing oscillatory current to flow. In Fig. 5.5 we try to illustrate pictorially the fields near the surface of a good conductor and contrast them with the fields that would exist there if the conductor were indeed perfect. The depths to which the fields and thus the real currents penetrate prove to be quite small. To obtain some feeling for the numbers involved, we quote formula (6.64) which says that the so-called "skin depth" δ is given by

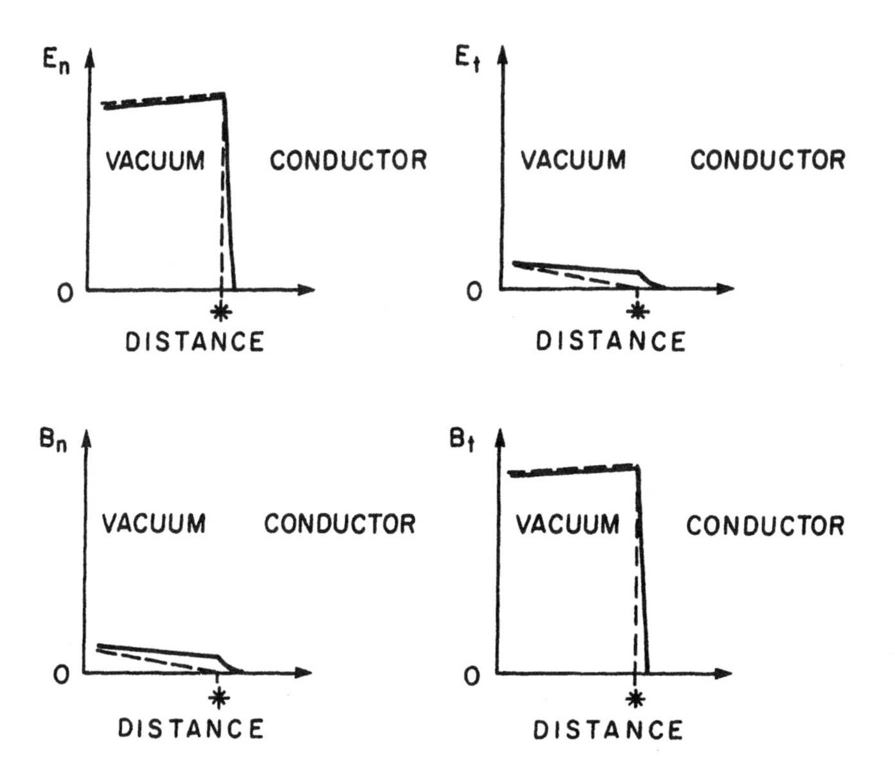

Fig. 5.5 Schematic of the fields near the surface of a conductor (the surface is marked with an asterisk *). The solid lines are for a good conductor, the dashed lines for a perfect conductor. The existence of the small tangential E_t (perpendicular to B_t) outside the surface means that there is power equal to $[\vec{E}_t \times \vec{B}_t/\mu_o]$ watt/m^2 flowing into the conductor. This is dissipated as joule heating in the bulk of the metal. One finds that the time-average power dissipated equals $(\omega\delta/4\mu_o)\, B_{ot}^2$ watt/m^2 where δ is the skin depth given by Eq. 5.11 and B_{ot} is the amplitude. [For a derivation of this result see section 5.4.]

$$\boxed{\delta = \sqrt{\frac{2}{\omega\mu_0\sigma}} \quad \text{meters}} \tag{5.11}$$

where ω is the angular frequency of the oscillations in radians per second and σ the electrical conductivity of the metal in units of $(\text{ohm-meter})^{-1}$. For copper, σ equals 5.81×10^7 $(\text{ohm-m})^{-1}$ and Eq. 5.11 then tells us that $\delta = 0.066\, f^{-1/2}$ meters where f is the frequency in hertz. Therefore, at a radio frequency of 1 MHz, $\delta = 6.6 \times 10^{-5}$ m. At a microwave frequency of 1000 MHz, $\delta = 2.08 \times 10^{-6}$ m; and at an optical frequency of 5×10^{14} Hz (wavelength equal to 6000 Å), $\delta = 2.95 \times 10^{-9}$ m or 29.5 Å. These are tiny distances indeed — observe that they are much smaller than the wavelength $\lambda = c/f$ of the electromagnetic fluctuation in question.

Equation 5.10 and Fig. 5.4 show that both J_z and B_t are parallel to the metal-vacuum interface, but are perpendicular to one another. If $\hat{n}$ is the unit normal directed outward from the metal, relation (5.10) can be written as $\hat{n} \times \vec{B} = \mu_0 \vec{J}_S$. Indeed, our four boundary conditions that must be satisfied at the surface of the metal* can be conveniently written in the following vector form:

$$\boxed{\left.\begin{aligned} \hat{n} \cdot \vec{E} &= \frac{\rho_S}{\epsilon_0} && \text{(a)} \\ \hat{n} \times \vec{E} &= 0 && \text{(b)} \\ \hat{n} \cdot \vec{B} &= 0 && \text{(c)} \\ \hat{n} \times \vec{B} &= \mu_0 \vec{J}_S && \text{(d)} \end{aligned}\right\} \text{at surface of a perfect conductor}} \tag{5.12}$$

In solving specific problems, we need only to satisfy explicitly conditions (5.12b) and (5.12c). It turns out that nature and Maxwell's

*General boundary conditions applicable at the interface separating any two media, be they imperfect metals or dielectrics, are given in the introduction to Chapter 7.

equations ensure that the surface charge density ρ_S and the surface current density $\vec{J}_S$ adjust themselves in magnitude and direction so that requirements (5.12a) and (5.12d) will be fulfilled automatically. We shall now apply our results to some useful problems.

A plane wave striking a metal at normal incidence

When a plane wave traveling in vacuum is incident normally on the surface of a perfect metal conductor, it is totally reflected from it. The reason is that neither $\vec{E}$ nor $\vec{B}$ can exist within the conductor so none of the incident energy can be absorbed by it and none can be transmitted through it. As a result, the amplitudes $\vec{E}_r$ and $\vec{B}_r$ of the reflected wave must equal the amplitudes $\vec{E}_i$ and $\vec{B}_i$ of the incident wave, a fact which we shall prove shortly. Suppose the conductor is situated in the z = 0 plane and a wave traveling from left to right is incident on it, as is shown in Fig. 5.6. Expressing the electric field

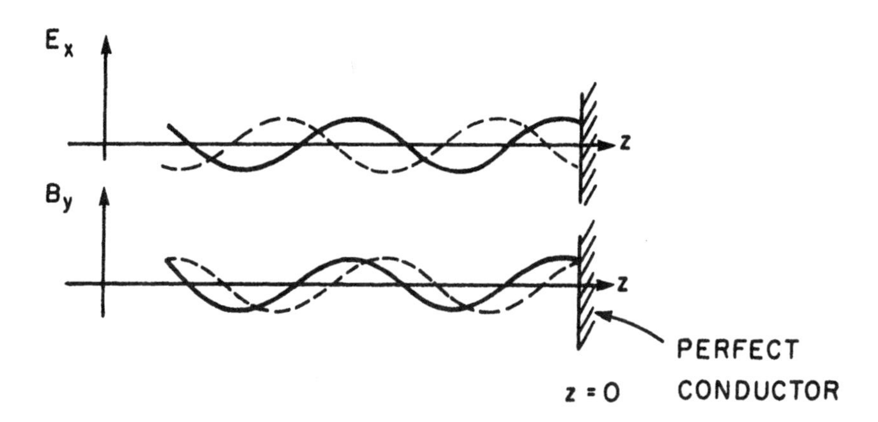

Fig. 5.6 Instantaneous picture of electric and magnetic fields near the surface of a conducting sheet. The solid lines represent the amplitudes of the incident plane, linearly polarized wave traveling from left to right. The dashed lines show the amplitudes of the reflected wave traveling from right to left. The total disturbance is obtained by summing the solid and dashed lines, giving a standing wave pattern like that shown in Fig. 3.16.

of the incident wave in complex notation

$$E_i = E_{oi}\, e^{j(\omega t - kz)} \tag{5.13}$$

the corresponding expression for the reflected wave will be

$$E_r = E_{or}\, e^{j(\omega t + kz)}. \tag{5.14}$$

We could use sines or cosines instead of complex exponentials as we did on earlier pages, but the present format is a little easier to manipulate. The amplitude E_{or} must be determined from boundary conditions. In accordance with Eq. 5.12b the total tangential electric field (incident plus reflected) vanishes at the metal surface; so, adding Eqs. 5.13 and 5.14 and setting $z = 0$ yields

$$E_{or} = -E_{oi} \tag{5.15}$$

which is the result we set out to prove. Notice that amplitudes of the two waves are equal but their phases are reversed — the reflected wave is 180° out of phase with the incident wave. The resultant electric field at any point distance z away from the surface is obtained by summing Eqs. 5.13 and 5.14 and imposing condition (5.15):

$$\left.\begin{aligned} E &= E_i + E_r \\ &= E_{oi}\, e^{j\omega t}\, [e^{-jkz} - e^{+jkz}] \\ &= -2jE_{oi}\, e^{j\omega t}\, [\sin kz] \end{aligned}\right\} \tag{5.16}$$

where, in deducing the last result we used the mathematical identity $\sin x = [\exp(jx) - \exp(-jx)]/2j$. As the final step we take the real part of the equation and obtain the real field

$$E = 2E_{oi} \sin(kz) \sin(\omega t). \tag{5.17}$$

This will be recognized immediately as the expression for a standing electromagnetic wave as discussed in section 3.5. The magnitude of the electric field varies sinusoidally with distance from the reflecting surface (see Fig. 3.16). The first electric node occurs at $z = 0$ and succeeding nodes are separated from one another by a distance of half a wavelength. The first antinode occurs at a distance of a quarter wavelength from the surface and succeeding antinodes are again separated from one another by half a wavelength. The positions of these electric nodes and antinodes are in full agreement with the experiments of Wiener (see section 2.3) who studied standing light waves above the surface of a plane mirror. In accord with Eq. 5.17 he found no blackening of the photographic emulsion at the mirror surface, showing that an electric node existed there. But blackening occurred at a quarter wavelength away where an antinode is expected. These findings proved to Wiener that the photochemical processes in the emulsion must have been triggered by the electric field of the wave.

In view of the fact that the electric field reverses phase on reflection, it follows that the magnetic field must be reflected without phase reversal – that is, $\vec{B}$ has an antinode at the metal surface. The reason is the following. The directions of the Poynting flux of the incident and reflected waves are given by the respective products, $\vec{E}_i \times \vec{B}_i$ and $\vec{E}_r \times \vec{B}_r$. These directions are of necessity reversed with respect to one another, as are the directions of $\vec{E}_i$ and $\vec{E}_r$. Hence $\vec{B}_i$ and $\vec{B}_r$ must remain in the same direction. The identical conclusion can be established more formally by inserting Eq. 5.13 in Eq. 3.73a and integrating to obtain the incident magnetic field B_i; doing the same with Eq. 5.14 to find the reflected magnetic field B_r, and then showing, subject to the requirement (5.15), that $B_{or} = +B_{oi}$.

Example

A traveling electromagnetic plane wave of angular frequency ω and amplitude E_o of the electric field is right-hand circularly polarized, i.e., the electric vector rotates clockwise (at constant z) when viewed in the direction of propagation, +z, and at z = 0 and t = 0 the electric vector has a y component only. The wave is incident on a polarizer, in the x-y plane, composed of closely spaced wires parallel to the y axis as shown.

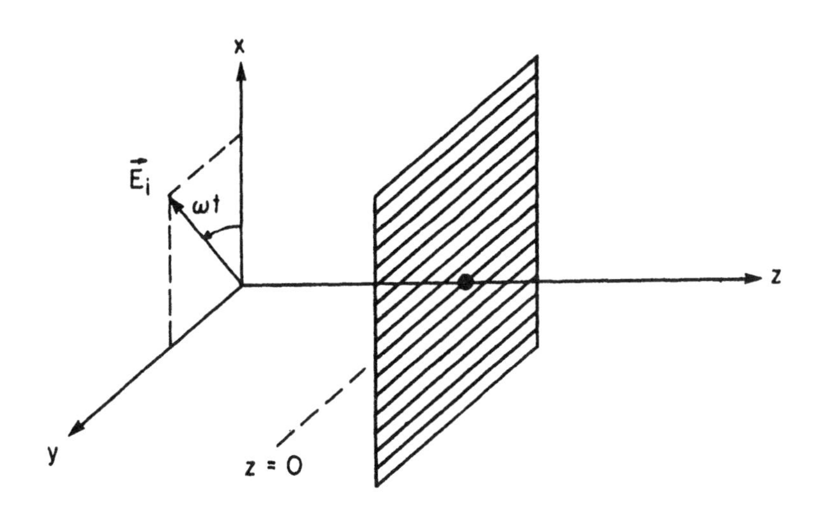

Such a device allows transmission of waves polarized along the x axis without appreciable attenuation and reflects waves polarized along the y axis. Write complete real vector expressions for the following:

a) The incident electric field $\vec{E}_i(z,t)$ for $z < 0$.

b) The transmitted electric field $\vec{E}_t(z,t)$ for $z > 0$, consistent with the answer to part (a).

c) The total electric field $\vec{E}_T(z,t)$ for $z < 0$, consistent with the answer to part (a).

d) The radiation pressure on the polarizer.

Solution

The incident wave is propagating along the positive z axis so the arguments of any sine and cosine functions must be $(\omega t - kz)$. The wave is circularly polarized so $|E_x|$ must equal $|E_y|$ but be 90° out of phase, i.e., one is a sine function and one is a cosine function. To insure that $\vec{E}_i$ has only a y component at $z = 0$ and $t = 0$ we make E_x proportional to $\sin(\omega t - kz)$ and E_y proportional to $\cos(\omega t - kz)$. Finally, to make $\vec{E}_i$ rotate clockwise (at $z = 0$, for example) we must have E_x increase negatively from $t = 0$. Thus the incident wave must be of the form

a) $$\vec{E}_i(z,t) = -E_o\hat{x}\sin(\omega t - kz) + E_o\hat{y}\cos(\omega t - kz).$$

An equally acceptable answer would be with both components reversed in sign since the sign of E_y at $z = 0$ and $t = 0$ was not given.

The nature of the polarizer is that the x component of the wave passes through without attenuation and that the y component is reflected. Thus the transmitted wave is simply the x component of the incident wave. Hence, consistent with part (a), we have

b) $$\vec{E}_t(z,t) = -E_o\hat{x}\sin(\omega t - kz).$$

The total electric field in the region $z < 0$ is composed of three parts:

(1) the x component of the incident wave,

(2) the y component of the incident wave, and

(3) the reflected wave which has only a y component since there is no reflection for the x component.

The reflected wave is of the argument $(\omega t + kz)$ as it is a traveling wave moving along the negative z axis. The amplitude of the reflected wave equals that of the y component of the incident wave but is of opposite sign so that complete cancellation of the y components occurs at the polarizer. Hence we can write for $z < 0$

c) $$\vec{E}_T(z,t) = -E_o\hat{x}\sin(\omega t - kz) + E_o\hat{y}\cos(\omega t - kz) - E_o\hat{y}\cos(\omega t + kz).$$

Alternatively, this can be written in the equivalent form

c) $$\vec{E}_T(z,t) = -E_o\hat{x}\sin(\omega t - kz) + 2E_o\hat{y}\sin kz \sin \omega t$$

which emphasizes that standing waves are set up in the y component due to the reflection at $z = 0$.

The radiation pressure is given by Eq. 3.129 for the case of a completely absorbed wave. For a perfectly reflecting surface the pressure is doubled, as implied in Eq. 3.128, for conservation of momentum. Only the y component is of importance here since the x component is unaffected by the polarizer. Thus the pressure is

$$\vec{P} = \frac{2}{\mu_o c}\left|E_y B_x\right|_{inc}\hat{z}.$$

Since $|B_x| = \frac{1}{c}|E_y|$

d) $$\vec{P} = \frac{2E_o^2}{\mu_o c^2}\cos^2(\omega t - kz)\,\hat{z}$$

Reflection from a metal surface at oblique incidence

When the electromagnetic wave impinges obliquely on the metal surface, it is necessary to consider separately two special cases, depending on the direction of polarization of the incident wave. In the first case the electric vector is parallel to the boundary surface, that is, $\vec{E}_i$ is perpendicular to the plane containing the propagation vector $\vec{k}_i$ and the normal to the surface $\hat{n}$. In the second case $\vec{E}_i$ lies in the plane containing $\vec{k}_i$ and $\hat{n}$. These two situations are illustrated in Fig. 5.7. We realize, of course, that the electric vector of the incident wave can have an arbitrary direction relative to the plane just defined and that the wave may not even be linearly polarized but could have, say, elliptical polarization. Such a more general situation of arbitrary elliptical polarization can be obtained by appropriate linear superposition of the two special cases discussed here, using the techniques described in section 3.4.

Let us begin with the situation in which the incident wave has its electric vector polarized perpendicular to the so-called "plane of incidence" defined by the vectors $\vec{k}_i$ and $\hat{n}$. Let the incident and reflected waves make an angle θ with respect to the y axis as is shown in Fig. 5.8. The electric vector of the incident wave can then be written in terms of our coordinates as

$$E_{ix} = E_{oi}\, e^{j(\omega t - \vec{k}_i \cdot \vec{r})} \tag{5.18}$$

where in accordance with section 2.4 (Eq. 2.53) the scalar product $\vec{k}_i \cdot \vec{r}$ is given by

$$\vec{k}_i \cdot \vec{r} = k_{ix}x + k_{iy}y + k_{iz}z. \tag{5.19}$$

But $k_{ix} = 0$, $k_{iy} = -k \cos \theta$, and $k_{iz} = k \sin \theta$, with the result that

$$E_{ix} = E_{oi}\, e^{j\omega t}\, e^{-jk(-y \cos \theta + z \sin \theta)} \tag{5.20}$$

Similarly, one finds for the reflected wave that

$$\begin{aligned} E_{rx} &= E_{or}\, e^{j\omega t - j\vec{k}_r \cdot \vec{r}} \\ &= E_{or}\, e^{j\omega t}\, e^{-jk(+y \cos \theta + z \sin \theta)}. \end{aligned} \tag{5.21}$$

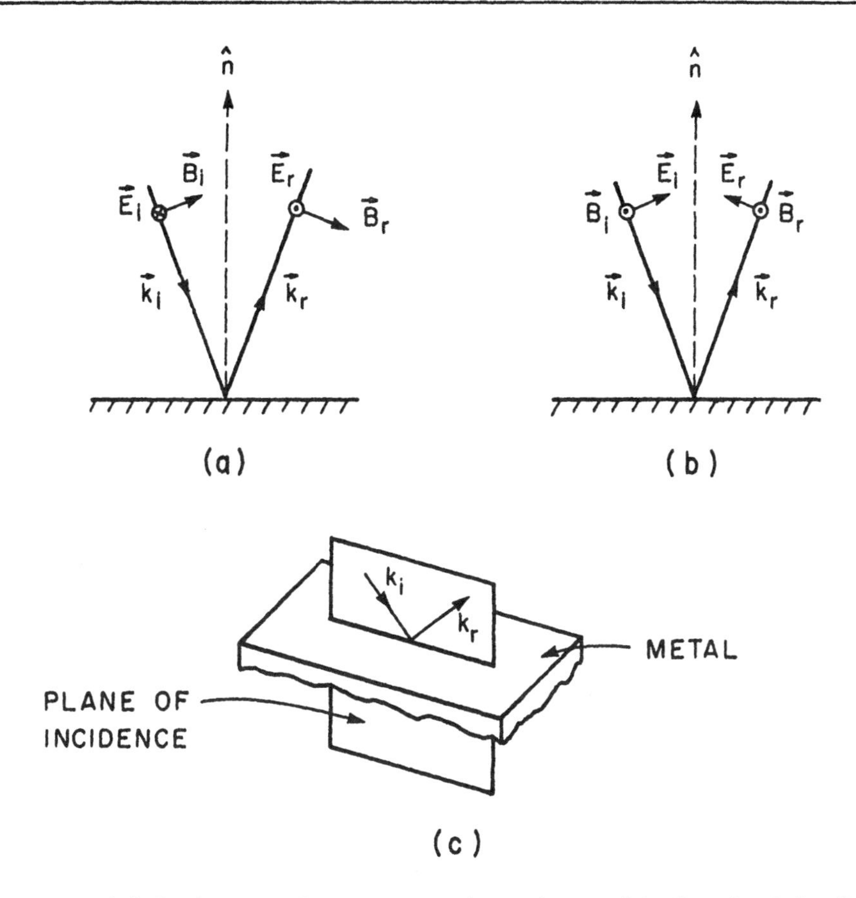

Fig. 5.7 (a) Reflection from a metal surface with the electric field polarized perpendicular to the plane of incidence. (b) Reflection for the case when the electric field lies in the plane of incidence (⊗ means that the vector points into the page; ⊙ means that the vector points out of the page). (c) Definition of the plane of incidence.

The total electric field above the conductor comes from summing Eqs. 5.20 and 5.21 subject to the boundary condition $E_{oi} = -E_{or}$. Hence

$$E_x = E_{oi}\, e^{j\omega t}\left[e^{-jk(-y\cos\theta + z\sin\theta)} - e^{-jk(y\cos\theta + z\sin\theta)}\right]$$
$$= 2jE_{oi}\, e^{j\omega t}\sin(ky\cos\theta)\, e^{-jkz\sin\theta} \qquad (5.22)$$

and on taking the real part, one obtains

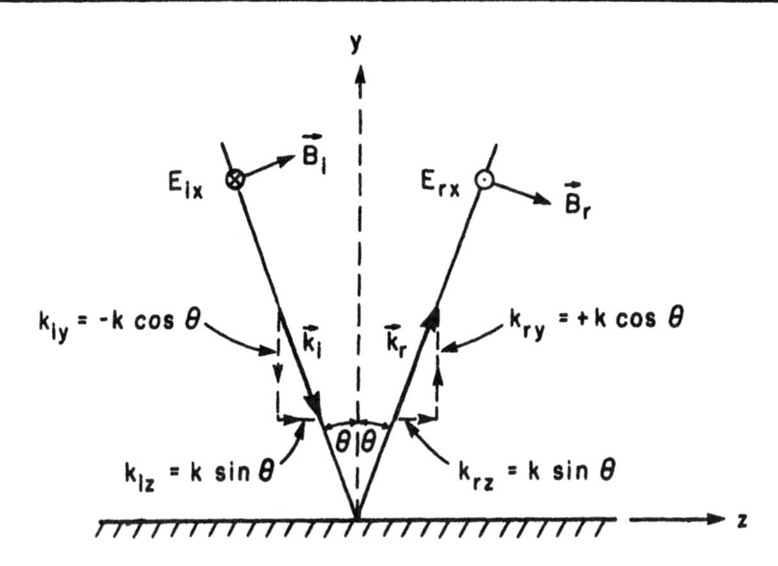

Fig. 5.8 Reflection from a metal surface with the electric field polarized perpendicular to the plane of incidence. Note that since both the incident and reflected waves travel in the same medium (vacuum), $|\vec{k}_i| = |\vec{k}_r| = k = 2\pi/\text{wavelength} = \omega/c$.

$$E_x = -2E_{oi} \sin(ky \cos \theta) \sin(\omega t - kz \sin \theta). \qquad (5.23)$$

This is a most interesting result. There is a standing wave pattern as one moves away from the surface in the direction of the y axis. The wavelength λ_y (defined as twice the distance between nodal points) equals $2\pi/k \cos \theta$ or $\lambda_i/\cos \theta$ and is therefore greater than the wavelength λ_i of the incident wave. In addition, the whole standing wave pattern moves parallel to the reflector, along the positive z axis, with a velocity $\omega/k \sin \theta$ or $c/\sin \theta$; the wavelength in this direction is $\lambda_i/\sin \theta$. Hence, the field $\vec{E}$ given by Eq. 5.23 is neither the field of a purely standing wave nor that of a purely traveling wave, but is a kind of hybrid thereof. This situation is also typical of propagation in waveguides, as we shall see in a later section.

In the case when the electric vector of the incident wave lies in the plane of incidence (see Fig. 5.7), a similar calculation can be performed with little difficulty. Now, however, $\vec{E}$ has two components

E_y and E_z, and for that reason it is more convenient to compute B_x which has but one component, and use the boundary condition $B_{oi} = +B_{or}$. We leave it to the reader to show that the total magnetic field (of the incident plus reflected wave) is given by

$$B_x = 2B_{oi} \cos(ky \cos \theta) \cos(\omega t - kz \sin \theta) \tag{5.24}$$

which exhibits a standing wave pattern traveling along the z axis, in exact analogy with the previous case. The associated electric field is easily deduced from either one of the Maxwell's equations (3.73a,b).

5.2 RADIO FREQUENCY TRANSMISSION LINES

A transmission line can be constructed, conceptually at least, out of a pair of plane parallel conducting sheets separated by a fixed distance a, infinite in extent in the y and z directions, as is illustrated in Fig. 5.9. Although this configuration is of little practical interest, it has the virtue of simplicity and thus serves as a prototype for more complex geometries. For the sake of concreteness, let z be the propagation direction. Undoubtedly, many different combinations of electric and magnetic field shapes can be fitted into the space between

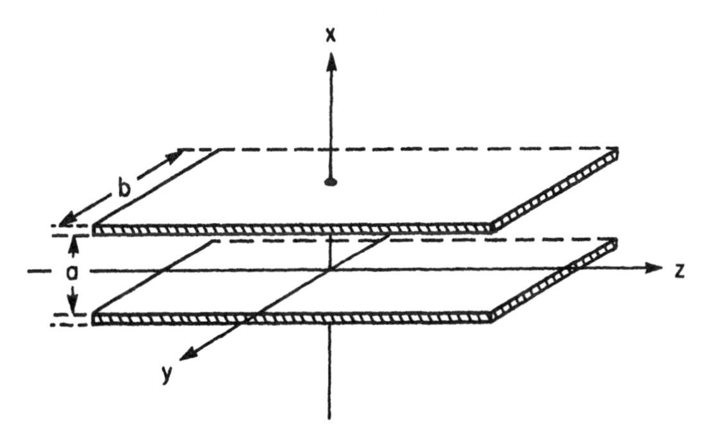

Fig. 5.9 Parallel plate transmission line constructed from two plane parallel sheets of metal, infinite in extent. In all our calculations, however, we shall be considering only a strip of width b.

the plates, and be made to satisfy boundary conditions (5.12b) and (5.12c). The simplest system that comes to mind is one in which $\vec{E}$ is entirely along the x axis and $\vec{B}$ entirely along the y axis. Since here there is no tangential $\vec{E}$ or normal $\vec{B}$, the aforementioned boundary conditions are fulfilled, so to say, by default. Since $\vec{E}$ is transverse to $\vec{B}$ and both are transverse to the assumed direction of propagation, what we have in fact constructed is the field configuration of a plane linearly polarized wave trapped between two conductors. Because we have simultaneously satisfied boundary conditions at the metal surfaces, we can feel confident that a wave must be able to propagate along this system, once it has been launched by some form of antenna situated between the plates. Therefore the primary purpose of the calculations we are about to make is not to prove the feasibility of propagation (which will be proved anyway in the course of the calculations) but to demonstrate that in describing the properties of the transmission line one can dispense with talk about electric and magnetic fields, and replace it with a discussion of the equivalent voltages and currents in the conductors. Because people have a better grasp of V and I compared with $\vec{E}$ and $\vec{B}$, this reformulation of the problem is both elegant and useful.

The electric field E_x and the magnetic field B_y are taken to vary only in the z direction and therefore they will have a uniform distribution in the x-y plane like that illustrated in Fig. 5.10. Observe that

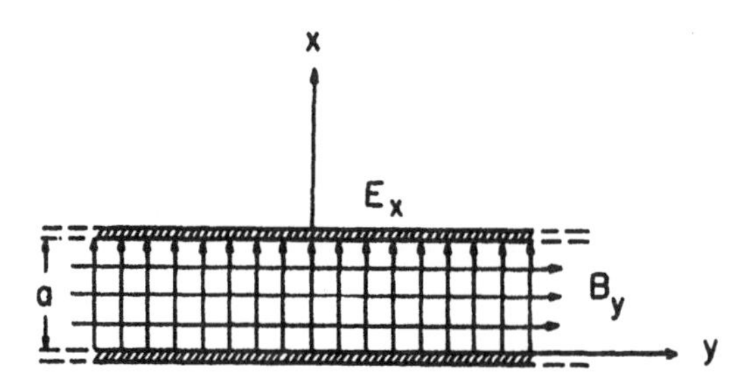

Fig. 5.10 Electric and magnetic fields of the principal or transverse electromagnetic mode TEM of a parallel-plate transmission line, as viewed in a plane perpendicular to the direction of propagation z.

with $\vec{E}$ pointing at a given instant of time in the direction shown, there is positive charge ρ_S induced on the inside of the bottom plate and negative charge on the inside of the top plate. In addition, there is current flow $\vec{J}_S$ in the plates: out of the page on the top plate and into the page at the bottom plate — the magnitudes of ρ_S and J_S being given by Eqs. 5.12a and 5.12d.

We now take a section of transmission line like that shown in Fig. 5.11a and focus our attention on a strip of width b (as measured along the y axis). We draw a closed rectangular path whose corners are labeled 1 through 4 and apply to it Faraday's law (3.16) with the result that,

$$\int_1^2 \vec{E}\cdot d\vec{s} + \int_2^3 \vec{E}\cdot d\vec{s} + \int_3^4 \vec{E}\cdot d\vec{s} + \int_4^1 \vec{E}\cdot d\vec{s} = -a\,dz\,\frac{\partial B_y}{\partial t} \tag{5.25}$$

where a is the separation between the conducting plates and dz is a small length measured along the propagation direction z. The first integral on the left of the equation equals minus the potential difference between the bottom and top plates at position z; we denote it $-V(z)$. The second and fourth integrals are zero because the tangential component of electric field at the surface of a perfect conductor is zero. And the third integral equals the potential difference between the plates at position z + dz, which we denote $V(z+dz)$. Therefore Eq. 5.25 reduces to $V(z+dz) - V(z) = -a\,dz(\partial B_y/\partial t)$. Dividing through by dz then yields

$$\frac{\partial V}{\partial z} = -a\,\frac{\partial B_y}{\partial t}. \tag{5.26}$$

Observe that V is the potential difference <u>between</u> the two conducting planes; it is a transverse voltage not to be confused with dc transmission lines where one talks of the voltage drop along the line. For the perfect conductors considered here, the voltage drop along the line is, of necessity, zero. We now avail ourselves of boundary condition (5.12d) which allows us to express B_y in terms of the surface current J_{Sz}

$$B_y = \mu_o J_{Sz} = \frac{\mu_o}{b} I \tag{5.27}$$

where I is the total current flowing through a strip of transmission line b meters wide. Inserting Eq. 5.27 in Eq. 5.26 gives a differential equation that involves voltages and currents only:

$$\frac{\partial V}{\partial z} = -\left(\mu_o \frac{a}{b}\right) \frac{\partial I}{\partial t}. \tag{5.28}$$

A second relation between V and I can be generated by turning to Fig. 5.11b and applying to the rectangular loop shown there, the Maxwell extension of Ampère's law (3.30). The result is

$$bB_y(z) - bB_y(z+dz) = \mu_o \epsilon_o b \, dz \left(\frac{\partial E_x}{\partial t}\right). \tag{5.29}$$

Replacing bB_y by $\mu_o I$ and E_x by V/a then yields the sought-for second equation connecting V and I:

$$\frac{\partial I}{\partial z} = -\left(\epsilon_o \frac{b}{a}\right) \frac{\partial V}{\partial t}. \tag{5.30}$$

Equations 5.28 and 5.30 are two partial differential equations in two unknowns V and I. To eliminate, say, I, differentiate the former with respect to z, substitute for $\partial I/\partial z$ from the latter and derive a wave equation for V

$$\frac{\partial^2 V}{\partial z^2} = \mu_o \epsilon_o \frac{\partial^2 V}{\partial t^2} \tag{5.31}$$

and, by similar means, a wave equation for I,

$$\frac{\partial^2 I}{\partial z^2} = \mu_o \epsilon_o \frac{\partial^2 I}{\partial t^2}. \tag{5.32}$$

Thus we have demonstrated that voltage and current waves can propagate down the transmission line. Their phase velocity is independent

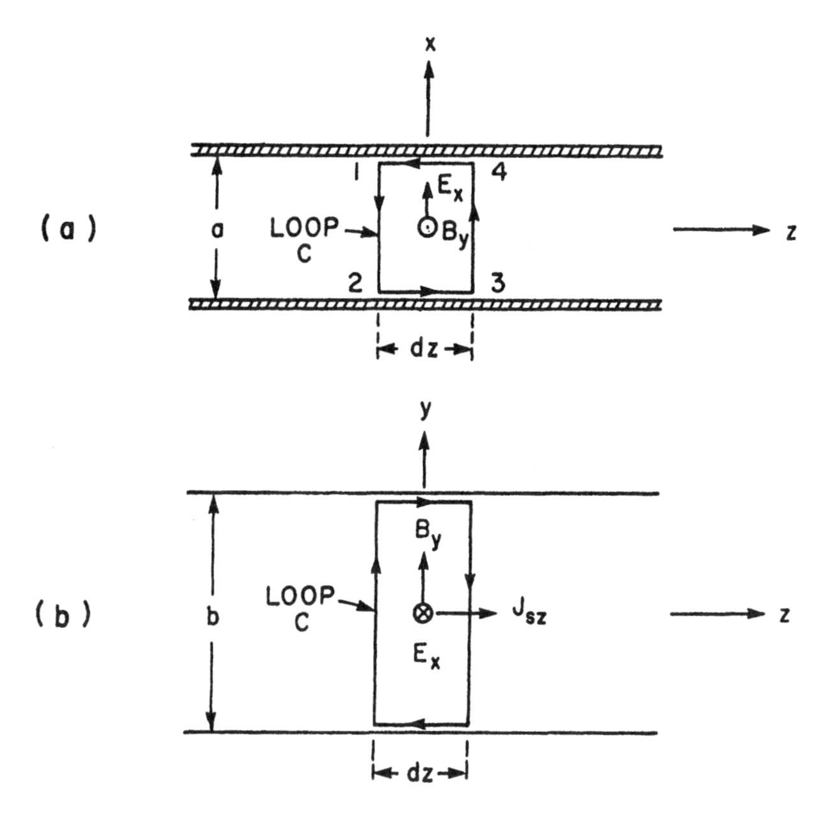

Fig. 5.11 Different views of the parallel-plate transmission line for use in evaluating integrals (3.16) and (3.30). The direction of the surface current density $\vec{J}_S$ is given by $\hat{n} \times \vec{B}$ where $\hat{n}$ is the unit normal pointing out of the metal surface. ⊙ means that the vector points out of the page; ⊗ means that the vector points into the page.

of the separation between the conducting plates and equals the speed of light in vacuum, $c = (\mu_o \epsilon_o)^{-1/2}$; indeed all geometric factors have dropped out of the expressions for the wave equations. This arises from the fact that the geometric term (a/b) of Eq. 5.28 appears as the reciprocal quantity (b/a) in Eq. 5.30.

What physical meaning can we ascribe to the factors in parentheses appearing in Eqs. 5.28 and 5.30? Let us take the term $(\epsilon_o b/a)$ and multiply it by a length ℓ of line, as measured off along the propagation direction z. This gives $(\epsilon_o b\ell/a)$ with $b\ell$ as the surface area of a piece of conducting plate ℓ long and b wide. Thus we see that $(\epsilon_0 b\ell/a)$ is nothing more than the capacity of a section of the transmission line as it would be calculated from electrostatics, assuming fixed surface charges $\pm\ \rho_S$ residing on the inside surfaces of the two conductors. It follows that $(\epsilon_0 b/a)$ is the capacity per unit length of line and we denote it by C_0 (farad/m). Similarly, the term $(\mu_0 a/b)$ of Eq. 5.28 is found to be the inductance per unit length of line L_0 (henry/m) as it would be calculated from magnetostatics, under the assumption that steady surface currents $\pm\ J_{Sz}$ flow in opposite directions in the two conducting plates (the inductance would be that of a strip of width b of plates). With these interpretations, we can rewrite Eqs. 5.28 and 5.30 to read

$$\boxed{\begin{aligned} \frac{\partial V}{\partial z} &= -L_0 \frac{\partial I}{\partial t} && \text{(a)} \\ \frac{\partial I}{\partial z} &= -C_0 \frac{\partial V}{\partial t}. && \text{(b)} \end{aligned}} \qquad (5.33)$$

These are the fundamental equations of transmission line theory. To be sure, we have proved them for the very special case of a parallel plate line, but they are applicable to any transmission line composed of two (or more) separate conductors whose geometrical cross section remains invariant as one proceeds along the line, in the propagation direction z. Let us summarize the general characteristics of these lines.

(i) On a two- (or more) wire transmission line one can always find a mode (that is, a configuration of electric and magnetic fields) where $\vec{E}$ and $\vec{B}$ are entirely transverse to the direction of energy flow (the z direction in our coordinate system). This so-called "principal" or transverse electromagnetic mode (TEM) is the one to which Eqs. 5.33 apply. There are other higher modes but they are all characterized by the fact that either $\vec{E}$ or $\vec{B}$ has a component along z. Such modes will not be discussed here any further but will be considered in the next section in connection with propagation in waveguides (which are essentially one-conductor systems and therefore cannot support a TEM mode).

(ii) The TEM wave will transmit energy at all frequencies down to and including zero cycles per second.

(iii) When losses in the conductor can be neglected, that is, in the limit of perfect conductivity, the phase velocity of the wave on the line is independent of frequency (there is no dispersion). Its magnitude is the same as the velocity in the absence of the conductors and is given by $v = c$. In terms of the inductance and capacity per unit length, $v = (L_o C_o)^{-1/2}$. [When the space between the conductors is filled with a dielectric, the phase velocity differs from c and is given by c/n where n is the refractive index of the material.]

(iv) When field penetration into the conductors can be neglected, which is permissible at high frequencies (see Eq. 5.11), the instantaneous electric and magnetic field configurations between the conductors are exactly the same as would exist there if only static charge and dc currents were properly distributed along the line. It is precisely this property which permits one to compute C_o and L_o of Eqs. 5.33 as if electrostatics and magnetostatics were applicable.

Let us now apply these considerations to an important practical situation — the familiar coaxial transmission line.

The coaxial line

The TEM mode of a coaxial line has an electric field which is purely radial and a magnetic field which is purely azimuthal as is illustrated in Fig. 5.12 — distributions which one may have guessed from symmetry arguments and boundary requirements at the surfaces of the two perfect conductors. To determine the characteristics of the coaxial line (or for that matter, any two-conductor transmission system) it is necessary to compute its inductance L_o and capacity C_o per unit length of conductor. Let us begin with the former.

The magnetic field B_θ at any radius r between the two conductors equals $\mu_o I/2\pi r$, a result which follows Ampère's law (3.25). Here I is the instantaneous axial current flowing down the center conductor in the direction of the positive z axis (a current of equal magnitude but opposite in direction flows at that instant and at that position z along

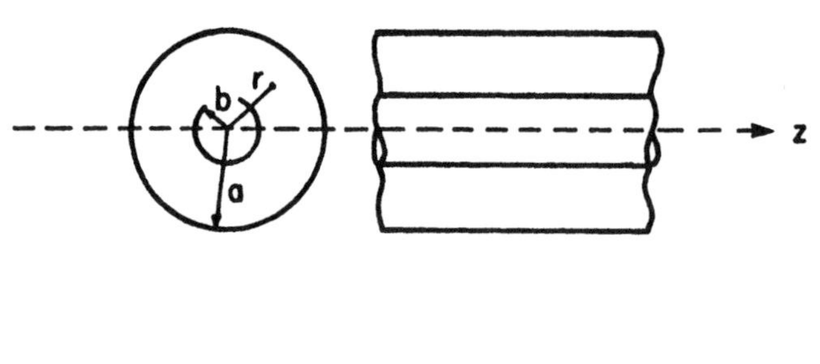

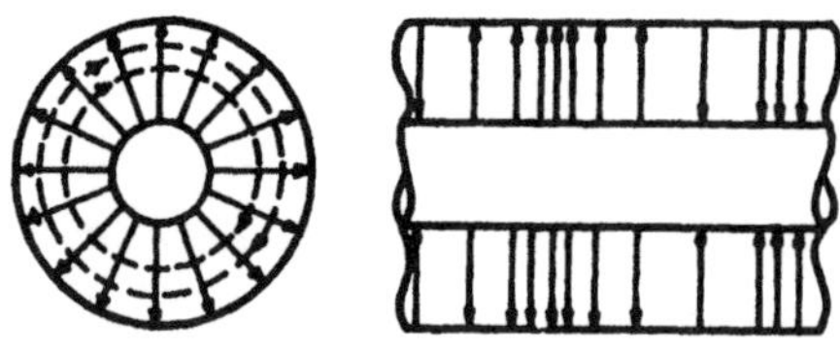

Fig. 5.12 Section of coaxial transmission line, illustrating the configuration of the electric [———] and magnetic [- - - -] fields of the principal (TEM) mode. Observe that although the fields are entirely transverse to the direction of propagation z, they are nonuniform and are thus quite unlike the plane waves on the parallel-plate transmission line shown in Fig. 5.10. Nevertheless, Eqs. 5.33 are fully applicable.

the outer conductor, but this current does not contribute to the magnetic field strength between the conductors). The magnetic flux $\Phi_B = \int \vec{B} \cdot d\vec{a}$ of length ℓ of line becomes

$$\begin{aligned}\Phi_B &= \frac{\ell\mu_o I}{2\pi} \int_b^a \frac{dr}{r} \\ &= \frac{\ell\mu_o I}{2\pi} \ell n \left(\frac{a}{b}\right).\end{aligned} \tag{5.34}$$

From Faraday's induction law and the definition of L, which is emf = $-d\Phi_B/dt = -L\, dI/dt$, it follows that $L = \Phi_B/I$, with the result that

$$\begin{aligned}L_o &= \frac{\mu_o}{2\pi} \ell n \left(\frac{a}{b}\right) \\ &= 4.61 \times 10^{-7} \log_{10}\left(\frac{a}{b}\right) \quad \text{henry/m}\end{aligned} \tag{5.35}$$

where, in obtaining the second form of the equation we substituted the numerical value for μ_o and changed bases on the logarithm.

The radial electric field at any radius r between the conductors is deduced from Gauss' law: if Q is the instantaneous charge at some position z along the inner conductor, then $E_r = Q/(2\pi\epsilon_o \ell r)$ (there is a charge -Q on the inner surface of the outer conductor, but this charge does not contribute to the electric flux existing between the conductors). The potential difference between the cylinders is

$$\begin{aligned} V &= -\frac{Q}{2\pi\epsilon_o \ell} \int_a^b \frac{dr}{r} \\ &= \frac{Q}{2\pi\epsilon_o \ell} \ln\left(\frac{a}{b}\right) \end{aligned} \tag{5.36}$$

and the capacitance per unit length $C_o = Q/\ell V$ is therefore

$$\begin{aligned} C_o &= \frac{2\pi\epsilon_o}{\ln\left(\frac{a}{b}\right)} \\ &= \frac{2.416 \times 10^{-11}}{\log_{10}\left(\frac{a}{b}\right)} \text{ farad/m.} \end{aligned} \tag{5.37}$$

The characteristic impedance Z_o

We excite one end of the transmission line by connecting the inner and outer conductors to an alternating, sinusoidal voltage source of frequency ω. This causes an electromagnetic wave to travel along the line with fields given by $E_r = E_{or} \exp[j\omega t - jkz]$, $B_\theta = B_{o\theta} \exp[j\omega t - jkz]$. The amplitudes E_{or} and $B_{o\theta}$ are both functions of the radial coordinate r. As was shown at the beginning of this section, the disturbance can also be represented in terms of equivalent voltage and current waves given by

$$\begin{aligned} V &= V_o\, e^{j(\omega t - kz)} \\ I &= I_o\, e^{j(\omega t - kz)} \end{aligned} \tag{5.38}$$

where ω/k is the phase velocity of magnitude

$$\frac{\omega}{k} = v = \frac{1}{\sqrt{L_o C_o}}. \tag{5.39}$$

On substituting Eqs. 5.35 and 5.37 in Eq. 5.39 we verify that the phase velocity is indeed equal to the speed of light in vacuum, c. V and I are of course related to one another through the two differential equations (5.33). To find this relationship, we insert Eqs. 5.38 in Eq. 5.33a, perform the designated differentiations and find that $I_o = [k/(\omega L_o)]V_o$. But, $k/\omega = (L_o C_o)^{1/2}$ so that $I_o = (C_o/L_o)^{1/2} V_o$ with the result that

$$V(z,t) = V_o \, e^{j(\omega t - kz)}$$
$$I(z,t) = \frac{V_o}{Z_o} e^{j(\omega t - kz)} \tag{5.40}$$

where

$$Z_o = \sqrt{\frac{L_o}{C_o}} \text{ ohm} \tag{5.41}$$

is the so-called "characteristic impedance." It proves to be an important parameter used in describing the propagation characteristic of the line. Recall that in section 2.2, Eq. 2.30 defines a corresponding quantity associated with sound propagation along a pipe. Voltage and current amplitudes are connected via the characteristic impedance, and if one is known, the other is obtained from the formula $I = V/Z_o$.

The instantaneous power transmitted is found by taking the real part of V and the real part of I of the quantities given by the right-hand sides of Eqs. 5.40 and multiplying one by the other:

$$P(z,t) = (\text{Re } V)(\text{Re } I)$$
$$= \frac{V_o^2}{Z_o} \cos^2(\omega t - kz) \text{ watt.} \tag{5.42}$$

The time-average power flow then becomes

$$\langle P \rangle = \frac{1}{2}\frac{V_o^2}{Z_o} \text{ watt.} \tag{5.43}$$

Observe that one cannot take the complex V and complex I of Eqs. 5.40, multiply one by the other and <u>then</u> take the real part of the product; the reason is that Re(VI) ≠ (Re V)(Re I). If one does wish to work with complex V and I, then the following is true: (Re V)(Re I) = $(1/4)[VI + V^*I^* + VI^* + V^*I]$ where the asterisk denotes the complex conjugate of the quantity in question (see Appendix A2).

In contrast to the phase velocity v which is geometry independent, the characteristic impedance is very much governed by the geometrical arrangement and size of the conductors. Thus, inserting Eqs. 5.35 and 5.37 in Eq. 5.41 we obtain for the coaxial transmission line the result

$$\begin{aligned} Z_o &= \frac{1}{2\pi}\sqrt{\frac{\mu_o}{\epsilon_o}}\, \ell n\left(\frac{a}{b}\right) \\ &= 138.1 \log_{10}\left(\frac{a}{b}\right) \text{ ohm.} \end{aligned} \tag{5.44}$$

As an example, take the rigid copper coaxial line in vogue during the early days of radar; the inner conductor is 3/8 inch OD; the outer conductor is 7/8 inch OD with a 1/32 inch thick wall. The space between conductors is air. This gives a ratio (a/b) equal to 2.17 and, in accordance with Eq. 5.44, a characteristic impedance of 46.4 ohm. What is the maximum power that can be carried by this line? In air, electrical breakdown occurs when the electric field E_r exceeds approximately 30,000 volts per centimeter. With a voltage V between conductors, the electric field E_r at any radius r is given by $E_r = V[r\, \ell n\, (a/b)]^{-1}$. This has its maximum value at r = b. Hence $E_r(\max) = V[b\, \ell n\, a/b]^{-1}$. With E_r set at 30,000 V/cm and with the values of a and b given above, the maximum permissible voltage V between the conductors equals 11,000 V. With Z_o = 46.4 ohm, the maximum time-averaged power, as given by Eq. 5.43, that can be carried by this line equals 1.31 MWatt. This is a theoretical maximum power-carrying capability, but in actual practice it is necessary to limit the powers to values considerably smaller than this. For example, small protrusions at joints cause undesirable field enhancements. In addition, any standing waves existing on the line cause an increase in V for a given net flow of power. The latter points up one of several reasons why so-called "impedance matching" is an important consideration in the transmission of electromagnetic energy, and we shall soon address ourselves to this problem.

Impedance matching

In writing Eqs. 5.38 we have permitted waves to travel in the positive z direction only, that is from the generator situated on the left to some receiving station or "load" on the right (see Fig. 5.13). Under more general conditions, a reflected wave will travel to the left so that the expressions for the voltage and current become

$$\begin{aligned} V &= V_i \, e^{j(\omega t - kz)} + V_r \, e^{j(\omega t + kz)} \\ I &= I_i \, e^{j(\omega t - kz)} + I_r \, e^{j(\omega t + kz)} . \end{aligned} \tag{5.45}$$

We wish to find how conditions at the load govern the amplitude of the reflected wave V_r, I_r. We found earlier that the amplitude I_i is given

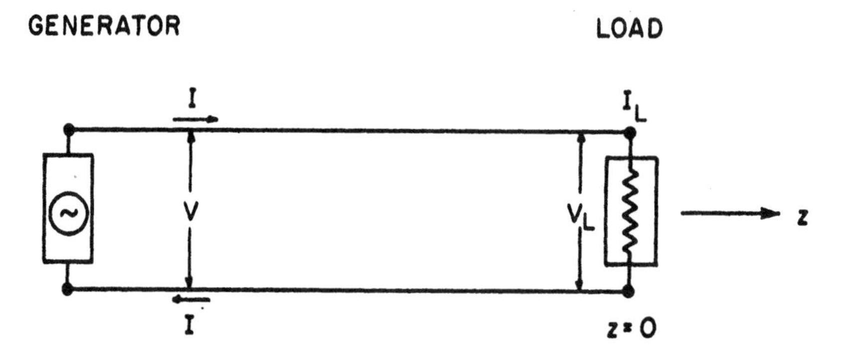

Fig. 5.13 Radio frequency transmission line showing the generator and the receiver (i.e., the load). Throughout we shall assume pure resistive loads.

by V_i/Z_o where Z_o is the characteristic impedance of the transmission line. In an exactly similar way it can be shown that for the reflected wave, $I_r = -V_r/Z_o$. Therefore Eqs. 5.45 can be rewritten in the form

$$V = V_i\, e^{j(\omega t - kz)} + V_r\, e^{j(\omega t + kz)}$$
$$I = \frac{V_i}{Z_o} e^{j(\omega t - kz)} - \frac{V_r}{Z_o} e^{j(\omega t + kz)}. \qquad (5.46)$$

Suppose the load is situated at $z = 0$ and that at this point the voltage takes on the value V_L and the current the value I_L. With these boundary conditions, Eqs. 5.46 become

$$V_L = (V_i + V_r)\, e^{j\omega t}$$
$$I_L = \left(\frac{V_i}{Z_o} - \frac{V_r}{Z_o}\right) e^{j\omega t}. \qquad (5.47)$$

Taking the ratio of V_L to I_L and denoting it Z_L, the "load impedance," we see that

$$Z_L \equiv \frac{V_L}{I_L}$$
$$= Z_o\left(\frac{V_i + V_r}{V_i - V_r}\right). \qquad (5.48)$$

This equation can be solved for the ratio V_r/V_i which represents the sought-after fraction of the incident wave reflected at the load. The ratio is called the voltage reflection coefficient and we designate it as R_V. Its value is

$$\boxed{R_V = \frac{V_r}{V_i} = \frac{Z_L - Z_o}{Z_L + Z_o}.} \qquad (5.49)$$

Since $I_i = V_i/Z_o$ and $I_r = -V_r/Z_o$ we can deduce the corresponding reflection coefficient for the current wave, which is

$$R_I = \frac{I_r}{I_i} = \frac{Z_o - Z_L}{Z_o + Z_L}. \quad (5.50)$$

Thus we conclude that when the impedance of the load Z_L equals the characteristic impedance of the line Z_o (that is, the two are "matched" to one another), the reflection coefficient is zero and no wave travels from the load back to the generator. Indeed, by making $Z_L = Z_o$, the transmission line, though of finite length, can be made to look infinitely long as far as its electromagnetic properties are concerned. Under all other conditions, the incident wave will in part be reflected back whenever it encounters an impedance other than the impedance of the line on which it travels, with the result that standing waves will be set up. The magnitude and phase of the reflections will depend on the impedance Z_L.

Several typical standing wave patterns that may be generated on a uniform, lossless transmission line are illustrated in Fig. 5.14. It is seen that when Z_L is zero or infinite, the magnitude of the reflection coefficients given above is unity and none of the energy incident on the load can be absorbed by it. Therefore the incident and reflected waves are of equal amplitude. It is apparent, however, that the standing wave pattern for the case $Z_L = 0$ differs from the standing wave pattern for the case $Z_L = \infty$. When the load impedance is zero, no voltage can be developed across its terminals and the two traveling waves of voltage (that is, the incident and reflected waves) must cancel one another at the position of the load, while the corresponding current waves will add. The reasons for this are clear: when the line is shorted, say by a metal bar, the tangential electric field at the short must be zero and so must the voltage which is proportional to the electric field. The situation is analogous to a plane wave striking a mirror, a case we studied in section 5.1.

The converse is true for a load of infinite impedance through which no current can flow. The current waves must therefore cancel

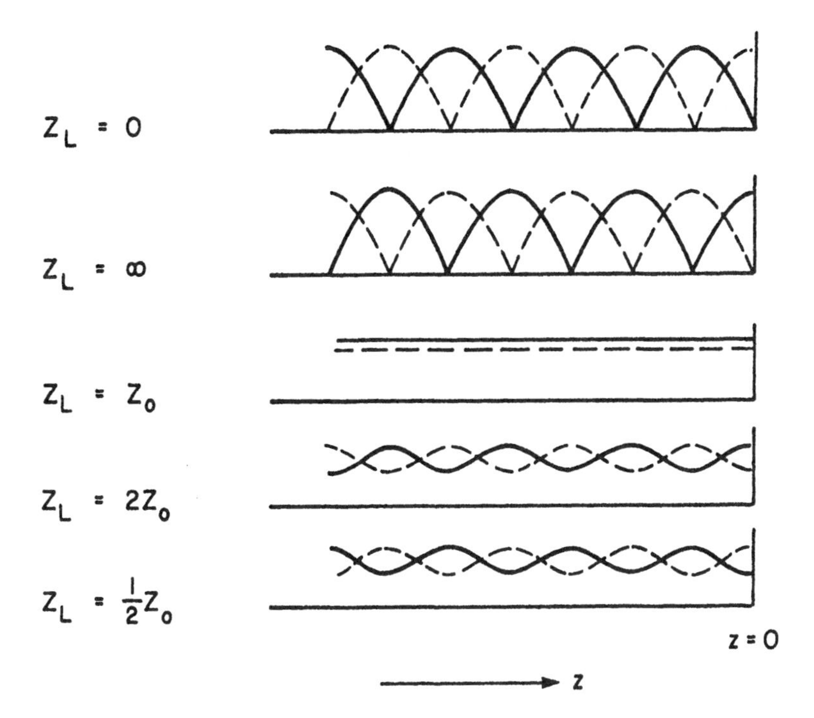

Fig. 5.14 Voltage [——] and current [----] distributions along a lossless transmission line with various load impedances Z_L. What is plotted is the magnitude of $|V|$ and the magnitude of $|I|$ as given by Eqs. 5.46 or the equivalent expressions (5.53). These are instantaneous pictures taken at a time when V and I are maximum. The reason for plotting magnitudes is that this is what voltmeters and ammeters measure with no regard to phase.

at the load, and the voltage waves add, giving an electric field maximum at this position twice that of the incident wave.

The reasons for attempting to match impedances are now becoming quite clear. One reason we stated already: in a matched line the electric field strength for a given power flow is at its minimum, there being no standing waves present. Thus the possibility of voltage breakdown at high power levels is reduced. More important, however, is the fact that if both the generator and the load are matched to the transmission line between them, the generator will deliver maximum power to the load whatever the length of the line. Let us calculate this power, assuming that the generator is matched but that the load is not necessarily matched. From Eq. 5.43 it follows that the time-averaged power flowing towards the load equals $V_i^2/2Z_o$; the time-averaged power traveling in the opposite direction is $V_r^2/2Z_o$, so that the power absorbed by the load is the difference between these two values, and is given by

$$\langle P_L \rangle = \frac{1}{2Z_o}\left[V_i^2 - V_r^2\right]. \tag{5.51}$$

The fraction absorbed is therefore

$$\begin{aligned} F_L &= \frac{V_i^2 - V_r^2}{V_i^2} \\ &= 1 - \left(\frac{Z_L - Z_o}{Z_L + Z_o}\right)^2 \end{aligned} \tag{5.52}$$

where the second form of the equation is obtained by use of the result (5.49). Figure 5.15 shows a plot of F_L as a function of Z_L/Z_o. We

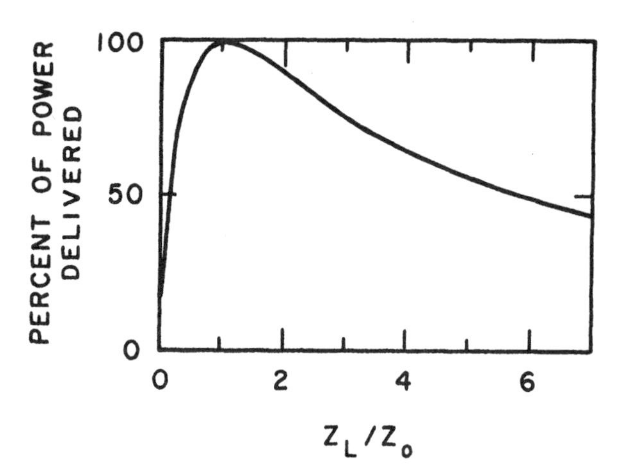

Fig. 5.15 Power transferred from a transmission line of impedance Z_o to a load of impedance Z_L. Z_L is assumed to be pure resistive.

see that when $Z_L = Z_o$, 100 percent of the power on the line flows into the load. On the other hand, if Z_L is, say, six times Z_o, only half the power is delivered, the remainder being reflected back to the generator. This is not only bad from the point of view of efficiency, but the power returned to the generator may cause damage to it and care must be taken to avoid this situation.

When the generator and the load do not match the line, the power delivered depends on the length of the transmission line, because (as we shall find in the next subsection) the effective impedance looking down a piece ℓ of unmatched transmission line depends on its electrical "length" $k\ell$, in addition to being a function of Z_o and Z_L.

Transmission lines as circuit elements

The transfer of energy from one point to another is one important function of a transmission line. Another is the use of pieces of line as high-frequency circuit elements such as capacitors, inductors (and resistors when energy loss is important). This can be understood qualitatively from the top picture in Fig. 5.14. Looking down the line toward the load, we can see at the input end a voltage maximum and zero current; that is, high electric and no magnetic field, just like in a capacitor. When, at the free end, we now chop off a piece of line a quarter of a wavelength long and look into the input end, we see a current maximum and zero voltage. That means a high magnetic field and zero electric field, like in an inductance.

In order to obtain quantitative results, let us take a section of transmission line of length ℓ as is illustrated in Fig. 5.16, and compute the voltage V_ℓ and current I_ℓ at its input end. All this means is that we need to set $z = -\ell$ in both Eqs. 5.46. But the amplitudes V_i and V_r can easily be replaced by V_L and I_L through the use of Eqs. 5.47 from which it is found that $2V_i = (V_L + Z_o I_L)\exp(-j\omega t)$ and $2V_r = (V_L - Z_o I_L)\exp(-j\omega t)$. Making this substitution, we obtain

$$
\begin{aligned}
V_\ell &= V_L \cos(k\ell) + jZ_o I_L \sin(k\ell) \\
I_\ell &= I_L \cos(k\ell) + j\frac{V_L}{Z_o}\sin(k\ell)
\end{aligned}
\qquad (5.53)
$$

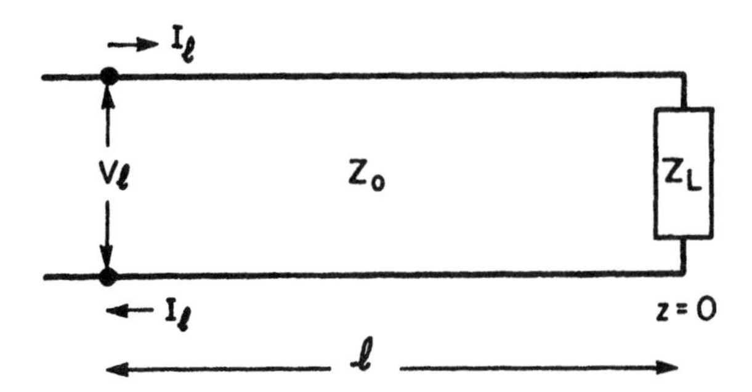

Fig. 5.16 A section of length ℓ of transmission line terminated in a load of impedance Z_L. The load is at position z = 0 and the generator is to the left of this reference point, beyond the page.

from which both the magnitudes and phases of V_ℓ and I_ℓ can be calculated for any desired length of line ℓ, given the values of Z_o, V_L and I_L. For purposes of illustration, let us consider a shorted line where $V_L = 0$. Then

$$\left.\begin{aligned} V_\ell &= jZ_o I_L \sin(k\ell) \\ I_\ell &= \quad I_L \cos(k\ell) \end{aligned}\right\} \text{ shorted line.} \tag{5.54}$$

The results are plotted in Fig. 5.17 where both the amplitudes and phases are depicted. The little complex vector diagrams help us in visualizing the situation. Observe that at all positions ℓ, the voltage and current are 90° out of phase. For line lengths less than 1/4 wavelength I_ℓ lags V_ℓ and therefore the system presents an impedance at its terminals which is inductive. For line lengths greater than $\lambda/4$ and less than $\lambda/2$, I_ℓ leads V_ℓ and the system presents an impedance that looks capacitive. The values of these impedances are given by the ratio of V_ℓ to I_ℓ. Using the expressions given by Eqs. 5.53 one obtains in the general case that

$$Z_\ell = Z_o \left[\frac{Z_L \cos(k\ell) + jZ_o \sin(k\ell)}{Z_o \cos(k\ell) + jZ_L \sin(k\ell)}\right] \tag{5.55}$$

and in the special case of the shorted line, that

$$Z_\ell = jZ_o \tan(k\ell) \quad \text{(shorted line).} \tag{5.56}$$

These are the so-called "input impedances" which the line presents as one looks down it, toward the load. Figure 5.18 shows some examples of these impedances and their equivalent circuits.

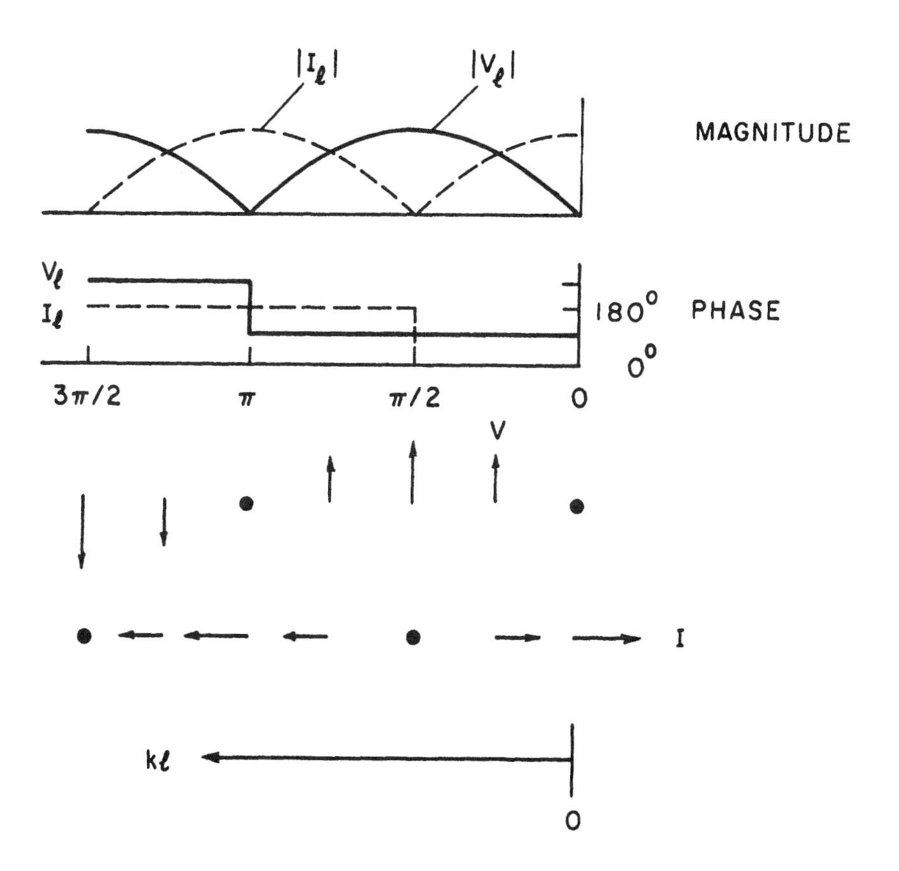

Fig. 5.17 The magnitudes and phases of the voltage V_ℓ current I_ℓ on a shorted transmission line. The short is at position $k\ell$ = 0. The arrows indicate amplitudes and phases at several positions along the line as shown in the complex plane. The horizontal axis is the real axis, the vertical axis is the imaginary axis. The figure is that of Eqs. 5.54.

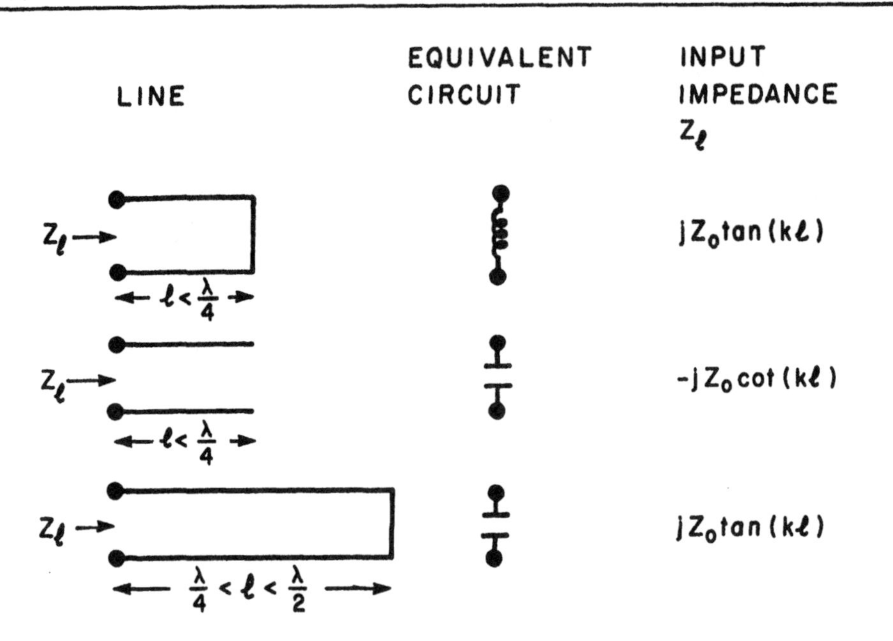

Fig. 5.18 Input impedances of different sections of lossless transmission lines.

The quarter-wave transformer

One of the important uses of such elements as we have described is in impedance matching where line, load and/or generator cannot in themselves be designed to have identical impedances. There are several different ways whereby matching can be accomplished (see for example E. C. Jordan and K. G. Balmain, Electromagnetic Waves and Radiating Systems (Prentice-Hall, Inc., Englewood Cliffs, New Jersey, 1968). One such is known as the quarter-wave transformer. Here a section of transmission line is interposed between the main line and the load, or between two lines of different characteristic impedances (Fig. 5.19). The aim is to eliminate reflections to the left of the transformer. We can guess how this is to be accomplished. A wave is reflected from the front edge of the transformer A and another wave is reflected from its back edge B (of course there are many successive reflections between the two surfaces, but for purposes of this qualitative argument we consider just one). Then, to

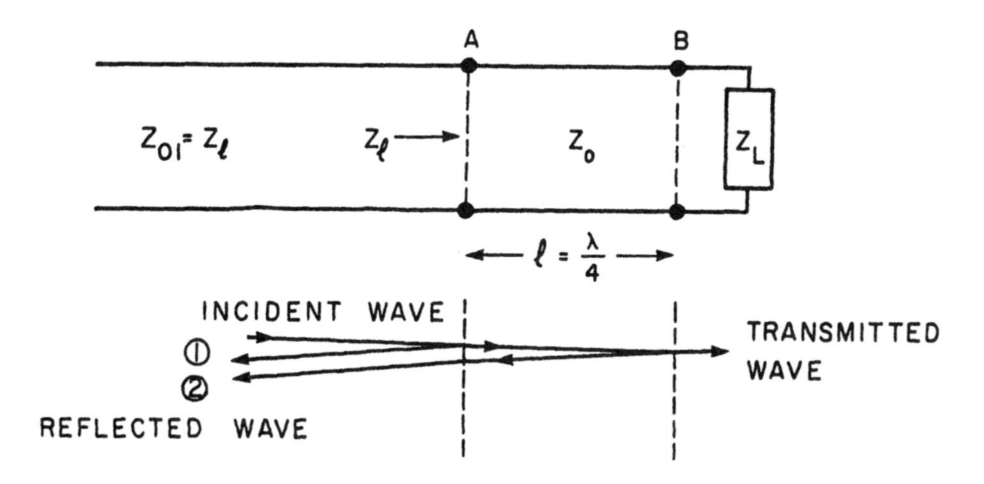

Fig. 5.19 A quarter-wave transformer with input impedance $Z_\ell = Z_o^2/Z_L$. It converts a high impedance Z_L (or Z_ℓ) into a low impedance Z_ℓ (or Z_L). To eliminate reflections to the left of point A, the transformer must have an impedance Z_o equal to $\sqrt{Z_{o1}Z_L}$, the geometric mean of the impedance on its left and right. By making $\ell = \lambda/4$, the path-length of wave (2) is one-half wavelength longer than that of wave (1) and thus the two reflected waves interfere destructively.

have zero net reflection at the front surface A, the two waves being considered must be 180° and of phase so as to interfere destructively. The argument suggests that the transformer be a quarter of a wavelength long. This is a necessary but not sufficient requirement. Setting $k\ell = \pi/2$ and substituting this value in Eq. 5.55 tells us the input impedance of the transformer,

$$Z_\ell = \frac{Z_o^2}{Z_L} \tag{5.57}$$

where Z_o is its characteristic impedance and Z_L is the impedance of the load situated behind it. Setting $Z_\ell = Z_{o1}$, the characteristic impedance of the line ahead of the transformer, yields the second requirement which must be met in designing this matching device, that $Z_o = \sqrt{Z_L Z_{o1}}$. Suppose, for example, that rf power from a generator is carried on a transmission line whose characteristic impedance Z_{o1} is 50 Ω. It is being delivered to a load whose impedance Z_L is 100 Ω. To eliminate standing waves on the line, we place near the load a quarter wavelength long line whose impedance must be adjusted to the value $\sqrt{(50) \times (100)}$ or 70.7 Ω. Often one needs to join rf cables of different impedances, say Z_{o1} and Z_{o2}. It follows from this discussion that in order to achieve this union without reflection, a quarter wavelength long cable having an impedance equal to $\sqrt{Z_{o1} Z_{o2}}$ must be interposed between the two principal lines.

What is somewhat unfortunate for this method of impedance matching (and this is true for almost all other methods) is that it is frequency sensitive; and having matched at one frequency, a mismatch occurs at other frequencies. In all such cases a crucial question is the bandwidth, that is, the range of frequencies over which good, if not perfect, matching can be achieved. Judicious combinations of the kind of circuit elements we have described can result in good and also fairly broadband matching. One way of obtaining broadband operation is by the use of several transformer sections that introduce a change in impedance gradually. Thus several quarter-wave sections of slightly different impedances may be employed. As a general rule, the greater the number of cascaded sections of some given type of matching device, and the smaller the discontinuity at each junction, the less sensitive to frequency is the overall device. Just a tapered section of line one or more wavelengths long may be used to join lines of different characteristic impedance, with little reflection at the junction, and good bandwidth properties.

Example

A lossless transmission line with a characteristic impedance Z_o of 50 Ω is terminated by a load impedance Z_L of 50 + j 20 Ω. The generator has an internal impedance of 50 Ω, hence is matched to the line, and an open-circuit output voltage given by

$V_s(t) = V_o \cos \omega_o t$ where V_o equals 15 volts. The length of line is 1/4 of a wavelength.

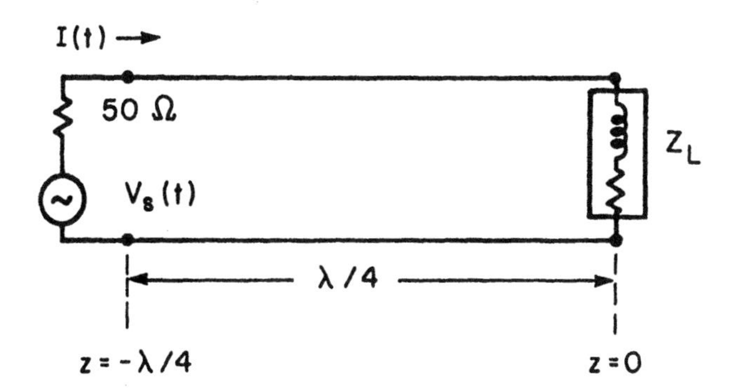

a) Determine the total current I(t) at the generator terminals.

b) Determine the time-average power dissipated in the load Z_L.

c) Find the impedance "looking into" the line if its length is half a wavelength.

Solution

The impedance of a line of length ℓ is given in general by Eq. 5.55. If $\ell = \lambda/4$ then $k\ell = \pi/2$ since $k = 2\pi/\lambda$, and Eq. 5.57, i.e., $Z_{\lambda/4} = Z_o^2/Z_L$. In our case, $Z_o = 50\ \Omega$ and $Z_L = 50 + j\,20\ \Omega$, hence we find

$$Z_{\lambda/4} = \frac{(50)^2}{50 + j20} = \frac{(50)^2(50 - j20)}{(50 + j20)(50 - j20)}$$

or

$$Z_{\lambda/4} = \frac{25}{29}(50 - j20) = 43.1 - j17.2\ \Omega.$$

Now the total impedance Z_T "seen by the generator" includes its own internal impedance, 50 Ω, thus

$$Z_T = 93.1 - j17.2\ \Omega.$$

The current flowing in the circuit I(t) is given by

$$I(t) = I_o e^{j\omega_o t} = \frac{V_s(t)}{Z_T} = \frac{V_o \cos \omega_o t}{Z_T} = \frac{V_o e^{j\omega_o t}}{Z_T}$$

where it is understood that the real part of all complex expressions must eventually be taken to give a physically possible answer. We now have

$$I_o = \frac{V_o}{Z_T} = \frac{15}{93.1 - j17.2} = 0.156 + j0.029.$$

Thus

$$I(t) = I_o e^{j\omega_o t} = (0.156 + j0.029)(\cos \omega_o t + j \sin \omega_o t).$$

Taking only the real part gives

a) $I(t) = 0.156 \cos \omega_o t - 0.029 \sin \omega_o t$ amps.

The time-average power dissipated in the load can be found from

$$\langle P \rangle = \frac{1}{2} R |I_o|^2$$

where the factor 1/2 comes from taking the time average of the current. The appropriate R is the real part of $Z_{\lambda/4}$ since this is the external resistance seen by the generator. (Remember, the current I(t) we have obtained is the current at the generator terminals.) Hence the time-average power dissipated in the load is

b) $\langle P \rangle = \frac{1}{2}(43.1)[(0.156)^2 + (0.029)^2] = 0.542$ watts.

If the line had been one-half wavelength then $k\ell = \pi$ and a simple substitution in Eq. 5.55 gives

c) $Z_{\lambda/2} = Z_L.$

This demonstrates that a half-wavelength line does not transform the impedance.

5.3 WAVEGUIDES

A waveguide is a hollow pipe customarily made from metal; brass for use at the lower microwave frequencies (below approximately 30 GHz or 1 cm wavelength), and from coin silver at extremely high frequencies (from around 30 GHz to 150 GHz) where damping of the waves caused by finite conductivity becomes a particularly serious problem. The waveguide, being a single-conductor system has several advantages over a two-wire transmission line like the coaxial cable. It is more rugged and also more easily constructed. The problems of how to mount and accurately position the center conductor of a coaxial cable, where misalignments cause mismatches, are no longer with us; and these difficulties become more and more acute as the wavelength becomes shorter and shorter. In addition, a waveguide generally causes less damping of a wave of given frequency than the coaxial cable used at the same frequency. However, matters are not entirely biased in favor of waveguides. They have one large drawback. They cannot support propagation of waves below a minimum frequency, in contrast to the two-wire transmission line which can operate down to zero values. To be sure, the larger the pipe the lower is their minimum, but the sizes become quite unwieldy for frequencies below approximately 500 MHz (which corresponds to a wavelength of 0.6 meters).

Let us think back for a moment on the parallel-plate system of Fig. 5.9, which was the two-conductor transmission line first discussed in section 5.2. To make this into a waveguide, all one needs to do is to close off the sides by two vertical walls, transforming the parallel plates into a rectangular pipe. This simple act has, however, quite profound effects on the propagation characteristics of the system. No longer can the pipe support a pure TEM wave which, as we remember, is characterized by the fact that the electric and magnetic fields are entirely transverse to the direction of propagation z. No longer does the wave propagate with a phase velocity c, and no longer can we transmit signals down to zero frequency. And the cause of all this? It is the additional boundary condition requiring the tangential component of the electric vector to vanish at the vertical sidewalls which we have just erected.

The electric and magnetic fields in waveguides are considerably more complex than the fields of the TEM wave in two-wire transmission lines. The associated currents flowing in the walls and the voltages across the walls have correspondingly more complicated patterns. As a result, the elegant description of the wave phenomena in terms of V and I employed in section 5.2 is now of little value to us and we shall no longer make use of it; rather, we shall revert back to the original language of $\vec{E}$ and $\vec{B}$ fields. Waveguides are made with rectangular, circular, or sometimes even elliptical cross sections. We shall now restrict ourselves entirely to the commonest type, the rectangular guide.

The rectangular waveguide

Consider a rectangular pipe of width a and height b made from perfectly conducting walls. Erect a Cartesian coordinate system like that shown in Fig. 5.20. Assume that a mode of propagation exists within the waveguide for which the electric vector is entirely along the y axis. This is a reasonable start because with this simple choice of $\vec{E} = \hat{y}E_y$, we are at least assured that the boundary condition

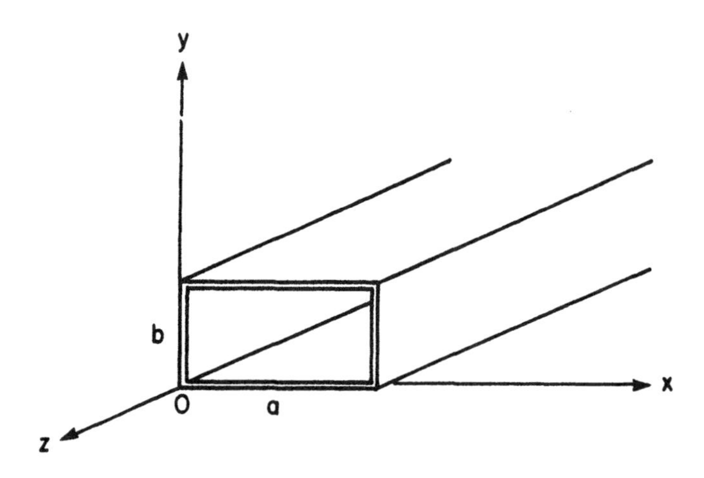

Fig. 5.20 Rectangular waveguide. The electric vector is assumed to be oriented along the y axis.

on $\vec{E}$ is satisfied along the top and bottom surfaces of the pipe. Now, let us be even bolder and assume that E_y is uniform in the y direction, that is, E_y is not a function of y. Of course, E_y cannot also be constant along the x direction; the requirement that the tangential $\vec{E}$ field vanish at a metal surface makes it mandatory that E_y be zero at $x = 0$ and $x = a$. There are many well-behaved functions that can be made zero at two positions and one of our tasks is to find this functional relationship; let us call it g(x). Thus E_y is proportional to the unknown function g(x). Because E_y is the electric field of a wave assumed to be traveling along the z axis, it is also proportional to $\exp[j\omega t - jk_z z]$ where k_z is the propagation constant of the wave, which is yet another unknown quantity to be determined. Therefore we write

$$E_y = g(x)\, e^{j(\omega t - k_z z)} \tag{5.58}$$

E_y must be a solution of the wave equation and since the field occupies a slice of three-dimensional space, E_y must be a solution of the three-dimensional wave equation (3.65):

$$\frac{\partial^2 E_y}{\partial x^2} + \frac{\partial^2 E_y}{\partial y^2} + \frac{\partial^2 E_y}{\partial z^2} = \frac{1}{c^2}\frac{\partial^2 E_y}{\partial t^2}. \tag{5.59}$$

Inserting Eq. 5.58 in 5.59, doing the differentiations and rearranging terms yields

$$\left[\frac{\partial^2 g}{\partial x^2} + \left(\frac{\omega}{c}\right)^2 g - k_z^2 g\right] e^{j(\omega t - k_z z)} = 0 \tag{5.60}$$

a result which can be satisfied only if the term in square brackets is zero, that is

$$\frac{\partial^2 g}{\partial x^2} + \left[\left(\frac{\omega}{c}\right)^2 - k_z^2\right] g = 0. \tag{5.61}$$

This is just the form of Eq. 2.38 whose solution is

$$g(x) = A\cos(k_x x) + B\sin(k_x x) \tag{5.62}$$

where k_x is just an abbreviation for the quantity

$$k_x = \pm\sqrt{\left(\frac{\omega}{c}\right)^2 - k_z^2}. \tag{5.63}$$

Since E_y must vanish at $x = 0$ and $x = a$, so must $g(x)$. This can be achieved only if the arbitrary constant A is zero, and if k_x equals $(m\pi/a)$, with m as an integer. We therefore have that $g(x) = B\sin(m\pi x/a)$, and for $m = 1$ that $g(x) = B\sin(\pi x/a)$, which is the case to which we shall restrict ourselves for the time being.* When we insert this value of $g(x)$ in Eq. 5.58, and take the real part, we arrive at the sought-after result for E_y,

$$E_y = E_{oy}\sin\left(\frac{\pi x}{a}\right)\cos(\omega t - k_z z)\ ; \qquad (m = 1) \tag{5.64}$$

where E_{oy} is the amplitude (which is just the constant B) and k_z the propagation constant along the guide given by

$$\begin{aligned} k_z^2 &= \left(\frac{\omega}{c}\right)^2 - \left(\frac{\pi}{a}\right)^2 \\ &= \left(\frac{\omega}{c}\right)^2\left[1 - \left(\frac{\pi c}{\omega a}\right)^2\right]. \end{aligned} \tag{5.65}$$

Application of Maxwell's equations (3.62b) and (3.62c) then yields the other electromagnetic components of this guided wave. They are

$$B_x = -\frac{E_{oy}}{c}\left(\frac{ck_z}{\omega}\right)\sin\left(\frac{\pi x}{a}\right)\cos(\omega t - k_z z) \tag{5.66}$$

$$B_z = -\frac{E_{oy}}{c}\left(\frac{\pi c}{\omega a}\right)\cos\left(\frac{\pi x}{a}\right)\sin(\omega t - k_z z). \tag{5.67}$$

---

*It is the principal mode of the rectangular waveguide and is usually designated as the TE_{10} mode. The TE stands for transverse electric. The subscript 10 designates the mode number, $m = 1$, $n = 0$ as is discussed in the subsection entitled "Higher waveguide modes."

The last four equations tell us several very interesting things. For one, the transverse electric field E_y and the transverse magnetic field B_x are exactly in spatial and temporal phase with one another, much like the situation of a plane wave propagating in unbounded space. But, quite unlike the last named, the guided wave is accompanied by a magnetic field component B_z oriented along the direction of propagation; it is out of phase by 90° with respect to the transverse component B_x. Observe the quantity $(\pi c/\omega a)$ which appears frequently in these equations. It can be written more simply as $(\lambda/2a)$ where λ is the free-space wavelength defined in the usual way as $2\pi c/\omega$. We see that when this ratio is very small, several things happen. The propagation constant k_z approaches ω/c which is just the value in free, unbounded space. The amplitude of the transverse magnetic field approaches the value E_{oy}/c just as in plane waves in far open spaces; and B_z goes to zero. All this is not at all surprising: we know that visible light has no difficulty passing through, say, a one-inch diameter pipe and we do not expect significant departures from plane wave propagation when the pipe is many millions of wavelengths across. In such large-sized pipes, the mode under discussion $(m = 1)$, together with all the higher modes that can propagate (see a later subsection) act together to produce the light radiation that allows one to see through the pipe.

Matters are quite different, however, when the wavelength becomes comparable to the guide width a (note that the height b does not enter any of our equations). Not only does the longitudinal field B_z become comparable in size with the transverse component B_x, but the phase velocity v of the guided wave (ω/k_z) departs significantly from its free-space value c. From Eq. 5.65 it follows that

$$\boxed{\begin{aligned} v &= \frac{\omega}{k_z} \\ &= c\,\frac{\omega}{\sqrt{\omega^2 - (\pi c/a)^2}}\,. \end{aligned}} \tag{5.68}$$

We see that at high frequencies such that $\omega \gg (\pi c/a)$, or equivalently when $\lambda \ll 2a$, the phase speed approaches that of the speed of light c. But as λ approaches 2a, v departs from c, and becomes imaginary when $\omega < (\pi c/a)$. This signifies that propagation is no longer possible for waves of these low frequencies. The limiting value $\omega = \omega_c$ where

$$\boxed{\omega_c = \frac{\pi c}{a}} \tag{5.69}$$

is known as the "cutoff frequency." Suppose the rectangular pipe is 3 inches wide. Inserting this value in Eq. 5.69 gives $\omega_c/2\pi$ as 1969 MHz which is the lowest frequency wave that this waveguide can support.

The field configuration of the wave is completely specified by Eqs. 5.64, 5.66 and 5.67. A pictorial representation of the electric and magnetic fields is shown in Fig. 5.21. Observe that the electric field is entirely in the y direction and the field lines terminate on the top and bottom walls. The field correctly vanishes at the side walls $x = 0$ and $x = a$ as is indicated by the low density of line shown in these two regions. On the other hand, the magnetic field must have no component normal to the metal walls and since $\nabla \cdot \vec{B} = 0$ the magnetic lines form closed loops. Surface currents flow on the inside surfaces of the guide walls and their magnitude and direction is given by Eq. 5.12d, $\mu_o \vec{J}_S = \hat{n} \times \vec{B}$. Since $\vec{B}$ is known, so is $\vec{J}_S$ and Fig. 5.22 gives a picture of its configuration. The fields are seen to be quite complicated compared to what we have encountered in two-wire transmission lines; so complicated that there is no unique way anymore of defining a current I or a voltage V across the guide; the values of I and V depend on where in the waveguide one attempts to compute them. And therefore the definition of characteristic impedance as the ratio V/I is likewise no longer a clearly definable and unique quantity.

Energy flow

The time-average energy crossing a unit area of waveguide per second is obtained from Poynting's vector which gives $\langle S_z \rangle = -\langle E_y B_x \rangle / \mu_o$. Inserting the appropriate values for E_y and B_x from

Eqs. 5.65 and 5.66 and time averaging the result, yields the value

$$\langle S_z \rangle = \frac{1}{2}\epsilon_o c E_{oy}^2 \left[\frac{ck_z}{\omega}\right] \sin^2\left(\frac{\pi x}{a}\right) \quad \text{watt/m}^2 \tag{5.70}$$

which can be compared with formula (3.106) applicable to plane waves

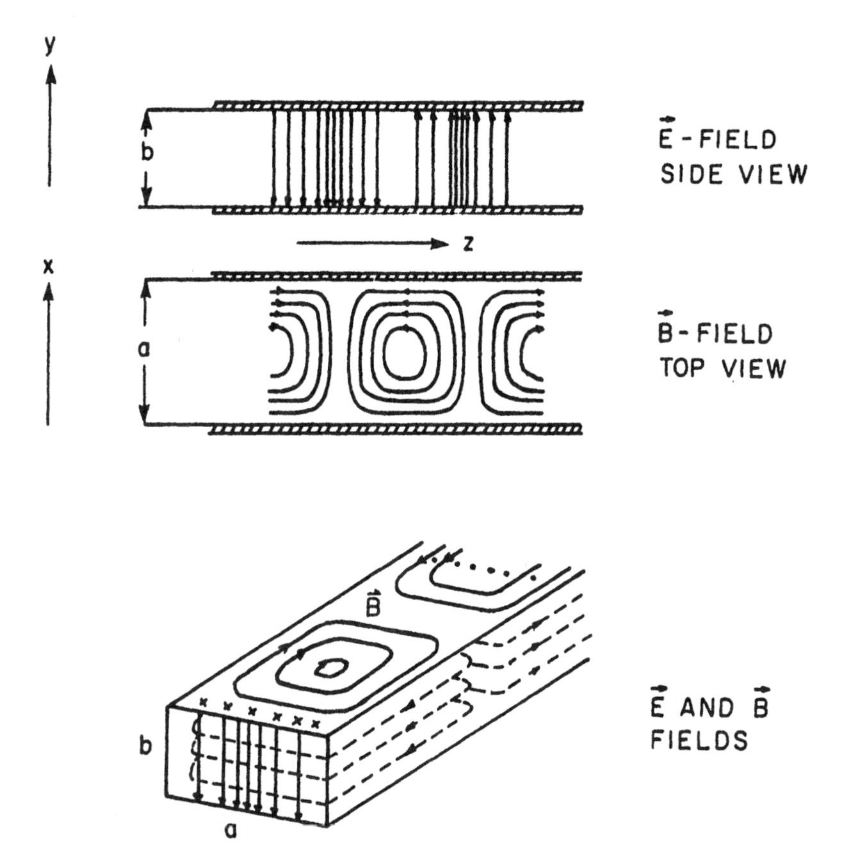

Fig. 5.21 Electric and magnetic field configurations of the TE_{10} wave of a rectangular waveguide, as seen from different directions. The density of lines is proportional to the field strengths. The nomenclature "TE" stands for transverse electric, which stresses the fact that for this mode the electric vector is entirely in the plane perpendicular to the direction of propagation. Observe, however, that the magnetic vector has components both perpendicular and along the propagation direction. The subscript 10 refers to the mode number, m = 1, n = 0.

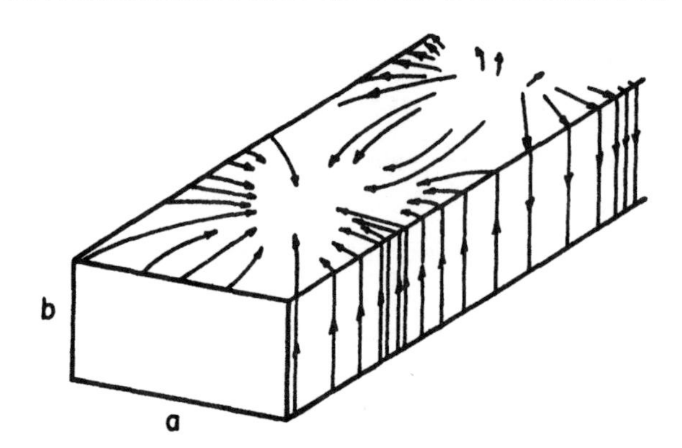

Fig. 5.22 Sketch of the surface current distribution $\vec{J}_S$ associated with the TE_{10} mode of a rectangular waveguide.

in unbounded space. To find the total power $\langle P \rangle$ flowing down the guide we must sum the above expression over the entire cross-sectional area of the guide. To this purpose, we split the area into elements of height b and width dx and by integrating over all the strips from x = 0 to x = a, obtain

$$\langle P \rangle = \frac{1}{2}\, \epsilon_o c E_{oy}^2 \left(\frac{ck_z}{\omega}\right) b \int_o^a \sin^2\left(\frac{\pi x}{a}\right) dx$$

$$= \frac{1}{4}\,(ab)\, \epsilon_o c E_{oy}^2 \left(\frac{ck_z}{\omega}\right) \text{ watt} \tag{5.71}$$

where the second formula comes from evaluating the integral whose value is a/2. The associated time-averaged energy density in electric and magnetic fields is given by

$$\langle U \rangle = \frac{1}{2}\,\epsilon_o \langle |E^2| \rangle + \frac{1}{2\mu_o} \langle |B|^2 \rangle . \tag{5.72}$$

On inserting the guide electric and magnetic fields, we obtain the energy per cubic meter as

$$\langle U \rangle = \frac{1}{4}\,\epsilon_o E_{oy}^2 \sin^2\left(\frac{\pi x}{a}\right) + \frac{1}{4\mu_o} \frac{E_{oy}^2}{c^2} \left(\frac{ck_z}{\omega}\right)^2 \sin^2\left(\frac{\pi x}{a}\right) +$$

$$+\frac{1}{4\mu_o}\frac{E_{oy}^2}{c^2}\left(\frac{\pi c}{\omega a}\right)^2\cos^2\left(\frac{\pi x}{a}\right) \quad \text{joule/m}^3. \tag{5.73}$$

On integrating this expression over the waveguide cross section and rearranging terms, we find the following result for the energy per unit length ℓ of guide:

$$\langle U\rangle_\ell = \frac{1}{4}(ab)\,\epsilon_o E_{oy}^2 \quad \text{joule/m}. \tag{5.74}$$

Comparison of the power flow $\langle P\rangle$ with the energy density $\langle U\rangle_\ell$ shows that the two quantities are proportional to one another. The constant of proportionality has dimensions of velocity, a fact that is familiar to us already from previous discussions in sections 2.2 and 3.6. It is the velocity at which the energy flows along the guide. Let us compute its value. The ratio of Eq. 5.71 to Eq. 5.74 yields the result that

$$\langle P\rangle = \left(\frac{c^2k_z}{\omega}\right)\langle U\rangle_\ell \tag{5.75}$$

with v_g as the velocity given by

$$\begin{aligned} v_g &= \frac{c^2k_z}{\omega} \\ &= c\sqrt{1-\left(\frac{\pi c}{\omega a}\right)^2}. \end{aligned} \tag{5.76}$$

The second form of the equation comes by substituting the value for k_z from Eq. 5.65. From this we see immediately that the energy propagates not at c, the speed of light in vacuum, as was the case in unbounded space (see Eq. 3.109) but at a slower speed. Nor does it propagate at the phase velocity v given by Eq. 5.68. The velocity in question proves to be the so-called group velocity which we shall now discuss.

Dispersion

We learned in section 2.1 that a convenient way of exhibiting dispersion is to plot the frequency ω as a function of the propagation constant k for the system in question. Let us do this for propagation in a waveguide. Equation 5.65 gives the desired relationship between ω and k_z; we can write it somewhat more conveniently as

$$k_z c = \pm \sqrt{\omega^2 - \omega_c^2} \tag{5.77}$$

where $\omega_c = (\pi c/a)$ is the cutoff frequency defined by Eq. 5.69. The plus and minus signs come from taking the square root of Eq. 5.65, but there is no ambiguity as to what the meaning of the signs is: the positive sign refers to waves traveling along the positive z axis, and the negative sign provides for the possibility of wave propagation in the opposite direction. Without loss of generality we shall restrict ourselves to the positive root of the equation.

The dashed line of Fig. 5.23 shows the dispersion characteristics of a wave propagating in evacuated unbounded space; the phase velocity

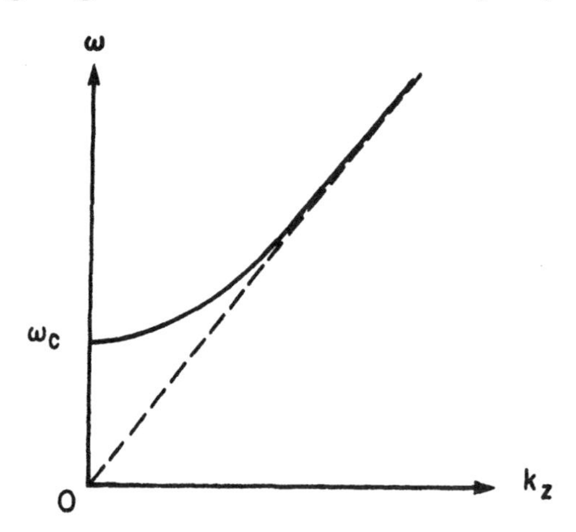

Fig. 5.23 The dispersion diagram for a wave in a waveguide; ω_c is the cutoff frequency. The solid line is typical of any waveguide mode, not only the TE_{10} mode. The dashed line represents the dispersion characteristic of a wave in empty unbounded space, and serves as a reference.

given by the ratio ω/k is the same for all frequencies and is precisely c. The system is dispersionless. The solid line represents a plot of Eq. 5.77, showing that now large dispersion occurs at low frequencies with ω near ω_c, but very little occurs at very high frequencies. Observe, however, that whatever the frequency is, the phase velocity of the guided wave is always greater than c.

So far we have assumed that ω is greater than ω_c. But what happens if it is less? Our plot of Fig. 5.23 tells us nothing about this regime of frequencies. It does not mean that it is unimportant. When $\omega < \omega_c$, k_z of Eq. 5.77 becomes purely imaginary and we can represent this situation by writing

$$k_z c = \pm j\sqrt{\omega_c^2 - \omega^2} \qquad (\text{for } \omega < \omega_c). \tag{5.78}$$

When we substitute this value in Eq. 5.58 we find that the electric field*

$$E_y = g(x) \exp\left[-\sqrt{\omega_c^2 - \omega^2}\,\frac{z}{c}\right] e^{j\omega t} \tag{5.79}$$

is still oscillatory in time, but it no longer oscillates harmonically with position z, as happens in a traveling wave. The field in fact no longer travels — it merely oscillates in time and decays exponentially in space, as is illustrated in Fig. 5.24. The wave is said to be "evanescent" or "beyond cutoff." And the smaller ω is relative to ω_c the faster is the exponential decrease. Let us take the concrete case of the three inch wide rectangular guide discussed earlier. We found that for this waveguide size the cutoff frequency $\omega_c/2\pi = 1969$ MHz. Suppose one sends into this guide a wave of frequency $\omega/2\pi = 1000$ MHz; in what distance $z = z_o$ will its amplitude fall to $(1/e)^{th}$ its original value? From Eq. 5.79 we see that the distance is obtained by solving the simple relation $z_o = c(\omega_c^2 - \omega^2)^{-1/2}$ which gives $z_o = 2.82$ cm.

*Care must be exercised in the choice of sign for the imaginary k_z. The negative sign gives an electric field that decays exponentially with z; the positive sign gives a growing field that becomes infinite at infinite distance and is thus unphysical.

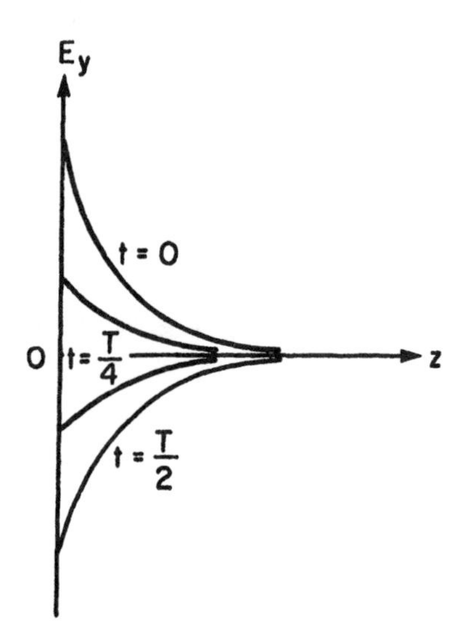

Fig. 5.24 The waveguide electric field as a function of position z under cutoff conditions $\omega < \omega_c$. The curves refer to different times of the oscillatory cycle. Not only E_y, but all the other components of the electromagnetic field exhibit a similar exponential decay. T is the period, $T = 2\pi/\omega$.

Therefore, in a piece of waveguide 15 cm long the amplitude falls to ~4.9 parts in a thousand of its original value and the intensity is down to approximately 2.4 parts in 100,000 (≈ 46 dB). We see that a waveguide beyond cutoff is a good highpass filter eliminating almost everything at low frequencies $\omega < \omega_c$ and passing everything at high frequencies, $\omega > \omega_c$. It is therefore not surprising that the waveguide is often used in this capacity as a simple high-frequency filter.

Wave velocities

The concept of phase velocity applies only to perturbations that are strictly periodic in space and time, and consequently the wave train in question must be of infinite duration. This allows transients generated during the creation of the wave to die out. Once this state has been reached the perturbation can be represented as $\exp[j(\omega t - kz)]$. The surfaces of constant phase defined as $(\omega t - kz)$ = constant (see section 2.4) are therefore propagated with a velocity, known as the phase velocity, given by

$$\boxed{v = \frac{dz}{dt} = \frac{\omega}{k}.} \tag{5.80}$$

This represents simply the velocity of propagation of a certain phase or a state, and does not necessarily coincide with the velocity with which the energy of a wave or a signal is propagated. In fact, as we have just seen for the case of the waveguide, v can exceed the speed of light in vacuum c, and it does so without violating Einstein's relativity postulate in any way. There are many other examples of phase velocities exceeding the value of c. One example is an ocean wave striking a vertical wall at oblique incidence. It is found that the speed of the bounding wave traveling along the wall depends upon the direction of approach of the ocean wave and increases without limit as normal incidence is approached. An analogous example is the reflection of light from a mirror, a case we treated in section 5.1. Equation 5.23 in fact shows that the phase velocity along the mirror surface (that is, the z direction) is given by $v_z = c/\sin\theta$, which goes to infinity as $\theta \to 0$.

Apart from v, there are other velocities associated with waves. To convey intelligence it is always necessary to "tag" or modulate the carrier frequency being transmitted (see section 1.6). When this is done, there is a group of frequencies $\Delta\omega$ centered around the carrier, as is illustrated in Fig. 1.33. This group forms a modulation envelope that does not necessarily travel at the phase velocity. Such a modulation envelope is shown in Fig. 1.32; our task here is to calculate the velocity at which the envelope travels. The modulated carrier is represented by (cf. Eq. 1.152)

$$E = [E_o + \alpha E_{om} \cos(\Delta\omega t)] \cos(\omega t). \tag{5.81}$$

Here E stands for, say, the electric field oscillations imposed upon a transmitting antenna positioned at z = 0 with E_o and ω as the carrier amplitude and frequency, respectively, and E_{om} and $\Delta\omega$ as the amplitude and frequency of the modulation; α is the modulation factor and its magnitude determines the modulation depth. This expression can be expanded through the use of standard trigonometric functions into three harmonic oscillations at the carrier and side-hand frequencies,

$$E = E_o \cos(\omega t) + \frac{1}{2}\alpha E_{om} \cos[(\omega + \Delta\omega)t] + \frac{1}{2}\alpha E_{om} \cos[(\omega - \Delta\omega)t]. \tag{5.82}$$

This signal is then radiated in the z direction through a region of space where the phase velocity is a function of frequency. This has the consequence that the phase of the oscillation represented by ωt at $z = 0$ becomes $(\omega t - kz)$ at a distance z away, and the phase represented by $(\Delta\omega)t$ becomes $[(\Delta\omega)t - (\Delta k)z]$ (see discussion in section 2.1). On account of this the resultant wave becomes

$$E = E_o \cos(\omega t - kz) + \frac{1}{2}\alpha E_{om} \cos[(\omega + \Delta\omega)t - (k + \Delta k)z] + \frac{1}{2}\alpha E_{om} \cos[(\omega - \Delta\omega)t - (k - \Delta k)z]. \tag{5.83}$$

This expression may be recombined to read

$$E = [E_o + \alpha E_{om} \cos(t\Delta\omega - z\Delta k)] \cos(\omega t - kz) \tag{5.84}$$

which gives the form of the amplitude-modulated wave progressing in the z direction. The surfaces of constant phase of the carrier are given by $\omega t - kz$ = constant, from which we deduce that they are propagated at a velocity $v = dz/dt = \omega/k$. This is just the phase velocity of an unmodulated sinusoidal wave. On the other hand, a given point on the envelope of the modulation is given by $t\Delta\omega - z\Delta k$ = constant, from which we infer that the envelope progresses in the z direction with a velocity $v_g = \Delta\omega/\Delta k$. If the frequency spread of the group is very small, we obtain in the limit

$$\boxed{v_g = \frac{d\omega}{dk}} \tag{5.85}$$

a quantity known as the group velocity. The series of snapshots shown in Fig. 5.25 illustrates the motion of the group relative to the motion of the individual waves in the group, plotted under the assumption that the group velocity v_g equals half the phase velocity. It is evident that the envelope is lagging behind with respect to the component waves. This can be easily seen in the case of water waves where it appears as if the component waves were slipping through, and past, the envelope.

To find the group velocity of guided electromagnetic waves we merely differentiate both sides of Eq. 5.77 with respect to ω, compute $dk_z/d\omega$ and invert the result:

$$\begin{aligned} v_g &= \frac{d\omega}{dk_z} \\ &= c\sqrt{1-\left(\frac{\omega_c}{\omega}\right)^2} \end{aligned} \tag{5.86}$$

from which we see that the group velocity is always less than the speed of light c in vacuum, and becomes zero at cutoff, $\omega = \omega_c$. Since the phase velocity v equals $c(1-(\omega_c/\omega)^2)^{-1/2}$, we obtain the interesting result that $vv_g = c^2$. The group velocity for these waves is less than their phase velocity, a fact which is particularly easy to understand from the diagram plotted in Fig. 5.26.

The dispersive properties of the waveguide can be put to good use. Imagine that a bunch of electrons is suddenly accelerated or decelerated, and as a result a burst of microwave radiation is generated [see Chapter 4]. This short microwave pulse can be thought of as being made up of a large number of harmonic wave trains each of different amplitude, and we would like to know the composition. Said differently, we would like to Fourier-analyze the burst and deduce the power spectrum – that is, find the radiated power as a function of frequency. To accomplish this we take a long piece of waveguide,

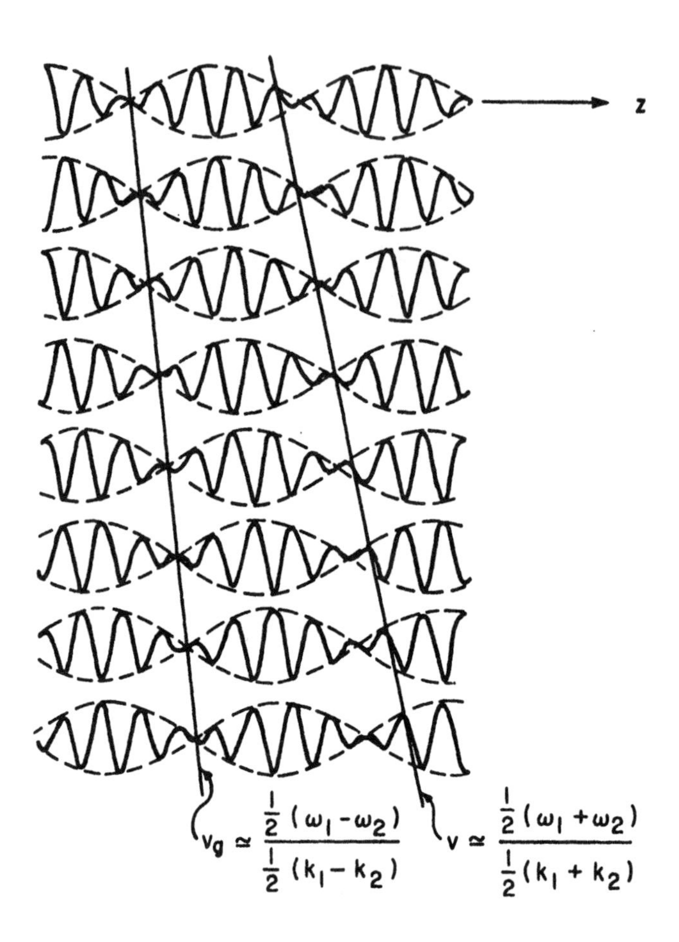

Fig. 5.25 Successive snapshots taken at equal intervals of time of two waves (no carrier) of frequencies ω_1 and ω_2 beating to give a resultant $E = E_0 \cos[\omega_1 t - k_1 z] + E_0 \cos[\omega_2 t - k_2 z] = 2E_0 \cos\left[\frac{1}{2}(\omega_1 - \omega_2)t - \frac{1}{2}(k_1 - k_2)z\right] \cos\left[\frac{1}{2}(\omega_1 + \omega_2)t - \frac{1}{2}(k_1 + k_2)z\right]$. In the picture $\omega_1 = (9/7)\omega_2$ and $k_1 = (5/3)k_2$ giving a phase velocity twice the group velocity. The sloping lines show the progress of the group and of an individual wave crest. [After F. S. Crawford, Jr., Waves (McGraw Hill Book Company, New York, 1965).] See also Appendix 1 which discusses the addition of two oscillations of nearly the same frequency.

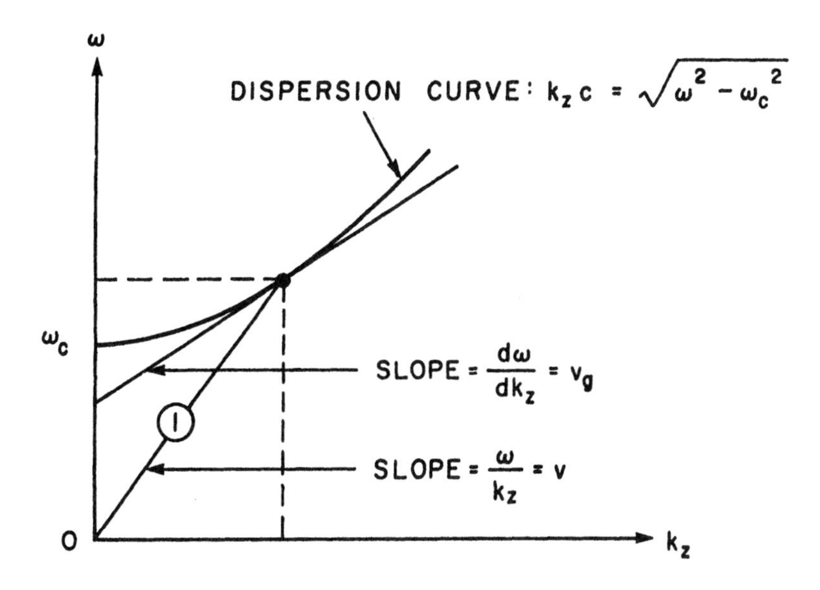

Fig. 5.26 Typical dispersion curve of a wave propagating in a waveguide. The slope of line (1) for this dispersion law is always greater than the slope of the tangent line $d\omega/dk_z$. Hence $v > v_g$.

maybe as long as a hundred meters, and allow the microwave pulse to pass through it. Since the different wavelengths travel at different speeds through the guide, some will be delayed compared with others, so that at the far (receiving) end of the waveguide, the different wavelengths of the pulse will, so to speak, be sorted out in accordance with their instance of arrival.* The waveguide therefore serves as a spectrum analyzer, and Fig. 5.27 shows the kind of experimental setup that has been used.

We found in an earlier subsection that electromagnetic energy travels in a waveguide with a speed equal to the group velocity [compare Eqs. 5.76 and 5.86]. This is not only true for propagation in waveguides but is true even when waves propagate through matter,

*Recall from section 2.5 that a pulse of radiation traveling in a nondispersive medium preserves its shape throughout its entire existence. No sorting out according to wavelengths takes place.

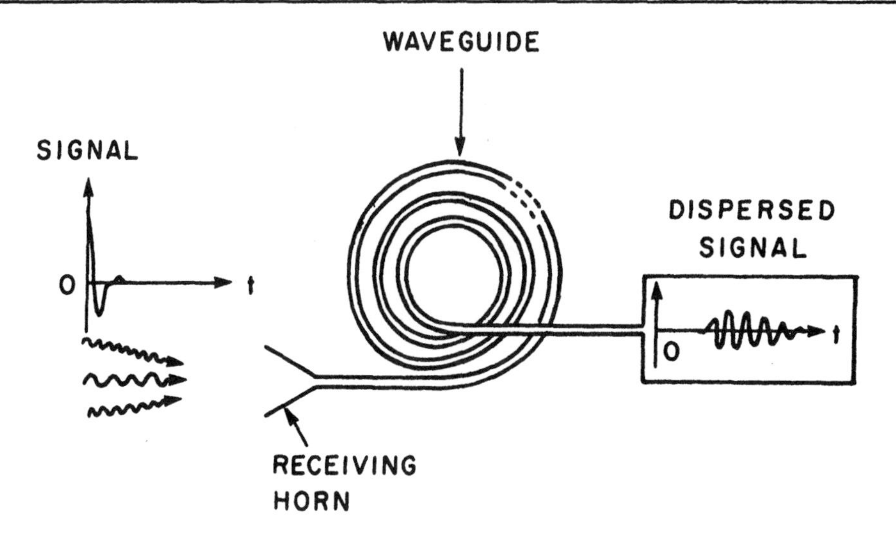

Fig. 5.27 A pulse of microwave radiation entering a horn and traveling through a long piece of waveguide. The dispersed signal can be "convolved" appropriately with the dispersion characteristics of the guide to obtain the power spectrum of the pulse entering the horn. [See, for example, M. Friedman and M. Herndon, Phys. Rev. Letters 28, 210 (1972); 29, 55 (1972). Dispersed microwave signals are shown in the paper by V. L. Granastein, M. Herndon, R. K. Parker, and P. Sprangle, IEEE J. Quant. Electronics, Vol. QE-10, p. 651, September 1974.]

or through other complex systems. The group velocity is less than c and so is the speed at which energy travels. Therefore, all is in harmony with Einstein's theory of relativity. But we mention in passing that there exist several anomalous cases when v_g is greater than c and this fact was used in the early days of relativity in attempts to refute its precepts. We now know* that it is not absolutely mandatory for energy to propagate at the group velocity and, even if $v_g > c$, the velocity of energy propagation never exceeds c. These and related problems are discussed in a most lucid way by L. Brillouin in a beautiful monograph entitled Wave Propagation

*As a rule of thumb, v_g is not a meaningful concept when the wave propagating in the medium is strongly damped; this is where the anomalous cases mentioned earlier occur; see also in section 6.5.

and Group Velocity (Academic Press, Inc., New York, 1960). He also defines other wave velocities in addition to v and v_g necessary to the understanding of the propagation of pulses.

From the relation $k = \omega/v = 2\pi/\lambda$, one can derive alternate expressions for the group velocity that are sometimes more convenient than formula (5.85):

$$\left.\begin{aligned} v_g &= \frac{d\omega}{dk} \\ &= v - \lambda\frac{dv}{d\lambda} \\ &= v + k\frac{dv}{dk}. \end{aligned}\right\} \tag{5.87}$$

The second form tells us immediately that in a dispersionless medium where v is not a function of λ the group and phase velocities are equal to one another. Also, we see that when the phase velocity v increases with increasing wavelength, v_g must be less than v. Such is the case for waveguides and for electromagnetic waves in matter. The anomalous cases occur in the opposite regime where $dv/d\lambda < 0$, giving $v_g > v$. We also wish to point out that the relationship $vv_g = c^2$ derived earlier is by no means universal; it applies only in those special circumstances in which ω and k are related by a dispersion law of the form

$$k^2c^2 = \omega^2 + \text{constant}. \tag{5.88}$$

Differentiating both sides with respect to k yields $2c^2k = 2\omega d\omega/dk$; but $v = \omega/k$ and $v_g = d\omega/dk$, with the result that $vv_g = c^2$.

The guided wave viewed as a problem of oblique reflection

Equation 5.23 represents the electric field of a plane wave incident obliquely on the surface of a metal. The field configuration is that of a standing wave which "stands" as one moves away from the surface, but propagates along the surface. Equation 5.64 represents the field configuration of the guided electromagnetic wave. The two equations are virtually identical, suggesting that the same physical mechanism, oblique reflection, operates in both situations. Consider our rectangular waveguide of width a and height b (Fig. 5.20). Let the bottom plate of the guide correspond to the single reflector discussed in section 5.1; it is illuminated obliquely as depicted in Fig. 5.8. This gives rise to a standing wave pattern sliding along the surface as was just mentioned. We now wish to put into place a second metal plate, parallel to the first, and make a waveguide, using the attractive idea of the sliding standing wave pattern as a means of locomotion of the wave along the surface. This can be done without perturbing the pattern by positioning the plate at a height where the incident and reflected waves interfere destructively. At such a height there exists a plane of electric nodes, and $\vec{E}$ being zero there already, the boundary condition requiring the tangential electric field to vanish is automatically fulfilled. Thus we have, so to speak, "trapped" the sliding standing wave pattern between two parallel metal walls. The remaining two walls orthogonal to the ones just mentioned and separated by a distance b from one another can be erected with no further ado, because $\vec{E}$ is perpendicular to these walls anyway, by prior assumption (see beginning of section 5.3).

An equivalent way to envisage the situation is to note that Eq. 5.64 for the sliding standing wave pattern can be decomposed by means of standard trigonometric identities into

$$\begin{aligned} E_y = {} & \frac{1}{2} E_{oy} \sin[\omega t - k_z z + k_x x] \\ & - \frac{1}{2} E_{oy} \sin[\omega t - k_z z - k_x x] \end{aligned} \tag{5.89}$$

thus showing that it is equivalent to a superposition of two plane traveling waves crisscrossing down the waveguide. In this equation we set $k_x = \pi/a$. The crisscrossing manifests itself in that the propagation constants $\vec{k} = \hat{z}k_z - \hat{x}k_x$ and $\vec{k} = \hat{z}k_z + \hat{x}k_x$ of the two respective waves have opposite x components. The successive wave fronts of the two waves are shown in Fig. 5.28. The two metal plates of the guide are placed along adjacent nodal planes of $\vec{E}$ and are thus separated by the distance

$$a = \frac{\lambda}{2 \sin \psi} = \frac{\pi c}{\omega \sin \psi} \tag{5.90}$$

where $\lambda = 2\pi c/\omega$ is the free-space wavelength and ψ is the angle between the propagation vector $\vec{k}$ and the surface of the metal waveguide. Now, the waveguide wavelength λ_z as determined by a probe moving along the z axis and measuring, say, the distance between two crests is

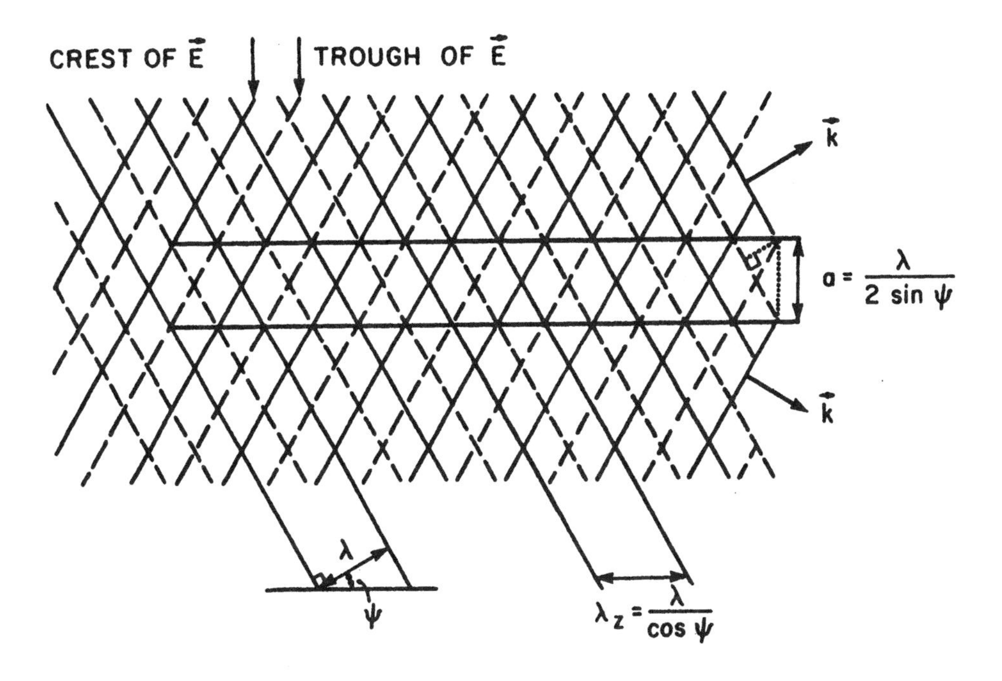

Fig. 5.28 Two plane linearly polarized waves of equal amplitude traveling at an angle to one another. Their electric vector is polarized perpendicular to the plane of the page. The oblique lines represent successive wave fronts of E. Wherever a solid line meets a dashed line the resultant amplitude is zero. The two horizontal lines represent the waveguide walls positioned along the adjacent nodal planes of the electric vector.

$$\lambda_z = \frac{\lambda}{\cos\psi} = \frac{2\pi c}{\omega\cos\psi}. \tag{5.91}$$

Eliminating the angle ψ from Eqs. 5.90 and 5.91 yields

$$\left(\frac{2\pi}{\lambda_z}\right)^2 = k_z^2 = \left[\omega^2 - \left(\frac{\pi c}{a}\right)^2\right] \Big/ c^2 \tag{5.92}$$

which is just the dispersion relation (5.65) or (5.77). The speed with which the standing wave pattern moves along the waveguide is $v = c/\cos\psi$, and since $\cos\psi = (1/\omega)[\omega^2 - (\pi c/a)^2]^{1/2}$, we arrive immediately at the formula for the phase velocity as given by (5.68).

The physical reason for the waveguide cutoff is now obvious. To satisfy boundary conditions, Eq. 5.90 must be obeyed. Suppose the guide size a is fixed; then as the frequency ω of the wave is decreased, $\sin\psi$ must increase: the crisscrossing waves strike the plates closer and closer to normal incidence. Eventually a frequency is reached for which $(\pi c/a) = \omega$ at which point $\sin\psi = 1$, $\psi = 90°$ and the waves bounce back and forth between the guide walls exactly at right angles to the surfaces; the "sliding" motion of the standing wave pattern is infinite but energy stopped flowing along the guide. The group velocity $v_g = c\cos\psi$ is zero but the phase velocity $v = c/\cos\psi$ is infinite.

Oblique incidence has one more important consequence. It is that if $\vec{E}$ is purely transverse, that is, perpendicular to the guide axis z, then $\vec{B}$ cannot be entirely in the x-y plane; but, as a result of the oblique angle of incidence, $\vec{B}$ of the crisscrossing plane waves must have a component along the z axis (see Eq. 5.67). Such waves are known as "transverse electric" or TE waves.

Higher waveguide modes

In separating the waveguide walls by one-half of the "effective" free-space wavelength ($\lambda/2\sin\psi$) of the crisscrossing waves, we have restricted ourselves to but one of many possible modes that the waveguide can support. More generally, Eq. 5.90 can be written as $a = m\lambda/2 \sin\psi = m(\pi c/\omega \sin\psi)$ with m as any positive integer. Therefore, the x-dependent term of Eq. 5.64 can take on the more general form $\sin(m\pi x/a)$. But one can go even further; our assumption that E_y is independent of y is a simplification that is not mandatory. All necessary boundary conditions can, in fact, be fulfilled provided that the electric vector varies harmonically with y as $\cos(n\pi y/b)$ where b is the waveguide height and n is a positive integer. Therefore the general form for E_y, replacing Eq. 5.64, is

$$E_y = E_{oy} \sin\left(\frac{m\pi x}{a}\right) \cos\left(\frac{n\pi y}{b}\right) \cos(\omega t - k_z z). \tag{5.93}$$

Here every set of integers (m,n) denotes a possible field configuration and thus a wave of the TE_{mn} type. This is still not the end! There exists a whole class of waves known as transverse magnetic modes TM_{mn} which are characterized by the fact that the magnetic field, rather than the electric field, is entirely transverse to z (now, of course, $\vec{E}$ must have a component E_z). And thus we are confronted with a doubly infinite set of possible waves that can travel along the rectangular waveguide. Each has its own electric and magnetic field shape (independent of all the other shapes) and each has its lowest frequency of propagation, the cutoff frequency ω_c. One finds that irrespective of whether it is a TE or TM mode the cutoff frequency is given by the formula

$$\omega_c(m,n) = \pi c \sqrt{\left(\frac{m}{a}\right)^2 + \left(\frac{n}{b}\right)^2}. \tag{5.94}$$

From this we see that when $a > b$ the wave with the lowest cutoff frequency is the one with $m = 1$ and $n = 0$, that is,

$$\omega_c(1,0) = \frac{\pi c}{a} \tag{5.95}$$

which is precisely the wave we have been studying throughout this section. Indeed, it is the dominant wave used in virtually all practical applications in which rectangular waveguides are used as transmission lines. Most waveguides are constructed with a width a approximately equal to twice the height b. Setting a = 2b in Eq. 5.94 gives

$$\begin{aligned} \omega_c(m,n) &= \frac{\pi c}{a}\sqrt{m^2 + 4n^2} \\ &= \omega_c(1,0)\sqrt{m^2 + 4n^2} \end{aligned} \qquad (5.96)$$

from which the cutoff frequencies of the various modes can be determined. In ascending order in frequency they are

$$\frac{\omega_c(m,n)}{\omega_c(1,0)} = 1;\ 2;\ 2.236;\ 2.828;\ 3;\ 3.606;\ 4;\ 4.123 \text{ etc.} \qquad (5.97)$$

From this we see that there is a frequency range from $\omega_c(1,0)$ to $2\omega_c(1,0)$ where only the TE_{10} mode can propagate, all other modes being below cutoff. Beyond the frequency $\omega = 2\omega_c(1,0)$ other modes begin to enter, and the higher the operating frequency, the more modes are possible. For purposes of communication it is very desirable that one and only one mode be used. This can be assured only if we operate with the principal TE_{10} mode and confine ourselves to frequencies such that $\omega_c(1,0) < \omega < 2\omega_c(1,0)$. When we transmit at a higher frequency than this, any discontinuities in the waveguide or other mismatches can result in the excitation of higher modes which are then free to propagate along the system.

Example

The y component of the electric field of the TE_{mn} mode of a rectangular waveguide is of the form (see Eq. 5.93)

$$E_y(\vec{r},t) = E_{oy}\sin\left(\frac{m\pi x}{a}\right)\cos\left(\frac{n\pi y}{b}\right)\cos(\omega t - k_z z).$$

a) Derive an expression for E_x.
b) If the frequency ω equals $2\pi c/a$, where c is the velocity of light in a vacuum, compute the phase velocity for the TE_{10} mode, i.e., m = 1, n = 0.
c) Compute the group velocity for the same mode.

Solution

Since the wave is the TE (transverse electric) wave propagating in the z direction there is no z component, i.e., $E_z = 0$. Hence from

$$\nabla \cdot \vec{E} = \frac{\partial E_x}{\partial x} + \frac{\partial E_y}{\partial y} + \frac{\partial E_z}{\partial z} = 0$$

we must have

$$\frac{\partial E_y}{\partial y} = - \frac{\partial E_x}{\partial x}.$$

Differentiating the given expression for E_y yields

$$\frac{\partial E_x}{\partial x} = \frac{E_{oy} n\pi}{b} \sin\left(\frac{m\pi x}{a}\right) \sin\left(\frac{n\pi y}{b}\right) \cos(\omega t - k_z z).$$

Integrating with respect to x gives

a) $$E_x = -E_{oy}\left(\frac{na}{mb}\right) \cos\left(\frac{m\pi x}{a}\right) \sin\left(\frac{n\pi y}{b}\right) \cos(\omega t - k_z z).$$

The components of the propagation vector $\vec{k}$ are related by

$$k_x^2 + k_y^2 + k_z^2 = \frac{\omega^2}{c^2}$$

and for the fields given above we have

$$k_x = \frac{m\pi}{a} \qquad k_y = \frac{n\pi}{b}.$$

For the TE_{10} mode, $m = 1$, $n = 0$, and we are also told that $\omega = 2\pi c/a$, hence

$$k_z^2 = \left(\frac{2\pi}{a}\right)^2 - \left(\frac{\pi}{a}\right)^2 = \frac{3\pi^2}{a^2}.$$

The phase velocity v is given by

b) $$v = \frac{\omega}{k_z} = \left(\frac{2\pi c}{a}\right)\left(\frac{a}{\sqrt{3}\,\pi}\right) = \frac{2c}{\sqrt{3}}.$$

To compute the group velocity v_g we write

$$k_z^2 = \left(\frac{\omega}{c}\right)^2 - k_x^2 = \left(\frac{\omega}{c}\right)^2 - \left(\frac{\pi}{a}\right)^2$$

or

$$\omega = c\sqrt{k_z^2 + \left(\frac{\pi}{a}\right)^2}.$$

The group velocity then follows as

$$v_g = \frac{d\omega}{dk_z} = \frac{k_z c}{\sqrt{k_z^2 + \left(\frac{\pi}{a}\right)^2}} = \frac{k_z c^2}{\omega},$$

but $k_z = \sqrt{3}\,\pi/a$ and $\omega = 2\pi c/a$ so we arrive at

c) $$v_g = \frac{\sqrt{3}}{2}c.$$

Note the product of v_g and v is c^2 and that $v_g < c$ whereas $v > c$.

Dielectric waveguides

The laser has stimulated much interest in developing efficient methods of long-distance transmission of light for the purpose of communication. The reason for wanting to go to optical frequencies is the potentially large frequency range that may be available. Take as an example the transmission of pictures in television. The bandwidth of the modulation imposed on the carrier (see section 1.6) must be approximately 10 MHz. The carrier frequencies used in television today range from ~50 MHz to 200 MHz and thus in this range of 150 MHz there is room for some 15 stations spread out over these frequencies. Actually about twice this number can be squeezed in by suppression of one of the sidebands so that 5 MHz instead of 10 MHz bandwidths per station are needed. If the carrier were visible light, the available range of wavelengths could be from, say, 6000 Å (red) to 4000 Å (blue). This corresponds to frequencies of 5×10^{14} Hz and 7.5×10^{14} Hz, respectively. The available range for communication would then be a span of 2.5×10^{8} MHz allowing some 2.5×10^{7} non-overlapping TV channels, each having a bandwidth of 10 MHz! However unpleasant such a prospect may be, research in optical communication is well under way and several promising contenders for long-distance transmission are being studied; amongst these are the dielectric waveguides.*

The name "dielectric waveguide" is reserved for those structures which have a dielectric cylinder of arbitrary cross section not enclosed by metallic walls. The dielectric (glass for example) can be solid or hollow and in essence can play the same role as the metal walls do in the metallic guides discussed earlier. Hollow dielectric pipes are particularly attractive because they can act both as waveguides for the light and as containers of the gaseous lasing medium of which the light source is composed. Systems making use of this dual role are described by P. W. Smith, Appl. Phys. Letters **19**, 132 (1971) and by T. J. Bridges, E. G. Burkhardt, and P. W. Smith, Appl. Phys. Letters **20**, 403 (1972); a recent very technical text *Optical Waveguides* by N. S. Kapany and J. J. Burke (Academic Press, New York, 1972) deals with all aspects of solid and hollow guiding structures.

The structure of the electric and magnetic fields in dielectric waveguides is considerably more complex than in metal guides because here the fields are not confined to the boundaries of the cylinder but extend off toward infinity. To be sure, they fall off rapidly with distance away from the cylinder axis; describing the fields subject to the proper boundary conditions at the dielectric and at infinity requires more complicated field configurations. A triply infinite set of modes can be supported on a dielectric waveguide. There are the TE_{mn} modes and TM_{mn} modes just like in a metal waveguide; the former

*See a series of articles in "Physics Today," Vol. 29, May, 1976.

has transverse fields E(transverse), B(transverse) (that is, transverse to the direction of propagation z), and a longitudinal magnetic field B_z. The latter has E(transverse), B(transverse), and a longitudinal electric field E_z. In addition, there are so-called "hybrid modes" not found in waveguides with perfectly conducting metal walls. The hybrids have all field components and a given mode therefore has E(transverse), B(transverse), E_z and B_z.

Figure 5.29 shows the electric fields of three modes of a hollow cylindrical dielectric waveguide: the TE_{10} mode which has a zero electric field on axis, and two hybrid modes designated EH_{11} and EH_{12}. The photographs on the right were taken from an experiment in which the light source was a He-Ne gas laser (wavelength 6328 Å). The light was propagated along hollow glass soda lime pipes (refractive index 1.512) some 5 cm long, having an inside diameter of approximately 0.03 mm. The mode patterns excited in the waveguides were then viewed with a microscope (40X objective, 10X eyepiece) and photographed. The glass waveguides have circular cross sections in contrast to the rectangular waveguides preferred at microwave frequencies. One reason is that construction of a precision rectangular bore a fraction of a millimeter across would be a formidable undertaking.

5.4 CAVITY RESONATORS

A conventional resonant circuit used at radio frequencies has a coil and a condenser. If L is the inductance of the coil and C the capacity of the condenser, the circuit will resonate at a frequency given by $\omega_0 \simeq 1/\sqrt{LC}$ rad/sec, provided that the losses represented by a resistance R are reasonably small (see Figs. 1.7 and 1.13). At microwave and higher frequencies ($\omega/2\pi \gtrsim 1000$ MHz) the components of such a circuit become physically too small to be of practical use and other ways must be sought to resonate the system. The answer to the problem is the cavity resonator; it is very efficient and is of convenient size physically.

Any closed box with conducting walls has associated with it an infinite number of discrete resonant frequencies, each corresponding

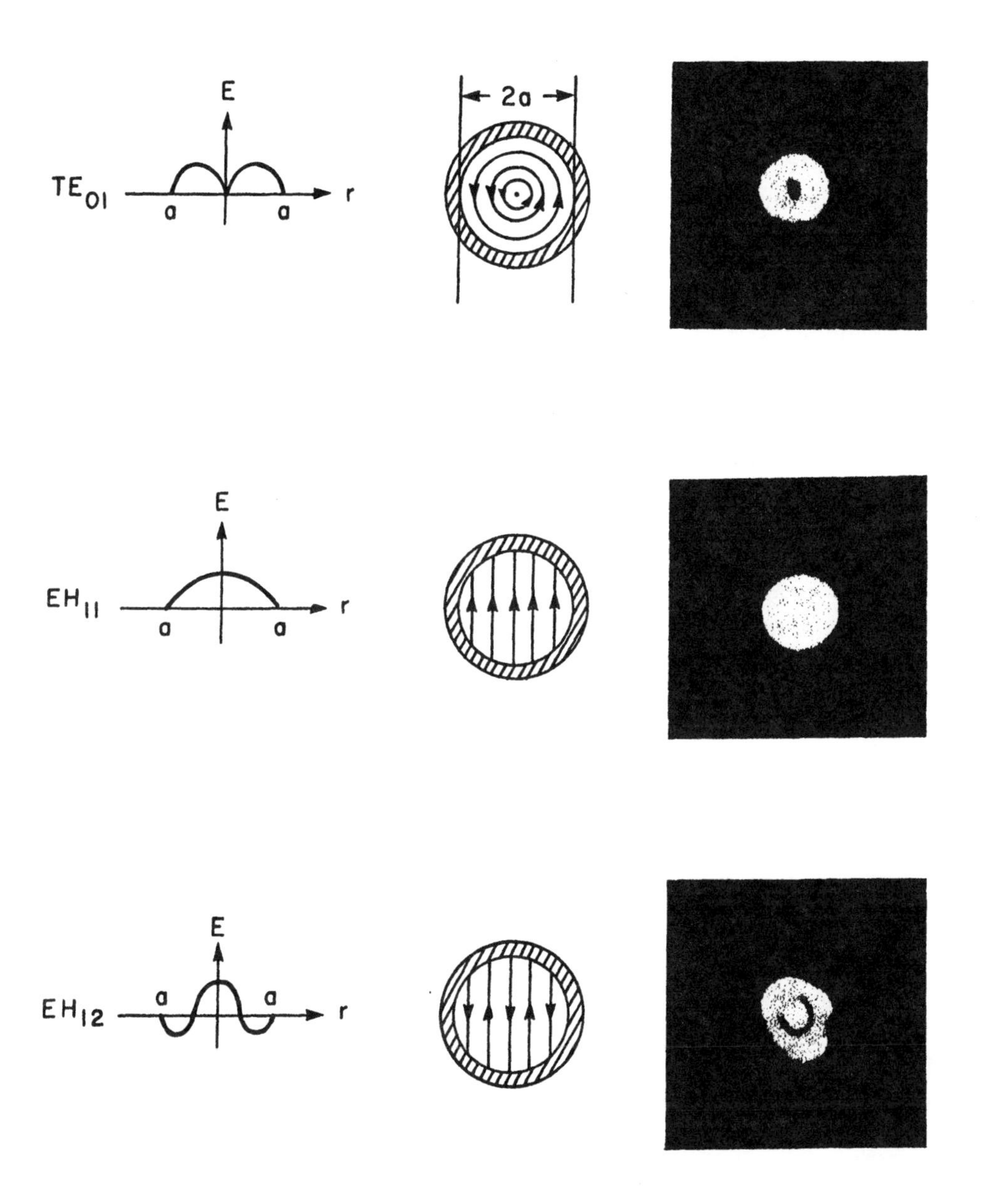

Fig. 5.29 The transverse electric field of three modes of a hollow dielectric waveguide. Notice in the photographs the positions of zero intensity where the electric field goes to zero. [From Optical Waveguides by N. S. Kapany and J. J. Burke (Academic Press, New York, 1972).]

to a different configuration of electric and magnetic fields. For a given cavity shape and a given mode, the resonant frequency ω_o depends only on the size of the resonator. Because the $\vec{E}$ and $\vec{B}$ fields are spatially distributed, however, it is often difficult to define ω_o in terms of lumped equivalent circuit elements L and C; and as we shall see, there is no need for it. Resonators come in all shapes – rectangular, cylindrical, spherical, and even ellipsoidal. Here we shall restrict ourselves to only two cases, the rectangular resonator for use at microwave frequencies, and the two-mirror resonator widely used with lasers at optical frequencies.

The rectangular microwave resonator

The rectangular prism resonator such as the one illustrated in Fig. 5.30 can be viewed simply as a section of rectangular waveguide closed off by two conducting end walls separated by a distance d. The resonance frequencies form a discrete set which can be determined graphically from the dispersion curve of the guide wave in question, by demanding that the wavenumber k_z be equal to $k_z = p\pi/d$ ($p = 1, 2, 3 \ldots$), as is illustrated in Fig. 5.31.

We know that in the absence of the end walls, the guide can support a doubly infinite set of TE_{mn} and TM_{mn} waves where the integers m and n designate the modes in question. The integer m represents the number of half-wave variations of the field as measured along the x axis and the integer n represents the number of half-wave variations along the y axis. The closing off of the guide along the z axis necessitates the specification of a third integer p, as we saw above, which determines the number of half-wave variations of the field along this coordinate axis. The resonant frequency $\omega_o(m,n,p)$ of the resonator can then be computed along the lines given in section 2.4 with the result that (cf. Eq. 2.59)

$$\omega_o(m,n,p) = \pi c \sqrt{\left(\frac{m}{a}\right)^2 + \left(\frac{n}{b}\right)^2 + \left(\frac{p}{d}\right)^2} \tag{5.98}$$

where $m,n,p = 0,1,2,3\ldots$ but where not more than one integer may

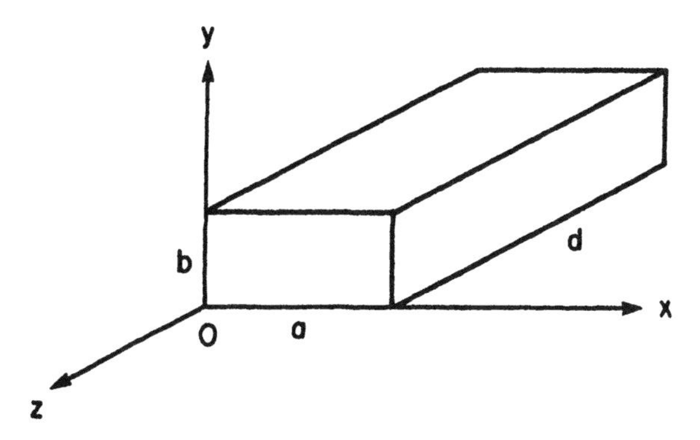

Fig. 5.30 A rectangular resonator with sides a, b, and d.

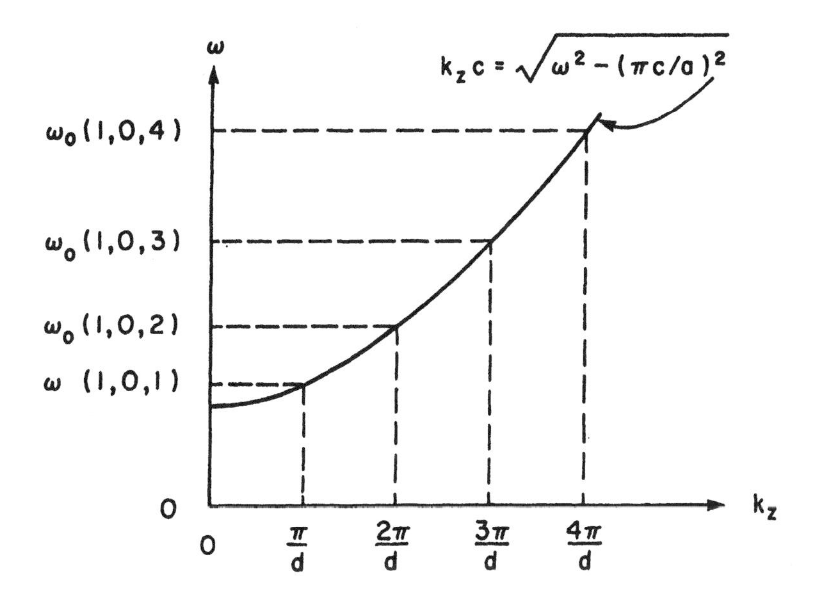

Fig. 5.31 Graphical method of determining the resonant frequencies of a resonator. As an example we took the $TE_{1,0}$ mode of the rectangular waveguide (see Fig. 5.23) and closed off the guide with walls separated by a distance d. Observe that the successive frequencies do not form a harmonic series. Similar diagrams can be drawn for all the higher TE and TM modes.

equal zero for fields to exist. When two integers are zero, it is not possible to satisfy boundary conditions on all conducting walls. For a cubical box with $a = b = d \equiv L$ the resonant frequencies are $\omega_o(m,n,p) = (\pi c/L)[m^2+n^2+p^2]^{1/2}$, and the lowest frequency is $\omega_o(1,1,0) = \sqrt{2}\ \pi c/L$. Thus, for example, an empty metal cavity with sides of 10 cm has its lowest resonant frequency at $\omega/2\pi = 2121.32$ MHz. Note that if the cavity were completely filled with a nonconducting medium of dielectric constant κ_e, Eq. 5.98 would have to be modified in that c would be replaced by $c/\sqrt{\kappa_e}$, which is the appropriate phase velocity in the medium in question (see section 6.5). Consequently, the resonant frequency shifts to lower frequencies when $\kappa_e > 1$ and to higher frequencies when $\kappa_e < 1$. Since small frequency changes can be measured very accurately, this property of the resonant cavity can be used in an accurate determination of κ_e, particularly for ionized and un-ionized gases for which κ_e is not very different from unity. [See G. Bekefi and S. C. Brown, Phys. Rev. 112, 159 (1958); also J. C. Ingraham and S. C. Brown, "Plasma Diagnostics," in R. F. Baddour and R. S. Timmons (Eds.), The Applications of Plasmas to Chemical Processing (The M.I.T. Press, Cambridge, Mass., 1967).] For example, the dielectric coefficient of oxygen at NTP equals 1.000542. Filling the aforementioned cavity with this gas shifts its frequency from 2121.32 MHz to 2120.75 MHz, which represents an easily measured frequency change of 0.57 MHz.

The physics of the frequency shift is readily understood: The cavity can be thought of as having an inductance L, a capacity C and a resonant frequency $\omega_o = 1/\sqrt{LC}$. Introduction of the dielectric into the capacitor changes its capacitance in proportion to the medium's dielectric constant κ_e; and ω_o therefore changes from $1/\sqrt{LC}$ to $1/\sqrt{\kappa_e LC}$.

The resonant frequency of a cavity oscillator is one quantity of prime importance; so is its "quality" Q. To compute the Q two quantities are needed. One is the time-averaged energy stored $\langle U \rangle$; the other, the time-averaged power dissipated $\langle P \rangle$. Then, in accordance with Eq. 1.108, $Q = \omega\langle U \rangle/\langle P \rangle$. The stored energy of an oscillation shifts back and forth in time between electric and magnetic forms and is equal to the peak energy in either the electric or the magnetic fields (compare this with potential and kinetic energy storage in mechanical oscillations, as discussed in section 1.1). In terms of B, then,

$$\langle U \rangle = \frac{1}{2\mu_o} \int_V B_o^2(x,y,z)\, dV \quad \text{joules} \tag{5.99}$$

where B_o is the amplitude. The integration extends over the entire volume V of the resonator. Note that B_o is the value of the magnetic field of the mode in question (m,n,p) as determined at an instant in time when the field has reached its maximum value. If one wishes to work with $\vec{E}$ rather than with $\vec{B}$ fields, the identical value of $\langle U \rangle$ can be obtained from the expression

$$\langle U \rangle = \frac{1}{2}\epsilon_o \int_V E_o^2(x,y,z)\, dV \quad \text{joules.} \tag{5.100}$$

In an empty evacuated cavity all energy dissipation comes from the joule heating $P = I^2R$ of the walls. Consider a small area of wall of length dz and width dx as is shown in Fig. 5.32. The current I flows in a thin layer of thickness $d\ell$ located at the metal-vacuum interface. Its value is related to the tangential magnetic field at the metal surface through Eq. 5.8. For the geometry of Fig. 5.32 it is given by $(B_t/\mu_o) = -J_z d\ell = -I_z/dx$. Hence the power dissipated can be written as $P = (B_t dx/\mu_o)^2 R$. But the resistance R of the strip of metal of conductivity σ equals $dz/(\sigma d\ell dx)$ with the result that $P = \left[B_t^2/\mu_o^2 \sigma d\ell\right] dxdz$. From this we see that the power dissipated per unit area of wall surface is just $B_t^2/\mu_o^2\sigma d\ell$. Integrating over the entire inner surface A of the resonator, and time averaging, yields the result

$$\langle P \rangle = \frac{1}{2\mu_o^2 \sigma d\ell} \int_A B_o^2(x,y,z)\, da \tag{5.101}$$

where B_o is the amplitude of the tangential magnetic field at the wall

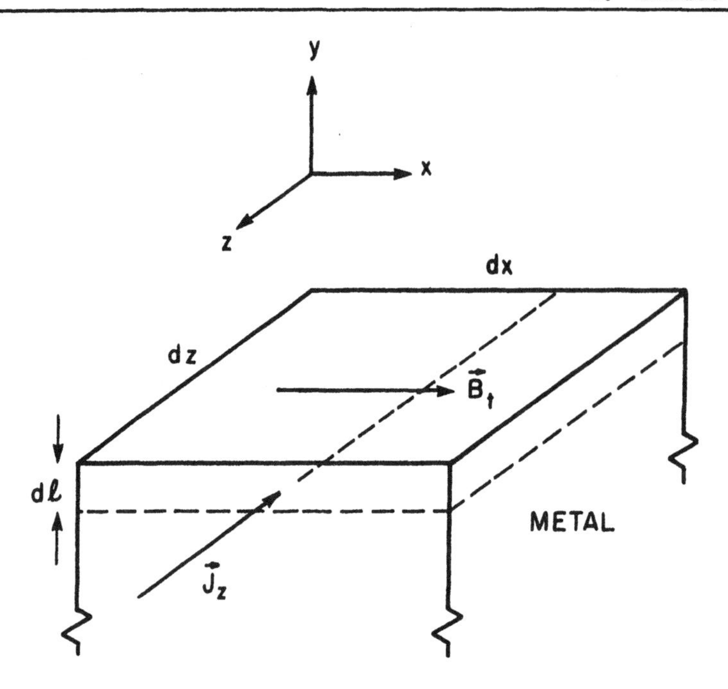

Fig. 5.32 The current flowing at the surface of a good conductor is related to the tangential magnetic field just above the surface through Eq. 5.8 or Eq. 5.12, $\hat{n} \times \vec{B} = \mu_o \vec{J}_z d\ell$ where $\hat{n}$ is the normal pointing out of the surface.

surface. Immediately above the surface of a good conductor B_t is in any case the dominant magnetic field and B_{normal} is vanishingly small, as is indicated in Fig. 5.5; and for that reason we simply drop the subscript t in B_t. Now, the thickness $d\ell$ through which most of the current flows equals the skin-depth $\delta = \sqrt{2/\omega\mu_o\sigma}$ given by Eq. 5.11. Thus, setting $d\ell = \delta$ and eliminating σ from Eq. 5.101 in favor of δ leads to the sought-after result

$$\langle P \rangle = \frac{\omega\delta}{4\mu_o} \int_A B_o^2(x,y,z)\, da \quad \text{watt.} \tag{5.102}$$

Combining Eqs. 5.99 and 5.102 gives the Q of the resonant cavity as

$$Q(m,n,p) = \frac{\omega\langle U \rangle}{\langle P \rangle} = \frac{2}{\delta} \frac{\int_V B_o^2(x,y,z)\, dV}{\int_A B_o^2(x,y,z)\, da}. \tag{5.103}$$

Since the magnetic field distribution $B_o(x,y,z)$ is a known function of position, the integrals can be evaluated for any desired mode (m,n,p), and the Q determined. In fact, Eq. 5.103 is of considerable generality and applies to resonators of any shape, not only rectangular.

To obtain an estimate of the Q we need not go through the rather tedious procedure of evaluating in detail the foregoing volume and surface integrals. If the magnetic field were uniform throughout the cavity, the ratio of the two integrals would simply be V/A, the ratio of cavity volume to surface area, and the Q would take on the value $2(V/A\delta)$. Because of the spatial variation of B_o, this result may be written as

$$Q(m,n,p) = 2\left(\frac{V}{A\delta}\right) \times F \qquad (5.104)$$

where F is a geometrical factor of order unity. We see that apart from the factor 2 and the geometrical factor F, the Q of the resonator equals the ratio of the volume V occupied by the electromagnetic fields, to the volume $(A\delta)$ of metal into which the fields penetrate as a result of finite conductivity. Thus, to maximize the Q for oscillations of a given frequency, it is usually desirable to use large cavities operated at one of the higher modes of oscillation. Still, the Q's of the lower modes are quite appreciable. Consider the cubical cavity discussed earlier with a side of length L = 10 cm and oscillating in its lowest mode $\omega_o(1,1,0)/2\pi$ = 2121 MHz. Then $V/A = L/6 = 1/60$ m. If the cavity is silvered, the skin depth $\delta = \sqrt{2/\omega\mu_o\sigma}$ for the frequency in question equals 1.395×10^{-6} m. Equation 5.104 with F set equal to unity then yields a Q of 23900. Because of imperfections, the Q of the real cavity will be somewhat lower. In fact, the introduction of an antenna coupling energy into and out of the cavity will in general cause a reduction in Q; thus great care must be exercised to perturb the interior of the resonator as little as possible if large Q's are desired.

Specially designed tunable cavity resonators called "wavemeters" are extensively used for the measurement of frequency of microwave signals. One type is illustrated in Fig. 5.33. An antenna couples the energy into the cavity and another samples the output. As the volume

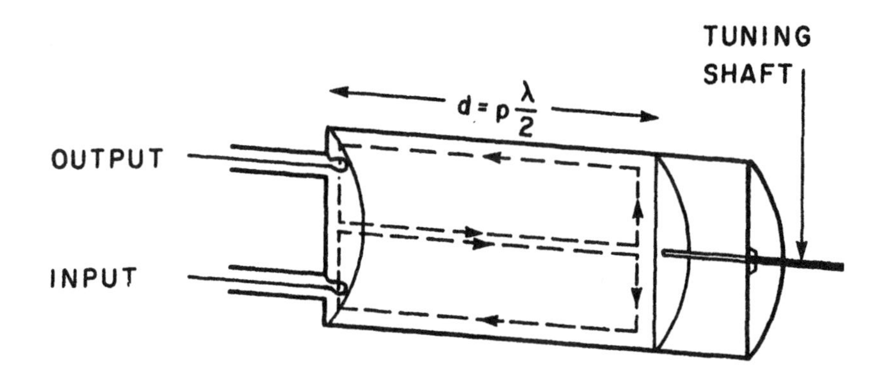

Fig. 5.33 Cutaway view of a cylindrical cavity. The input and output coupling loops couple to the magnetic field of the TE_{011} mode of this cavity. The dashed lines indicate the magnetic field. The electric field lines (not shown) are concentric circles perpendicular to the cylinder axis and concentric with it. By changing the plunger distance d one tunes for different frequencies. With this mode and at a frequency of ~10 GHz, Q's of 100,000 and higher can be achieved.

of the cavity is changed by means of the movable plunger, an appreciable signal is received only when the resonant frequency ω_o coincides (to within the width of the resonance $\Delta\omega$) with the signal frequency ω. The plunger position is normally calibrated in units of frequency or wavelength. Good accuracy can be achieved in determining ω; suppose $\omega/2\pi = 2000$ MHz and $Q = 20,000$. Then the uncertainty in frequency will be $\Delta\omega/2\pi \approx \omega_o/2\pi Q = 10^5$ Hz (cf. Eq. 1.111).

Optical resonators

The cavity resonators used in lasers generally have two flat or slightly curved mirrors facing one another, as is illustrated in Fig. 5.34. The lasing medium, for example, a tenuous gas, is interposed between the two. The light generated in the medium can bounce back and forth between the mirrors and after a few bounces a quasi-planar standing wave is established in the cavity. For resonance to occur, the distance L between mirrors must equal an integral number of half wavelengths $L = p(\lambda/2)$ where $p = 1, 2, 3, \ldots$ is the axial mode number.

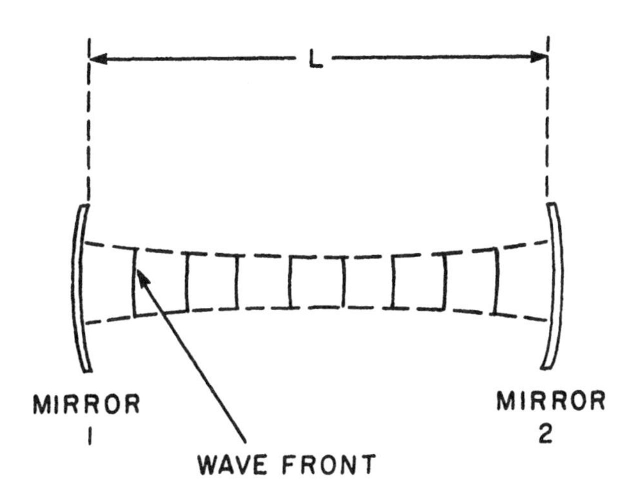

Fig. 5.34 Optical resonator formed by two mirrors facing each other. Using curved rather than plane mirrors as shown leads to stable optical oscillations. A well-studied arrangement is the confocal resonator in which the foci of the two mirrors coincide. If their radii of curvature are R_1 and R_2, then for this configuration $R_1 = R_2 = L$. Observe that the wave fronts of this quasi-plane wave are slightly curved. Also, the beam cross section (dashed line) expands somewhat toward the mirror surfaces.

Written in terms of the frequency ω of the light, the resonance condition becomes

$$\omega_o = \frac{p\pi c}{L}. \tag{5.105}$$

Since L is typically 10-100 cm and λ is 10^{-4}-10^{-5} cm, the axial mode numbers p at which optical resonators are operated is in the hundreds of thousands; this is to be contrasted with microwave resonators which operate typically with mode numbers of order one to ten. Consequently the Q of the optical cavity can be expected to be considerably larger. Let us compute it.

Suppose $\langle U \rangle$ is the average energy density (i.e., energy per unit volume) stored in the cavity. Half of this stored energy may be associated with a wave traveling from left to right and half with the wave traveling from right to left. In accordance with Eq. 3.109, the time-averaged power flow in each wave is $\langle S\rangle = c\langle U\rangle/2$ watt/m^2. This is the power per unit area striking one of the mirrors; if r represents the fraction of power lost per bounce, because of losses at the mirror, then the time-averaged power lost equals $c\langle U\rangle rA/2$ watts, where A is the cross-sectional area of the laser beam. The total power lost at both mirrors is $c\langle U\rangle rA$. And, the time-averaged energy stored equals $\langle U\rangle LA$. Hence the Q becomes

$$\left.\begin{aligned} Q &= \frac{\omega_o\langle U\rangle LA}{c\langle U\rangle rA} \\ &= \frac{\omega_o L}{rc} \\ &= \frac{2\pi}{r}\frac{L}{\lambda}. \end{aligned}\right\} \tag{5.106}$$

As an example, consider a one meter long helium-neon gas laser operated at the well-known transition corresponding to a wavelength of 6328 Å. Suppose the loss per bounce is 1% ($r = 10^{-2}$). Inserting these values in Eq. 5.106 gives $Q = 0.99 \times 10^9$ (!) which is a large Q indeed. A common misconception is that when r is, say, 0.01 per bounce, the corresponding Q is approximately $Q \approx 1/r$ or 100. Although there is a 1% loss per bounce, the cavity is many wavelengths long, and thus many, many cycles of the oscillations are needed to complete the bounce.

As a result of the large Q of the laser resonator, the resonant cavity modes have a narrow spectral width $\Delta\omega = \omega/Q$. For the case we have just considered, $\Delta\omega/2\pi = 4.8 \times 10^5$ Hz. This is much narrower than the linewidth of the atomic transition $\Delta\omega_a/2\pi$. In the case of the helium-neon laser, the atomic line is broadened because of the thermal agitation of the radiating atoms (Doppler broadening) and at room temperature this has a value $\Delta\omega_a/2\pi \approx 1500$ MHz. Within this bandwidth a number of cavity modes can easily be accommodated and, unless special precautions are taken, the laser will oscillate simultaneously at several discrete cavity modes, as is illustrated in Fig. 5.35. The modes can be observed in the following way. Shine

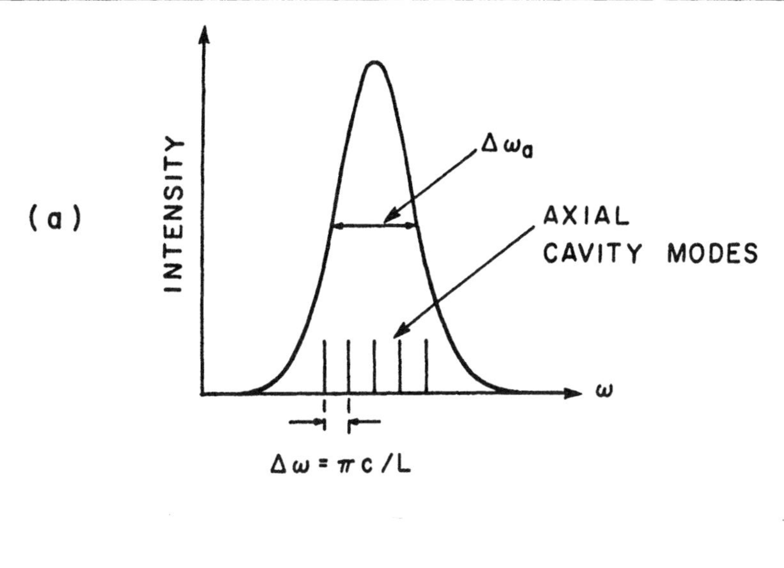

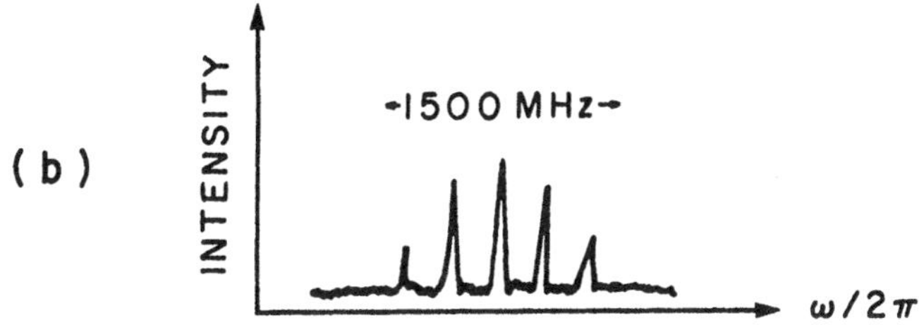

Fig. 5.35 (a) Doppler-broadened laser transition of half-power width $\Delta\omega_a$, together with five axial cavity modes separated by a frequency $\Delta\omega = \pi c/L$. (b) Observation made on a helium-neon 6328 Å laser showing five simultaneously oscillating axial modes. Note that one way to obtain single-mode operation is to make the laser sufficiently short so that $\Delta\omega \gg \Delta\omega_a$, but this reduces the gain.

the laser light on a photomultiplier or photodiode. This detector is a square-law device; hence the output current generated by the light contains not only a dc output proportional to the total light intensity, but also the beat frequencies $\omega_1 - \omega_2$, $\omega_1 - \omega_3$, $\omega_2 - \omega_3$, etc. between the various oscillation components present (see section 1.6). These beats can be seen on an oscilloscope provided, of course, that the photodetector response is fast enough to follow the different frequencies $\omega_1 - \omega_2$, $\omega_1 - \omega_3$ ... present. For example, Eq. 5.105 tells us that two adjacent (p) and (p+1) axial modes are separated by a frequency $\Delta\omega/2\pi = c/2L$, which equals 150 MHz for a one-meter long laser. Such a frequency can be detected fairly readily with a reasonably good photomultiplier.

We have seen that there are many axial nodes and antinodes along the length L of the optical cavity. Are there any in the transverse direction, over the cross section of the laser beam? The answer is, yes indeed. The lowest mode and one that is desired in most practical applications is the single circular spot as illustrated in Fig. 5.36.

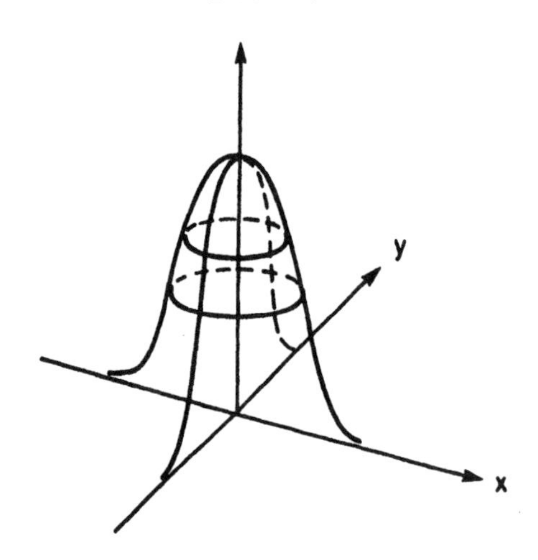

Fig. 5.36 Intensity distribution across a laser beam oscillating in the lowest transverse mode $[TEM_{00p}]$. The distribution is a Gaussian with an intensity profile that varies as $\exp\left[-2(x^2+y^2)/w_o^2\right]$; w_o is the spot size. An aperture of radius equal to w_o will transmit approximately 86% of the total laser beam; an aperture of radius equal to 1.5 w_o will transmit about 99% of the beam. Diffraction is responsible for the Gaussian profile. [For a review of the subject at a quite advanced level see H. Rogelnik, "Modes in Optical Resonator," in A. K. Levine (Ed.), Lasers: A Series of Advances (Marcel Dekker, New York, 1966).]

But it is noteworthy that even in this simplest of modes the amplitude of the electric (and magnetic) fields is not uniform over the cross section (x,y) as in a truly plane wave. For mirrors that are large compared with the laser beam the amplitude distribution is Gaussian,

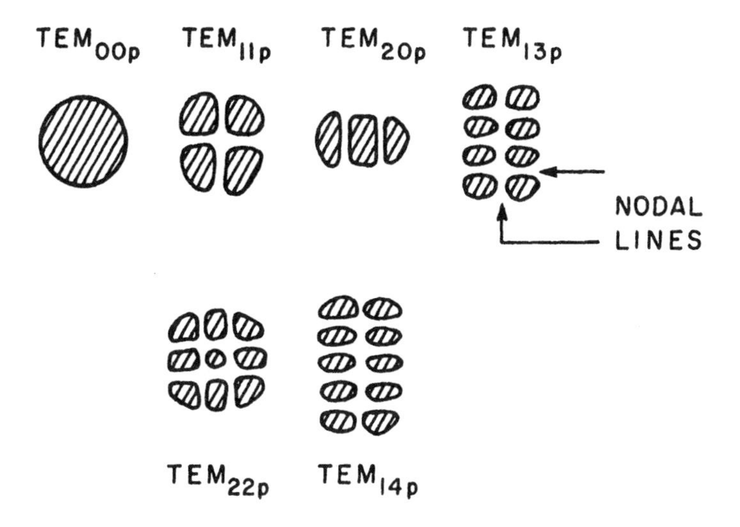

Fig. 5.37 Examples of transverse modes of an optical resonator.

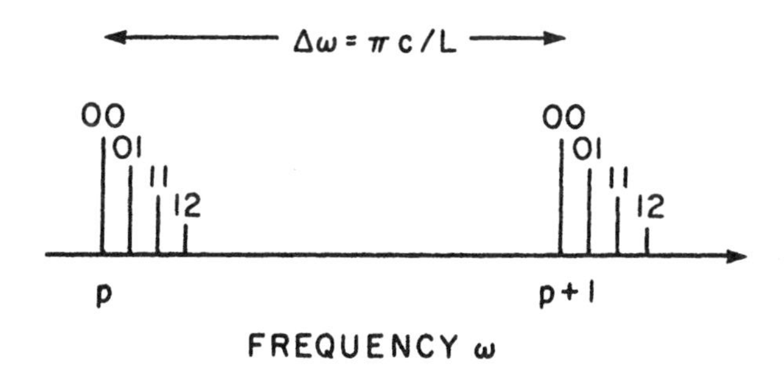

Fig. 5.38 The frequency spectrum $\omega(m,n,p)$ of the higher order transverse modes of an optical cavity with near-planar mirrors; $\Delta\omega = \pi c/L$ is the axial mode spacing.

$$\left.\begin{matrix} E \\ H \end{matrix}\right\} \propto e^{-(x^2+y^2)/w_o^2} \cos(\omega t - kz) \tag{5.107}$$

where w_o is called the "spot size." The mode of this nonuniform plane wave is commonly designated as TEM_{00p}, indicating that here we are dealing with transverse electric and magnetic waves. They may be likened to the TEM waves of two-wire transmission lines (section 5.2) but they differ from the TE and TM waves of hollow waveguides (section 5.3). Higher order TEM_{mnp} modes are often seen in laser beams. The integers m and n give the number of nodal lines in the x and y direction across the emerging beam. A few of these are illustrated in Fig. 5.37. Their theoretical derivation is quite difficult but the interested reader is referred to the following very good exposition: A. E. Sigman, An Introduction to Lasers and Masers (McGraw Hill Book Company, New York, 1971). Each of the higher modes is associated with a different natural frequency $\omega_o(m,n,p)$ as is shown schematically in Fig. 5.38.

CHAPTER 6

INTERACTION OF WAVES WITH MATTER

The interaction of waves with matter concerns the interplay of electromagnetic radiation with electrons, atoms and molecules. The story is an interesting one which asks – and answers – many questions about the workings of nature and man. Why is the sky blue and gold yellow? Why are x-rays refracted oppositely to light rays? How does a laser differ from a flashlight, and why and when is there communication blackout between Earth and an astronaut? These questions and many others like it are the type of problem with which we shall be concerned in Chapters 6 and 7. In our attempt to answer them, we shall replace matter by an ensemble of classical oscillators, electrons in particular, sometimes held by spring-like forces to the nuclei, and oscillating in response to the electromagnetic force $q[\vec{E}+\vec{v}\times\vec{B}]$ acting on them. As we learned in Chapter 1, the equation of motion appropriate to each such oscillator (cf. Eq. 1.121) is

$$\boxed{\frac{d^2\vec{r}}{dt^2}+\beta\frac{d\vec{r}}{dt}+\omega_o^2\vec{r}=\frac{q}{m}\left[\vec{E}+\frac{d\vec{r}}{dt}\times\vec{B}\right]} \tag{6.1}$$

where $\vec{r}$ is its displacement vector, $\omega_o=\sqrt{K/m}$ its natural frequency of oscillation and β is a phenomenological damping factor. The approach, which takes Eq. 6.1 as its basis, is generally known as the "classical electron theory of matter," and acquires two forms. In the one, the electrons are considered to be essentially free; that is, not attached to molecules or nuclei, as is the case, for example, in metals or in ionized gases (plasmas); here the "spring constant" K and hence ω_0 are of necessity zero. In the other case the electrons are bound to atoms as is typical of dielectrics like air or glass.

We know, of course, that a classical equation of motion like that written above cannot adequately describe the real behavior of atoms and molecules, and that a rigogous theory of the interaction of the electromagnetic field $\vec{E}, \vec{B}$ with matter requires quantum mechanics for its complete solution. However, the truth of the matter is that, quite remarkably, there is a large number of physical phenomena which agree very well with the classical model, as we shall find out. Furthermore, the classical model clarifies effects such as scattering, dispersion, emission, and absorption, and illustrates their interrelationship in a way that the much more involved quantum mechanical calculations often do not.

The interaction of waves with matter can be treated on two levels. There is the microscopic level just discussed to which one must ultimately resort for a full understanding of the problem. And then there is the macroscopic level where the interaction is described in terms of certain large-scale phenomenological quantities which express the response of the bulk medium to the electromagnetic perturbation. (The electrical conductivity of a metal or the refractive index of a dielectric are two such parameters.) By viewing these quantities as known from experiment, for example, their great virtue lies in the fact that they can be readily incorporated into Maxwell's equations and then manipulated subject to a few simple rules. By using them, we obtain information about the interaction of electromagnetic radiation with matter in bulk. Since these macroscopic quantities are model-independent, they are meaningful in the classical as well as in the quantum-mechanical context. The ultimate aim of theory is to make the necessary connection between the macroscopic measurable world of matter and the detailed equation of motion of the constituent atomic oscillators, to the extent that we understand them. We shall begin with the macroscopic description.

6.1 POLARIZATION

Maxwell's equations (3.38) govern electromagnetic phenomena of all types. The electric and magnetic fields $\vec{E}$ and $\vec{B}$ at any position and any time are fully specified provided that the complete charge and current distributions ρ and $\vec{J}$ are given. In problems involving but a few idealized point charges and a few current elements, the equations in the form given are tractable and amenable to fairly straightforward solution (see, for example, Chapter 4). However, when we come to problems involving fields in the presence of matter a complete specification of $\vec{E}$ and $\vec{B}$ in terms of the individual charges (and currents) would be impossible; a cubic centimeter of matter contains anywhere from $\sim 10^{19}$ to $\sim 10^{24}$ charges all of them undergoing complicated motions caused by thermal agitation of the molecules. Fortunately, in many problems of interest one does not ask for the detailed microscopic electric and magnetic fields but for averages over small volume elements ΔV. By averaging over regions with characteristic dimensions $\sim 10^{-4}$ cm ($\Delta V \sim 10^{-12}$ cm^{-3}) or greater, one encompasses 10^7 or more molecules, which is sufficient to average out most of the microscopic fluctuations. The remaining "macroscopic" fields averaged in the sense

$$\langle \vec{E} \; ; \; \vec{B} \rangle = \frac{1}{\Delta V} \int_{\Delta V} (\vec{E} \; ; \; \vec{B}) \; dV \tag{6.2}$$

are what this chapter is all about. In order not to complicate the notation we omit the sign $\langle \; \rangle$. Henceforth, however, all quantities will be assumed implicitly to have been averaged in the way we have just outlined. The resulting modified forms of Maxwell's equations produced by this averaging process are usually referred to as the "macroscopic field equations."

The movement of an electron of charge q from a point A to a point B is equivalent to leaving the original electron undisturbed at A and adding a dipole as is shown in Fig. 6.1, so as to cancel the original charge at A. The resulting dipole moment $\vec{p}$ is given by

$$\vec{p} = q\vec{\ell} \quad \text{coulomb-meter} \tag{6.3}$$

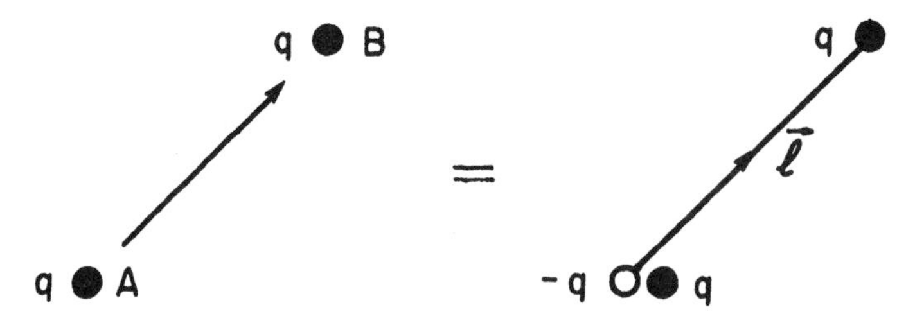

Fig. 6.1 Movement of a charge q from A to B is equivalent to the addition of a dipole $\vec{p} = q\vec{\ell}$. Its electric field in the x-z plane and at a large distance r is given by

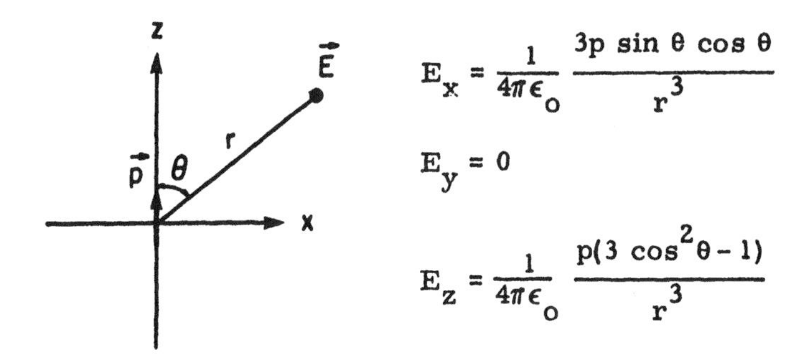

where $\vec{\ell}$ is the vector displacement of the charge; the direction of the vector $\vec{\ell}$ is taken from $-q$ to $+q$. When an electric field is applied to a collection of molecules (typically to the molecules of a dielectric) it causes the aforesaid movement of electrons and thus creation of an ensemble of dipoles. This is known as the "polarization" of matter. It arises in two ways as is illustrated in Fig. 6.2:

(a) The applied electric field distorts the charge distribution of the atom (or molecule), which in the absence of the field may be perfectly symmetrical. Such is the case in so-called nonpolar substances, as for example, in H, He, C, Na or in a molecule like methane (CH_4). With zero electric field, these substances have no net $\vec{p}$ of their own.

(b) The applied field tends to line up the initially randomly oriented permanent dipole moments of the molecules. Such is the situation in so-called polar substances like water (H_2O), hydrogen chloride (HCl), ammonia (NH_3), carbon monoxide (CO), methanol (CH_3OH). Of course, not every polar molecule will be aligned precisely by the field

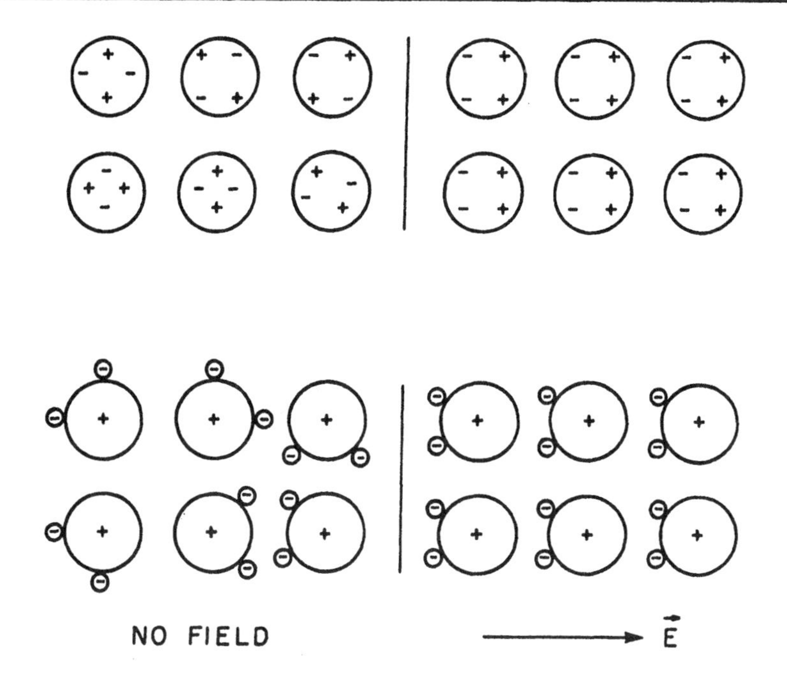

Fig. 6.2 Action of an electric field on a symmetrical molecule (above) inducing a dipole moment; and its action on an asymmetrical molecule (below) causing an alignment of the permanent dipoles.

since thermal agitation partly destroys complete polarization.

No matter whether the effect occurs by process (a) or (b), or a combination thereof, the substance will experience some net polarization. If there are N atoms (or molecules) per unit volume, the dipole moment per unit volume averaged over many such particles (cf. Eq. 6.2) is given by

$$N\langle \vec{p} \rangle = \frac{1}{\Delta V} \int_{\Delta V} N\vec{p}\, dV \quad \text{coulomb/m}^2. \tag{6.4}$$

The quantity $N\langle \vec{p} \rangle$ is known as the "polarization" and is designated by the letter $\vec{P}$. It is the fundamental macroscopic quantity associated with the polarization process.

Let us now consider an elementary volume of dielectric (Fig. 6.3) shaped like a slanted box of base da and height $\vec{\ell}$, where $\vec{\ell}$ is the same vector displacement of the charge as is given by Eq. 6.3. The

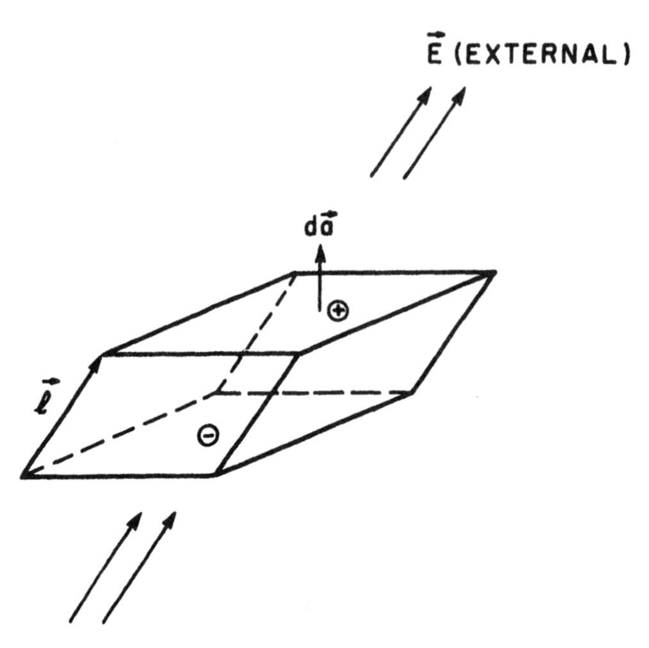

Fig. 6.3 An element of dielectric polarized by an externally applied electric field.

medium is polarized by an externally applied electric field acting in the direction shown. As a result, charges will be pushed across $d\vec{a}$; the average amount of charge pushed being just that which was originally contained in the box of base da and slant height $\vec{\ell}$, that is, $Nq\vec{\ell} \cdot d\vec{a}$. Since $Nq\vec{\ell}$ equals $\vec{P}$, we conclude that the charge crossing da equals $\vec{P} \cdot d\vec{a}$. If $d\vec{a}$ should coincide with the outer surface of the dielectric, then $Nq\vec{\ell} \cdot d\vec{a}$ represents the surface charge induced on this element of area. Denoting ρ_S(pol) as the polarization surface charge density, we arrive at the important result that the charge induced on the surface of a polarized dielectric is equal to the normal component of the polarization vector at that point. Stated in symbols,

$$\boxed{\rho_S(\text{pol}) = \vec{P} \cdot \hat{n} \quad \text{coulomb/m}^2} \tag{6.5}$$

where $\hat{n}$ is the outward normal to the surface.

Equation 6.5 allows us to explain quantitatively some familiar

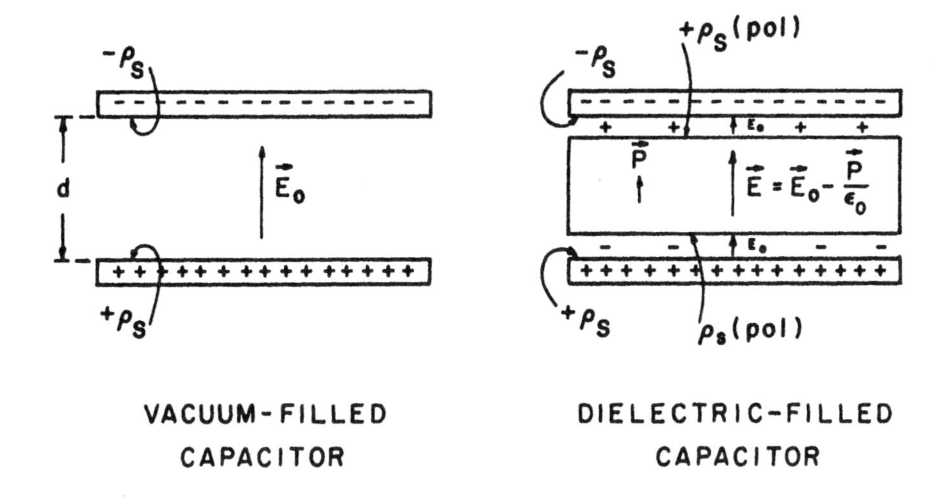

Fig. 6.4 The field in a dielectric-filled capacitor is the superposition of the vacuum field E_0 and the field $E' = -P/\epsilon_0 = -\rho_S(\text{pol})/\epsilon_0$ (acting from top to bottom in our figure), due to the polarization charge induced on the dielectric slab surfaces.

results, as for example, why insertion of a dielectric between the plates of a capacitor increases its capacity. Let us first take two parallel metal plates in vacuum of area A with charge density $+\rho_S$ on the bottom plate and $-\rho_S$ on the top plate (Fig. 6.4). The vacuum electric field $\vec{E}_0$ due to the two charged sheets is, by Gauss' law, equal to ρ_S/ϵ_0, and the potential V is equal to $V = E_0 d = \rho_S d/\epsilon_0$ where d is the plate separation. Thus, the capacity of the empty capacitor is $C_0 = \rho_S A/V$, or

$$C_0 = \frac{\epsilon_0 A}{d} \text{ farad.} \tag{6.6}$$

We now entirely fill the space between the plates with an isotropic, uniform dielectric. The electric field polarizes the medium giving rise to the polarization $\vec{P}$. But, by Eq. 6.5 this polarization $\vec{P}$ is fully equivalent to removing the dielectric and replacing it by an induced surface charge $-\rho_S(\text{pol})$ residing on the bottom surface and

$+\rho_S$(pol) residing on the top surface. These two induced charged sheets themselves produce an electric field $E' = -\rho_S(\text{pol})/\epsilon_o = -P/\epsilon_o$ acting downward, as is shown in Fig. 6.4. By the principle of superposition the net electric field within the medium equals

$$\begin{aligned} \vec{E} &= \vec{E}_o + \vec{E}' \\ &= \vec{E}_o - \frac{\vec{P}}{\epsilon_o} \text{ volt/m.} \end{aligned} \tag{6.7}$$

We recall that this field $\vec{E}$ is the macroscopic field averaged over many polarized molecules. The local microscopic field can depart hugely from this value. Its magnitude depends sensitively on the precise place where the tiny electric field probe is placed (if this could be done) relative to a given atom; in principle there would be nothing wrong even in sticking the probe into the atom itself. Such probing followed by averaging over volume then leads to $\vec{E}$.

Observe that while the charge $\rho_S A$ on the capacitor is still the same, the electric field between the plates is reduced compared with the vacuum field $\vec{E}_o$, by an amount $[E_o - (P_o/\epsilon_o)]/E_o$. And since the potential is reduced by a similar factor, it follows that the capacitance is changed by the reciprocal amount,

$$\begin{aligned} C &= C_o \frac{E_o}{E_o - (P/\epsilon_o)} \\ &= C_o\left[1 + \frac{P}{\epsilon_o E}\right]. \end{aligned} \tag{6.8}$$

The second form of this equation is derived from the first by eliminating E_o in favor of the net electric field E through use of the relation (6.7), which can be written as $E_o = E + (P/\epsilon_o)$. Since the dimensionless quantity $(P/\epsilon_o E)$ is usually positive, the capacity C exceeds the vacuum capacity C_o. The quantity $1 + (P/\epsilon_o E)$ is known as the dielectric constant of the polarized medium, a quantity that will be discussed later in this chapter.

In the paragraph preceding Eq. 6.5 we showed that as a consequence of the polarization phenomenon, the magnitude of charge crossing an elementary area da equals $\vec{P} \cdot d\vec{a}$. Now suppose S is a closed surface entirely immersed within the bulk of the dielectric. If the polarization is uniform (that is, $\vec{P}$ is constant throughout the volume enclosed by the area S), then $\int P \cdot d\vec{a}$ integrated over S is identically equal to zero because as much charge must leave some points of S as enters other points. We see this clearly in the case of the capacitor of Fig. 6.4: the electric field pointing up the page pushes electrons downward, and having pushed $-\rho_S(\text{pol})A$ charge onto the bottom surface, it left $+\rho_S(\text{pol})A$ of positive charge behind at the top surface. However, if $\vec{P}$ is not uniform as a result of, say, nonuniformities in the dielectric, an incomplete cancellation of the dipole "end charges" would produce a net volume charge density, as is illustrated schematically in Fig. 6.5. Hence, denoting $\rho(\text{pol})$ as the volume polarization charge density, it follows that

$$\int_{\text{closed } S} \vec{P} \cdot d\vec{a} = -\int_V \rho(\text{pol})\, dV \qquad (6.9)$$

where V is the volume enclosed by the surface S. The negative sign comes from the fact that $\vec{P} \cdot d\vec{a}$ gives the magnitude of charge crossing S and <u>escaping</u> the volume V; so, by charge conservation, an equal and opposite polarization charge is left behind within V. On using the

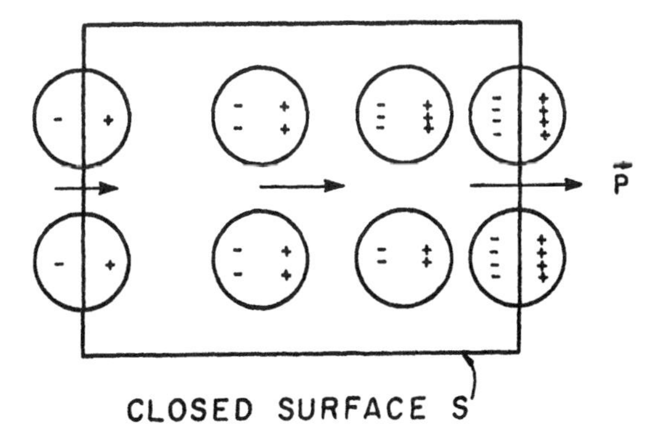

Fig. 6.5 Nonuniform polarization of a dielectric showing more positive "end" charges sticking out of S than negative "end" charges sticking out of S.

divergence theorem (3.13) one can readily cast Eq. 6.9 into the differential form

$$\nabla \cdot \vec{P} = -\rho(\text{pol}). \tag{6.10}$$

Note that there is nothing fictitious about ρ(pol). It is a real charge density and we just call it "polarization charge" to remind ourselves where it came from.

We are now in a position to rewrite the first Maxwell equation (3.38a), that is, Gauss' law, $\nabla \cdot \vec{E} = \rho(\text{total})/\epsilon_0$ to allow for the presence of matter. By writing ρ(total) we wish to stress that all of the charge density including the part arising from polarization must be inserted on the right-hand side of Gauss' equation. We therefore write

$$\rho(\text{total}) = \rho(\text{pol}) + \rho \tag{6.11}$$

where the last term ρ stands for any other, so-called "free" charge density, not included in ρ(pol). Thus, for example, if extra charge is introduced into the medium, it is contained in ρ; or if the molecules carry net charge (they are usually neutral), that also is included in ρ. On substituting for ρ(pol) from Eq. 6.10 in Eq. 6.11 and inserting ρ(total) into Gauss' equation we get

$$\boxed{\nabla \cdot \vec{E} = \frac{\rho}{\epsilon_0} - \frac{\nabla \cdot \vec{P}}{\epsilon_0}} \tag{6.12}$$

as the sought-for modification caused by the presence of matter. The equation is as applicable to time-stationary as to time-varying charge distributions. In the latter case $\vec{P}$, of course, becomes a function of time.

Next we inquire what modifications the presence of matter causes in the second Maxwell equation (3.38b). The answer is, none whatever, so that

$$\nabla \times \vec{E} = -\frac{\partial \vec{B}}{\partial t}. \tag{6.13}$$

The reason is simply that the equation contains no statement about real charges or, for that matter, currents and it must therefore be equally valid in vacuum as it is in a substance. The same argument applies to the last equation (3.38d) and therefore

$$\nabla \cdot \vec{B} = 0. \tag{6.14}$$

The situation is different in regard to the remaining Maxwell equation, Eq. 3.38c. Whenever the polarization of matter changes in time (as a result of time-varying electric fields) there occurs a genuine motion of charge, and thus a current. Each oscillating charge q contributes a current equal to q multiplied by its velocity. If N is the number of bound charges per unit volume, then the current density $\vec{J} = Nq\vec{v} = Nq\,d\vec{\ell}/dt$ is just $d\vec{P}/dt$. Therefore, the current associated with the time-varying polarization is given by the general result

$$\boxed{\vec{J}(\text{pol}) = \frac{d\vec{P}}{dt} \ \text{amp/m}^2} \tag{6.15}$$

and represents part of the total current $\vec{J}$(total) appearing on the right-hand side of Eq. 3.38c. Our task concerning $\vec{J}$(total) is now similar to that we undertook above in regard to ρ(total) appearing in Gauss' law; that is, to break up $\vec{J}$(total) into convenient, readily identifiable parts. To this purpose we write

$$\vec{J}(\text{total}) = \frac{\partial \vec{P}}{\partial t} + \vec{J}(\text{cond}) + \vec{J}(\text{mag}). \tag{6.16}$$

The first term is just the current associated with the time-varying polarization of the bound charges as discussed above. $\vec{J}$(cond) is generally associated with the current carried by the "free" electrons of conductors or semiconductors, but it may also include other currents, as for example, the current of free electrons or ions flowing in vacuum. And, finally, $\vec{J}$(mag) is intended to represent the circulating currents associated with the elementary atomic magnets which account for any magnetic properties of the material in question. We shall discuss the last named in the following subsection, but before doing so, let us just write down the last of Maxwell's equations with $\vec{J}$(total) as given by Eq. 6.16:

$$\nabla \times \vec{B} = \mu_o \epsilon_o \frac{\partial \vec{E}}{\partial t} + \mu_o \frac{\partial \vec{P}}{\partial t} + \mu_o \vec{J}(\text{cond}) + \mu_o \vec{J}(\text{mag}). \tag{6.17}$$

It is an equation for the most general type of material one can envisage: one which exhibits dielectric properties associated with its bound electrons, conduction by its free electrons and magnetism. In addition there is the fictitious but nonetheless important displacement current given by $\epsilon_o\ \partial\vec{E}/\partial t$. Fortunately, there is virtually no substance in which all four currents are equally prominent, and in most problems of interest drastic simplifications can be made on the right-hand side of Eq. 6.17.

6.2 MAGNETIZATION

The magnetic properties of materials come from tiny circulating currents within atoms – either from spinning electrons or from the orbital motion of electrons in the atoms. Since a circulating current I has a magnetic moment $\vec{m}$ given by

$$\vec{m} = \vec{A}I \ \text{amp-m}^2 \tag{6.18}$$

where $\vec{A}$ is the vector area of the current loop (see Fig. 6.6), we can take the alternate view in which every atom or molecule is a tiny magnetic dipole. The material is then said to be magnetized if there is some net alignment of these magnetic dipoles. This viewpoint has the advantage that it parallels the story of the electric dipoles in electric fields discussed earlier. And, just like in the case of $\vec{P}$ of Eq. 6.4 we define a macroscopic averaged quantity $\vec{M}$, the magnetization, which equals the magnetic dipole moment per unit volume, that is

$$\begin{aligned}\vec{M} &= N\langle\vec{m}\rangle \\ &= \frac{1}{\Delta V}\int N\vec{m}\,dV \ \text{amp/m}\end{aligned} \tag{6.19}$$

with N as the density of atoms or molecules. We know of three major classes of magnetized matter:

(a) There are paramagnetic substances like sodium (Na), aluminum (Al), copper chloride ($CuCl_2$) nickel sulfate ($NiSO_4$), and liquid oxygen (O_2) where the alignment is in the direction of the applied

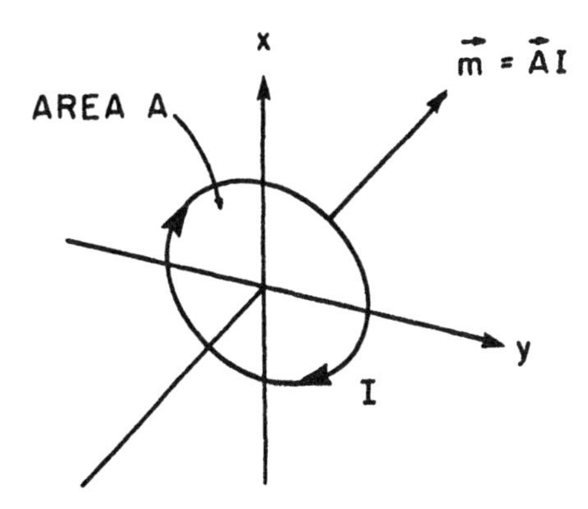

Fig. 6. 6 The magnetic dipole $\vec{m}$ of a small current loop is defined as the vector of magnitude IA directed along the normal to the plane of the loop. The sense of the vector $\vec{m}$ is such that a right-hand screw rotation is in the direction of the current. If $\vec{m}$ is located at the origin of coordinates and points along the z axis, one can show that the magnetic field $\vec{B}$ in the x-z plane at large r is given by

$$B_x = \frac{\mu_0}{4\pi} \frac{3m \sin\theta \cos\theta}{r^3}; \quad B_y = 0; \quad B_z = \frac{\mu_0}{4\pi} \frac{m(3\cos^2\theta - 1)}{r^3}$$

just like the $\vec{E}$ field of an electric dipole (see Fig. 6. 1); θ is the angle between $\vec{m}$ and $\vec{r}$.

magnetic field $\vec{B}$ but is fairly weak. Thus here the internal magnetic field is increased but not by very much.

(b) There are ferromagnetic materials like iron (Fe), magnetite (Fe_3O_4) and certain alloys like Mu metal in which the alignment is in the direction of the magnetic field but is very strong. Once again the internal magnetic field is increased compared with the imposed field but the effect is one to five orders of magnitude greater than for the paramagnetic substances discussed above.

(c) And finally there are diamagnetic materials, as for example water (H_2O), copper (Cu), lead (Pb), quartz (SiO_2), and graphite (C), where the alignment is weak and is opposed to the direction of the external $\vec{B}$ field. Here the internal field is just slightly lower than the applied field. We point out that all matter is intrinsically diamagnetic, but the diamagnetism in substances mentioned in (a) and (b) is

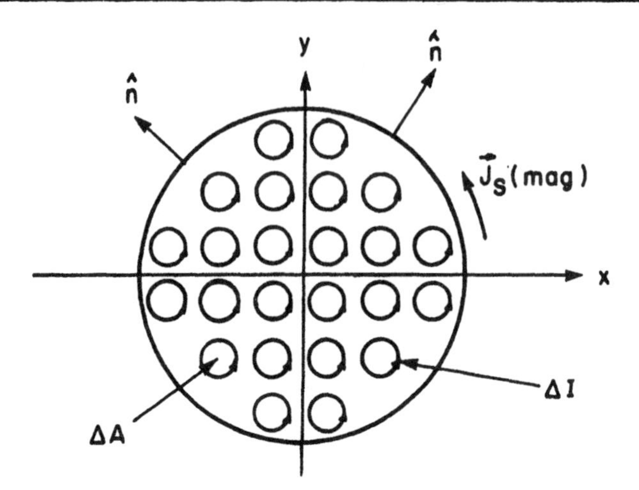

Fig. 6.7 The elementary "Ampèrian" currents ΔI of a uniformly magnetized cylindrical rod of area $A = \Sigma\, \Delta A$. The magnetization is along the z axis. The totality of these current loops is a surface current $\vec{J}_S$ flowing around the outer curved surface of the rod in the direction shown. $\vec{J}_S$ is the current per unit length Δz; $\hat{n}$ is a unit vector drawn perpendicular to the curved surface. $\vec{J}_S(\text{mag}) = \vec{M} \times \hat{n}$.

overcome by para- and ferromagnetic effects.

In accordance with Eq. 6.18 every elementary atomic dipole $\vec{m}$ is associated with an elementary circulating current ΔI. We must now inquire about the current associated with the overall, bulk magnetization $\vec{M}$. To begin with, consider a homogeneous cylindrical rod of thickness Δz uniformly magnetized. Let the magnetization be entirely along the axis of the rod, that is, along the z axis of the coordinate system shown in Fig. 6.7. Physically, we interpret this to mean that in the material there is a uniform distribution of elementary currents all circulating in the x-y plane as is illustrated schematically in the figure. All the atomic currents go around in little circles; their magnitudes are the same, as is their circulating direction. Now, what is the total effective current of this system? Clearly, in the bulk of the bar there is no effect at all because next to each little current there is another current going in the opposite direction. Note, however, that at the curved surface of the material this cancellation does not take place and so there is a net current after all. We can relate it to the magnetization M_z per unit volume. By Eqs. 6.18 and 6.19 $M_z = Nm_z = \Sigma m_z/V = \Sigma \Delta A \Delta I/V$ where the summation is over all the elementary circulating currents ΔI, each enclosing the elementary area ΔA. But, $\Sigma \Delta A \Delta I = \Delta I A$ where A is the cross-sectional area of the rod, and the volume $V = A\Delta z$. Hence

$$M_z = \frac{\Delta I}{\Delta z}. \tag{6.20}$$

Now, $\Delta I/\Delta z$ is the surface current density, that is, current per unit length as measured along the surface.

Denoting this quantity by $J_S(\text{mag})$ [amp/m] we have the result that $M_z = J_S(\text{mag})$ where the subscript S denotes the fact that it is a surface current, and the word (mag) is to remind us that it is a circulation current associated with the magnetic properties of the material. In a more general situation where the magnetization is not necessarily in the z direction the vector expression

$$\vec{J}_S(\text{mag}) = \vec{M} \times \hat{n} \text{ amp/m} \tag{6.21}$$

gives the full relation between the magnetization $\vec{M}$, the surface current density $\vec{J}_S(\text{mag})$, and the unit vector $\hat{n}$ drawn normal to the surface and pointing outward from the volume of the magnetized material. This equation plays the same role in magnetic effects as Eq. 6.5 plays in dielectric effects.

Besides the surface currents of Eq. 6.21 there will appear in general a volume distribution of such currents whenever there is any nonuniformity in the magnetization $\vec{M}$. This is precisely analogous to the volume density of polarization (Eq. 6.10) which we found to occur whenever there was any nonuniformity in the polarization $\vec{P}$ of the dielectric. It is our task to determine the relationship between $\vec{M}$ and the volume current density which we shall denote by $\vec{J}(\text{mag})$. Consider a region of inhomogeneous magnetization but assume at first that $\vec{M}$ is oriented entirely in the z direction. Figure 6.8a shows two neighboring elements, M_z at position x and the slightly different magnetization $(M_z + \Delta M_z)$ at position $(x + \Delta x)$. The circulating current around M_z is termed ΔI_1 and that around $(M_z + \Delta M_z)$ is termed ΔI_2.

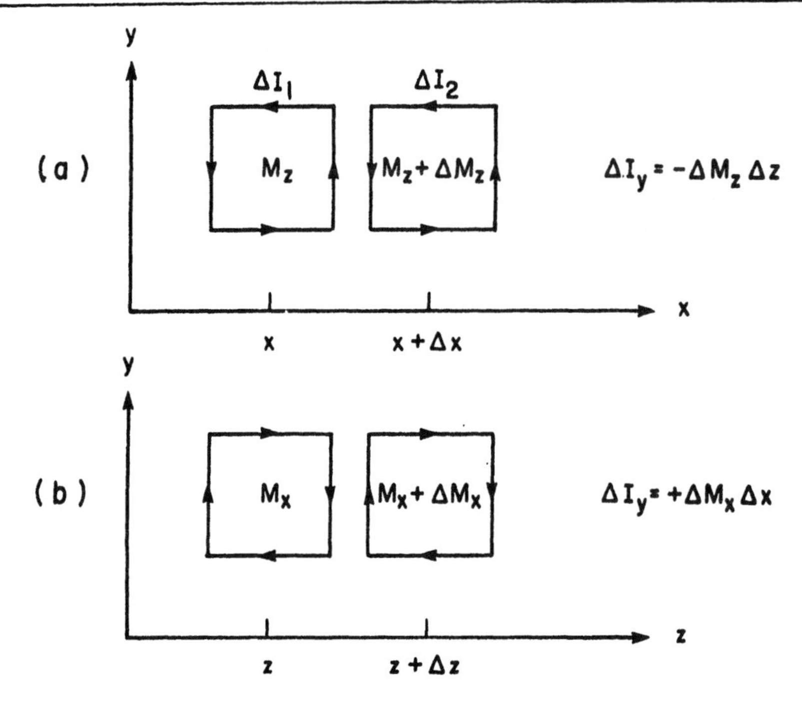

Fig. 6.8 Geometry for the case of inhomogeneous magnetization, with components of $\vec{M}$ along the z axis (above) and x axis (below). Square Ampèrian current loops are used for simplicity.

Because of the inhomogeneity in M_z, the two adjacent currents do not cancel but instead give rise to a net surface current in the y direction of magnitude $I_y(\text{mag}) = \Delta I_1 - \Delta I_2$. Now, ΔI is is related to M_z through Eq. 6.20, so that

$$
\begin{aligned}
I_y(\text{mag}) &= \Delta I_1 - \Delta I_2 \\
&= M_z \Delta z - [M_z + \Delta M_z]\, \Delta z \\
&= -\Delta M_z \Delta z.
\end{aligned}
\tag{6.22}
$$

We can write ΔM_z as the derivative of M_z with respect to its spatial variation x, that is $\Delta M_z = [\partial M_z/\partial x]\,\Delta x$ with the result that

$$
I_y(\text{mag}) = -\left(\frac{\partial M_z}{\partial x}\right) \Delta x \Delta z. \tag{6.23}
$$

Since $I_y(\text{mag})/\Delta x \Delta z$ is just the average volume current density $J_y(\text{mag})$,

$$J_y(\text{mag}) = -\left(\frac{\partial M_z}{\partial x}\right). \qquad (6.24)$$

However, this is not the sole contribution to the current density in the y direction. As is seen from Fig. 6.8b, a z-directed inhomegeneity in the component M_x gives an additional contribution to $J_y(\text{mag})$ equal to $+(\partial M_x/\partial z)$, so that the total current in the y direction becomes

$$J_y(\text{mag}) = \frac{\partial M_x}{\partial z} - \frac{\partial M_z}{\partial x}. \qquad (6.25)$$

We note that the right-hand side of the equation has the structure of the curl of a vector, specifically the y component of the curl. Indeed, when one works out the expressions for the remaining currents $J_x(\text{mag})$ and $J_z(\text{mag})$ and combines these with Eq. 6.25, one can verify that

$$\boxed{\vec{J}(\text{mag}) = \nabla \times \vec{M} \ \ \text{amp/m}^2} \qquad (6.26)$$

which is the result needed to complete the right-hand side of Eq. 6.17.

6.3 MAXWELL'S MACROSCOPIC EQUATIONS IN THE PRESENCE OF MATTER

We have now established all of the fundamental field equations appropriate to conditions in the presence of matter. They are Eqs. 6.12, 6.13, 6.14, and 6.17. They can be put into a somewhat more elegant form by moving $\nabla \cdot \vec{P}$ and $\nabla \times \vec{M}$ of Eqs. 6.12 and 6.17 to the respective left-hand sides, with the result that

$$\left.\begin{aligned}
&\nabla \cdot \left[\vec{E} + \frac{\vec{P}}{\epsilon_0}\right] = \frac{\rho}{\epsilon_0} && \text{(a)}\\
&\nabla \times \vec{E} = -\frac{\partial \vec{B}}{\partial t} && \text{(b)}\\
&\nabla \times [\vec{B} - \mu_0 \vec{M}] = \mu_0 \epsilon_0 \frac{\partial}{\partial t}\left[\vec{E} + \frac{\vec{P}}{\epsilon_0}\right] + \mu_0 \vec{J}(\text{cond}) && \text{(c)}\\
&\nabla \cdot \vec{B} = 0. && \text{(d)}
\end{aligned}\right\} \qquad (6.27)$$

It has become a tradition, ever since the days of Maxwell and the other early researchers to pretty up the equations even further by defining two new vectors

$$\vec{D} = \epsilon_o \vec{E} + \vec{P}$$

and

$$\vec{H} = [(\vec{B}/\mu_o) - \vec{M}]$$

and thus rid the equations of $\vec{P}$ and $\vec{M}$. Even today, most text books travel that route. However, this road has been fraught with misunderstanding and some confusion and for that reason we shall keep the equations as shown — in terms of the physical quantities as we understand them today: the macroscopic electric and magnetic fields $\vec{E}$ and $\vec{B}$, and the macroscopic polarization and magnetization $\vec{P}$ and $\vec{M}$.

We have worked hard in arriving at Eqs. 6. 27 and one may well ask what it is that we have gained. In bringing in the unknown vectors $\vec{P}$ and $\vec{M}$, it looks, on the surface, as if we added to our problems. The answer is that $\vec{P}$ is a function of $\vec{E}$ and $\vec{M}$ is a function of $\vec{B}$ and that the functional relationships

$$\vec{P} = f(\vec{E}) \qquad (6.28)$$
$$\vec{M} = \psi(\vec{B})$$

are intrinsic to the material in question. Therefore, if we know enough about the material, either from experiment or from the microscopic theory, to determine these relationships, $\vec{P}$ and $\vec{M}$ can be eliminated from Eqs. 6. 27 and solutions for the $\vec{E}$ and $\vec{B}$ fields obtained. In general, the relationships (6. 28) can be extremely complicated. For example, when $\vec{E}$ is very large (of the order of 10^8 V/cm) $\vec{P}$ may not be proportional to $\vec{E}$, as might be expected instinctively, but it may have a power dependence of the form $\vec{P} = a\vec{E} + b\vec{E}^2 + c\vec{E}^3 + \ldots$. (For a discussion of this problem see sections 1. 5 and 1. 6.) In ferromagnetic materials $\vec{M}$ is not linearly related to $\vec{B}$ even in very weak magnetic fields and, moreover, $\vec{M}$ depends on the previous history of achieving the magnetization.

When we exclude large electric and magnetic fields and

ferromagnetic substances and deal with so-called linear systems, a profound simplification results, in that Maxwell's equations (6.27) now become linear, partial differential equations and these are the kind of equations we know very well how to handle (for example, superposition of fields now becomes applicable). But we must still be a little careful. When simple substances (gases, amorphous glasses, many liquids, etc.) are subjected to a weak electric field, not only is the magnitude of $\vec{P}$ proportional to the magnitude of $\vec{E}$, but the direction of $\vec{P}$ is exactly lined up with the direction of $\vec{E}$. Such substances are said to be isotropic. In more complicated, so-called anisotropic substances, as, for example, crystals like quartz or calcite, $\vec{P}$ does not line up with $\vec{E}$. But, this should not be considered as too surprising. A crystal presents different electronic configurations along its different crystallographic axes and when an electric field is applied along these axes, the "stiffness" to perturbation characterized by the quantity $\vec{P}$ is expected to be different along each of them. This has the consequence that when $\vec{E}$ is in some arbitrary direction, the proportionality between $\vec{P}$ and $\vec{E}$ is not a simple number, a scalar, but rather a set of numbers. These numbers express the linear dependence of one component of $\vec{P}$ on all three components of $\vec{E}$, namely, $P_x = \alpha_{xx}E_x + \alpha_{xy}E_y + \alpha_{xz}E_z$ and so on for the remaining components of the vector $\vec{P}$. The coefficients α are the elements of a tensor. A similar tensor relationship can exist between $\vec{M}$ and $\vec{B}$. However, the subject of fields in anisotropic matter lies outside the scope of this book and we shall pursue it no further.

In summary, then, the extent of our discussion so far is the following. In linear, isotropic matter $\vec{P} = \alpha\vec{E}$ and $\vec{M} = \beta\vec{B}$ where the proportionality constants α and β are simple scalar quantities intrinsic to the material being considered. When $\vec{E}$ is a steady field, $\vec{P}$ is a steady response, proportional to $\vec{E}$ and aligned with it. This is also true for the relation between $\vec{M}$ and $\vec{B}$. When $\vec{E}$ varies slowly in time this picture is still expected to be all right, at least approximately. But it cannot be true for any kind of time variation. Matter cannot respond instantaneously to an arbitrarily fast change of $\vec{E}$ – there is, after all, inertia even of the light electrons. We therefore expect that $\vec{P}(t)$ as determined at a time t will not only depend on $\vec{E}$ at time t but will also be a function of $\vec{E}$ at all previous times which we shall call t'. In other words, what we are saying is that the response $\vec{P}$ very likely depends on the history of the event.* We can readily put these words into useful mathematical form. $\vec{P}(t)$ depends linearly on all $\vec{E}(t')$ that existed at all earlier times $t' \leq t$. The constant of proportionality α must clearly be a function of time also, and in particular, it must be a function of the time difference $(t-t')$ between the application of the perturbation at t' and the response observed at t. Thus we write that $\alpha = \alpha(t-t')$. Since the substance is assumed to be the same at all times, α cannot depend on absolute time; it would however depend on absolute time if for instance the perturbing electric field burned holes in the material or ionized some of the atoms, and so forth. Hence $\vec{P} = \Sigma_{t'} \alpha(t-t') \vec{E}(t') \Delta t'$ where the summation is over all the previous time intervals $\Delta t'$. Replacing the summation by an integration we get

$$\begin{aligned} \vec{P}(t) &= \int_{t'=-\infty}^{t} \alpha(t-t') \vec{E}(t') \, dt' \\ &= \int_0^{\infty} \alpha(\tau) \vec{E}(t-\tau) \, d\tau. \end{aligned} \tag{6.29}$$

The second form of the equation is merely a mathematical manipulation; it is a little trick in which we change variables, by making the substitution $t - t' = \tau$. In this way we eliminate t' in favor of the dummy variable τ.

We seem to be faced with pretty unpleasant prospects: when we substitute for $\vec{P}$ from Eq. 6.29 in Maxwell's equations (6.27) and then demand to know the electric (and magnetic) fields, say, now, at time t, we must trace their historical development and first find $\vec{E}$ and $\vec{B}$ at all earlier times $t' < t$. These are indeed the facts, when matter is perturbed by some general, rapidly varying electromagnetic field. Fortunately, when the perturbations are <u>sinusoidal</u> (but of arbitrary rapidity) the problem simplifies dramatically and all unpleasantness

*Our discussion will apply equally well to the response of $\vec{M}$ to $\vec{B}$ as long as we deal with linear, nonferromagnetic material.

vanishes. The reason is clear. We know from Fourier analysis of section 2.5 (Eq. 2.99) that a purely sinusoidal oscillation is one that has gone on for ever and ever. Therefore, the atoms exposed to the field oscillations have always vibrated, which is enough time for all transient responses to have died out. Therein lies the root of the simplification. We shall illustrate this right now.

Let us write the sinusoidal electric field in complex notation as

$$\vec{E}(t) = \vec{E}(x, y, z)\, e^{j\omega t} \tag{6.30}$$

where ω is the angular frequency of the oscillations. Then, $\vec{E}(t-\tau) = \vec{E}(x, y, z) \exp[j\omega(t-\tau)]$ and inserting this form under the integral of Eq. 6.29 yields

$$\begin{aligned} \vec{P}(t) &= \vec{E}(x, y, z) \int_0^\infty \alpha(\tau)\, e^{j\omega(t-\tau)}\, d\tau \\ &= \vec{E}(x, y, z)\, e^{j\omega t} \left[\int_0^\infty \alpha(\tau)\, e^{-j\omega\tau}\, d\tau\right]. \end{aligned} \tag{6.31}$$

The second equation follows directly from the first and is a consequence of the fact that the quantity $\exp[j\omega t]$ is not a variable of integration and can be pulled out from under the integral sign. But now we see that $\vec{P}(t)$ at time t is in fact proportional to $\vec{E}(x, y, z) \exp(j\omega t)$ which is just the total electric field $\vec{E}(t)$ at the <u>same</u> time t. The factor in square brackets is the new proportionality constant and we denote it $a(\omega)$; it is itself time independent. Thus we have the simple result

$$\vec{P}(t) = a(\omega)\, \vec{E}(t) \tag{6.32}$$

and a corresponding result for magnetized matter

$$\vec{M}(t) = b(\omega)\, \vec{B}(t). \tag{6.33}$$

In writing $a(\omega)$ rather than a, and $b(\omega)$ rather than b we try to stress here that the new coefficients are in general expected to be frequency dependent, a fact which follows directly from an inspection of the integral contained in the square brackets of Eq. 6.31. Indeed, we shall find in a later section that the coefficients are not only frequency dependent but they may also be complex functions of frequency — that is, they can have both real and imaginary parts. Thus we have shown that matter subjected to a weak sinusoidal electromagnetic perturbation of arbitrarily high frequency responds in a simple linear fashion to the perturbation. The coefficients of proportionality are what the experimenter wants to measure and the theorist wants to deduce from an appropriate microscopic theory of matter; the coefficients are generally frequency-dependent and often complex. The frequency can, of course, be zero and we then talk of the dc values of $\vec{P}$ and $\vec{M}$, that is, values obtained in response to steady electric and magnetic fields.

By tradition the proportionalities shown in Eqs. 6.32 and 6.33 are generally written in a slightly modified, though physically equivalent way,

$$\vec{P}(t) = \epsilon_o[\kappa_e(\omega) - 1]\,\vec{E}(t)$$
$$\vec{M}(t) = \frac{1}{\mu_o}\left[1 - \frac{1}{\kappa_m(\omega)}\right]\vec{B}(t) \qquad (6.34)$$

which introduces into the equations two dimensionless numbers $\kappa_e(\omega)$ and $\kappa_m(\omega)$. The dimensionless coefficient $\kappa_e(\omega)$ is known as the dielectric constant,* sometimes referred to as the "relative permittivity" of the medium; for vacuum κ_e equals unity so that $\vec{P}(t)$ becomes identically equal to zero. The dimensionless coefficient $\kappa_m(\omega)$ is the "relative permeability." It is unity for vacuum, greater than unity for paramagnetic (and ferromagnetic) substances, and less than unity for diamagnetic materials. Table 6.1 lists the experimentally determined values of κ_e and κ_m of a few substances. Observe the weak frequency dependence exhibited by some materials and the extremely large frequency variations shown by others. The results are taken from the Handbook of Chemistry and Physics and from Dielectric Materials and Applications, A. R. von Hippel, Editor (M.I.T. Press, 1954).

The electric and magnetic susceptibilities, χ_e and χ_m, respectively, are additional parameters that can be defined from Eqs. 6.34.

*When the first Eq. 6.34 is inserted in Eq. 6.8 we get that $C = \kappa_e C_o$. This is just a statement of the well-known fact that the capacity is proportional to the dielectric constant of the material filling the capacitor.

Table 6.1 The dielectric constant κ_e and the relative permeability κ_m of a few substances at different frequencies $\omega/2\pi$ hertz.

		$\omega/2\pi$ (Hz)				
		0	10^5	10^8	10^{10}	5×10^{14}
Fused Quartz (S_iO_2)	κ_e	3.78	3.78	3.78	3.78	2.14
	κ_m	1	1	1	1	1
Sodium Chloride (rock salt)	κ_e	5.9	5.9		5.9	2.38
	κ_m	1	1		1	1
Water	κ_e	78.5	78.2	77.5	55	1.77
(at 25°C)	κ_m	1	1	1	1	1

		$\omega/2\pi$ (Hz)				
		10^3	10^5	10^7	10^8	10^{10}
Ferrite	κ_e	9.82	9.13	8.87	8.5	8.5
(Ferramic A)	κ_m	19.6	19.5	24.6	7.2	1.0
Ferrite	κ_e	12000	930	26	13.6	
(Ferramic I)	κ_m	890	890	215	10.7	

These are

$$\vec{P}(t) = \epsilon_o \chi_e \vec{E}(t)$$
$$\vec{M}(t) = \frac{\chi_m}{\mu} \vec{B}(t) \tag{6.34a}$$

from which it immediately follows that the susceptibilities and the relative permittivity (dielectric constant) and relative permeability are related as

$$\epsilon = \epsilon_o(1+\chi_e) \qquad \mu = \mu_o(1+\chi_m)$$

$$\kappa_e = \frac{\epsilon}{\epsilon_o} = 1+\chi_e \qquad \kappa_m = \frac{\mu}{\mu_o} = 1+\chi_m. \tag{6.34b}$$

In materials that are electrically conducting, the current generated by the local electric field is given by Ohm's law. If the conduction current varies linearly with field and if the field is sinusoidal, Ohm's law can be written in the form

$$\vec{J}(t;\text{cond}) = \sigma(\omega)\vec{E}(t) \tag{6.35}$$

where $\vec{J}$ is the current density (in amp/m^2) and $\sigma(\omega)$ $(\text{ohm}-\text{m})^{-1}$ is the conductivity that may well be frequency dependent; our discussion concerning the frequency dependence of $\vec{P}$ and $\vec{M}$ also applies to σ.

Table 6.2 The dc conductivities ($\omega = 0$) of several substances. [The measurements were made at room temperature — σ for most substances is very temperature-dependent.]

	Substance	Conductivity σ $(\text{ohm-m})^{-1}$
Conductors	Silver	6.14×10^7
	Copper (annealed)	5.80×10^7
	Plasma of 1 keV temperature	$\sim 3 \times 10^7$
	Steel (304 stainless)	1.38×10^6
	Mercury	1.04×10^6
Semiconductors	Carbon (graphite)	3.3×10^4
	Germanium	2×10^{-1}
	Silicon	6×10^{-3}
Dielectrics	Water (distilled)	2×10^{-4}
	Glass	2×10^{-12}
	Quartz (fused)	2×10^{-17}

Table 6.2 lists the dc conductivities of some representative materials. All substances exhibit conductivity to some degree but the range of observed values is indeed tremendous as one goes from the so-called "conductors" to the "dielectrics." Note that a hot plasma has an electrical conductivity nearly as good as that of copper.

We are now in a position to write Maxwell's equations in the final form that will prove to be the most useful to us. Inserting Eq. 6.34 and 6.35 in Eqs. 6.27 and eliminating from the latter $\vec{P}$, $\vec{M}$, and $\vec{J}$(cond) in favor of κ_e, κ_m, and σ we obtain the following elegant set of equations

$$\left.\begin{aligned} &\nabla \cdot (\kappa_e \vec{E}) = \frac{\rho}{\epsilon_o} && \text{(a)} \\ &\nabla \times \vec{E} = -\frac{\partial \vec{B}}{\partial t} && \text{(b)} \\ &\nabla \times \left(\frac{\vec{B}}{\kappa_m}\right) = \mu_o \epsilon_o \kappa_e \frac{\partial \vec{E}}{\partial t} + \mu_o \sigma \vec{E} && \text{(c)} \\ &\nabla \cdot \vec{B} = 0 && \text{(d)} \end{aligned}\right\} \quad (6.36)$$

which express in differential form the electromagnetic field ($\vec{E}$, $\vec{B}$) solely in terms of the three physical constants of the matter in question: κ_e, κ_m, and σ. Note that we have left κ_e and κ_m under the operator sign ∇. This allows for the possibility that the substance may be inhomogeneous, in which case κ_e, κ_m, and σ would be functions of position (x, y, z). Finally, we wish to reiterate that if κ_e, κ_m, and σ can be assumed to be frequency independent (usually they cannot except at low frequencies or some narrow band of frequencies), then $\vec{E}$ and $\vec{B}$ can be allowed to have any kind of time variation — for example, they can be in the form of a pulse or series of pulses of arbitrary shape. If, however, κ_e κ_m, and σ are functions of frequency then $\vec{E}$ and $\vec{B}$ of Eqs. 6.36 must be sinusoidal. Since $\vec{E}$ varies as $\exp[j\omega t]$, $\partial \vec{E}/\partial t$ of Eq. 6.36c is in fact

$$\frac{\partial \vec{E}(x, y, z, t)}{\partial t} = j\omega \vec{E}(x, y, z)\, e^{j\omega t} \qquad (6.37)$$

which is also the case for $\partial\vec{B}/\partial t$ of Eq. 6.36b.

6.4 PROPAGATION OF PLANE WAVES IN MATTER – GENERAL CONSIDERATIONS

In Chapter 3 we saw that the basic feature of Maxwell's equations for the electromagnetic field is the existence of traveling wave solutions which represent transport of energy in vacuum from one point to another. The modified Maxwell equations (6.36), modified to accommodate the presence of matter, also exhibit this basic feature. But, the propagation characteristics are now changed and it is our purpose here to discuss the changes. To proceed with our task, we suppose that the medium is infinite in extent and uniform throughout. A plane linearly polarized electromagnetic wave

$$\vec{E}_x = \hat{x}E_{ox}\, e^{j\omega t - jkz}$$
$$\vec{B}_y = \hat{y}B_{oy}\, e^{j\omega t - jkz} \tag{6.38}$$

is taken to propagate in the z direction. If the medium were vacuum ($\kappa_e = \kappa_m = 1$) one knows that the amplitude of $\vec{B}$ is related to the amplitudes of $\vec{E}$ and that the frequency ω is related to the propagation constant k as

$$\frac{B_{oy}}{E_{ox}} = \sqrt{\mu_o \epsilon_o} = \frac{1}{c}$$
$$\omega = \frac{k}{\sqrt{\mu_o \epsilon_o}} = ck. \tag{6.39}$$

To obtain corresponding formulas for waves in matter, proceed as follows: substitute Eqs. 6.38 in Eqs. 6.36b and 6.36c, perform the designated differentiations with respect to time and coordinates (using, say, Eq. 3.52 for the curl) and find that

$$kE_{ox}\, e^{j\omega t - jkz} = \omega B_{oy}\, e^{j\omega t - jkz}$$
$$jkB_{oy}\, e^{j\omega t - jkz} = [j\omega\mu_o\epsilon_o\kappa_e\kappa_m + \mu_o\kappa_m\sigma]\, E_{ox}\, e^{j\omega t - jkz}. \tag{6.40}$$

The first equation gives immediately the ratio of amplitudes,

$$\frac{B_{oy}}{E_{ox}} = \frac{k}{\omega}. \tag{6.41}$$

The second equation (6.40) can be combined with Eq. 6.41, E_{ox} and B_{oy} eliminated, and the following result found:

$$k^2 = \omega^2 \mu_o \epsilon_o \kappa_e \kappa_m - j\omega\mu_o \kappa_m \sigma. \tag{6.42}$$

This is the dispersion relation giving k in terms of ω; knowledge of κ_e, κ_m, and σ yields virtually all one wishes to know about the wave. It is immediately apparent that k must be complex as Eq. 6.42 has both real and imaginary parts. Thus we write

$$k = k_r - jk_i, \tag{6.42a}$$

substitute in Eq. 6.42 and equate separately the real and imaginary parts to get

$$k_r^2 - k_i^2 = \frac{\omega^2 \kappa_e \kappa_m}{c^2}$$

$$2k_r k_i = \omega \mu_o \kappa_m \sigma$$

where we have explicitly assumed (for the moment) that all the parameters have only real parts. Solving these equations for k_r and k_i yields

$$\begin{aligned} k_r &= \frac{\omega}{c}\sqrt{\frac{\kappa_e \kappa_m}{2}}\left[\left\{1 + \left(\frac{\sigma}{\omega\epsilon_o\kappa_e}\right)^2\right\}^{1/2} + 1\right]^{1/2} \\ k_i &= \frac{\omega}{c}\sqrt{\frac{\kappa_e \kappa_m}{2}}\left[\left\{1 + \left(\frac{\sigma}{\omega\epsilon_o\kappa_e}\right)^2\right\}^{1/2} - 1\right]^{1/2} \end{aligned} \tag{6.42b}$$

These are rather cumbersome expressions and we shall be interested in several limiting cases to be discussed in subsequent sections. One such limiting case we point out now: when σ goes to zero we find

$k_r = (\omega/c)\sqrt{\kappa_e\kappa_m}$ and $k_i = 0$. Thus k becomes real when $\sigma = 0$, a result easily anticipated from Eq. 6.42.

The significance of an imaginary component to the propagation constant has profound consequences on the electromagnetic fields. Combining Eqs. 6.38 and 6.42a we see that the fields now decay as $e^{-k_i z}$ as they propagate along z. Thus we conclude that the imaginary part of the propagation constant leads to absorption of the electromagnetic wave in its passage through matter. Note: our choice of signs in Eqs. 6.42a and 6.42b has been such as to exclude the case of a wave exponentially growing as it propagates along z.

We have encountered a complex propagation constant k in our discussion of propagation in a waveguide but there k was either entirely real, corresponding to propagation along the waveguide, or entirely imaginary, corresponding to no propagation (cf. Eqs. 5.65 and 5.78). We now see that the presence of matter can lead to the concept of a k with both real and imaginary parts.

One might conclude from this discussion that the absorption of electromagnetic energy is due entirely to the conductivity of the material since $k_i \to 0$ as $\sigma \to 0$. However, dielectric materials may be modeled quite well by assuming them to have zero conductivity, yet experiments show that there is still attenuation of a wave in traversing a dielectric. Has our theory failed us? No, it has not, and the reason can be traced back to our assumption that the electrical parameters, κ_e, κ_m, and σ had only real parts. It can be seen from Eq. 6.42 that even if σ is zero the propagation constant k will have an imaginary part if, for example, the dielectric constant κ_e is complex. This is exactly the case in dielectrics and it arises because of the frictional forces suffered by the oscillating charges.

We can explain this a little differently as follows. Friction, as represented by the coefficient β of Eq. 6.1, is the ultimate cause of all wave damping. And it is this frictional coefficient that can make its way into Maxwell's macroscopic equations in one of two ways. Either we model matter by assuming the polarization $\vec{P}$ and hence κ_e to be complex and let σ be identically equal to zero, thus allowing for the presence of β in the complex value of κ_e, or we view κ_e as purely real and account for friction by allowing σ to be finite. The final result of course is independent of the model we adopt, because we are merely putting all the physics concerning the microscopic electron motions we know into the right-hand side of Eq. 6.27c, and it does not really matter whether we put our knowledge into the term $\vec{P}$ or the term $\vec{J}$ (or into both as is sometimes done).

Our comments concerning friction apply equally well to every detail of the electron's motion. We know that an oscillating charge can be viewed in one of two ways, as an oscillating dipole or as an oscillating current. When we take the viewpoint that an oscillating charge is a dipole, we then insert this information into Maxwell's macroscopic equation through the polarization $\vec{P}$ and the dielectric coefficient κ_e. This is the customary procedure for dielectrics in which the oscillating charges bound to the lattice points are better pictured as dipoles. On the other hand, when we take the viewpoint that an oscillating charge creates a current, we then insert this information into Maxwell's macroscopic equation via the current density $\vec{J}$ and the conductivity σ. This is the typical procedure adopted for metals and plasmas which contain a large number of free (conduction) electrons. In semiconductors, however, the contribution from the

Table 6.3 Modeling of nonmagnetic matter with κ_e and σ. (κ_m is taken to be unity.)

Matter	κ_e	σ
Good dielectrics	complex; $\text{Im}\,\kappa_e \ll \text{Re}\,\kappa_e$	0
Good conductors	–	complex; usually $\text{Im}\,\sigma \ll \text{Re}\,\sigma$
Plasmas	–	complex; $\text{Im}\,\sigma \gtrless \text{Re}\,\sigma$
Semiconductors	contribution from lattice electrons	contribution from free carrier (conduction) electrons

lattice electrons is as important as the contribution from the conduction electrons and here we insert the dynamics of the motions both into $\vec{P}$ and $\vec{J}$. Table 6.3 summarizes the conventional procedure of modeling matter. Now that we understand the overall picture we shall proceed with details of wave propagation in matter. We begin with dielectrics.

6.5 WAVES IN DIELECTRICS

In a perfect dielectric (if such a substance existed) not only is σ identically zero, but the coefficients κ_e and κ_m are purely real numbers. Hence Eq. 6.42 becomes $k^2 = \omega^2 \mu_o \epsilon_o \kappa_e \kappa_m$, from which we deduce that the phase velocity ω/k of the wave is

$$v = \frac{\omega}{k} = \frac{1}{\sqrt{\mu_o \epsilon_o \kappa_e \kappa_m}}. \qquad (6.43)$$

Since most dielectrics have virtually no magnetic properties, κ_m is very close to unity, and $v \approx [\mu_o \epsilon_o \kappa_e]^{-1/2}$, showing that knowledge of the dielectric constant κ_e suffices to determine the speed of the wave. Now, the ratio of the propagation velocity in vacuum to the propagation velocity in the medium defines the familiar "index of refraction" $n = c/v$. But, $c = (\mu_o \epsilon_o)^{-1/2}$ and insertion of this quantity in Eq. 6.43 yields a most remarkable relationship

$$n = \sqrt{\kappa_e \kappa_m} \approx \sqrt{\kappa_e} \qquad [\kappa_m \approx 1] \qquad (6.44)$$

between the electric and magnetic responses of matter as defined by κ_e and κ_m on the one hand and the optical constant n on the other hand. Thus an experimental or theoretical determination of κ_e (and κ_m) yields the refractive index, with one proviso, that this determination be made at the same frequency ω as the frequency at which knowledge of n is desired. The reason is, of course, that κ_e is usually frequency-dependent as is shown in Table 6.1.

To exhibit the frequency dependence of $\kappa_e(\omega)$ theoretically, and thus also of $v(\omega)$ and of $n(\omega)$, we adopt the simple classical model described at the beginning of this chapter (and also in section 1.5). We single out one atom of a tenuous substance, a gas; dense matter will be treated later. The equation of motion of one of its electrons is then given by Eq. 6.1 with $\vec{E}$ equal to $\vec{E}(x, y, z) \exp[j\omega t]$. The magnetic force $q(d\vec{r}/dt) \times \vec{B}$ is small compared with the electrical force $q\vec{E}$ since the velocity $d\vec{r}/dt$ acquired by the particle is tiny compared with c, a fact which can be proved a postiori once one has solved for the particle motion with $q(d\vec{r}/dt) \times \vec{B}$ neglected. The resulting differential equation is solved by proceeding in the manner described in some detail in section 1.4. Postulate a trial solution of the form $\vec{r} = \vec{r}_o \exp[j\omega t]$ and insert it in the differential equation (6.1). The resulting steady-state motion (after transients have died out) is

$$\vec{r} = \frac{(q/m)}{\left(\omega_o^2 - \omega^2\right) + j\beta\omega} \vec{E}_o(x, y, z)\, e^{j\omega t}. \tag{6.45}$$

The induced dipole moment $\vec{p}$ is then given by $\vec{p} = q\vec{r}$, or

$$\vec{p} = \frac{(q^2/m)}{\left(\omega_o^2 - \omega^2\right) + j\beta\omega} \vec{E}_o(x, y, z)\, e^{j\omega t}. \tag{6.46}$$

Now suppose there are N electrons per unit volume. Not all electrons, however, have the same "spring constant" and hence the same natural frequency of oscillation ω_o. Let f_1 be the fraction with frequency ω_{o1}, f_2 be the fraction with frequency ω_{o2}, and so on. The total dipole moment per unit volume is then given by $\vec{P} = \sum_i Nf_i\vec{p}_i$, or

$$\vec{P} = \sum_i \left[\frac{Nf_i(q^2/m)}{\omega_{oi}^2 - \omega^2 + j\beta_i\omega}\right] \vec{E} \tag{6.47}$$

where β_i is the damping factor of the ith group of bound, oscillating electrons. Thus, having determined the constant of proportionality between $\vec{P}$ and $\vec{E}$, we have accomplished our assigned task. In accordance with Eq. 6.34 $\vec{P} = \epsilon_o[\kappa_e(\omega) - 1]\vec{E}$, and comparison of this relation with Eq. 6.47 yields $\kappa_e(\omega)$ directly. Additionally, because $\kappa_e(\omega) = n^2(\omega)$ we also know the refractive index, and since on the basis of Eq. 6.42, $k^2c^2 = \omega^2\kappa_e$, we also know the dispersion characteristics of the wave:

$$\boxed{\begin{aligned} n^2(\omega) &= \kappa_e(\omega) = k^2c^2/\omega^2 \\ &= 1 + \sum_i \frac{Nf_i(q^2/m\epsilon_o)}{\omega_{oi}^2 - \omega^2 + j\beta_i\omega} \end{aligned}} \qquad (6.48)$$

The imaginary term $j\beta_i\omega$ in the denominators of Eqs. 6.47 and 6.48 tells us something that we have anticipated already in our discussion of $\vec{P}$, that $\vec{P}(\omega)$ can be complex due to a complex dielectric constant κ_e. This comes about as a result of the inevitable damping suffered by all systems. It leads to the somewhat strange looking result that $n^2(\omega)$ and therefore the refractive index $n(\omega)$ are complex numbers. What does that really mean? To see the implications we need only to recall that $n(\omega) = kc/\omega$ and refer to the discussion following Eq. 6.42b. We can write

$$n(\omega) = n_r - jn_i = kc/\omega = (c/\omega)(k_r - jk_i) \qquad (6.49)$$

and we see that a complex index of refraction simply implies a complex propagation constant. Insertion of these values into the exponential space-time variation of the wave

$$\begin{aligned} \vec{E}_x &= \hat{x}E_{ox}\, e^{j\omega t - jkz} \\ &= \hat{x}E_{ox}\, e^{-(\omega n_i/c)z}\, e^{j\omega t - j(\omega n_r/c)z} \end{aligned} \qquad (6.50)$$

shows that the imaginary part of n simply represents spatial damping of the wave. The amplitude falls off exponentially with distance as is illustrated in Fig. 6.9 and drops to one $(e)^{th}$ of its value in a distance

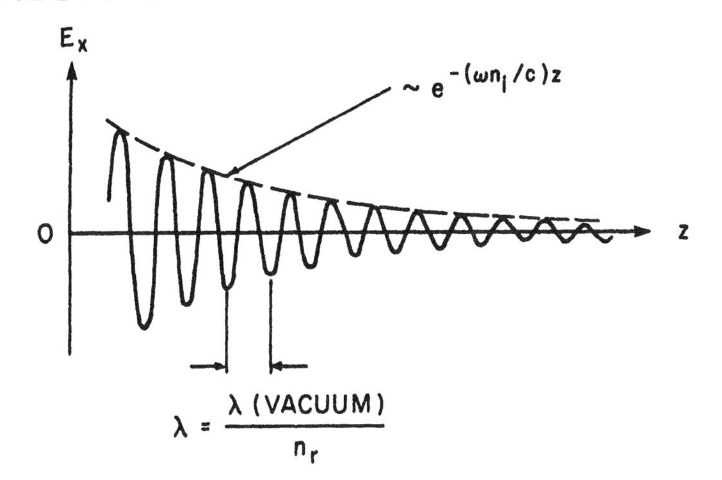

Fig. 6.9 Exponential attenuation of a wave propagating in a lossy ($n_i \neq 0$) dielectric.

$$z_0 = \frac{c}{\omega n_i} \text{ meter.} \tag{6.51}$$

The intensity of the wave varies as $\exp[-(2\omega n_i/c)z]$ and the quantity $(2\omega n_i/c)$ is known as the absorption coefficient. The real part of $n(\omega)$, on the other hand, determines the phase velocity of the wave in accordance with the relation

$$\left.\begin{aligned} v &= \frac{\omega}{k_r} \\ &= \frac{\omega}{(\omega n_r/c)} \\ &= \frac{c}{n_r}. \end{aligned}\right\} \tag{6.52}$$

This points up the fact that it is n_r which is identical to the quantity which we originally called the index of refraction n for the case of zero damping (see Eqs. 6.43 and 6.44).

Eq. 6.48 can be rationalized to obtain the real $(\kappa_e)_r$ and imaginary $(\kappa_e)_i$ parts of the dielectric constant,

$$\kappa_e(\omega) = (\kappa_e)_r - j(\kappa_e)_i$$

$$= 1 + \sum_i \frac{Nf_i(q^2/m\epsilon_o)\left[\omega_{oi}^2 - \omega^2 - j\beta_i\omega\right]}{\left(\omega_{oi}^2 - \omega^2\right)^2 + (\beta_i\omega)^2}. \qquad (6.52a)$$

These can now be related to the real and imaginary parts of the index of refraction as follows:

$$n(\omega) = n_r - jn_i = \sqrt{\kappa_e} = \sqrt{(\kappa_e)_r - j(\kappa_e)_i}$$

$$= \sqrt{\kappa_o\, e^{-j\phi}} = \sqrt{\kappa_o}\, e^{-j\phi/2}$$

$$= \sqrt{\kappa_o}\cos\phi/2 - j\sqrt{\kappa_o}\sin\phi/2,$$

from which we have

$$n_r = \sqrt{\kappa_o}\cos\phi/2 \qquad n_i = \sqrt{\kappa_o}\sin\phi/2$$

$$\kappa_o = \sqrt{(\kappa_e)_r^2 + (\kappa_e)_i^2} \qquad \tan\phi/2 = \frac{(\kappa_e)_i}{(\kappa_e)_r}. \qquad (6.52b)$$

A much simpler approach is possible when κ_e is near unity, as is frequently the case for gases. Then $\kappa_e - 1 \ll 1$ and we can write

$$n = \sqrt{\kappa_e} = \sqrt{1 + (\kappa_e - 1)} \approx 1 + \frac{1}{2}(\kappa_e - 1).$$

This immediately gives us

$$n_r = \frac{1 + (\kappa_e)_r}{2} \qquad n_i = \frac{(\kappa_e)_i}{2}. \qquad (6.52c)$$

Using Eq. 6.52a, we get

$$n_r = 1 + \sum_i \frac{Nf_i q^2}{2m\epsilon_o} \frac{\omega_{oi}^2 - \omega^2}{\left(\omega_{oi}^2 - \omega^2\right)^2 + (\beta_i\omega)^2}$$

$$n_i = \sum_i \frac{Nf_i q^2}{2m\epsilon_o} \frac{\beta_i\omega}{\left(\omega_{oi}^2 - \omega^2\right)^2 + (\beta_i\omega)^2}. \qquad (6.52d)$$

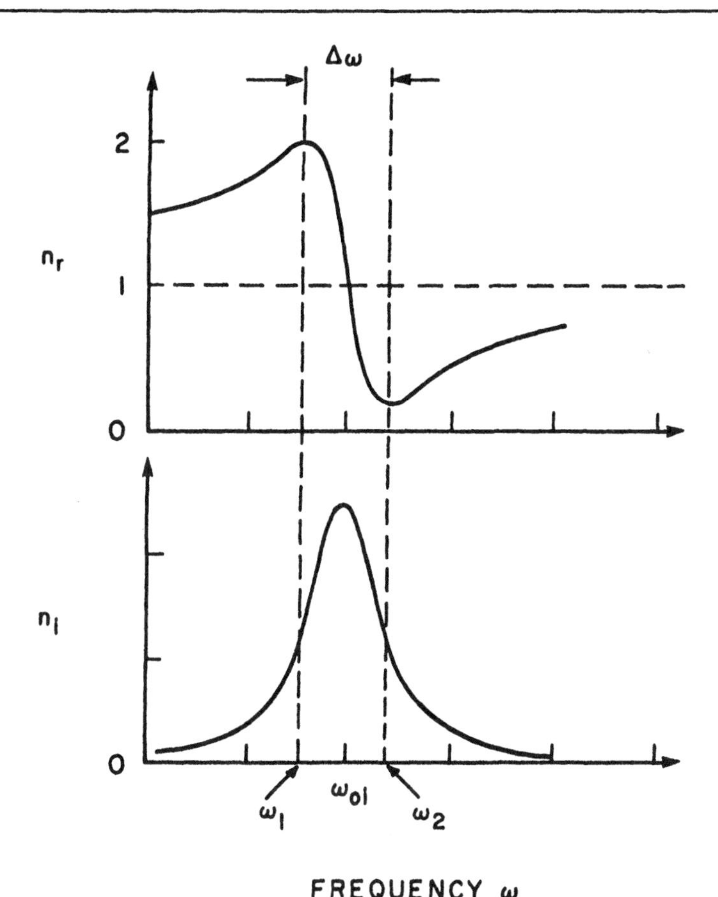

Fig. 6.10 Real and imaginary parts of the refractive index n of a dielectric near one of its resonances ω_{oi}. The damping characterized by n_i is maximum at $\omega = \omega_{oi}$. Note that the real part of n is greater than unity for $\omega < \omega_1$ and less than unity for $\omega > \omega_2$.

In the case of sharp resonances, i.e., small damping, the resonance term in the second equation above can be simplified in the manner used to give Eq. 1.105.

The quantities n_r and n_i are shown plotted in Fig. 6.10 for a single value i in the summation indicated in Eq. 6.52d. We see that at frequencies near the resonant frequency, both n_r and n_i undergo violent changes. In the frequency range labeled $\Delta\omega = \omega_2 - \omega_1$ the index of refraction n_r decreases as the frequency of the wave increases. Since in this region $dn_r/d\omega < 1$, rays of shorter wavelength are refracted less than those of longer wavelength. This results in a reversal of the usual ordering of prismatic colors that would be normally observed if, say, a prism were used to disperse white light. For that reason the term "anomalous dispersion" is used to describe the behavior of the dielectric. Outside the frequency interval $\Delta\omega$ (both at $\omega < \omega_1$ and $\omega > \omega_2$) the dispersive behavior is normal in that a shorter wavelength (blue, say) is refracted more than a longer wavelength (red). Moreover, here damping is small and n is almost purely real.

Figure 6.10 also shows that the anomalous behavior of the refractive index n_r is accompanied by large absorption which becomes maximum at a frequency ω equal to the natural frequency of oscillation ω_{oi}. If the absorption factor β of Eq. 6.52d is large, the medium may, in fact, become quite opaque to radiation in the frequency band $\Delta\omega$. Colorless, transparent gases, liquids, and solids have their characteristic frequencies outside the visible range of the spectrum (which is why they are, in fact, transparent and colorless). For example, hydrogen, oxygen, and air have their resonant frequencies at 3.40×10^{15} Hz, 3.55×10^{15} Hz, and 3.98×10^{15} Hz, respectively, all of which lie in the ultraviolet. Glasses also have their resonances in the ultraviolet region, and for this reason the anomalous phenomena occuring at $\omega \approx \omega_{oi}$ cannot be observed visually in any of these substances. At frequencies well below ω_{oi} and for sufficiently weak damping, Eq. 6.52d, written for a single term of the summation, can be simplified readily to read

$$n_r = A + \frac{B}{\lambda^2} + \frac{C}{\lambda^4} + \dots \qquad [\omega \ll \omega_{oi}] \tag{6.53}$$

where $\lambda = 2\pi c/\omega$ is the wavelength of the radiation, and where A, B, and C are constants involving N, q, m, and ω_o. This formula of Cauchy agrees remarkably well with index of refraction measurements of gases made at optical wavelengths.

The phase and group velocities (Eqs. 5.80 and 5.87) written in terms of the refractive index

$$v = \frac{c}{n_r(\omega)}$$

$$v_g = \frac{c}{[n_r(\omega) + \omega(dn_r/d\omega)]}$$

can be less than, equal to, or greater than c, depending on the magnitude of n_r and on its derivative $dn_r/d\omega$. At optical wavelengths n_r is greater than unity for the majority of substances; here also the dispersion is normal ($dn_r/d\omega > 1$) with the result that $v_g < v < c$. But, in the anomalous region shown in Fig. 6.10, $dn_r/d\omega$ can become large and negative and v_g can exceed both v and c. Note that this does not mean that energy propagates faster than c; rather it implies that v_g is no longer a meaningful concept when $dn_r/d\omega$ and hence $dk/d\omega$ are large; see also discussion in section 5.3 (wave velocities). Figure 6.11 illustrates qualitatively the frequency dependence of v and v_g in the vicinity of the resonance $\omega \sim \omega_{oi}$. Note that at high frequencies

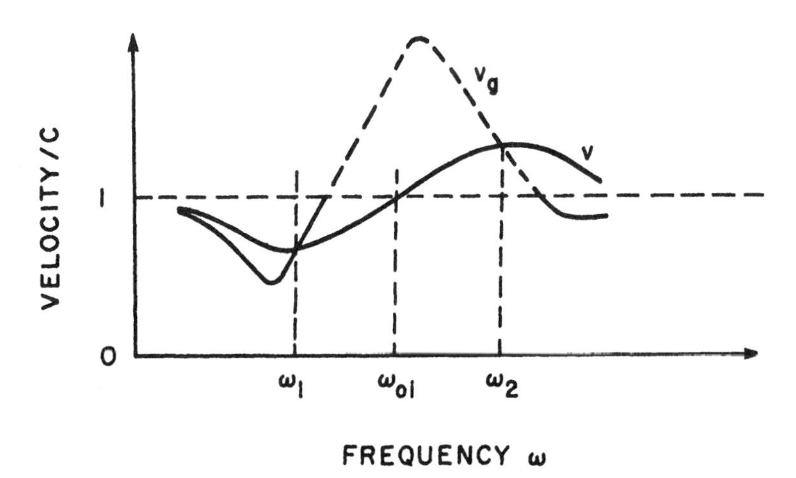

Fig. 6.11 The phase velocity $v = \omega/k$ and the group velocity $v_g = d\omega/dk$ near the region of anomalous dispersion. The frequencies ω_1, ω_{oi}, and ω_2 are the same as those shown in Fig. 6.10. The group velocity v_g is not meaningful at frequencies where it is larger than c (the dotted portions of the curve).

such that $\omega > \omega_0$ (for air this would correspond to a frequency $\omega/2\pi \gtrsim 4 \times 10^{15}$ Hz) the phase velocity is greater than c and therefore the refractive index is less than unity, quite unlike its behavior at optical wavelengths where its index is greater than unity. This has some curious ramifications, for example, how an x ray refracts at a vacuum-medium boundary compared with how a light ray refracts at the same boundary, but we shall leave this problem for the reader to work out.

Figure 6.10 illustrates the dispersion characteristics near a single natural frequency ω_0 of the oscillating atom. In general, there are many such resonances, and around each the indicated behavior repeats, as is shown in Fig. 6.12. For optically transparent substances all of these so-called "electronic resonances" occur, as we pointed out already, in the ultraviolet range. However, there are other groupings of such resonances at infrared and also at microwave frequencies, which are typical of ionic rather than electronic oscillations. Each such group of resonant frequencies is associated with a different effective mass of the microscopic bodies which contribute to the polarization effect. As a rule of thumb, the greater the effective mass, the

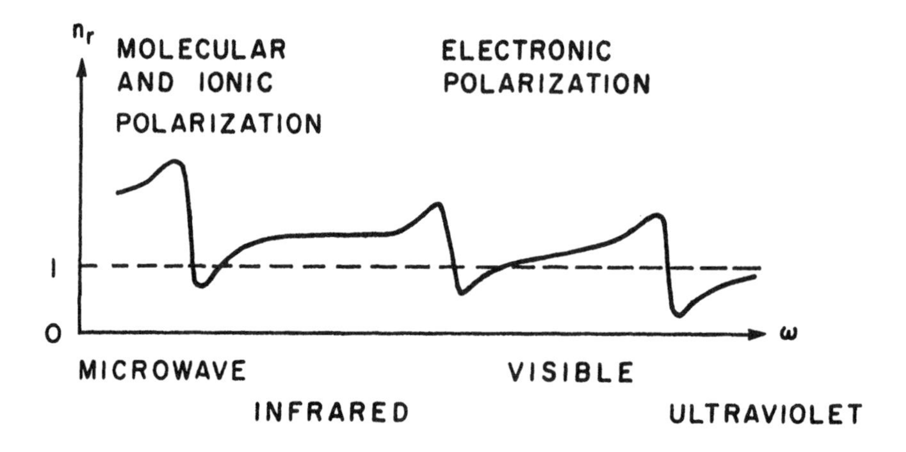

Fig. 6.12 Variation of the refractive index with frequency. In the case of molecules, there are strong contributions to n from the alignment of permanent dipoles in polar substances. These contributions are most prominent at microwave and lower frequencies. That is why water, for example, has such a large n at low frequencies (see Table 6.1).

lower the resonant frequency (see also discussion in section 1.5, Electrical polarizability of atoms).

It is worth pointing out that quantum mechanics yields essentially the same formula as that given by Eq. 6.52d for the dispersion of a gas, with the exception that some of the terms appearing in the equation must be given a different interpretation. Accordingly, each ω_{oi} represents one of the allowed frequencies at which the atom can absorb (or emit) a photon; and f_i is no longer the fraction of electrons associated with the oscillation frequency but rather a factor that tells the strength with which the atoms can exhibit their oscillation. It is termed the "oscillator strength."

One might think that our master equation (6.52d) is good not only for gases but also for dense dielectrics like liquids and solids. But this is not so because the field $\vec{E}$ in Eq. 6.1 acting on a given atom must include the electric field produced by all neighboring atoms, and in dense matter this can be large. To proceed with the problem, imagine then that the medium has become polarized by the presence of an electric field (polar substances having permanent dipole moments are excluded from our consideration); denote the spatially averaged electric field at any point of the dense medium by $\vec{E}$. To find the field $\vec{E}$(site) at the site of a given atom remove the atom from its site, otherwise the tiny imaginary field probe placed there will measure a superposition of the desired $\vec{E}$(site) and the self-field of the polarized particle. This excision will leave a vacuum hole and our task is to calculate $\vec{E}$(site) within it. Note that $\vec{E}$(site) is greater than $\vec{E}$ just because in the hole there are no dipoles that oppose the applied field. If the void left by the atom were in the shape of a long slot oriented perpendicular to the field, then Eq. 6.7 derived for the parallel plate capacitor would give us the answer, which is that $\vec{E}(\text{site}) = \vec{E} + (\vec{P}/\epsilon_o)$. This, however, is an unreasonable assumption. The removal of a dipole from within a random distribution of neighboring dipoles is better approximated by a spherical hole as is illustrated in Fig. 6.13. Calculations, which we shall not attempt to make here, then show that

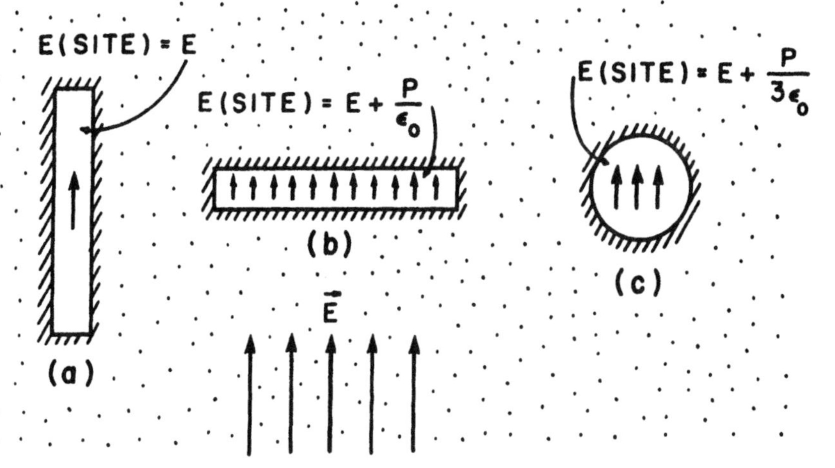

Fig. 6.13 The electric field $\vec{E}$(site) in various shaped holes cut in a uniform dielectric medium containing an electric field $\vec{E}$. When the slot is long and very thin as in (a), virtually no polarization charge is induced on its upper and lower surfaces, and therefore E(site) = E. When the slot is wide as in (b), polarization charge $+\rho_S$ is induced on the bottom surface and $-\rho_S$ of charge is induced along the top surface of the slot. Hence the field inside is, by Gauss' law, E(site) = E + (P/ϵ_0). The case of a spherical hole, (c), gives an E(site) = E + $(P/3\epsilon_0)$ which lies between the results for (a) and (b).

one has, instead, the result*

$$\vec{E}(\text{site}) = \vec{E} + \frac{\vec{P}}{3\epsilon_0}. \tag{6.55}$$

Now, the remaining task is simple. Wherever we find in Eqs. 6.45 through Eqs. 6.47 the field $\vec{E}$, we must replace it by the field $\vec{E}$(site) given by Eq. 6.55. In particular, Eq. 6.47 then becomes

$$\begin{aligned} \vec{P} &= \sum_i \left[\frac{Nf_i(q^2/m)}{\omega_{oi}^2 - \omega^2 + j\beta_i\omega}\right]\left(\vec{E} + \frac{\vec{P}}{3\epsilon_0}\right) \\ &= \chi\left(\vec{E} + \frac{\vec{P}}{3\epsilon_0}\right), \end{aligned} \tag{6.56}$$

*See, for example, Electricity and Magnetism, E. M. Purcell (McGraw Hill Book Company, New York, 1965), sections 9.10, 9.11.

where the symbol χ is simply the entire expression in the brackets. This result can now be rearranged to read

$$\vec{P} = \frac{\chi}{1 - (\chi/3\epsilon_o)} \vec{E} \tag{6.57}$$

demonstrating once again that the polarization $\vec{P}$ is directly proportional to the electric field $\vec{E}$. Comparing this result with the first equation (6.34) yields the dielectric constant in terms of χ, namely $\epsilon_o(\kappa_e - 1) = \chi[1 - (\chi/3\epsilon_o)]^{-1}$. On reshuffling terms it is found that $(\kappa_e - 1)/(\kappa_e + 2) = (\chi/3\epsilon_o)$ and since $\kappa_e = n^2$,

$$\frac{n^2 - 1}{n^2 + 2} = \frac{1}{3} \sum_i \frac{Nf_i(q^2/m\epsilon_o)}{\omega_{oi}^2 - \omega^2 + j\beta_i\omega} \tag{6.58}$$

which is known as the Clausius-Mossotti equation. Observe that for gases where n^2 is not far removed from unity, Eq. 6.58 becomes identical with our previous result (6.48).

Equation 6.58 proves to be remarkably accurate for a wide variety of substances. The right-hand side of the equation is proportional to N, the number of electrons per unit volume, and so it is proportional to the number of atoms per unit volume, and hence to the ratio of mass density ρ to the molecular weight W of the material. This means that

$$\left(\frac{n^2 - 1}{n^2 + 2}\right) \frac{W}{\rho} = \text{constant } A \tag{6.59}$$

where A is known as the molar refractivity. This equation says that if one knows the refractive index n_1 of a substance when its density is ρ_1, one can determine its index n_2 when it is in a different state ρ_2 (that is, when it is expanded or compressed), simply by solving the equation $(A/W) = \left(n_1^2 - 1\right)/\rho_1\left(n_1^2 + 2\right) = \left(n_2^2 - 1\right)/\rho_2\left(n_2^2 + 2\right)$. This works remarkably well, so well that one of the states can be a liquid and the other its vapor. Table 6.4 shows the quite extraordinary constancy of A, despite the tremendous disparities of n or ρ associated with the two states of the substance in question. Observe also that Eq. 6.58 can be used to predict the refractive index of a mixture. Suppose unit

Table 6.4 The molar refractivities A of different substances in their liquid and gaseous states. The measurements of n were made at the wavelength of sodium D light. The damping is so light that $n = n_r$ (where not shown otherwise, the density of the vapor was calculated from the ideal gas law).

Substance		n_r	ρ (gm/cc)	$A = W(n_r^2-1)/\rho(n_r^2+2)$
O_2	liquid	1.221	1.124	4.00
(W = 32)	gas	1.000271	0.001429	4.05
H_2O	liquid	1.334	1.00	3.71
(W = 18)	gas	1.000249	—	3.70
CS_2	liquid	1.628	1.264	21.34
(W = 76)	gas	1.00147	—	21.78

volume of the first substance contains N_1 molecules and of the second substance N_2, then the molar refractivity of the mixture will be

$$A = \frac{N_1A_1 + N_2A_2}{N_1 + N_2}, \tag{6.60}$$

a result which follows directly from the "summability" of terms on the right-hand side of Eq. 6.58.

Example

A plane electromagnetic wave of angular frequency ω is incident normally on a slab of dielectric material of thickness d and dielectric constant κ_e (at frequency ω). Assume the dielectric is lossless, i.e., κ_e is real and σ is zero.

a) At any instant of time, how many wavelengths of the field are there in the dielectric?

b) What is the phase difference $\Delta\phi = \phi_1 - \phi_2$ between a wave which passes through the dielectric and one which did not?

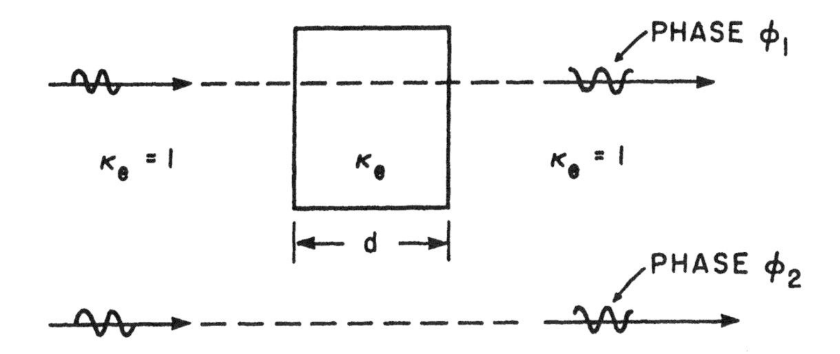

Solution

The propagation constant in the dielectric is given by Eqs. 6.43 and 6.44 as

$$k = \frac{\omega}{v} = \frac{\omega}{c} n = \frac{\omega}{c}\sqrt{\kappa_e} = \frac{2\pi}{\lambda_d}$$

where λ_d is the wavelength in the dielectric

$$\lambda_d = \frac{2\pi c}{\omega\sqrt{\kappa_e}} = \frac{\lambda_o}{\sqrt{\kappa_e}}.$$

In this expression λ_o is the wavelength in vacuum. Hence we have

a) $$\frac{d}{\lambda_d} = \text{number of } \lambda\text{'s in dielectric} = \frac{\sqrt{\kappa_e}\,\omega d}{2\pi c} = \frac{\sqrt{\kappa_e}\,d}{\lambda_o}.$$

The phase difference between a wave that passed through the dielectric and one that did not is given by

$$\Delta\phi = \phi_1 - \phi_2 = 2\pi\left(\frac{d}{\lambda_d} - \frac{d}{\lambda_o}\right).$$

Using this expression, we have

b) $$\Delta\phi = \frac{\omega d}{c}(\sqrt{\kappa_e} - 1) = \frac{2\pi d}{\lambda_o}(\sqrt{\kappa_e} - 1).$$

Example

We have presented two equations for the real part of the index of refraction n_r: Eq. 6.52d based on a classical, microscopic model of bound electrons in molecules, and Eq. 6.53, an empirical relation due to Cauchy which agrees well with experiment, which is valid for $\omega \ll \omega_{oi}$ and for small damping, i.e., $\beta_i \omega \ll \omega_{oi}^2$.

a) Considering only the first two terms in Cauchy's relation (Eq. 6.53), derive the constants A and B in terms of the parameters of Eq. 6.52d. Assume only one resonance frequency in Eq. 6.52d and take $f_i = 1$.

Solution

Equation 6.52d, written for one resonance and with negligible damping, yields

$$n_r = 1 + \frac{Nq^2}{2m\epsilon_o}\frac{1}{\omega_o^2 - \omega^2} = 1 + \left(\frac{Nq^2}{2m\epsilon_o\omega_o^2}\right)\frac{1}{1 - \left(\frac{\omega}{\omega_o}\right)^2}$$

$$n_r = 1 + \left(\frac{Nq^2}{2m\epsilon_o\omega_o^2}\right)\left[1 + \left(\frac{\omega}{\omega_o}\right)^2 + \ldots\right] \qquad \omega \ll \omega_o$$

$$n_r = \left(1 + \frac{Nq^2}{2m\epsilon_o\omega_o^2}\right) + \left(\frac{2\pi^2 Nq^2 c^2}{m\epsilon_o\omega_o^4}\right)\frac{1}{\lambda^2} + \ldots$$

$$n_r = A + \frac{B}{\lambda^2} + \ldots.$$

Thus we can easily identify A and B as follows:

$$A = 1 + \frac{Nq^2}{2m\epsilon_o\omega_o^2}$$

$$B = \frac{2\pi^2 Nq^2 c^2}{m\epsilon_o\omega_o^4}.$$

One could just as easily have obtained an expression for the constant C in Eq. 6.53 by keeping higher order terms in the expansion $[1-(\omega/\omega_o)^2]^{-1}$.

6.6 WAVES IN CONDUCTORS AND PLASMAS

Metals and semiconductors are characterized by the fact that dispersed in between the lattice of fixed positive ions and electrons is, so to speak, a gas of almost freely circulating electrons moving randomly hither and thither. The application of an external electric field induces a general drift of the electrons antiparallel to the field. This results in a current density $\vec{J}$ proportional to the field strength $\vec{E}$, with the conductivity σ being the constant of proportionality, in accordance with Ohm's empirical law $\vec{J} = \sigma\vec{E}$.

Equation 6.42 is the dispersion relation applicable to such a conducting medium. The equation may be simplified somewhat. It is known that for metallic atoms the resonance frequencies of the bound electrons lie far out in the ultraviolet region of the spectrum so that for frequencies in the visible, infrared, and radio, $\omega_{oi} \gg \omega$. This has the consequence that n and hence κ_e associated with the bound electrons approaches unity (see Fig. 6.10). If, moreover, the metal is nonferromagnetic κ_m is also approximately equal to one with the result that Eq. 6.42 becomes

$$k^2 \approx \frac{\omega^2}{c^2}\left[1 - j\,\frac{\sigma(\omega)}{\omega\epsilon_o}\right] \tag{6.61}$$

where the conductivity $\sigma(\omega)$ can in general be complex and can also be a function of the frequency ω. However, we shall see momentarily that at frequencies below about 10^{12} Hz (wavelength $\lambda \gtrsim 3 \times 10^{-2}$ cm) σ is not only real but its magnitude is very well approximated by its dc value as given, for example, in Table 6.2. Thus, for now at least, we take σ of Eq. 6.61 to be a fixed real constant.

The first term on the right-hand side of Eq. 6.61 originates from the displacement current $\epsilon_o\, \partial\vec{E}/\partial t$ and the second term from the conduction current $\sigma\vec{E}$ of the Maxwell formula (6.36c). In regard to wave propagation, therefore, good conduction is synonymous with the requirement that $\sigma/\omega\epsilon_o \gg 1$, an inequality which imposes an upper limit on the frequency of the wave. For a good conductor, like silver, for which $\sigma = 6.14 \times 10^7$ (ohm-m)$^{-1}$, $\omega(\max)/2\pi \ll 1.1 \times 10^{18}$ Hz which is readily satisfied for just about any frequency of interest to us. Even for a poor conductor like carbon for which $\sigma = 3 \times 10^4 (\text{ohm-m})^{-1}$, $\omega(\max)/2\pi \ll 6 \times 10^{14}$ Hz. But note that above approximately 10^{12} Hz, σ is no longer given by its dc value! Assuming that we are in this high conductivity regime, Eq. 6.61 becomes $k^2 \simeq -j(\omega/c)^2 [\sigma/\omega\epsilon_o]$. Taking the square root and noting that $\sqrt{-j} = (1-j)/\sqrt{2}$, we obtain

$$k = \pm(1-j)\frac{1}{c}\sqrt{\frac{\omega\sigma}{2\epsilon_o}} \tag{6.62}$$

showing that k is complex with a large imaginary part. This means damping. The wave propagating as $E_x = E_{ox} \exp[j\omega t - jkz]$ now has the form

$$E_x = E_{ox} \underbrace{\exp\left[-\frac{z}{c}\sqrt{\frac{\omega\sigma}{2\epsilon_o}}\right]}_{\substack{\text{damping}\\ \text{term}}} \underbrace{\exp j\left[\omega t - \frac{z}{c}\sqrt{\frac{\omega\sigma}{2\epsilon_o}}\right]}_{\substack{\text{wave}\\ \text{term}}} \tag{6.63}$$

and therefore an electromagnetic wave entering a conductor is damped to $(1/e)^{\text{th}} = 0.369$ of its initial amplitude in a distance

$$\boxed{\begin{aligned} \delta &= \sqrt{\frac{2\epsilon_o c^2}{\omega\sigma}} \\ &= \sqrt{\frac{2}{\omega\mu_o\sigma}} \end{aligned} \quad \text{meter.}} \tag{6.64}$$

This is precisely the "skin depth" which we discussed in some detail in section 5.1. Observe that the real part of k, $k_r \equiv 2\pi/\text{wavelength}$, is just $1/\delta$ from which we conclude that the amplitude falls to 0.369 of its initial value in a distance $z = \lambda/2\pi$, which is a fraction of a wavelength. It is a rapid falloff, indeed, and one can hardly call this a wave.

The relation between the magnetic field $\vec{B}$ and the electric field $\vec{E}$ is noteworthy. Inserting Eq. 6.62 in Eq. 6.41 and rearranging terms, we obtain

$$\begin{aligned}\frac{B_{oy}}{E_{ox}} &= \frac{1}{c}(1-j)\sqrt{\frac{\sigma}{2\epsilon_o\omega}} \\ &= \left(\frac{B_{oy}}{E_{ox}}\right)_{vacuum} \times (1-j)\sqrt{\frac{\sigma}{2\epsilon_o\omega}}\end{aligned} \tag{6.65}$$

where the second form of the equation comes from the fact that for a plane wave in vacuum $B_{oy}/E_{ox} = 1/c$ (see Eq. 6.39). We therefore see that in a very good conductor $[\sigma/\epsilon_o\omega \gg 1]$, the ratio B/E is large compared to what it is in vacuum and, moreover, B is out of phase* with the electric field by 45°. This is in complete contrast to the situation in vacuum (or in a good dielectric), where the $\vec{E}$ and $\vec{B}$ fields are in phase and their relative amplitudes are given by $B_{oy}/E_{ox} = n/c$.

In one of the simplest microscopic models of conduction in metals due to Drude (1863-1906) the so-called free "conduction electrons" acquire directed motion in the applied field $\vec{E}$. If these electrons were completely free to move, they would keep accelerating and reach arbitrarily large velocities. But this does not occur because they lose forward momentum as a result of several processes, one of the most important being collisions with the lattice ions. Calculations show that if an average electron makes ν collisions per second, it is being deflected sidewise and thus loses $mv\nu$ amount of forward momentum per second. This constitutes a frictional force proportional to velocity – a form of damping with which we are well acquainted. The resulting equation of motion

$$\frac{d^2\vec{r}}{dt^2} + \nu\frac{d\vec{r}}{dt} = \frac{q}{m}\vec{E}(\vec{r},t) \tag{6.66}$$

is just Eq. 6.1 with the restoring force $\omega_o^2\vec{r}$ set equal to zero. (There

*This phase angle derives from $(1-j)/\sqrt{2} = \exp[-j\pi/4] = \cos(\pi/4) - j\sin(\pi/4)$.

is no restoring force acting on the free conduction electrons, unlike the case of bound electrons discussed in connection with dielectrics.) The quantity ν is known as the "collision frequency for momentum transfer" and represents an intrinsic quantity of the conductor in question.

In a sinusoidal field $\vec{E}(x,y,z)\exp(j\omega t)$ the electrons acquire an oscillatory displacement which can be obtained by solving Eq. 6.66. Alternatively, by setting $\omega_o = 0$ and $\beta = \nu$ in Eq. 6.45 we find that

$$\vec{r} = \frac{(q/m)}{-\omega^2 + j\nu\omega}\vec{E}(x,y,z)\,e^{j\omega t}. \tag{6.67}$$

Each electron creates an oscillatory current and if there are N such free electrons per unit volume, the net current density is $\vec{J} = Nq\vec{v} = Nq(d\vec{r}/dt)$, or

$$\vec{J} = \left(\frac{Nq^2}{m}\right)\left(\frac{1}{\nu + j\omega}\right)\vec{E}(x,y,z)\,e^{j\omega t}. \tag{6.68}$$

But, by Ohm's law, $\vec{J} = \sigma(\omega)\vec{E}$ and thus the conductivity $\sigma(\omega)$ becomes

$$\sigma(\omega) = \left(\frac{Nq^2}{m}\right)\left(\frac{1}{\nu + j\omega}\right)\ (\text{ohm-m})^{-1} \tag{6.69}$$

which is the basic formula of this theory. We see immediately that when the frequency ω is much smaller than ν, the conductivity

$$\sigma(\text{dc}) = \frac{Nq^2}{m\nu}\ (\text{ohm-m})^{-1} \tag{6.70}$$

becomes independent of frequency. It is in fact just the dc conductivity like that listed in Table 6.2. To see what frequencies this corresponds to, we calculate ν. We know the experimental dc conductivity σ, and the mass m and charge q of an electron. The conduction electrons N are supplied by the "valence electrons" of the substance. For many conductors the valence is unity which says that there is one conduction electron for every metallic atom (approximately). Thus knowledge of the valence, the atomic weight and the

Table 6.5 The estimated empirical collision frequency ν for several conductors.

Substance	σ $(\text{ohm-m})^{-1}$	Valence	N (m^{-3})	ν $(\sec^{-1})$
Silver	6.14×10^{7}	1	5.8×10^{28}	3×10^{13}
Copper	5.80×10^{7}	1	8.5×10^{28}	4×10^{13}
Carbon	3.3×10^{4}	4	4.01×10^{29}	3×10^{17}

mass density of the substance allows one to deduce from Eq. 6.70 the collision frequency ν. This is shown in Table 6.5. The results shown there confirm our earlier statement that at wave frequencies ω below approximately 10^{13} rad/sec ($\sim 1.5 \times 10^{12}$ Hz which lies in the long infrared) a good conductor like silver or copper will exhibit its "low frequency" behavior characterized by the fact that σ is a purely real, frequency-independent number. Carbon will exhibit this behavior even at very much higher frequencies, $\omega \ll 10^{17}$ rad/sec.

Now, what happens when $\omega \gg \nu$, that is when the wave frequency is very much greater than the collision frequency ν? Equation 6.69 tells us that σ becomes purely imaginary:

$$\sigma(\omega) \simeq -j \frac{Nq^2}{m\omega} \tag{6.71}$$

To find out what this means physically, in regard to wave propagation, we must insert this value of σ in the dispersion relation (6.61). This yields

$$\begin{aligned} kc &= \pm \sqrt{\omega^2 - \frac{Nq^2}{m\epsilon_o}} \\ &= \pm \sqrt{\omega^2 - \omega_p^2} \end{aligned} \tag{6.72}$$

where $\omega_p = \sqrt{Nq^2/m\epsilon_o}$ is the so-called plasma frequency; a quantity

Table 6.6 The frequencies above which a metal becomes transparent.

Metal	$\omega_p/2\pi$ (Hz)	Wavelength $\lambda_p = 2\pi c/\omega_p$ (Å)
Silver	2.16×10^{15}	1386
Copper	2.6×10^{15}	1154
Lithium	1.94×10^{15}	1550
Rubidium	9.32×10^{14}	3220

which we met already in section 1.2 (Eq. 1.63). This general form of dispersion is precisely like the one we deduced for waves propagating in a waveguide! Comparing Eq. 5.77 with the above, we see that ω_p plays the same role as the cutoff frequency ω_c plays in waveguide propagation. And so the dispersion curve of Fig. 5.23 tells the story: at wave frequencies $\omega > \omega_p$ the metal permits free propagation of waves without attenuation. The metal becomes transparent and when a wave is incident on it, it behaves like a dielectric and not like a mirror. Table 6.6 gives the plasma frequencies for several metals, showing that this occurs at wavelengths corresponding to the ultraviolet.

As the wave frequency is decreased from $\omega > \omega_p$ to $\omega = \omega_p$, the phase velocity $v = \omega/k$ goes to infinity, the group velocity $v_g = d\omega/dk$ goes to zero, and the wave ceases to propagate. The wave is said to be "cutoff" in the range $\nu \ll \omega \leq \omega_p$. The electric (and magnetic) fields decay exponentially with distance from the surface, as is illustrated in Fig. 5.24. This differs from the low frequency behavior $\omega \ll \nu$ where we found that while the wave is damped (it loses energy) it nonetheless exhibits oscillatory behavior both in space and time. In the cutoff regime, on the other hand, $\nu \ll \omega \leq \omega_p$ there simply is no wave and no energy dissipation, merely a sinusoidal exponentially decaying field. A wave incident from outside cannot penetrate and is totally reflected. The three regimes typical of propagation in conducting media are illustrated schematically in Fig. 6.14.

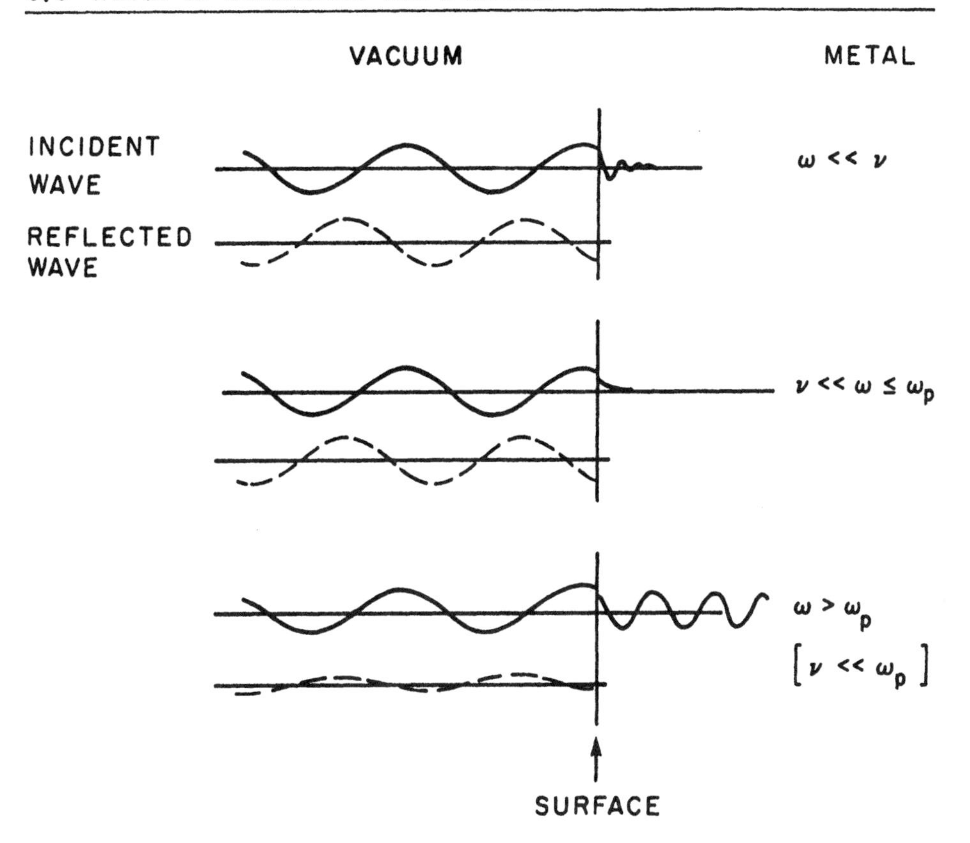

Fig. 6.14 Three possible regimes of waves in a metal. When $\omega \ll \nu$, almost all of the incident energy is reflected. The little that penetrates gets dissipated as heat as a result of electron-lattice ion collisions. When $\nu \ll \omega \leq \omega_p$ all incident energy is reflected; there is no wave in the metal – just exponentially decaying fields. When $\omega > \omega_p$ (but $\nu \ll \omega_p$), the metal is transparent and behaves like a good dielectric. There is a little reflection, as there is from any dielectric. (However, the overall behavior is different in materials like carbon where $\nu \gtrsim \omega_p$.)

We mention that the simple (but very satisfactory) model of a plasma composed of free electrons, ions and possibly atoms is governed by very much the same processes as we have just described for the metal. Equation 6.69 appropriately defines the RF conductivity of a plasma except that now ν defines the collisions of electrons with

Table 6.7 The plasma frequency of some ionized gases. Values of N are only order of magnitude estimates.

Medium	N (m^{-3})	$\omega_p/2\pi$ (Hz)
Interstellar gas	10^6	8.9×10^3
Ionized Nebula	10^9	2.8×10^5
Ionosphere	10^{11}	2.8×10^6
Fluorescent lamp	10^{17}	2.8×10^9
Solar atmosphere	10^{18}	8.9×10^9
Fusion plasma	10^{20}	8.9×10^{10}
Laser-generated plasma (or exploded wire)	10^{26}	8.9×10^{13}

the free ions and atoms of this gaseous medium. Also the electron densities of gas plasmas are much smaller than the density of conduction electrons in the metallic "solid state plasmas" and accordingly ω_p is much smaller. Typical values of some naturally occuring and man-made gas plasmas are given in Table 6.7. From this we see, for example, that radio waves beamed vertically up at the ionosphere will be reflected from it if their frequency is less than about 3 MHz but will pass readily through the ionosphere at frequencies above 3 MHz. Also when a plasma is illuminated with intense radiation of frequency $\omega < \omega_p$ it bounces off as from a solid mirror and imparts pressure to the medium (see section 3.7). When the radiation is applied from all directions, the inward directed force opposes the outwardly directed plasma pressure; the plasma after all wants to expand like any other gas. In this way the plasma could in principle be kept away from the container walls. This method of confinement of hot, fusionlike plasmas has at one time been suggested as a possible schema.

When a spacecraft reenters the Earth's atmosphere, it ionizes the gas around it by collisions with neutral atoms and thus becomes enveloped by a fairly dense plasma with $N \gtrsim 10^{14}\ m^{-3}$ ($\omega_p/2\pi \lesssim 8.9 \times 10^7$ Hz). This results in a temporary "communication blackout" whenever the plasma frequency ω_p exceeds the transmitter frequency ω.

Example

While we have given the impression that most materials may be treated as very good dielectrics or as very good conductors, there are some materials which are not in either category. One of these is human bone and fat tissue which have relatively low water content. At a frequency ν of 3 GHz (3×10^9 Hz) the dielectric constant k_e is approximately 5.5, $k_m \simeq 1$, and the conductivity σ is 0.150 mho/meter. What is the wavelength in the tissue and what is the "depth of penetration" or "skin depth" of the electromagnetic fields and of the power?

Solution

For this problem we must evaluate the real and imaginary parts of the propagation constant k using Eqs. 6.42b. Since we have

$$\omega = 2\pi \times 3 \times 10^9 = 1.88 \times 10^{10}\ \text{sec}^{-1} \qquad k_e = 5.5$$

$$\sigma = 1.50 \times 10^{-1}\ \text{mho/m} \qquad k_m = 1$$

$$\epsilon_o = 8.85 \times 10^{-12}\ \text{farad/m} \qquad c = 3 \times 10^8\ \text{m/sec}$$

we find

$$\frac{\sigma}{\omega \epsilon_o \kappa_e} = 0.164$$

and

$$k_r = 1.48 \times 10^2\ m^{-1} \qquad k_i = 8.26 \times 10^{-2}\ m^{-1}.$$

Thus the wavelength in the tissue λ_t is given by

$$\lambda_t = 2\pi/k_r = 4.24\ \text{cm}$$

which can be compared with the vacuum wavelength λ_o of 10 cm.

The penetration of the fields into the tissue is given as

$$d_{fields} = (k_i)^{-1} = 12.1 \text{ cm}.$$

Since the power varies as $E^2 \sim e^{-2k_i z} = e^{-z/d_{power}}$ the penetration of the power into the tissue is

$$d_{power} = (2k_i)^{-1} = 6.05 \text{ cm}.$$

Thus microwave energy can penetrate human (and animal) tissue to appreciable depths and this fact forms the basis for the principle of microwave ovens and microwave thermography, the sensing of subsurface body temperatures by measuring the body's thermal radiation.

Example

Pulsating galactic radio sources, called pulsars, were discovered in 1968 and are believed to be rotating neutron stars, i.e., stars of such density that they consist entirely of neutrons. The existence of such pulsating radio sources has allowed radio astronomers to derive the integrated electron density, $\int N_e \, d\ell$, between the observer and the pulsar with incredible accuracy. Since the pulses propagate through the interstellar gas, a tenuous ionized gas, the difference in the arrival time of a pulse at two different frequencies is a measure of the dispersion along the line of sight to the pulsar. Show that a measurement of the difference in arrival times of a pulse at frequencies ω_1 and ω_2 is directly proportional to the integrated electron density. Assume the ω's are much greater than the plasma frequency ω_p.

Solution

The transit time T of a pulse from the source to the observer, a distance L, is given by

$$T = \int_0^L \frac{d\ell}{v_g}$$

where the group velocity v_g is given by

$$v_g = \frac{d\omega}{dk} \qquad \text{and} \qquad k^2c^2 = \omega^2 - \omega_p^2$$

from Eq. 6.72. The plasma frequency ω_p is

$$\omega_p^2 = N_e q^2/m\epsilon_o$$

where N_e is the electron density. With these values, the group velocity v_g is

$$v_g = c\sqrt{1 - \left(\frac{\omega_p}{\omega}\right)^2}$$

so the transit time for frequency ω_1 is

$$T_1 = \frac{1}{c}\int_0^L \frac{d\ell}{\sqrt{1 - \left(\frac{\omega_p}{\omega_1}\right)^2}} \approx \frac{1}{c}\int_0^L \left(1 + \frac{\omega_p^2}{2\omega_1^2}\right) d\ell.$$

The difference in transit times $\Delta T = T_2 - T_1$ becomes

$$\Delta T = \frac{1}{2c}\int_0^L \left(\frac{\omega_p^2}{\omega_2^2} - \frac{\omega_p^2}{\omega_1^2}\right) d\ell = \frac{q^2}{2m\epsilon_o c}\left(\frac{1}{\omega_2^2} - \frac{1}{\omega_1^2}\right)\int_0^L N_e\, d\ell.$$

Since time and frequency can be measured to very high accuracy we can determine the integrated electron density

$$\int_0^L N_e\, d\ell = \frac{2m\epsilon_o c}{q^2}\left(\frac{\omega_1^2\omega_2^2}{\omega_1^2 - \omega_2^2}\right)\Delta T.$$

Assuming ω_1 and ω_2 are emitted simultaneously at the source, we find that ΔT is the difference in arrival times at the radio telescope.

CHAPTER 7

REFLECTION, REFRACTION AND SCATTERING

Reflection and refraction occur when a wave passes from one medium to another, or whenever it encounters inhomogeneities in the medium through which it propagates. An understanding of these phenomena is needed in the solution of many diverse problems, including the design of optical instruments, reflection from radar targets, refraction of radio waves from the ionosphere, and many more. In section 5.1 we examined the special problem of wave reflection from a perfect conductor, and to solve it our first task was to determine the boundary conditions that $\vec{E}$ and $\vec{B}$ must obey at the vacuum-metal interface (Eqs. 5.12). In this section we shall study the behavior of waves at the interface between two dielectrics or at the boundary of a good but not perfect conductor. And, just as in section 5.1, we shall need to know the appropriate boundary conditions. The method of derivation is the same as was discussed already in that section and for this reason it will not be repeated here. In summary, the procedure is as follows. Consider a segment of the interface like that shown in Fig. 7.1. Take each of the four Maxwell equations (6.36) (which apply to waves in bulk matter) and rewrite them in integral form; draw appropriate pillboxes and contours exactly like those illustrated in Figs. 5.2 through 5.4 and evaluate the four equations – one at a time. This will lead to the following four boundary conditions which we write both in words and in symbols:

(a) The normal component of the quantity $(\epsilon_o \kappa_e \vec{E})$ is continuous at the interface provided that there is no <u>free</u> surface charge density. Otherwise $(\epsilon_o \kappa_e \vec{E})$ is discontinuous by the amount of surface charge density ρ_S present:

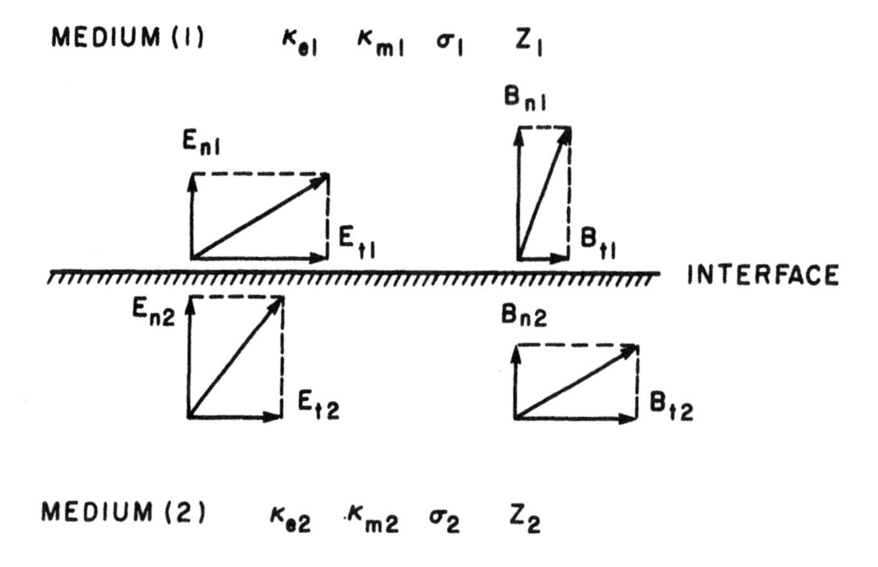

Fig. 7.1 The fields at the interface between two media. (cf. Fig. 5.1.)

$$[\epsilon_o \kappa_e E]_{n1} - [\epsilon_o \kappa_e E]_{n2} = \rho_S \tag{7.1a}$$

(b) The tangential component of $\vec{E}$ is continuous across the interface; that is, it is the same just outside the surface as it is just inside the surface:

$$E_{t1} - E_{t2} = 0 \tag{7.1b}$$

(c) The normal component of $\vec{B}$ is continuous across the interface:

$$B_{n1} - B_{n2} = 0 \tag{7.1c}$$

(d) The tangential component of the quantity $\vec{B}/(\mu_o \kappa_m)$ is continuous across the interface except at the interface of an ideal perfect conductor. In the latter case, the tangential component of $\vec{B}/(\mu_o\kappa_m)$ is discontinuous by the amount of surface current density $\vec{J}_S$ (cf. section 5.1):

$$\left[\frac{B}{\mu_o\kappa_m}\right]_{t1} - \left[\frac{B}{\mu_o\kappa_m}\right]_{t2} = 0 \qquad\qquad (7.1d)$$
$$= -J_S \quad \text{(perfect conductor).}$$

It can be verified easily that in the special case when medium 2 is a perfect conductor ($E_2 = B_2 = 0$) and medium 1 is vacuum ($\kappa_{e1} = \kappa_{m1} = 1$), relations (a)-(d) reduce to those given by Eqs. 5.12.

We have now provided the basic physical input necessary to explain the laws of reflection and refraction of electromagnetic waves. Much of what follows will be algebraic manipulations. To help us in these manipulations it is convenient, but by no means mandatory, to avail ourselves of the useful concept of "wave impedance" Z, which relates the electric and magnetic field strengths associated with a plane wave. We know already one such relation given by Eq. 6.41, that $E/B = (\omega/k)$. Multiplying both sides of this equation by $\mu_o\kappa_m$ defines for us a quantity whose units are ohms and which is termed the wave impedance:

$$Z \equiv \mu_o\kappa_m \frac{E}{B}$$

$$= \frac{\mu_o\kappa_m\omega}{k} \text{ ohm.} \qquad\qquad (7.2)$$

That some such quantity with dimensions of ohms, relating $\vec{E}$ and $\vec{B}$, must exist is evident from the fact that $\vec{E}$ has units of volts/m and $\vec{B}$ is proportional to current. Note that in the special circumstance of a wave in vacuum, $\omega/k = c$, $\kappa_m = 1$, and the impedance becomes

$$\left.\begin{aligned} Z_o(\text{vacuum}) &= \mu_o c \\ &= \sqrt{\mu_o/\epsilon_o} \\ &= 377 \text{ ohm.} \end{aligned}\right\} \qquad\qquad (7.3)$$

We stress that the wave impedance of an optical system, like the impedance of a transmission line (Eq. 5.44) does in no way ascribe power dissipation to the wave; rather it tells us something about wave reflection and matching characteristics, as we shall see momentarily. (However, since in lossy dielectrics or in metals the propagation constant k can be complex, the impedance Z of Eq. 7.2 can likewise become complex. It is the imaginary part of Z which then tells us something about power dissipation from the wave.)

7.1 REFLECTION FROM A PLANE DIELECTRIC INTERFACE – NORMAL INCIDENCE

Consider a beam of radiation incident at right angles on a surface separating two media. If the surface irregularities are small compared with the wavelength of the radiation, a sharply defined reflected beam and a sharply defined transmitted beam emerge. It is our purpose here to find what fraction is reflected, and thus what remaining fraction is transmitted through the surface. If the surface is smooth,

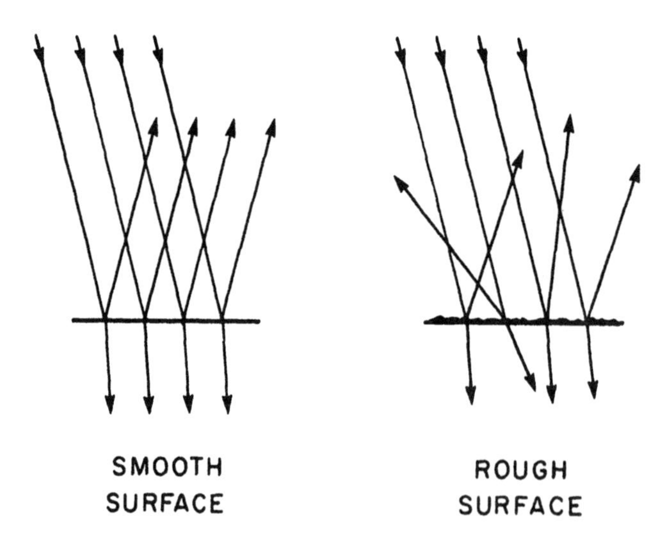

Fig. 7.2 Reflection from and transmission through, smooth and rough surfaces. The reflected and refracted rays represent schematically the directions of energy flow, i.e., directions of the time-averaged Poynting's vector.

as was just mentioned, the wave is said to be reflected specularly; if it is relatively rough, the reflection is diffuse (see Fig. 7.2). In the last-named case the laws of reflection and refraction still hold exactly over any region which is small enough so as to be considered smooth and planar, but large enough to contain many atoms.

Figure 7.3 illustrates the electric and magnetic fields of the incident (i), reflected (r), and transmitted (t) waves near the interface (z = 0) between medium (1) and medium (2). The waves are taken to be linearly polarized and monochromatic of frequency ω; the field amplitudes are given by

$$\left.\begin{aligned} E_i(1) &= E_{oi}(1)\, e^{j\omega t - jk_1 z} \;; & B_i(1) &= B_{oi}(1)\, e^{j\omega t - jk_1 z} \\ E_r(1) &= -E_{or}(1)\, e^{j\omega t + jk_1 z} \;; & B_r(1) &= B_{or}(1)\, e^{j\omega t + jk_1 z} \\ E_t(2) &= E_{ot}(2)\, e^{j\omega t - jk_2 z} \;; & B_t(2) &= B_{ot}(2)\, e^{j\omega t - jk_2 z} \end{aligned}\right\} \qquad (7.4)$$

The numbers in parentheses (1) and (2) specify whether the fields* in question are in medium (1) or in medium (2). The existence of boundary conditions at z = 0 permits determination of the amplitudes of the reflected and transmitted waves in terms of the amplitude of the incident wave (absolute values of the amplitudes cannot be specified without some prior knowledge of the source strength which determines E_i and B_i). Applying boundary requirement (7.1b) leads to $E_{oi}(1) - E_{or}(1) = E_{ot}(2)$, and applying requirement (7.1d) leads to $(B_{oi}(1) + B_{or}(1))/(\mu_o \kappa_{m1}) = B_{ot}(2)/(\mu_o \kappa_{m2})$. But, use of Eq. 7.2 permits us to eliminate B in favor of E and Z. After carrying out this elimination of B we obtain

*In the reflected wave, either $\vec{E}$ or $\vec{B}$ must have a change in sign to insure that $(\vec{E}\times\vec{B})$ point along the negative axis. We chose to change the sign of $\vec{E}$. If this phase reversal of 180° was chosen incorrectly, the consequence will show up in the final equations.

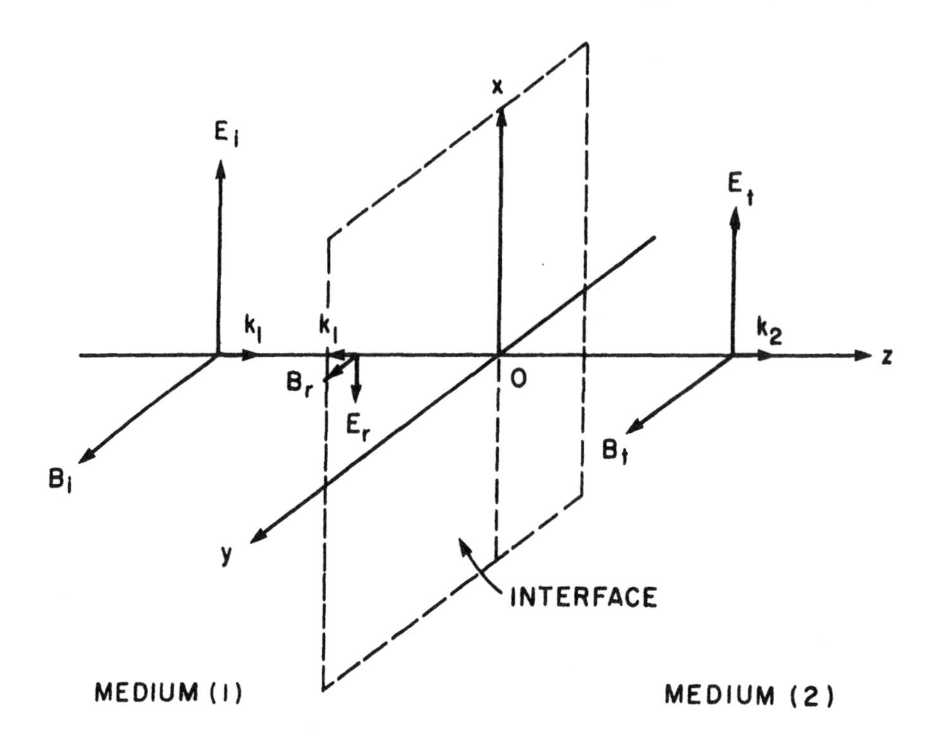

Fig. 7.3 The incident (i), reflected (r), and transmitted (t) waves near the interface between two media. The electric field of the reflected wave at z = 0 is drawn 180° out of phase with respect to the electric field of the incident wave (at z = 0).

$$E_{oi}(1) - E_{or}(1) = E_{ot}(2)$$

$$\frac{E_{oi}(1)}{Z_1} + \frac{E_{or}(1)}{Z_1} = \frac{E_{ot}(2)}{Z_2} \tag{7.5}$$

where $Z_1 = (\mu_o \kappa_{m1} \omega / k_1)$ is the wave impedance of medium (1) and $Z_2 = (\mu_o \kappa_{m2} \omega / k_2)$ is the wave impedance of medium (2). Solution of the two equations yields E_r and E_t in terms of the incident field E_i,

$$E_r(1) = E_i(1) \left[\frac{Z_1 - Z_2}{Z_1 + Z_2}\right]$$

$$E_t(2) = E_i(1) \left[\frac{2Z_2}{Z_1 + Z_2}\right], \tag{7.6}$$

a result which is remarkably similar to the transmission line equation (5.49). The coincidence is, of course, not fortuitous. When $Z_1 = Z_2$ the two media are "matched" and there is no reflection.

The impedance Z is an intrinsic property of the medium. For a lossless, nonmagnetic dielectric, k is purely real, $\kappa_m = 1$, and $Z = \mu_o v$, with $v = \omega/k$ as the phase velocity of the wave. If n denotes the refractive index, then $v = c/n$ and $Z = \mu_o c/n$. As a result, Eqs. 7.6 become

$$E_r(1) = E_i(1)\left[\frac{n_2 - n_1}{n_2 + n_1}\right]$$
$$E_t(2) = E_i(1)\left[\frac{2n_1}{n_2 + n_1}\right] \tag{7.7}$$

and the reflection properties are fully specified in terms of a single quantity, the refractive index of the media on the two sides of the interface. Suppose medium (2) is glass with $n \approx 1.5$ and medium (1) is vacuum (or air) with $n \approx 1$; then $E_r/E_i \approx (1/5)$ which says that the amplitude of the reflected wave is one fifth that of the incident wave. The Poynting flux* for a wave in vacuum is $\vec{E} \times \vec{B}/\mu_o$, and hence its intensity is proportional to $|E|^2$. Thus, the fraction of the incident power which is reflected from the surface (that is, the reflection coefficient R) is given by $R \equiv |E_r/E_i|^2 \approx 0.04$. Approximately 4% of the intensity of the light beam incident normally on a clean glass surface is reflected from it.

The first equation (7.7) shows us that when $n_2 > n_1$, that is, the wave is incident on an optically denser medium, the electric vector

*In a nondispersive medium the Poynting flux and the wave energy density are, respectively, $\vec{S} = \vec{E} \times \vec{B}/\mu_o\kappa_m$, $U = (\epsilon_o\kappa_e E^2/2) + (B^2/2\mu_o\kappa_m)$. But when dispersion is important, these results must be modified. For an advanced discussion of this subject see the advanced text Electrodynamics of Continuous Media, L. D. Landau and E. M. Lifshitz (Addison-Wesley Publishing Company, Inc., South Reading, Mass., 1960), p. 253.

at the surface undergoes a phase change of 180° on reflection and the magnetic vector undergoes no change, precisely as is shown in Fig. 7.3, and as we originally guessed in drawing the figure. On the other hand, if $n_2 < n_1$, as would be the case of a beam traveling from, say, glass to air, the electric vectors of the incident and reflected waves are in phase at the interface, whereas the magnetic vectors are flipped in phase by 180°. The consequence of this phase-reversal phenomenon can be observed as follows. Shine collimated light at normal incidence on a thin flat soap film produced by dipping a plane wire grid into a detergent. Look at the reflected light (Fig. 7.4). At first, many colored fringes will be seen, due to interference of waves reflected from the front and back of the film (see section 8.2). As the film drains and becomes thinner and thinner, the fringes disappear and become replaced by complete darkness. The explanation is the following: the $\vec{E}$ vector of the wave reflected from the front surface of

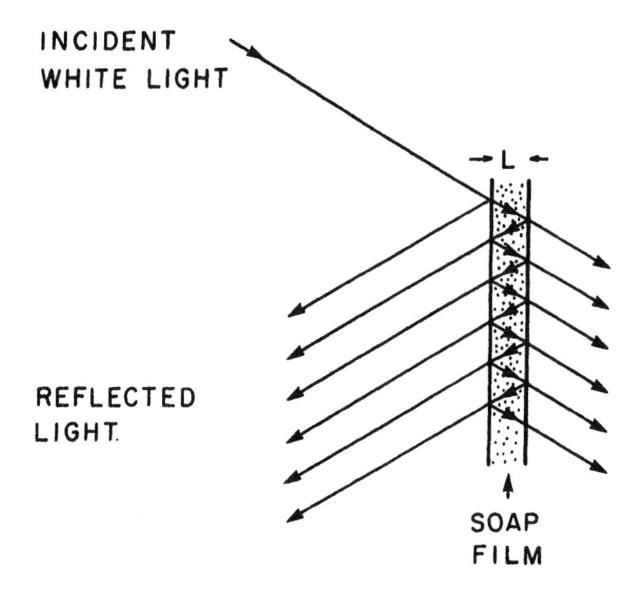

Fig. 7.4 White light incident on a soap film. When the thickness L is small compared with the wavelength in the film, reflected light waves interfere destructively giving zero reflected intensity. The light is incident nearly at right angles; for purposes of illustration, the angles are greatly exaggerated.

the film has suffered a 180° phase change. The $\vec{E}$ vector of the wave reflected from the back surface of the soap film suffered zero phase change. The two waves interfere destructively and there is darkness, because now the film has drained to a small thickness compared with a wavelength and there are no longer any additional phase (or path length) differences to produce successive destructive and constructive interferences, and thus fringes.

Example

Using Eqs. 7.7, compute the energy flux density reflected and transmitted when a plane wave in a medium of index n_1 is incident normally on a nonmagnetic dielectric slab of index n_2, and show that energy is conserved.

Solution

For convenience we take the electric field polarized along x and the magnetic field along y. Then for the incident wave we have

$$\langle \vec{S}_i \rangle = \frac{1}{\mu_o} \langle \vec{E}_i \times \vec{B}_i \rangle = \frac{1}{\mu_o} \langle E_x B_y \rangle_i \, \hat{z} = \frac{\epsilon_o c}{2} E_{ox}^2 n_1 \hat{z}$$

where use has been made of $B_y = E_x n_1/c$, $c^2 = (\epsilon_o \mu_o)^{-1}$ and the factor of 1/2 comes from the time average.

For the reflected fields we have

$$\langle \vec{S}_r \rangle = \frac{1}{\mu_o} \langle -E_x B_y \rangle_r \, \hat{z} = -\frac{\epsilon_o c}{2} E_{ox}^2 \left(\frac{n_2 - n_1}{n_2 + n_1} \right)^2 n_1 \hat{z}.$$

Finally, for the transmitted fields:

$$\langle \vec{S}_t \rangle = \frac{1}{\mu_o} \langle E_x B_y \rangle_t \, \hat{z} = \frac{\epsilon_o c}{2} E_{ox}^2 \left(\frac{2n_1}{n_2 + n_1} \right)^2 n_2 \hat{z}.$$

For conservation of energy we must have

$$\langle |\vec{S}_i| \rangle = \langle |\vec{S}_r| \rangle + \langle |\vec{S}_t| \rangle$$

$$\frac{\epsilon_o c}{2} E_{ox}^2 n_1 \hat{z} = \frac{\epsilon_o c}{2} E_{ox}^2 \left[\left(\frac{n_2 - n_1}{n_2 + n_1} \right)^2 n_1 + \left(\frac{2n_1}{n_2 + n_1} \right)^2 n_2 \right] \hat{z}$$

$$= \frac{\epsilon_o c}{2} E_{ox}^2 n_1 \hat{z}.$$

Thus the results of Eqs. 7.7 are consistent with the conservation of energy as, of course, they must be.

One can use these expressions to define power reflection (R) and transmission (T) coefficients for propagation from medium n_1 to n_2. These are

$$R = \frac{\langle |\vec{S}_r| \rangle}{\langle |\vec{S}_i| \rangle} = \left(\frac{n_2 - n_1}{n_2 + n_1} \right)^2$$

$$T = \frac{\langle |\vec{S}_t| \rangle}{\langle |\vec{S}_i| \rangle} = \frac{4n_1 n_2}{(n_2+n_1)^2}.$$

7.2 REFLECTION FROM DIELECTRIC LAYERS

From the previous discussion it can be seen that the thickness of the dielectric matters a great deal and that unless somehow all the radiation incident on the back surface is absorbed, multiple reflections can occur, giving rise to interference phenomena. But to do the calculations, we need not consider the progress of each individual reflected wave and then sum over an infinity of such waves to obtain the net effect. That is a messy calculation; instead, we sum the waves first and let $[E(2), B(2)]^+$ be the resultant of all waves traveling to the right in medium (2) and let $[E(2), B(2)]^-$ be the resultant of all waves traveling to the left in medium (2) as is indicated in Fig. 7.5. At the first interface $z = 0$, in accordance with our boundary conditions that the tangential $\vec{E}$ and $\vec{B}$ fields be contiuous at the boundary, we see that

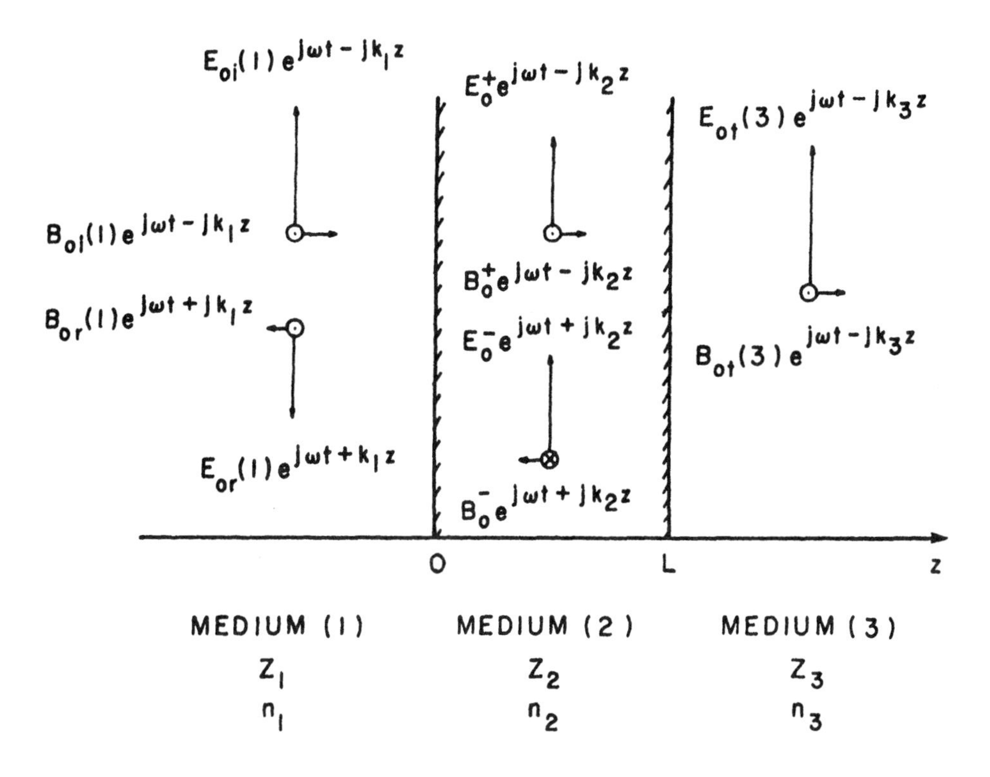

Fig. 7.5 The waves in a dielectric layer of thickness L, separating two media (1) and (3). The propagation constants k_1, k_2, and k_3 are given in terms of the refractive indices by

$$k_1 = \left(\frac{\omega}{c}\right) n_1 \quad ; \quad Z_1 = \mu_o \omega / k_1$$

$$k_2 = \left(\frac{\omega}{c}\right) n_2 \quad ; \quad Z_2 = \mu_o \omega / k_2$$

$$k_3 = \left(\frac{\omega}{c}\right) n_3 \quad ; \quad Z_3 = \mu_o \omega / k_3 .$$

$$E_{oi}(1) - E_{or}(1) = E_o^+(2) + E_o^-(2)$$

$$\frac{E_{oi}(1)}{Z_1} + \frac{E_{or}(1)}{Z_1} = \frac{E_o^+(2)}{Z_2} - \frac{E_o^-(2)}{Z_2} . \qquad (7.8)$$

Similarly, at the second (back) interface z = L we have

$$E_o^+(2)\, e^{-jk_2 L} + E_o^-(2)\, e^{+jk_2 L} = E_{ot}\, e^{-jk_3 L}$$

$$\frac{E_o^+(2)}{Z_2}\, e^{-jk_2 L} - \frac{E_o^-(2)}{Z_2}\, e^{+jk_2 L} = \frac{E_{ot}}{Z_3}\, e^{-jk_3 L} \tag{7.9}$$

where the phase factors $\exp[\pm jkL]$ account for the phase changes suffered by the fields in traversing the distance L. This ends the physics of the problem; what remains are somewhat lengthy algebraic manipulations aimed at eliminating E_o^+, E_o^- and E_t with the view of obtaining a relation between E_{or}, E_{oi} and the wave impedance Z. The outcome is

$$\frac{E_{or}(1)}{E_{oi}(1)} = \frac{[Z_2 Z_3 - Z_1 Z_2] \cos(k_2 L) + j\left[Z_2^2 - Z_1 Z_3\right] \sin(k_2 L)}{[Z_2 Z_3 + Z_1 Z_2] \cos(k_2 L) + j\left[Z_2^2 + Z_1 Z_3\right] \sin(k_2 L)}. \tag{7.10}$$

Multiplying this result by the complex conjugate (see Appendix 2) leads to the reflection coefficient R

$$R \equiv \left|\frac{E_{or}(1)}{E_{oi}(1)}\right|^2$$

$$= \frac{[Z_2 Z_3 - Z_1 Z_2]^2 \cos^2(k_2 L) + \left[Z_2^2 - Z_1 Z_3\right]^2 \sin^2(k_2 L)}{[Z_2 Z_3 + Z_1 Z_2]^2 \cos^2(k_2 L) + \left[Z_2^2 + Z_1 Z_3\right]^2 \sin^2(k_2 L)} \tag{7.11}$$

which gives the fraction of the incident power reflected from the front surface of the dielectric sheet. This formidable looking equation exhibits a very remarkable property. The minimum value of R occurs when $k_2 L = \pi/2$, at which thickness it takes on a value

$$R = \left[\frac{Z_2^2 - Z_1 Z_3}{Z_1^2 + Z_1 Z_3}\right]^2 \qquad (k_2 L = \pi/2). \tag{7.12}$$

More important, R can now be made identically zero provided that

$$Z_2 = \sqrt{Z_1 Z_3} \tag{7.13}$$

which is precisely the condition imposed on the quarter wave transformer described in section 5.2 in order to achieve perfect impedance matching on a high frequency transmission line. This result can be expressed in terms of refractive indices since Z and n are related through $Z = \mu_o c/n$:

$$n_2 = \sqrt{n_1 n_3}. \tag{7.14}$$

Thus, a dielectric film one quarter wave thick, $L = \lambda_2/4$, (with λ_2 as the wavelength in the film) acts as an impedance matching device and constitutes the basis of all antireflection coatings. When n_2 and L do not satisfy exactly the aforementioned criteria, a reduction in R can nonetheless be achieved as is shown in Fig. 7.6 for the case of a film interposed between a water substrate and air.

In practice medium (1) is generally air ($n_1 \approx 1$) and medium(3) is a glass substrate ($n \approx 1.5$). This requires that the coating have a refractive index $n_2 \approx 1.22$. Common coatings are sodium aluminum fluoride ($n_2 = 1.35$) and magnesium fluoride ($n_2 = 1.38$). Although these indices are somewhat too large to exactly satisfy equality (Eq. 7.14), a single layer of magnesium fluoride reduces the reflection coefficient from 4% to ~1% over a good portion of the visible spectrum. It is common practice to coat optical instruments with such antireflection films. On camera lenses they reduce haziness caused by stray internally reflected light and lead to an increase in image brightness. For greater effectiveness, double, or even triple quarter-wavelength layers are used on high quality optical elements. However, it is important to note that for multiple layers none of the equations (7.10) to (7.14) are applicable any more. New equations, one for each of the interfaces, must then be generated and the problem solved anew. A convenient matrix method of solution is available which greatly eases the problem of solution [see, for example, Optics by E. Hecht and A. Zajac, (Addison-Wesley, 1974); also M. Born and

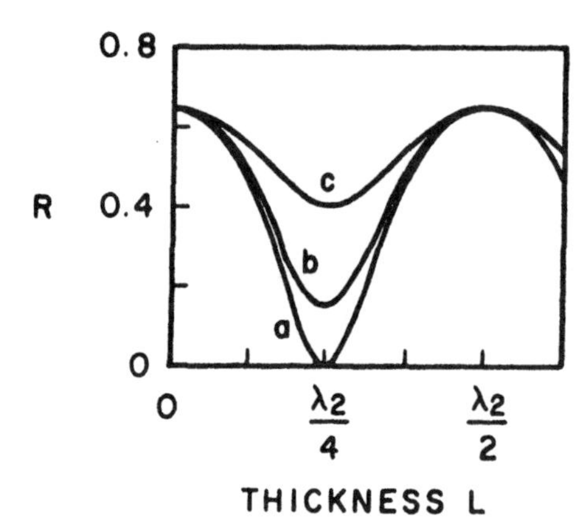

Fig. 7.6 Reflection from a thin dielectric film of thickness L interposed between two media: vacuum ($n_1 = 1$) and water ($n_3 = 9$). Each curve refers to a different value of refractive index n_2 of the film and was calculated from Eq. 7.11: (a) $n_2 = 3$, (b) $n_2 = 2$, (c) $n_2 = \sqrt{2}$. In (a) the "matching" condition given by Eq. 7.14 is obeyed exactly and therefore when $L = \lambda_2/4$, $R = 0$. Note that only for low-frequency waves is the dielectric coefficient of water $\kappa_e \approx 80$, and hence $n_3 = \sqrt{\kappa_e} \approx 9$ (see Table 6.1). From Electromagnetic Theory, J. A. Stratton (McGraw Hill Book Company, New York, 1941).

E. Wolf, Principles of Optics (Pergamon Press, 1964)].

It is fairly evident that the stacking on a substrate of alternate layers of high and low refractive index of appropriate n, a variety of effects can be achieved other than just the reduction of reflectivity. Notably, multiple films of dielectric can be arranged to achieve very high reflectivity, greater than 99 percent! (see Fig. 7.7). Such mirrors are commonly used for resonant cavities of lasers. High reflectivity or high transmission can be wavelength "tailored"; that is, bandpass, lowpass, and highpass filters can be manufactured. Here the successive layers are arranged so that the composite film stack permits reflection or transmission of designated portions of the frequency spectrum (see Fig. 7.8).

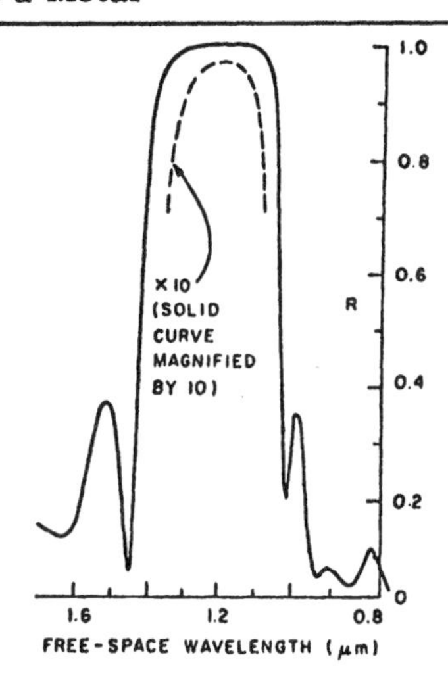

Fig. 7.7 Reflection coefficient of a multilayer dielectric stack peaked for maximum reflection at a free-space wavelength of approximately 1.2 micrometers (1.2×10^{-6} meter). The maximum R is ~99.6 percent.

7.3 REFLECTIONS FROM A METAL

When a wave is incident on a metal having an infinite conductivity, all of the incident energy is reflected from the surface (see section 5.1). In practical situations of high but finite conductivity, some of the incident energy is absorbed. How much? To find this out all we need to do is to calculate the wave impedance of the material, substitute it in the first Eq. 7.6 and obtain the amplitude of the reflected wave.

The main difference between the impedance of a lossless dielectric and the impedance of a metal is that the former is a purely real number but the latter is a complex one. This comes from the fact that the propagation constant κ and hence $Z \equiv \mu_o \kappa_m \omega/k$ are complex. Inserting κ from Eq. 6.62 in Eq. 7.2 it follows that the impedance of a nonmagnetic conductor ($\kappa_m = 1$) is

$$Z(\text{conductor}) = (1+j)\,\mu_o c \sqrt{\frac{\omega \epsilon_o}{2\sigma}}. \tag{7.15}$$

Also, the impedance of vacuum equals $\mu_o c$; hence at a vacuum-metal

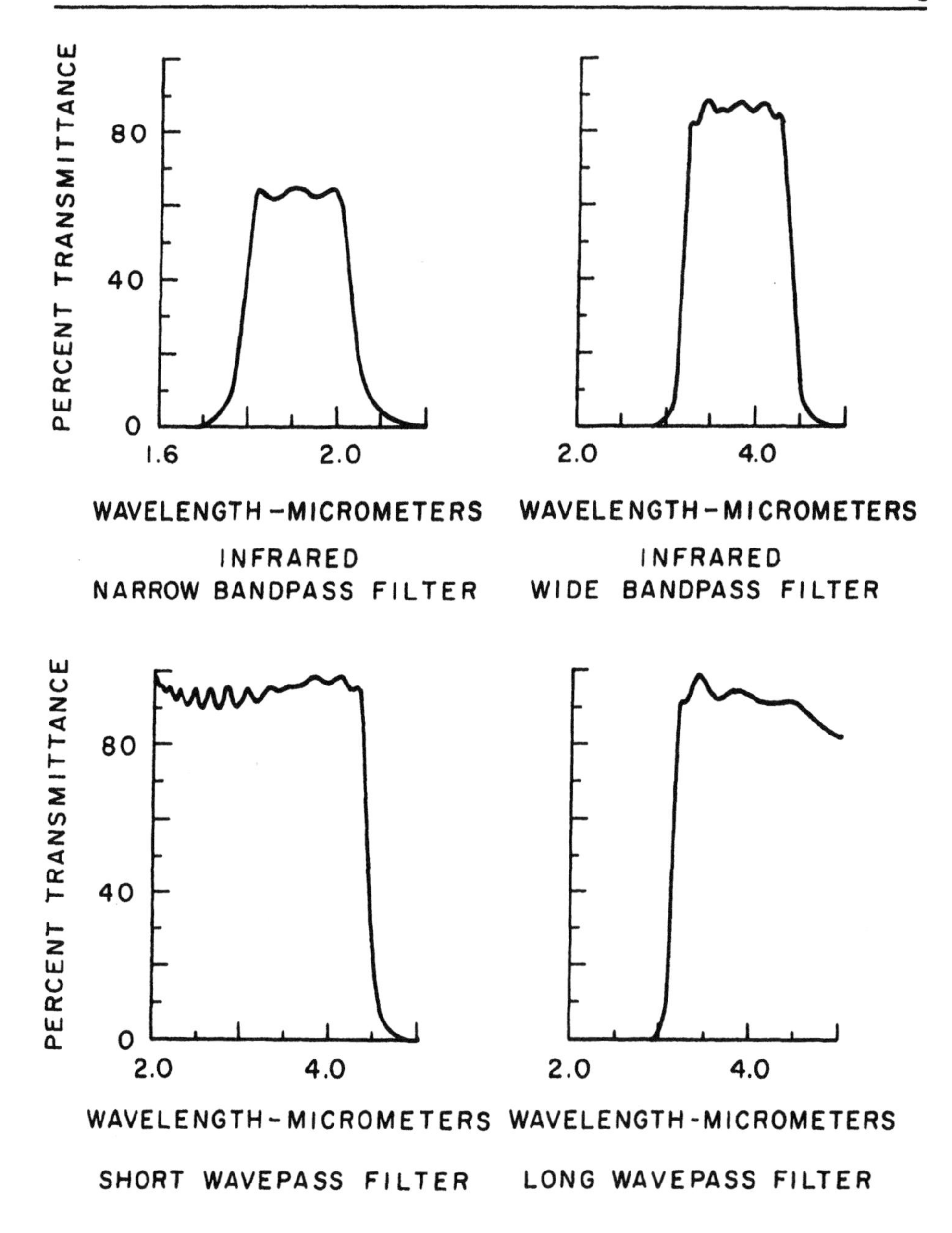

Fig. 7.8 Various infrared lowpass, bandpass, and highpass multilayer dielectric filters [from Heliotek, Division of Textron, Inc., Sylmar, California].

interface use of these impedances in Eq. 7.6 yields

$$\frac{E_r(1)}{E_i(1)} = \frac{1 - (1+j)\sqrt{\omega\epsilon_0/2\sigma}}{1 + (1+j)\sqrt{\omega\epsilon_0/2\sigma}} \qquad \text{(a)}$$

$$\approx 1 - 2\sqrt{\frac{\omega\epsilon_0}{2\sigma}} - 2j\sqrt{\frac{\omega\epsilon_0}{2\sigma}}. \qquad \text{(b)} \qquad (7.16)$$

The second formula is an approximation valid under conditions of good conduction, $\sqrt{\omega\epsilon_0/2\sigma} \ll 1$. When this inequality is satisfied, the denominator of Eq. 7.16a can be expanded in a binomial series; then, to lowest order in the parameter $\sqrt{\omega\epsilon_0/2\sigma}$, Eq. 7.16b follows.

To obtain the fraction R of the incident power reflected from the metal surface, one must multiply Eq. 7.16b by its complex conjugate; and again, to lowest order in the aforementioned parameter, it is found that

$$R = 1 - \sqrt{\frac{8\omega\epsilon_0}{\sigma}} \qquad [\sqrt{\omega\epsilon_0/2\sigma} \ll 1] \qquad (7.17)$$

which is the sought-for result. Let us see what this gives. Take silver whose dc conductivity is 6.14×10^7 (ohm-m)$^{-1}$ and let $\omega = 7.39 \times 10^{13}$ rad/sec (free space wavelength $\lambda = 25.5\ \mu$m). Then, by Eq. 7.17, $(1-R) = 9.23 \times 10^{-3}$. This says that close to 1% of the incident energy is absorbed by the metal. Measurements made at this wavelength by Hagen and Rubens [Ann. der Physik 11, 873 (1903)] give a value of $(1-R) = 11.3 \times 10^{-3}$ which is in fair agreement with the theoretical prediction. Let us remember, however, not to push such calculations to too high frequencies and still expect good agreement with measurements. The use of the dc conductivity instead of the rf conductivity is limited to relatively low frequencies (see section 6.6) of infrared and longer wavelengths, where ω is small compared with the collision frequency ν. At high frequencies $\omega \gg \nu$, we recall that the propagation constant k is given by, approximately, $k^2c^2 \approx \omega^2 - \omega_p^2$ with ω_p as the plasma frequency. Under these conditions, the impedance of the metal conductor becomes

$$Z(\text{conductor}) = \frac{\mu_0 c}{\sqrt{1 - (\omega_p/\omega)^2}} \qquad (\omega \gg \nu). \tag{7.18}$$

It is left to the reader to verify that in this regime of frequencies the reflection coefficient R is exactly unity when $\omega \leq \omega_p$, and is small when $\omega > \omega_p$, where the metal becomes quite transparent and behaves like a dielectric (see Fig. 6.14). Of course, a real metal surface does not follow the above model exactly. For one thing, the simple Drude theory (section 6.6) on which Eq. 7.18 is based neglects the role of the bound electrons — bound to the lattice ions. These electrons can resonate when light of the right frequency is incident on them and thus they effectively absorb the radiant energy. Indeed, it is precisely this excitation of bound electrons by the incident light which gives the metal its characteristic color through the process of selective reflection. Figure 7.9 shows the measured wavelength dependence of R for silver and gold. Silver is seen to reflect equally well over the entire visible spectrum: there are no important resonances

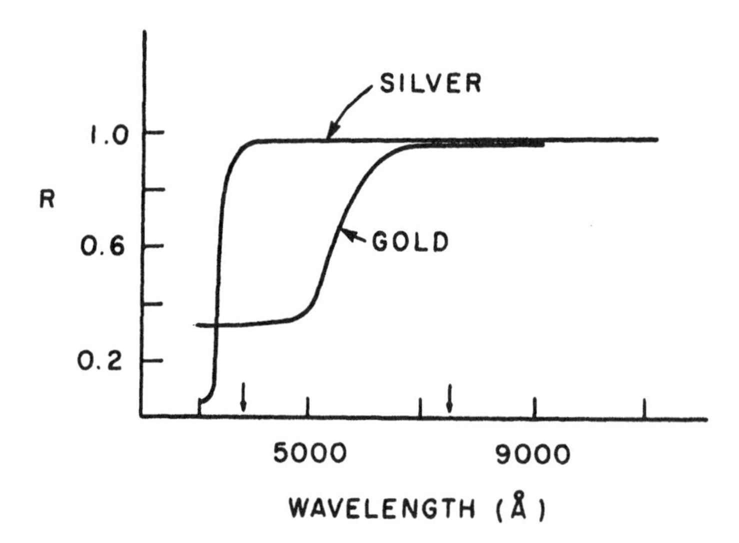

Fig. 7.9 The reflection coefficient at normal incidence of evaporated films of silver and gold. The region between the two arrows delineates the visible spectrum.

in this wavelength range. That is why it has the silvery-gray appearance. The same is true for other metals like aluminum, steel, tin, and sodium. Gold, on the other hand, is seen to be a good reflector of yellow but transmits fairly well at wavelengths in the green region of the spectrum and shorter. Thus, when white light is incident on its surface, it selectively reflects the yellow giving it its characteristic color.

As yet we have not explained the connection between good absorption and good reflection. The general rule is that if any material is a very good absorber of waves of a certain frequency, these same waves are also strongly reflected at the surface of the material and little energy penetrates to be absorbed. One can see the truth of this by coating a glass plate with red ink, say, and letting it dry. Now, when white light is shone on its surface, the light transmitted through the film is red and that reflected from it is green. The concentrated pigment absorbs most effectively in its resonant frequency range (green), and by the general rule it preferentially reflects in the same resonant range. In solution, of course, the bottle of red ink looks red in either reflected or transmitted light.

The aforementioned rule is well founded in theory. We know that a wave strongly absorbed in matter at some given frequency must have large imaginary parts to its propagation constant, because it is the imaginary part of k which gives rise to the exponential damping of the wave in the bulk of the material (see Eq. 6.50). Now, writing that $k = k_r + jk_i$ and inserting this form into expression (7.2) for the impedance, we obtain the real and imaginary parts of Z:

$$\begin{aligned} Z &= Z_r + jZ_i \\ &= \frac{\mu_o \omega k_r}{k_r^2 + k_i^2} - j\frac{\mu_o \omega k_i}{k_r^2 + k_i^2} \end{aligned} \tag{7.19}$$

which shows that large k_i leads to large Z_i. Use of Eq. 7.6 then yields the reflection coefficient R at the metal-vacuum interface expressed in terms of the real and imaginary parts of Z:

$$R = \frac{(Z_r - Z_o)^2 + Z_i^2}{(Z_r + Z_o)^2 + Z_i^2} \tag{7.20}$$

with $Z_o = \mu_o c$ as the free-space impedance of 377 ohm. Thus, we see immediately that when Z_i (and thus k_i) are large, R tends towards unity and becomes exactly unity in the limit as Z becomes pure imaginary. This proves what at first sounded like a paradox — that very high absorption is concomitant with high reflection.

7.4 REFLECTION AND REFRACTION FROM A PERFECT INSULATOR AT OBLIQUE INCIDENCE

A wave of arbitrary polarization, incident obliquely on a plane interface, can always be considered as a superposition of two waves, one with the electric vector polarized perpendicular to the plane of incidence and the other with the electric vector lying in the plane of incidence, as is illustrated in Fig. 7.10 (see section 5.1 for the analogous situation of oblique incidence on a perfect conductor). Therefore, it is sufficient to consider these two cases separately as we shall now do; the general situation may be obtained by suitable linear superposition.

There are two main questions to be answered. First, what are the relationships between the angle of incidence θ_1, the angle of refraction θ_2 and the angle of reflection θ_3; and second, what are the relative amplitudes of the incident E_i, reflected E_r and transmitted E_t waves. The answer to the first question is readily obtained by noting that in order to satisfy boundary conditions at all times and at every point of the interface $z = 0$ the phases of the three waves $\vec{E}_{oi} \exp[j\omega t - j\vec{k}_i \cdot \vec{r}]$, $\vec{E}_{or} \exp[j\omega t - j\vec{k}_r \cdot \vec{r}]$, $\vec{E}_{ot} \exp[j\omega t - j\vec{k}_t \cdot \vec{r}]$ must be the same. This means that

$$[\vec{k}_i \cdot \vec{r}]_{z=0} = [\vec{k}_r \cdot \vec{r}]_{z=0} = [\vec{k}_t \cdot \vec{r}]_{z=0}. \tag{7.21}$$

But, $\vec{k} \cdot \vec{r} = [k_x x + k_y y + k_z z]$, and therefore in accordance with the geometry of Fig. 7.10, $(\vec{k}_i \cdot \vec{r})$ at $z = 0$ equals $k_1 y \sin \theta_1$; $(\vec{k}_r \cdot \vec{r})$ at

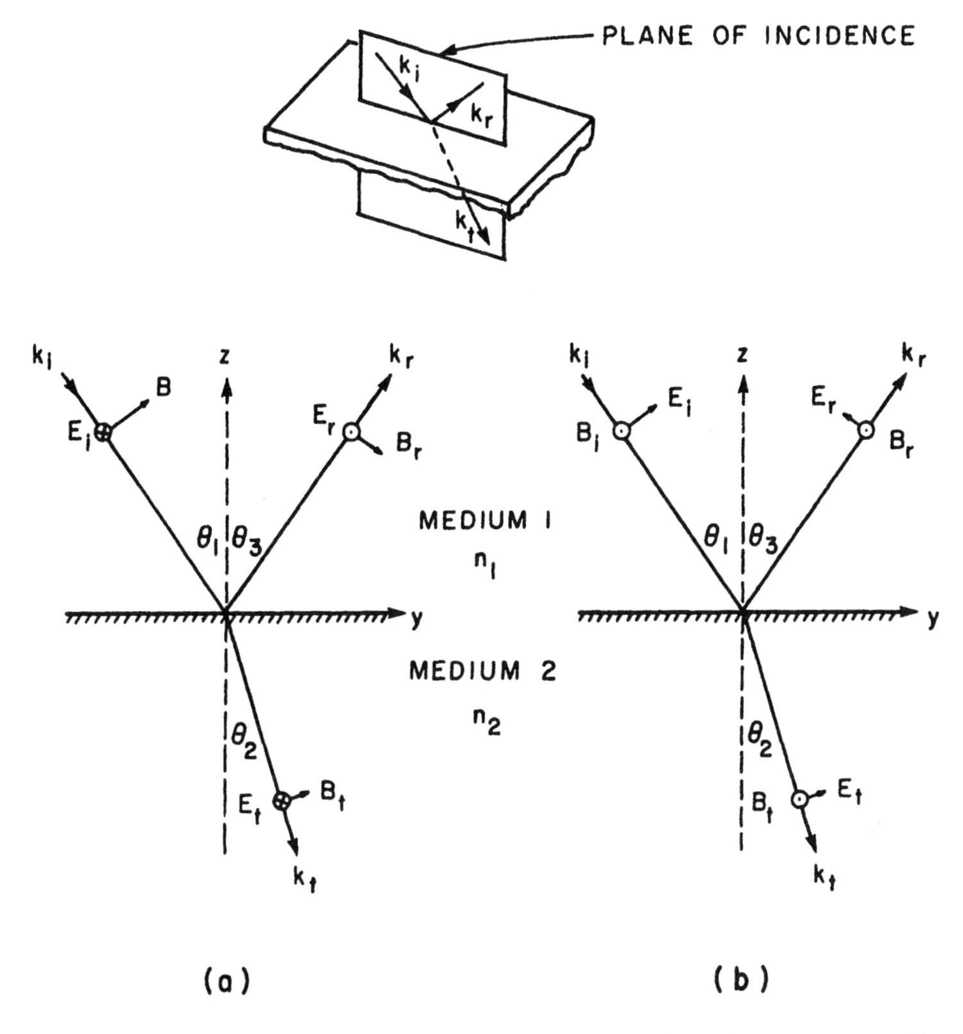

Fig. 7.10 Wave incident obliquely on the interface between two dielectrics, with (a) the electric vector polarized perpendicular to the plane of incidence, and (b) in the plane of incidence.

$z = 0$ equals $k_1 y \sin \theta_3$, and $(\vec{k}_t \cdot \vec{r})$ at $z = 0$ equals $k_2 y \sin \theta_2$, where $k_1 = n_1 \omega/c$ and $k_2 = n_2 \omega/c$ are the propagation constants in medium (1) and medium (2), respectively. After inserting these values in Eq. 7.21, the laws of reflection and refraction follow directly:

$$\boxed{\begin{array}{l}\theta_1 = \theta_3 \\ n_1 \sin\theta_1 = n_2 \sin\theta_2.\end{array}} \qquad (7.22)$$

The latter, known as the "law of sines" or "Snell's Law" is generally credited to Willebrord Snell who apparently discovered it in the year 1621. It is noteworthy that in deriving Eqs. 7.22 it was unnecessary for us to use specific boundary conditions at the interface $z = 0$. Thus, the two laws are not restricted to electromagnetic waves but are equally valid for, say, sound or water waves, with the understanding, of course, that $n \propto 1/\text{phase velocity}$ is just another way of speaking of the wave's velocity in the medium. This freedom from detailed boundary conditions unfortunately does not carry over to the next problem on hand: What are the relative amplitudes of the incident, reflected, and refracted waves?

Case (a): Electric field polarized perpendicular to the plane of incidence

In this situation the electric vector lies parallel to the interface, and at right angles to the plane of incidence, as is illustrated in Fig. 7.10a. Using boundary conditions (7.1b) and (7.1d) we obtain

$$\begin{aligned} &E_{oi}(1) - E_{or}(1) = E_{ot}(2) \\ &\frac{E_{oi}(1)\cos\theta_1}{Z_1} + \frac{E_{or}(1)\cos\theta_1}{Z_1} = \frac{E_{ot}(2)\cos\theta_2}{Z_2}. \end{aligned} \qquad (7.23)$$

We eliminate the amplitude E_{ot} between the two equations, replace Z by the appropriate refractive index n, use Snell's law to remove θ_2 in favor of θ_1, and find that

$$\frac{E_{or}(1)}{E_{oi}(1)} = -\frac{\cos\theta_1 - \sqrt{(n_2/n_1)^2 - \sin^2\theta_1}}{\cos\theta_1 + \sqrt{(n_2/n_1)^2 - \sin^2\theta_1}} \qquad (7.24)$$

which is the desired expression for the field strength of the reflected wave. Once E_{or}/E_{oi} is known, the amplitude of the transmitted wave follows from the relation

$$\frac{E_{ot}(2)}{E_{oi}(1)} = 1 - \frac{E_{or}(1)}{E_{oi}(1)}. \tag{7.25}$$

Case (b): Electric field polarized parallel to the plane of incidence

In this case the magnetic field vector $\vec{B}$ is parallel to the interface, with $\vec{E}$ lying in the plane containing the propagation vector $\vec{k}_i$ and the normal to the surface $\hat{n}$ (Fig. 7.10b). Boundary requirements (7.1b) and (7.1d) yield

$$\begin{aligned} &E_{oi}(1)\cos\theta_1 - E_{or}(1)\cos\theta_1 = E_{ot}(2)\cos\theta_2 \\ &\frac{E_{oi}(1)}{Z_1} + \frac{E_{or}(1)}{Z_1} = \frac{E_{ot}(2)}{Z_2} \end{aligned} \tag{7.26}$$

from which, employing methods similar to those described above, we find that

$$\begin{aligned} &\frac{E_{or}(1)}{E_{oi}(1)} = \frac{(n_2/n_1)^2\cos\theta_1 - \sqrt{(n_2/n_1)^2 - \sin^2\theta_1}}{(n_2/n_1)^2\cos\theta_1 + \sqrt{(n_2/n_1)^2 - \sin^2\theta_1}} \\ &\frac{E_{ot}(2)}{E_{oi}(1)} = \frac{n_1}{n_2}\left[1 + \frac{E_{or}(1)}{E_{oi}(1)}\right]. \end{aligned} \tag{7.27}$$

It would now be in order to undertake a detailed analysis of the two principal equations (7.24) and (7.27), and compute the concomitant reflection coefficients as a function of (n_2/n_1) and angle θ. Rather, we shall concentrate on just two aspects of these equations, which are both interesting and useful.

Total internal reflection

Suppose a wave, traveling in a medium like glass with a refractive index n_1, encounters a boundary with an optically less dense medium like air whose index $n_2 < n_1$. By Snell's law (Eq. 7.22) the angle of refraction θ_2 exceeds the angle of incidence θ_1; a situation often expressed by saying that the "refracted wave bends away from the normal." Now, as θ_1 is increased (Fig. 7.11) a situation is reached where $\theta_2 = \pi/2$ at which point the angle of incidence reaches its so-called critical value given by $\theta_1 = \theta^*$ where

$$\boxed{\theta^* = \sin^{-1}\left(\frac{n_2}{n_1}\right).} \tag{7.28}$$

For waves incident on the interface at $\theta_1 = \theta^*$, the refracted wave is propagated parallel to the interface. There can be no energy flow across it. As an example, consider a glass-air interface for which $n_2/n_1 = 1/1.5$; then, by Eq. 7.28, $\theta^* = 41.8°$. For a water-air

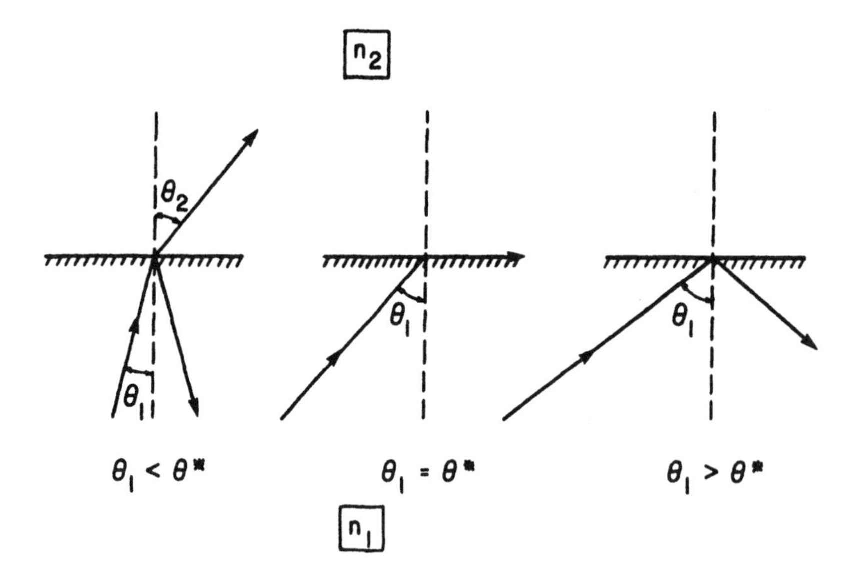

Fig. 7.11 A wave incident from an optically denser to a less dense medium $[n_1 > n_2]$, showing total internal reflection at an angle $\theta_1 > \theta^*$.

interface where $n_2/n_1 = 1/1.33$, $\theta^* = 48.8°$.

What happens when θ_1 is greater than the critical angle θ^*, that is, when $\sin\theta_1 > n_2/n_1$? Both Eqs. 7.24 and 7.27 for the reflected amplitudes then become complex numbers of the form $E_{or}/E_{oi} = (A+jB)/(A-jB)$. But numbers of this form have unit magnitude, as is readily checked by multiplying by the complex conjugate $(A-jB)/(A+jB)$. In other words, the reflection is total. It is noteworthy that total reflection does not imply that there is no field whatever in the less dense medium (2). There is a field and we can compute it. In conformity with Fig. 7.10, the electric field of the refracted wave has the form $E_t = E_{ot}\exp[j\omega t - jk_2 y\sin\theta_2 - jk_2 z\cos\theta_2]$. But, $\cos\theta_2 = (1-\sin^2\theta_2)^{1/2}$ and by Snell's law, $\sin\theta_2 = (n_1/n_2)\sin\theta_1$. Moreover, by Eq. 7.28, $(n_1/n_2) = 1/\sin\theta^*$ and therefore

$$\cos\theta_2 = \pm\sqrt{1-\left(\frac{\sin\theta_1}{\sin\theta^*}\right)^2} = \pm j\sqrt{\left(\frac{\sin\theta_1}{\sin\theta^*}\right)^2 - 1} \tag{7.29}$$

where the second relation follows from the realization that under conditions of total internal reflection, $\sin\theta_1/\sin\theta^*$ is greater than unity and $\cos\theta_2$ becomes imaginary. The fact that the cosine of an angle is imaginary has, in itself, no physical content. But it takes on physical significance as soon as this value of $\cos\theta_2$ is inserted in the phase factor $jk_2 z\cos\theta_2$ appearing in the equation for the transmitted wave. Then

$$E_t = E_{ot}\exp\left[-k_2 z\sqrt{(\sin\theta_1/\sin\theta^*)^2 - 1}\right] e^{j\omega t - jk_2 y(n_1/n_2)\sin\theta_1} \tag{7.30}$$

which is the sought-after electric field in the less dense medium. Observe its interesting properties: it falls off exponentially (see Fig. 7.12) with distance z from the interface and, except when $\theta_1 = \theta^*$, the amplitude becomes negligibly small at a distance of a few

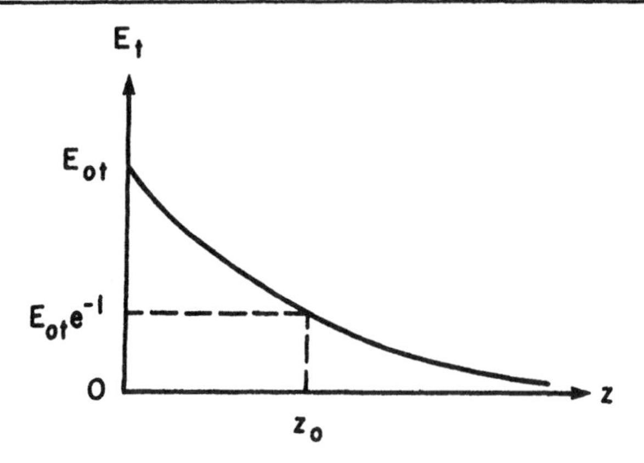

Fig. 7.12 Amplitude of the surface wave as a function of distance from the interface between two dielectrics n_1 and n_2, under conditions of total internal reflection, $\theta_1 > \theta^* = \sin^{-1}(n_2/n_1)$. When medium (1) is glass ($n_1 = 1.50$) and medium (2) is air ($n_2 = 1$), $\theta^* = 41.8°$. From Eq. 7.30 the e-folding distance z_o is given by

$$z_o = \frac{\lambda_2}{2\pi}\left[(\sin\theta_1/\sin\theta^*)^2 - 1\right]^{-1/2}$$

where λ_2 is the wavelength in medium (2). For $\theta_1 = 45°$ and an air-glass interface, $z_o = 0.45\ \lambda_2$; when $\theta_1 = 80°$, $z_o = 0.15\ \lambda_2$.

wavelengths from the boundary between the two media. Thus the electric field is confined to the immediate neighborhood of the surface. However, the field exhibits a phase progression along the interface and represents a wave which propagates along the y direction. It is for this reason that the disturbance is known as a "surface wave." Its phase velocity has the value

$$v = \frac{\omega}{k_2(n_1/n_2)\sin\theta_1} \qquad (7.31)$$

which under conditions of total internal reflection is less than the phase velocity ω/k_2 of a "regular" plane wave propagating in medium (2). Consequently this surface wave is also a "slow wave."

The existence of a surface wave in the optically less dense medium has important consequences. Suppose some material, a piece of dielectric, for example, is brought near the surface where total internal reflection occurs. If this distance is not many wavelengths, the surface wave is strong enough to drive electrons in the dielectric causing them to radiate. The net effect of all these elementary radiations will be a new wave, as shown in Fig. 7.13. Thus some light is transmitted across the gap where before there was no outward energy flow. This radiation comes, of course, at the expense of the internally reflected wave. In other words, total internal reflection is no longer total; in modern terminology "frustrated total internal reflection" (FTIR) is the somewhat amusing wording used to describe the effect. The phenomenon is similar to the "tunneling" or "barrier penetration" familiar in quantum mechanics by which a material particle can escape from a classically inescapable potential well. It is fairly obvious that the smaller the gap between the two dielectrics the

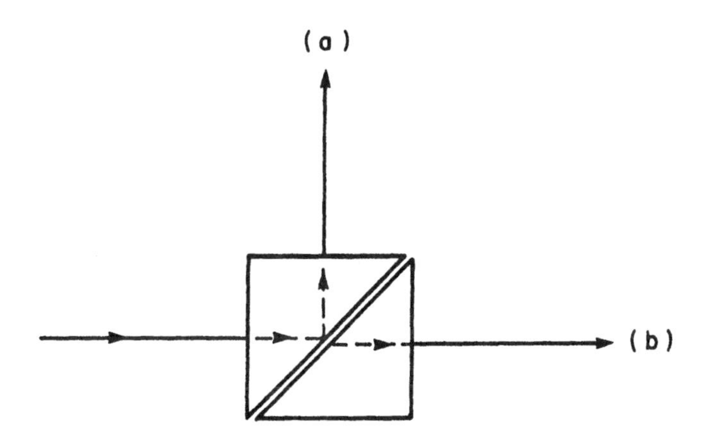

Fig. 7.13 Demonstration of "frustrated total internal reflection" (FTIR) using two 45°-45°-90° prisms with an air gap between them. As the width of the gap is increased, the amplitude of the transmitted wave (b) falls off exponentially in accord with expectations based on Eq. 7.30 and Fig. 7.12. [For details of the experiment using the green line from mercury as the incident light, see D. D. Coon, Am. J. Phys. 34, 240 (1966).]

larger will be the fraction of light transmitted across it, and with the gap closed completely, reflection becomes zero. We see therefore that FTIR can be put to practical use: light can be split and the amount of reflected and transmitted energy can be governed by the gap size. The gadget becomes an optical beam splitter.

The phenomenon of total internal reflection is exploited extensively when one wishes to deflect or guide radiation without loss of intensity. For example, for a glass-air interface $\theta^* \approx 42°$. Hence, light incident normally on a 45°-45°-90° prism as shown in Fig. 7.14 will have a $\theta_1 > 42°$ and will therefore be totally reflected. This is a convenient way of deflecting light without the use of mirrors whose

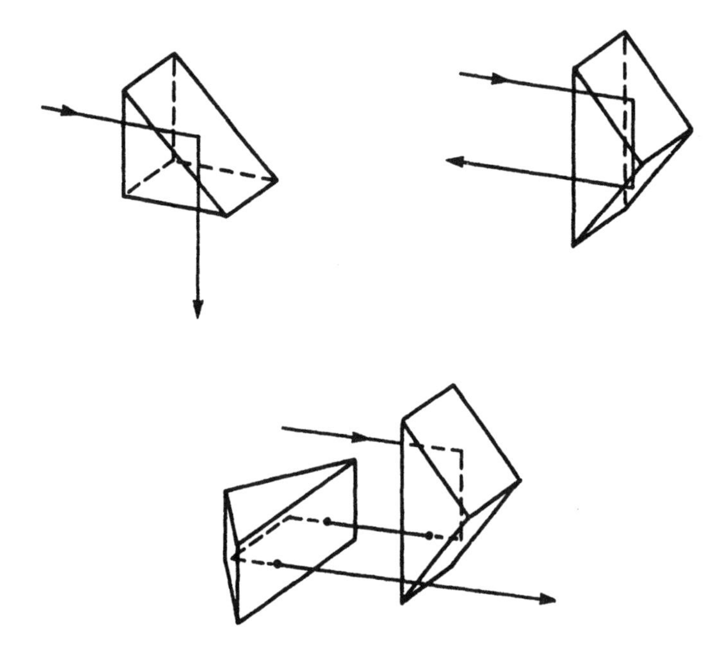

Fig. 7.14 Three examples of using total internal reflection to deflect a beam of radiation from 45°-45°-90° prisms. The lower picture is the characteristic mounting used in binoculars where the two prisms are interposed between the objective lens and the eyepiece, thus shortening the instrument. The four successive reflections cause the emergent image to be rotated through 180° and have the same "handedness" as the object. Thus, the two prisms also serve to erect what would otherwise be an inverted image of the binocular system.

metallic surfaces are prone to deterioration. Binoculars use this method of light deflection. Recent years have also seen a rapid development in the technology of so-called "fiber-optics" in which dielectric rods or pipes are used in efficient guidance of light from one point to another. When the fiber cross section is not many wavelengths in size, the propagation characteristics can be understood only by detailed consideration of the precise geometry of the fiber. That is, the problem is one of waveguide propagation as discussed in section 5.3. On the other hand, when the fibers are many wavelengths across, the above considerations of waves at a plane interface are approximately valid. Then, the entire picture of fiber propagation is one in which the light skips back and forth and is trapped by total internal reflections within the material of the fiber (see Fig. 7.15). Hundreds to thousands of reflections per foot length of fiber are common. When a single fiber or a single light pipe (which is just a very thick fiber) are

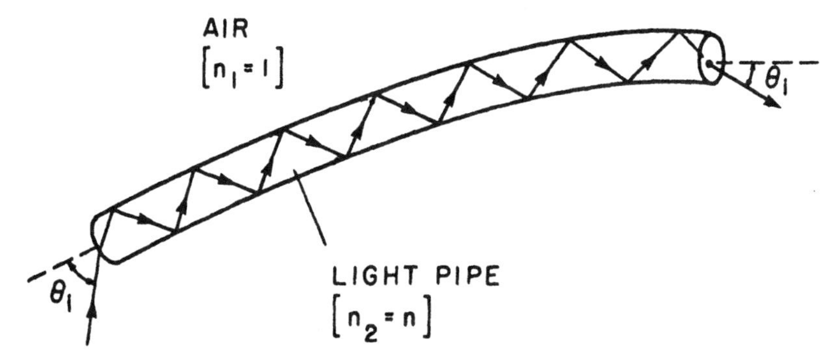

Fig. 7.15 A beam of light undergoing successive total internal reflections in a light pipe. The number of reflections per unit length of light pipe suffered by a ray which is co-planar with the fiber axis is approximately

$$\mathcal{N} \approx \frac{\sin\theta_i}{D\left[n^2 - \sin^2\theta_i\right]^{1/2}} \qquad (\text{for } \mathcal{N} \gg 1)$$

where D is the pipe diameter. Note that a beam incident at any angle $\theta_i \leqslant 90°$ on the face of the light pipe will be totally internally reflected. (Use Snell's law to prove this.)

used, their smooth surface must be kept scrupulously clean of dirt and moisture. Otherwise light leakage occurs via the phenomenon of frustrated total internal reflection. When a large number of fibers is packed into a bundle, FTIR causes light leakage between fibers. Such undesirable "cross talk" is avoided by cladding each fiber by a sheath of dielectric of lower refractive index. A typical bundle may be composed of fibers 50 μ in diameter having an index of ~1.6, with each fiber surrounded by a sheath of ~20 μ thick dielectric having an index ~1.5.

Example

Red and blue light are incident on a glass-air interface, from the glass side, at an angle of incidence θ_i. The index of refraction for red light is 1.50 and 1.52 for blue light. If θ_i is greater than some angle θ_c the transmitted beam contains only red light. What is the minimum angle θ_r for which only red light emerges?

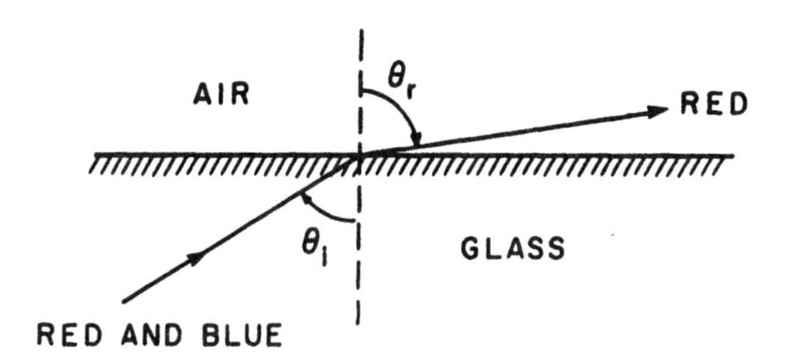

Solution

Total internal reflection takes place when

$$n_{blue} \sin \theta_c = 1$$

or

$$\sin \theta_c = 1/n_{blue} = 1/1.52.$$

Since this is now the minimum angle of incidence for the red light, the minimum refracted angle θ_r is given by

$$n_{red} \sin \theta_i = \sin \theta_r$$

$$\theta_i = \theta_c$$

$$\sin \theta_r = 1.50/1.52 = 0.987$$

$$\theta_r = 80.7 \text{ degrees.}$$

Brewster's angle θ_B

In 1811 the Scottish physicist Sir David Brewster discovered that when a wave is incident at a particular angle on the interface between two dielectrics, a state of zero reflection can be achieved. Let us see what our two principal equations (7.24) and (7.27) can tell us in regard to this effect. In order that $E_{or}(1)$ be zero in Eq. 7.24 its numerator must be zero. To achieve this $n_1 = n_2$, which is a trivial result implying that the two media are optically identical. Hence, for a wave polarized perpendicular to the plane of incidence, the condition of zero reflection cannot be achieved. Matters are different for a wave polarized in the plane of incidence. Here E_{or} of Eq. 7.27 can be made equal to zero at an angle $\theta_1 = \theta_B$ for which $(n_2/n_1)^2 \cos \theta_B = [(n_2/n_1)^2 - \sin^2\theta_B]^{1/2}$. On squaring this result and performing a few simple trigonometric manipulations, one finds

$$\boxed{\theta_B = \tan^{-1}\left(\frac{n_2}{n_1}\right)} \tag{7.32}$$

as that angle of incidence for which no reflection occurs for the stated polarization. If the incident wave is randomly polarized, only that portion of the wave with the electric vector perpendicular to the plane of incidence will be reflected, whereas the remaining portion with the electric vector parallel to the plane of incidence will be entirely transmitted (Fig. 7.16). Since the reflected wave is now linearly polarized, Brewster's angle is sometimes also called the polarizing angle. This behavior has been utilized to generate linearly polarized radiation. However, since the total amount of reflected light is small, the

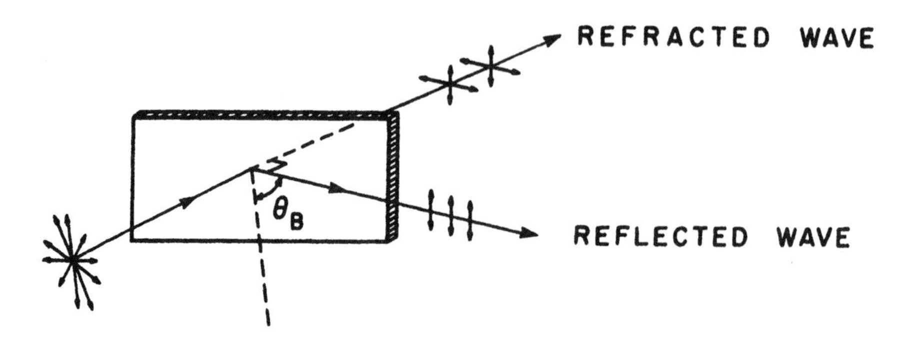

Fig. 7.16 Generation of linearly polarized radiation when a wave is incident obliquely on the interface between two dielectrics at the Brewster angle $\theta_B = \tan^{-1}(n_2/n_1)$. At this angle, the reflected and refracted rays are perpendicular to one another.

scheme is not too efficient. The effect can be greatly improved by stacking a large number of glass plates one on top of another and then illuminating this "pile of plates" with unpolarized light at the Brewster angle θ_B. The reflections from each surface then add up giving a greatly intensified linearly polarized beam of light. When the incident beam travels from air ($n_1 \approx 1$) and strikes glass ($n_2 \approx 1.5$), the angle of incidence must be adjusted so that $\theta_1 = \theta_B = \tan^{-1} 1.5$ which equals about 56°. At angles on either side of θ_B the radiation is still polarized but only partially, rather than 100 %. This tendency for reflected light to be predominantly polarized perpendicular to the plane of incidence, even when $\theta_1 \neq \theta_B$ is exploited in Polaroid sun glasses which transmit preferentially only one direction of polarization. Figure 7.17 shows a plot of the reflection coefficient as a function of the angle of incidence for the two principal states of polarization when light is incident on an air-glass interface ($n_1 = 1$, $n_2 = 1.5$). To make quantitative the concept of degree of polarization we define the quantity

$$V = \frac{R_\perp - R_\parallel}{R_\perp + R_\parallel} \times 100 \text{ percent.} \tag{7.33}$$

Using this definition and the figure, it can be seen that $V = 0\%$ when $\theta_1 = 0$ or 90° and $V = 100\%$ at the Brewster angle $\theta_B = 56°$. However,

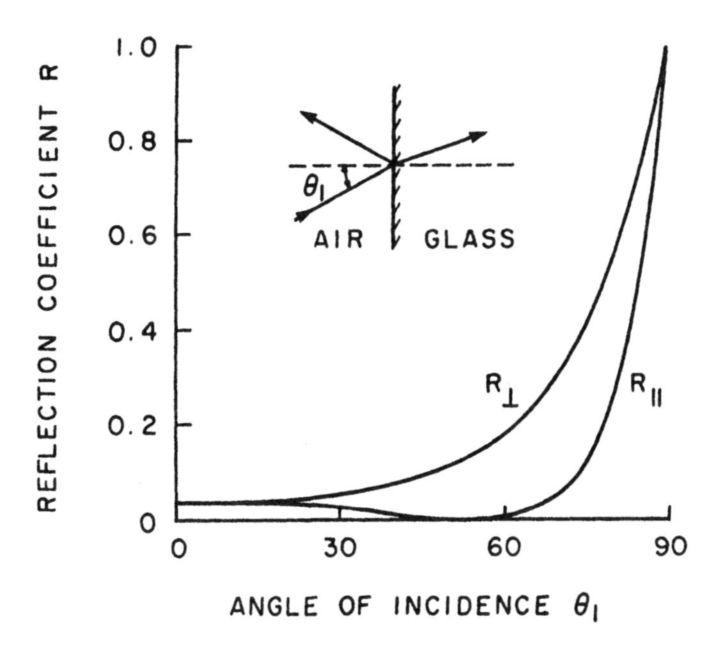

Fig. 7.17 Reflection at an air-glass interface ($n_1 = 1$, $n_2 = 1.5$) as a function of the angle of incidence θ_1. The symbols $\perp$ and $\parallel$ denote that the wave is polarized perpendicular and parallel to the plane of incidence, respectively. When $\theta_1 = 0$, $R_\perp = R_\parallel = 0.04$ as obtained from Eq. 7.7. For other angles $R_\perp$ and $R_\parallel$ were computed from Eqs. 7.24 and 7.27.

even when θ_1 deviates appreciably from θ_B, V can be substantial. Thus, for example, at $\theta_2 = 80°$ one finds from the figure and Eq. 7.33 that $V \approx 40\%$.

So-called "Brewster angle windows" are extensively used to close off the discharge tubes of gas lasers. When light traverses the optical window at normal incidence, approximately 4% of light gets reflected out of the beam at each surface and thus only about 92% of the incident light is transmitted through its two faces (see section 7.1). This is acceptable in many optics applications but it is not tolerable in the case of lasers having their high reflectance mirrors external to the lasing medium. The reason is that the light is required to make tens to hundreds of passes through the lasing medium and loss at the windows greatly reduces the Q of the resonant laser cavity (with 92% transmission the beam on its hundredth traversal is reduced to $0.92^{100} = 0.00024$ of its initial intensity!). However, by tilting the windows so that the radiation is incident at the Brewster angle as shown in Fig. 7.18, that component of the laser light that is polarized parallel to the plane of incidence is fully transmitted with no loss whatever. The other polarization component is reflected out of the beam at each pass, so that eventually after a sufficient number of passes the laser radiation becomes 100% linearly polarized. Some lasers do not employ Brewster windows and their emission is therefore unpolarized.

We have talked at some length about the manifestations and uses of Brewster's law. But, can we explain the physics of this phenomenon in some simple way other than just solving Eq. 7.27 for no reflection? The answer is yes. When the electromagnetic wave is incident on the interface between, say, vacuum and a dielectric, it sets into motion the atomic charges of the material. The accelerated charges reradiate (i.e., scatter) in accordance with precepts developed in section 4.1. Addition of the "wavelets" from each individual charge gives rise, through constructive interference, to the reflected and

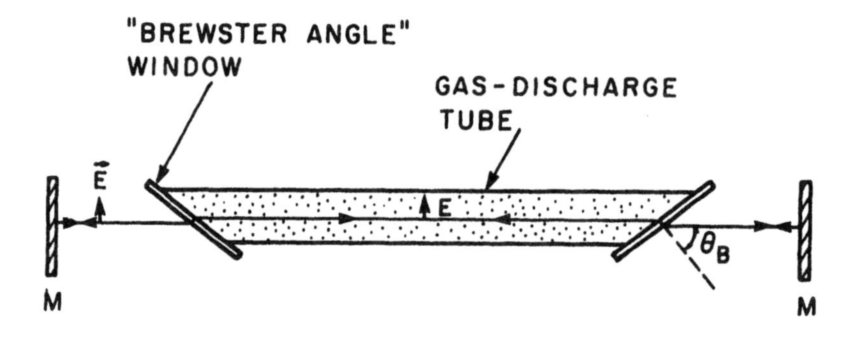

Fig. 7.18 Schematic of a gas laser provided with Brewster angle windows. M are highly reflecting mirrors which form the resonant laser cavity. The laser light is polarized entirely in the plane of incidence as shown in the figure. The other polarization is reflected out of the system as a consequence of the Brewster windows.

transmitted waves as we know them. The radiation in directions other than the reflected and transmitted ones is much, much weaker since the waves do not have special phase relationships to cause them to add constructively. Now, the direction of motion of the accelerated electrons lies along the direction of the $\vec{E}_i$ vector of the incident wave. But there is no reradiation in the direction of $\vec{E}_i$ because, as we learned in section 4.1, the emission has a sin θ dependence with θ as the angle between the acceleration vector and the direction of the propagation vector. It follows, then, that at the special angle where the reflected and transmitted waves are perpendicular to one another [i.e., $\theta_1 + \theta_2 = \pi/2$], the reflected wave receives no energy from electrons vibrating in the plane of incidence. This is illustrated in Fig. 7.19. Does this situation coincide with the requirement for Brewster

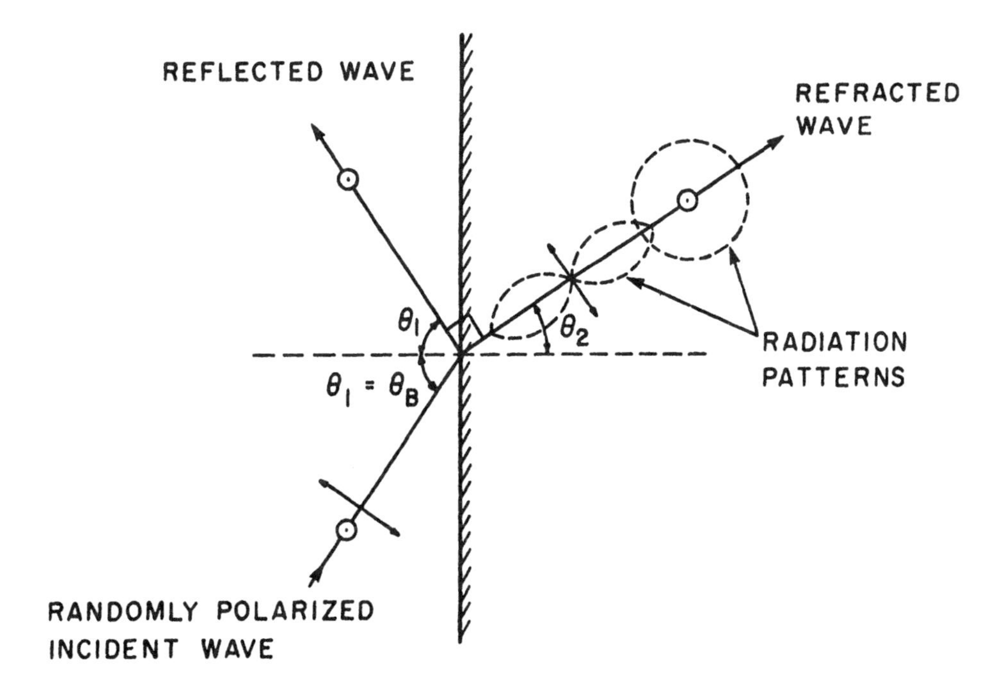

Fig. 7.19 At the Brewster angle, where $\theta_1 + \theta_2 = 90°$, the radiation patterns of the electrons vibrating in the plane of the page are such that no radiation travels in the direction of the reflected wave. The only contribution to the reflected wave comes from electrons vibrating perpendicular to the page.

reflection? It does indeed. Snell's law states that $n_1 \sin \theta_1 = n_2 \sin \theta_2$, but $\theta_2 = (\pi/2) - \theta_1$; eliminating the angle θ_2 from these two equations we have that, $n_1 \sin \theta_1 = n_2 \sin [(\pi/2) - \theta_1]$ from which it follows that $\tan \theta_1 = n_2/n_1$. But, electrons vibrating perpendicular to the plane of incidence radiate uniformly in the plane of the figure (remember that the radiation pattern of an accelerated nonrelativistic charge is a doughnut, as shown in Fig. 4.8), and it is these electrons, then, that do, in fact, contribute to the reflected wave.

7.5 SCATTERING

When a beam of radiation passes through matter, the electric field of the wave causes the charges (electrons) to move. The moving electrons in turn radiate (see Fig. 7.20) in various directions because, as we have learned in Chapter 4, any accelerated charge must emit electromagnetic waves. This process of reradiation is known as scattering. As a result of it, the forward-directed beam of radiation is somewhat weakened; and the energy lost from the forward direction is redistributed (i.e., scattered) into other angles. Although the magnitude of this process is very tiny, it is an important and widespread phenomenon in nature. Here we shall study a few examples of it.

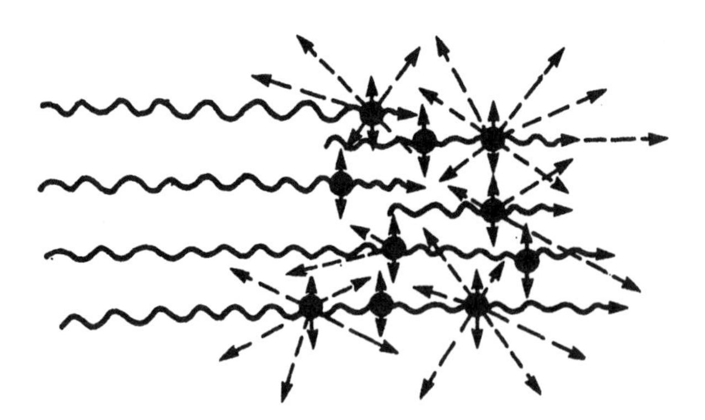

Fig. 7.20 A beam of light enters an assembly of electrons which scatter the radiation in all directions. The incident beam is shown here polarized vertically in the plane of the paper, causing the electrons to jiggle up and down the page.

Scattering by free electrons — Thomson scattering

A beautiful manifestation of scattering by free electrons is the solar corona. The corona is a tenuous outer atmosphere of the Sun. Even close to the Sun the corona contains only about 10^9 hydrogen atoms per cubic centimeter. But it is hot ($\sim 10^6$ °K) and the gas is therefore almost completely ionized into the constituent protons and electrons. The corona is not self-luminous, and therefore cannot be seen by its own emitted light. It is visible, however, by virtue of the fact that its electrons scatter the light emitted by the solar photosphere. Figure 7.21 shows the coronal light during an eclipse when the direct light of the sun is largely eliminated from the observer's view.

To calculate the amount of scattered light, we begin by considering a single electron of charge q and mass m. It is stationary, except insofar as it vibrates under the influence of a plane, monochromatic wave whose electric field is given by (see Fig. 7.22)

$$E_x = E_o \cos(\omega t - kz). \tag{7.34}$$

The charge, assumed to be at the origin z = 0, experiences an acceleration

$$a_x = \frac{q}{m} E_o \cos(\omega t). \tag{7.35}$$

The amplitude E_o of the incident wave is taken to be sufficiently small so that the charge never attains relativistic velocities. Because of the acceleration, the charge radiates in a doughnut shaped pattern shown in Figs. 4.7 and 4.8. The total power radiated into all directions is given by Larmor's formula (4.13):

$$\begin{aligned} P(t) &= \frac{q^2 a^2(t')}{6\pi\epsilon_o c^3} \qquad \left(t' = t - \frac{r}{c}\right) \\ &= \frac{q^4}{6\pi\epsilon_o m^2 c^3} E_o^2 \cos^2\left[\omega\left(t - \frac{r}{c}\right)\right] \text{ watt} \end{aligned} \tag{7.36}$$

Fig. 7.21 The solar corona seen during a total eclipse of the Sun on March 7, 1970 at "High Point" San Carlos, Yautepec, Mexico. (Courtesy of the High Altitude Observatory/NCAR. The National Center for Atmospheric Research is funded by the National Science Foundation).

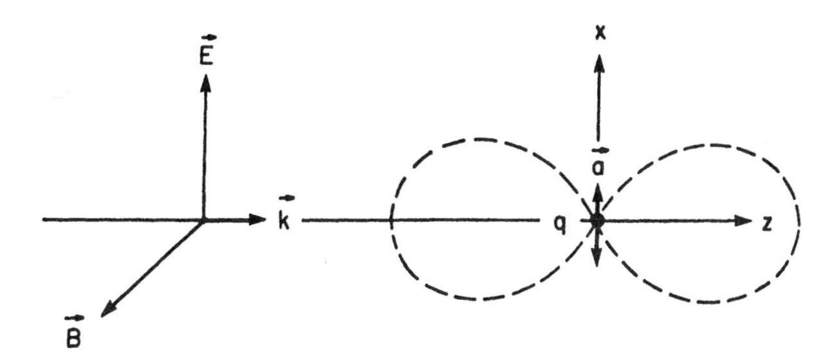

Fig. 7.22 An electron of charge q is illuminated by a plane, linearly polarized electromagnetic wave. It undergoes driven simple harmonic oscillations which cause it to radiate in the typical dipole radiation pattern shown by the dashed line. The acceleration $\vec{a}$ is due to the electric field, $\vec{a} = (q/m)\vec{E}$; the acceleration due to the magnetic field $(q/m)(\vec{v} \times \vec{B})$ is negligibly small as long as the induced motion is nonrelavistic ($v/c \ll 1$).

where the second form of the equation comes from substituting for the acceleration from Eq. 7.35. This then is the formula for the scattered power by one electron; if there are N electrons in the incident beam, the total power will be N times greater.

One often desires to know the time-averaged scattered power, averaged over a period of the oscillation. This is given by

$$\langle P \rangle = \frac{q^4 E_o^2}{12\,\pi\epsilon_o m^2 c^3} \text{ watt} \tag{7.37}$$

a result which follows from the fact that $\langle \cos^2(\ldots.) \rangle = 1/2$. To find what fraction of the total incoming light is scattered, it is convenient to eliminate the amplitude squared $\left(E_o^2\right)$ in favor of the incident flux of radiation. The last-named quantity is, of course, the Poynting flux S which is related to E_o^2 through Eq. 3.106:

$$\langle S \rangle = \frac{1}{2}\epsilon_o c E_o^2 \text{ watt/meter}^2. \tag{7.38}$$

Substituting this value of E_o^2 in Eq. 7.37 gives a result that can be written in the form

$$\underbrace{\langle P \rangle_{\text{scattered}}}_{\text{watts}} = \underbrace{\frac{8\pi}{3}\left(\frac{q^2}{4\pi\epsilon_o mc^2}\right)^2}_{\text{area } \sigma} \underbrace{\langle S \rangle_{\text{incident}}}_{\text{watts/area}}. \tag{7.39}$$

Since the total power scattered is in watts, and since the incident wave is characterized by a flux density (watts/m^2), the relationship between $\langle P \rangle$ and $\langle S \rangle$ must involve a quantity of dimensions of an area. It is known as the Thomson scattering cross section and denoted by σ:

$$\sigma = \frac{8\pi}{3}\left(\frac{q^2}{4\pi\epsilon_o mc^2}\right)^2 \ m^2 \tag{7.40}$$

$$= 6.65 \times 10^{-29} \ m^2.$$

Physicists love to express processes in terms of cross sections, and do so frequently whenever some phenomenon occurs in proportion to the intensity of a beam, for example, a beam of atoms or mesons. The advantage is that the cross section is not a function of the strength of the beam, but is a fundamental quantity of the process itself: the larger the cross section (for a given fixed beam intensity) the larger the amount of the scattering. Crawford [Waves (McGraw Hill, 1968)] uses the following mechanical analogy which helps one to visualize the meaning of cross section. Suppose you had a billiard ball of radius R sitting in a broad uniform beam of steel BB's. Those BB's that hit the ball are elastically scattered out of the beam. The total number of BB's scattered per unit time is the product of the incident number flux, in BB's per square meter per second, times the cross section $\sigma = \pi R^2$ of the ball. In other words,

$$\text{Scattered BB's per second} = \sigma \times \text{incident particle flux.} \tag{7.41}$$

Since the BB's are assumed to be scattered elastically, that is, without loss of energy, we can multiply both sides of Eq. 7.41 by the energy of one BB. Then, Eq. 7.41 becomes

Scattered power = $\sigma \times$ incident energy flux. (7.42)

This analogy must not be carried too far. Whereas in our mechanical example $\sigma = \pi R^2$ is a real surface and R a real radius, no such physical quantities must be ascribed to the oscillating electron shaken by the incident electromagnetic waves. One does not know the real radius of the electron (if indeed it has a radius); nonetheless, the quantity (cf. Eq. 7.40)

$$r_o \equiv \frac{q^2}{4\pi\epsilon_o mc^2} \tag{7.43}$$

$$= 2.82 \times 10^{-15} \text{ meter}$$

having the dimensions of a length, is usually called the "classical electron radius."

Knowledge of the cross section σ allows one to compute how much a light beam is degraded in energy in its passage through the material. Take a slice of matter like that shown in Fig. 7.23. The incident power is $\langle S \rangle A$; it is reduced by an amount $Ad\langle S \rangle$ in passing through a thickness dz. The reduction $Ad\langle S \rangle$ is due to the combined scattering from all the scatterers contained in the slice. If the density of

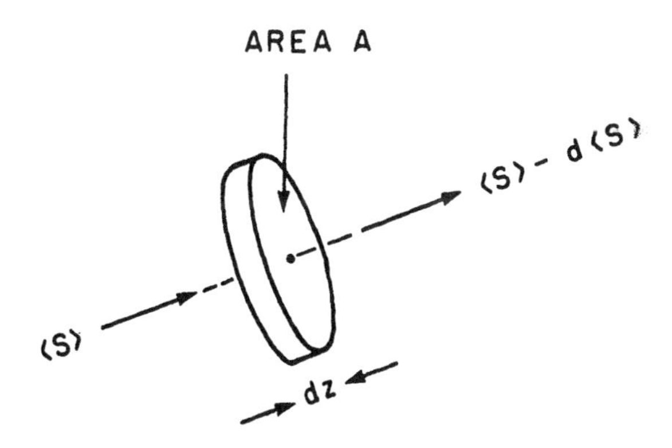

Fig. 7.23 The radiation flux $\langle S \rangle$ is diminished by an amount $d\langle S \rangle$ in passing through a slice of scattering material of thickness dz.

scatterers is N per (meter)^3, then there is NAdz of scatterers in the slice, each having a scattering "area" equal to σ. This assumes that the density N is sufficiently low that the scatterers do not shadow each other. The total power reduction is therefore

$$A d\langle S \rangle = -NAdz\sigma\langle S \rangle \tag{7.44}$$

where the minus sign indicates that as dz increases $\langle S \rangle$ decreases. Rearranging terms, it follows that $d\langle S \rangle / \langle S \rangle = -N\sigma dz$ which can be integrated to give

$$\langle S \rangle = \langle S_o \rangle e^{-N\sigma z} \tag{7.45}$$

where $\langle S_o \rangle$ is the energy flux of the light at the entrance into the medium, $z = 0$.

As an example, take light passing through the solar corona assumed to be uniform, and having an average density $N = 10^{15}$ electrons per cubic meter. By how much is the light intensity degraded in passing through 7×10^8 meters of corona? (this is a distance equal to about one solar radius). The quantity $N\sigma z$ of Eq. 7.45 equals $10^{15} \times 6.65 \times 10^{-29} \times 7 \times 10^8$ or $N\sigma z \approx 4.66 \times 10^{-5}$ so that

$$\begin{aligned} \langle S \rangle &= \langle S_o \rangle e^{-4.66 \times 10^{-5}} \\ &= 0.999953 \langle S_o \rangle \end{aligned} \tag{7.46}$$

which is a very tiny intensity loss, indeed. Nonetheless, the amount of scattering is enough to make the corona visible, albeit only during an eclipse, at which time the overpowering direct light from the Sun is sufficiently dimmed.

Observations of scattering from a medium can yield a great deal of information about the physical characteristics of that medium. For that reason many experiments are being made on the scattering from laboratory-produced plasmas, and from the ionosphere. [See, for example, G. Bekefi, Radiation Processes in Plasmas (John Wiley & Sons, New York, 1966), Chapter 8. Observations of scattering from interstellar grains is described in "Interstellar Grains" by J. M. Greenberg, Scientific American, October 1967. For a very advanced text, see H. C. Van de Hulst, Light Scattering by Small Particles (Wiley, 1957).]

Scattering by atoms – why the sky is blue – Rayleigh scattering

When light passes through a gas of neutral atoms, the bound electrons are set into vibration, and scattering takes place once again. However, the acceleration experienced by a bound electron is different from that of a free electron and as a result the details of the scattering are changed greatly. Consider the equation of motion (6.1) of a bound electron, but, for the sake of simplicity, neglect the damping term β. We know from section 6.5 that this neglect is not serious as long as the frequency of the wave ω is sufficiently far removed from the resonance frequency ω_o; that is, ω must not be in the region of anomalous dispersion illustrated in Fig. 6.10. Under these conditions the equation of motion of the electron is

$$\frac{d^2x}{dt^2} + \omega_o^2 x = \frac{q}{m} E_o \cos(\omega t) \tag{7.47}$$

which gives a displacement

$$x = \frac{q/m}{\omega_o^2 - \omega^2} E_o \cos(\omega t) \tag{7.48}$$

and an acceleration

$$a = \frac{d^2x}{dt^2} = -\frac{(q/m)\omega^2}{\omega_o^2 - \omega^2} E_o \cos(\omega t). \tag{7.49}$$

We now proceed exactly as in the previous subsection. We insert the acceleration a in the first equation (7.36) and time-average the result. Elimination of E_o^2 in favor of the time-average Poynting flux $\langle S \rangle$ then yields

$$\underbrace{\langle P \rangle_{\text{scattered}}}_{\text{watts}} = \underbrace{\frac{8\pi}{3}\left(\frac{q^2}{4\pi\epsilon_0 mc^2}\right)^2 \left[\frac{\omega^2}{\omega_0^2 - \omega^2}\right]^2}_{\text{cross-section } \sigma} \underbrace{\langle S \rangle_{\text{incident}}}_{\text{watts/area}} \qquad (7.50)$$

which is the sought-for result for the scattering from a bound electron, equivalent to Eq. 7.39 for the free electron. Indeed, when ω_0 is set identically equal to zero, the two equations are the same, as they must be.

Observe that when light is scattered from atoms, the cross-section σ is a strong function of frequency ω of the light. (For the free electron, σ is independent of ω.) Take, for example, the scattering of visible light from air atoms and molecules. We recall that for atmospheric gases the resonant frequencies ω_0 of the bound electronic oscillators lie in the ultraviolet and are therefore much higher than the frequency of visible light. Thus, $\omega_0 \gg \omega$ and one can approximate Eq. 7.50 by

$$\langle P \rangle_{\text{scattered}} \approx \frac{8\pi}{3}\left(\frac{q^2}{4\pi\epsilon_0 mc^2}\right)^2 \left(\frac{\omega}{\omega_0}\right)^4 \times \langle S \rangle_{\text{incident}} \qquad (7.51)$$

which shows that the scattered power is proportional to the fourth power of the light frequency. That is to say, blue light is scattered much more than red light. Take, for example, the two limits of visible light: dark blue of 3900 Å and dark red at 7600 Å. The former will be scattered more than the latter in the ratio $(7600/3900)^4$ or 14.4 times. This means that when white sunlight is incident on the Earth's atmosphere, the blue end of the spectrum is scattered to a far greater extent than the red end. That is why the sky is blue. Sunsets are red because the blue has been largely removed from the direct beam, leaving mainly red. In contradistinction, the scattering from free electrons is frequency-independent. That is why the sunlight scattered from the solar corona (Fig. 7.21) is white.

Example

The nitrogen (N_2) molecules in the terrestrial atmosphere have an electronic transition in the ultraviolet at a wavelength of approximately 750 Å. The total number of N_2 molecules in a column of 1 cm^2 cross section in a path through the atmosphere at the zenith is 1.68×10^{25} cm^{-2}.

a) Compute the fraction of blue sunlight, $\lambda \approx 4500$ Å, scattered out of the direct path from the Sun when the Sun is directly overhead.
b) Repeat the calculation for the case when the Sun is setting. Take the Sun to be at an elevation angle of 5° and assume a planar Earth and atmosphere, i.e., neglect the curvature of the Earth.

Solution

This problem involves Rayleigh scattering and since $\omega_o \approx 6\omega$ we are justified in using Eq. 7.51. Comparing this equation with Eq. 7.39 we see that the scattering cross section σ_R is given by

$$\sigma_R = \frac{8\pi}{3}\left(\frac{q^2}{4\pi\epsilon_o mc^2}\right)^2 \left(\frac{\omega}{\omega_o}\right)^4 = \sigma_T\left(\frac{\omega}{\omega_o}\right)^4 = \sigma_T\left(\frac{\lambda_o}{\lambda}\right)^4$$

where σ_T is the Thomson cross section given by Eq. 7.40 and equals 6.65×10^{-29} m^2.

The fractional reduction in intensity F can be found from Eq. 7.45 as

$$F = 1 - \frac{\langle S\rangle}{\langle S_o\rangle} = 1 - e^{-N\sigma z}.$$

We are told that $Nz = 1.68 \times 10^{25}$ $cm^{-2} = 1.68 \times 10^{29}$ m^{-2} for a path through the zenith. Thus

$$N\sigma_R z = Nz\sigma_T\left(\frac{\lambda_o}{\lambda}\right)^4 = (1.68\times 10^{29})(6.65\times 10^{-29})\left(\frac{750}{4500}\right)^4$$

$$N\sigma_R z = 8.62 \times 10^{-3}.$$

Hence the fraction scattered out of the direct path is

$$F = 1 - e^{-8.62\times10^{-3}} \approx 1 - (1 - 8.62\times10^{-3})$$

$$F = 0.0086 = 0.86 \text{ percent.}$$

When the Sun is at an elevation angle of 5° (zenith angle of 85°) the path length through the atmosphere is increased by 1/cos 85° = 1/0.0872 = 11.47. Thus

$$N\sigma_R z_{5°} = (8.62\times10^{-3})(11.47) = 0.0989.$$

Therefore

b) $$F = 1 - e^{-9.89\times10^{-2}} = 0.0989 = 9.89 \text{ percent.}$$

Polarization of scattered light

Although the direct light from the Sun is randomly polarized, the light scattered by the Earth's atmosphere is polarized, as can be easily verified by looking at a section of blue sky through a piece of polaroid held close to one's eyes.

The explanation for this polarizing characteristic possessed by elementary scatterers comes from the fact well known to us already: that an accelerated charge radiates primarily in a direction perpendicular to the acceleration vector $\vec{a}$, with no emission taking place along $\vec{a}$. Thus, as the observer of Fig. 7.24 sweeps his head along the arc of a circle, he sees the following: at zero angle of scattering ($\theta_{sc} = 0$) the radiations from both the horizontal and vertical components of the acceleration vectors contribute equally — the radiation is unpolarized at this angle of observation. As θ_{sc} increases from zero, the emission associated with the vertical component of $\vec{a}$ remains unchanged, but the emission intensity associated with the horizontal component of $\vec{a}$ decreases as $\cos^2\theta_{sc}$ (cf. Eq. 4.11 and Figs. 4.7 and 4.8). Ultimately, when $\theta_{sc} = 90°$, there is zero radiation associated with the horizontal component of $\vec{a}$; the observer then sees 100% vertically polarized light.

In section 7.4 we found that specular reflection is one way of pro-

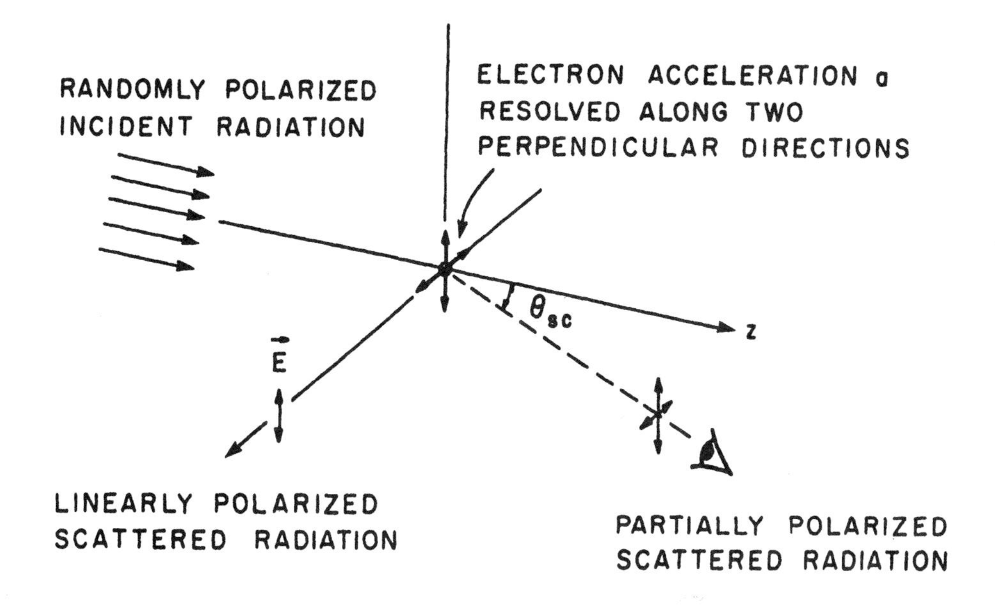

Fig. 7.24 As the angle of scattering θ_{sc} increases from zero, light becomes more and more polarized, and is 100% linearly polarized when $\theta_{sc} = 90°$. This is true as long as the density of scatterers is sufficiently small so that the radiation escapes from the medium without suffering further scattering events. Multiple scattering in dense media results in depolarization, and eventually the scattered light becomes randomly polarized.

ducing polarized light. Scattering offers another means of doing this.

A slight dilemma

On previous pages we computed the scattering from a single stationary electron and a single stationary atom. (By stationary we mean that in the absence of the incident wave there is no movement. The wave induces vibrations.) The implication was that if we had NV such scatterers randomly distributed throughout a volume V (N is the particle density), the scattered power would be simply $NV\sigma\langle S\rangle$ with σ as the cross section and $\langle S\rangle$ as the incident energy flux. That is, one simply multiplies the effect by the total number of scatterers present. This conclusion is erroneous and the total scattered power from such a system would, in fact, be zero. This is readily demonstrated as follows.

Take a very thin plane slab of electrons oriented perpendicularly to the vector $\vec{K}$ defined as $\vec{K} = \vec{k}_i - \vec{k}_s$ where $\vec{k}_i$ and $\vec{k}_s$ are the propagation vectors of the incident and scattered waves, respectively (see Fig. 7.25). The slab is so thin that all the electrons within it contribute in phase to the radiation emitted in the direction $\vec{k}_s$. Now, choose a second slab parallel to the original one, and separated from it by a distance $\pi/|\vec{K}|$. Thus the signals from the two slabs arrive exactly out of phase with one another. Since the medium is perfectly homogeneous, the two slabs contain the same number of electrons, and the two signals cancel exactly. This is true for any such pair of slabs. There is no scattered radiation!

We got ourselves into this apparent dilemma by taking account of the relative phases of the radiations. Destructive interference gave the null result. That scattering does, in fact, take place is entirely due to <u>density fluctuations</u> that produce an unbalance in the number of scatterers in the respective slabs. Statistical mechanics shows that the root mean square density fluctuation is $\sim\sqrt{N/V}$. Thus, if V is the volume of each of the two slabs, the difference in their electron populations is proportional to $\sqrt{N/V}$. Therefore the net amplitude of the scattered signal is proportional to $\sqrt{N/V}$. The power is proportional to the square of this value and is therefore proportional to the average number density N. Note that we were right after all by taking the scattered power from each particle and multiplying by the number of particles! The reason, however, is quite subtle, as we have just shown.*

We can now understand better why scattering from transparent liquids and solids is usually much smaller than from a gas, although the density of scatterers can be several orders of magnitude greater. In these substances the oscillators are arrayed in a more orderly

*There is another consequence. Fluctuations imply that the scatterers are in motion; their random velocities cause the scattered light to be Doppler-shifted. Thus the scattered light is not at a single frequency ω but there is a spread of frequencies about ω. A measurement of this spread is a favorite way of determining the mean-square velocity of the scatterers, i.e., their temperature.

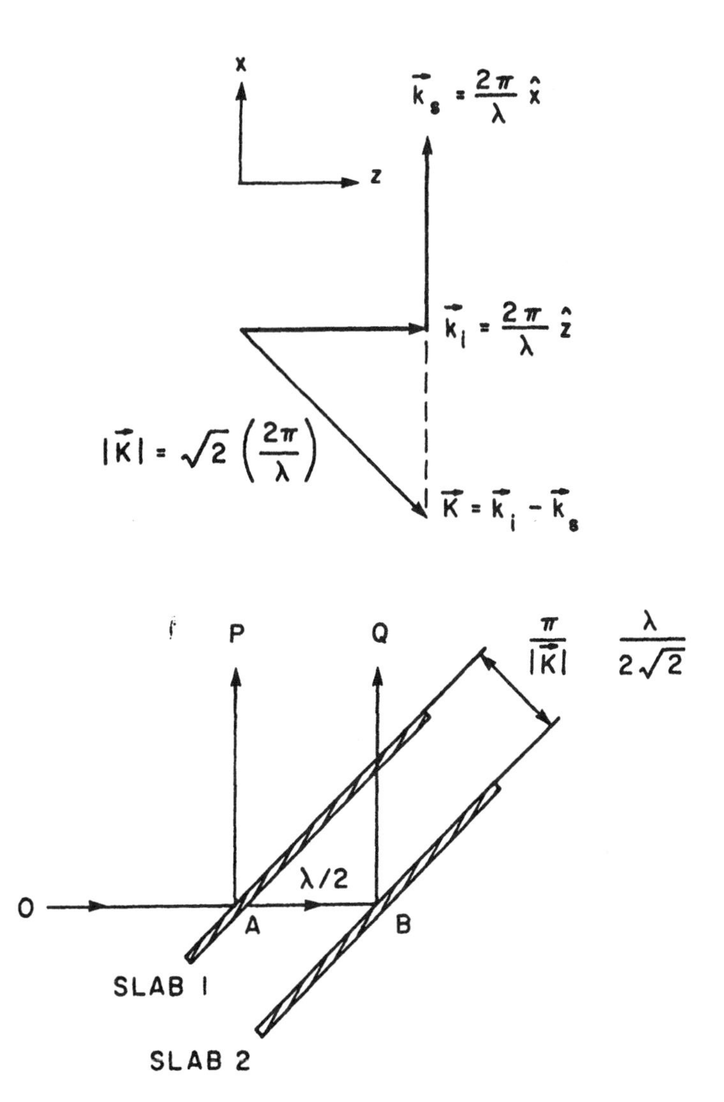

Fig. 7.25 For simplicity we take 90° scattering, with the incident wave along the z axis and the scattered wave along the x axis, as shown in the top figure. The two slabs are separated by a distance $\pi/|\vec{K}| = \lambda/2\sqrt{2}$; they lie perpendicular to $\vec{K}$. The difference in the optical path lengths 0AP and 0ABQ is AB = $\lambda/2$ and the signals cancel destructively.

manner and the fluctuations are also smaller. Hence the reemitted waves tend to cancel one another to a high degree. An exception is in the forward direction, that is, the direction of incidence ($\theta_{sc} = 0$). In only this direction the reemitted waves add up in phase to give a beam in the same direction as the incident beam but of somewhat different phase. The beam is precisely what we call the "transmitted ray through the medium." The phase difference, obtained by adding the contributions of all of the scatterers, is the origin of the index of refraction discussed in previous sections. A detailed proof of this will be found, for example, in The Magneto-Ionic Theory and Its Applications to the Ionosphere, J. A. Ratcliffe (Cambridge University Press, London, 1959), Chapter 3. For a less exhaustive discussion, see The Feynman Lectures on Physics(Addison-Wesley, 1963), Volume I, section 31.

Radiation damping

Let us consider once again the equation of motion

$$m\frac{d^2x}{dt^2} = qE_o \cos \omega t \tag{7.52}$$

of a free electron subjected to an external oscillatory force $F_{ext} = qE_o \cos \omega t$. The electron acquires a velocity

$$v = \frac{qE_o}{m\omega} \sin \omega t \tag{7.53}$$

and an acceleration

$$a = \frac{qE_o}{m} \cos \omega t. \tag{7.54}$$

Because of the acquired acceleration, the electron will radiate an average power given by Eq. 7.37:

$$\langle P \rangle = \frac{q^4 E_o^2}{12\, \pi\epsilon_o m^2 c^3} \text{ watt.} \tag{7.55}$$

Conservation of energy demands that the work done ($F_{ext}v$) by the applied field be equal to the radiated energy. Let us check whether or not this is so. Substituting for F_{ext} and v from the above equations and time-averaging over a period $T = 2\pi/\omega$ gives

$$\langle Fv \rangle = \frac{q^2E_o^2}{m\omega}\frac{1}{T}\int_0^T \cos(\omega t)\sin(\omega t)\,dt \quad \text{watt} \tag{7.56}$$

$$= 0.$$

We see that the work done by the applied electric field averages out to zero. If this is so, where does the radiated energy (7.55) come from? Herein lies the inconsistency of the equation of motion (7.52) (note that the magnetic force of the wave $q(\vec{v} \times \vec{B})$ which has not been included in Eq. 7.52 is not the answer to our dilemma, since this force does not contribute to the work done).

These considerations impel us to conclude that the basic equation of motion should contain some additional force doing work on the electron, which does not time-average to zero. For this to be so, the velocity v needs to have a component in phase with the externally applied force. So, with a little guessing and a little intuition we add a correction term to the right-hand side of Eq. 7.53

$$v = \frac{qE_o}{m\omega}\sin\omega t + A\cos\omega t \tag{7.57}$$

and see where this is going to lead us. Specifically, our task is to determine the unknown coefficient A. With the new term added, the work done per second by the external force $qE_o \cos\omega t$ becomes

$$\langle Fv \rangle = \frac{1}{T}\int_0^T [qE_o\cos(\omega t)]\left[\frac{qE_o}{m\omega}\sin(\omega t) + A\cos(\omega t)\right]dt \tag{7.58}$$

$$= \frac{1}{2}qE_oA.$$

Setting this power equal to the power radiated as given by Eq. 7.55 gives the value of A

$$A = \frac{q^3E_o}{6\pi\epsilon_o m^2c^3} \tag{7.59}$$

and hence the new value of the corrected electron velocity:

$$v = \frac{qE_o}{m\omega}\sin(\omega t) + \frac{q^3E_o}{6\pi\epsilon_o m^2c^3}\cos(\omega t.) \tag{7.60}$$

Differentiation with respect to time allows us to write an equation in the form of F = ma,

$$\underbrace{m\frac{d^2x}{dt^2}}_{\substack{\text{mass} \times \\ \text{acceleration}}} = \underbrace{qE_o\cos(\omega t)}_{\substack{\text{external force} \\ F_{ext}}} - \underbrace{\frac{q^3\omega}{6\pi\epsilon_o mc^3}E_o\sin(\omega t)}_{\substack{\text{radiative reaction} \\ \text{force } F_{rad}}}. \tag{7.61}$$

The new force, F_{rad}, can be viewed as a radiative reaction force which accounts for the emission of electromagnetic waves. Its value

$$F_{rad} \equiv \frac{q^3\omega}{6\pi\epsilon_o mc^3}E_o\sin(\omega t) \tag{7.62}$$

is normally much, much smaller than the value of most externally applied fields. Taking the ratio $|F_{rad}/F_{ext}|$ gives

$$\left|\frac{F_{rad}}{F_{ext}}\right| = \frac{q^2\omega}{6\pi\epsilon_o mc^3} = 6.26 \times 10^{-24}\,\omega \tag{7.63}$$

which is small for all frequencies well below ~10^{23} rad/sec. This means that with the exception of very energetic cosmic rays, the radiative reaction force is but a small correction factor to the equation of motion of the electron.

What we have said about the free electron applies also to the bound electron driven by an external force $qE_o\cos(\omega t)$. The radiative energy loss now manifests itself as a line broadening of the natural frequency ω_o, because, as we know from section 1.4, a damped, driven harmonic oscillator exhibits a line broadening like that illustrated in Fig. 1.21.

Writing the equation of motion of such as oscillator in the form

$$m \frac{d^2x}{dt^2} + m\omega_o^2 = qE_o \cos(\omega t) - m\beta \frac{dx}{dt} \tag{7.64}$$

with $m\beta\, dx/dt$ as the damping force (cf. Eqs. 1.90 and 6.1), allows us to identify the radiative reaction force F_{rad}

$$m\beta \frac{dx}{dt} \rightarrow \frac{q^3\omega}{6\pi\epsilon_o mc^3} E_o \sin(\omega t) \tag{7.65}$$

simply by comparing Eqs. 7.61 and 7.64. But from Eq. 7.57 we know that $v = dx/dt \approx (qE_o/m\omega) \sin \omega t$ (the second term $A \cos \omega t$ is but a small correction). Inserting this value of dx/dt in Eq. 7.65 leads to the sought-for value of the damping constants β:

$$\beta = \frac{q^2\omega^2}{6\pi\epsilon_o mc^3}. \tag{7.66}$$

An examination of Fig. 1.21 or Eq. 1.103 tells one that β equals the linewidth $\Delta\omega$, or more precisely, "the full linewidth at half-maximum power"; that is,

$$\Delta\omega = \frac{q^2\omega^2}{6\pi\epsilon_o mc^3} \text{ rad/sec.} \tag{7.67}$$

It is more convenient to express the broadening in wavelength units $\Delta\lambda$, rather than frequency unit $\Delta\omega$. Since $\omega\lambda = 2\pi c$, it follows that $\omega\Delta\lambda + \lambda\Delta\omega = 0$ or $|\Delta\lambda| = |\lambda\Delta\omega/\omega| = (2\pi c/\omega^2)\ \Delta\omega$. Hence

$$\begin{aligned} \Delta\lambda &= \frac{q^2}{3\epsilon_o mc^2} \\ &= 1.2 \times 10^{-4}\ \text{Å} \end{aligned} \tag{7.68}$$

and is a universal constant. This width is known as the "natural linewidth" of a spectral line. It is an intrinsic irreducible broadening which remains after all other broadening mechanisms (Doppler broadening, collisional broadening, etc.) are eliminated. Quantum-mechanical calculations give a somewhat different result for $\Delta\lambda$, and they show that the natural linewidth is not a universal constant, but is, in fact, a function of wavelength of the spectral line. For a good discussion, see V. F. Weisskopf, Rev. Mod. Phys. 21, 305 (1949).

There are two other, related pieces of information that can be extracted from these results. The one has to do with the Q of the driven oscillator, damped by radiation alone. From Eq. 1.110

$$Q = \frac{\omega}{\Delta\omega} = \frac{6\pi\epsilon_o mc^3}{q^2\omega} \tag{7.69}$$

with $\omega \sim 10^{15}$ rad/sec (which corresponds to the visible part of the electromagnetic spectrum), the Q takes on a value $\sim 10^8$ which is a high value indeed. Second, one can compute the typical lifetime τ of the oscillator. Adopting the classical picture of the simple harmonic oscillator set into vibration and then allowed to decay freely, we see from Eqs. 1.79 and 7.66 that

$$\tau = \frac{1}{\beta} = \frac{6\pi\epsilon_o mc^3}{q^2\omega^2}. \tag{7.70}$$

Substituting numbers into this expression, we find that when ω lies in the optical frequency range, τ lies between $\sim 10^{-8}$ and $\sim 10^{-9}$ seconds, which is a well-known and much quoted result.

In conclusion we must warn the reader: The superficially plausible derivation of the radiative reaction force F_{rad} given here is beset by a host of difficulties which touch upon some of the most fundamental aspects of the nature of the electron and other fundamental particles. Unfortunately, not even quantum mechanics has yet been able to resolve the difficulties.

7.6 STIMULATED EMISSION – THE LASER

One of the most exciting discoveries of the last two decades, with innumerable applications to science and technology, is the laser, a word which stands for "light amplification by stimulated emission of radiation," and its next of kin, the maser (microwave amplification by stimulated emission of radiation). Ever since the demonstration of the first microwave ammonia maser in 1954 by Townes, and the first light ruby laser in 1960 by Maiman, there has been an extraordinary growth in the variety of laser and maser devices. Today, they operate at frequencies from the audio to the ultraviolet, and with output power levels from milliwatts to gigawatts. Yet, despite the almost bewildering variety, the devices are all based on a few rather simple physical principles. As the name suggests, there is amplification. That is, when an electromagnetic wave passes through matter (gas, liquid, or solid) which was initially prepared in a very special way, the wave can be made to grow exponentially with distance as

$$E = E_o e^{\alpha z} \cos(\omega t - kz) \qquad \alpha > 0 \qquad (7.71)$$

where α is a constant that is a function of the medium properties. The growth of energy in the wave [which is proportional to $E^2 \sim \exp(2\alpha z)$] must, by conservation of energy, come at the expense of the energy degradation of the lasing medium, that is, at the expense of the energy of the electrons, atoms, or molecules.

The form of Eq. 7.71 is not unfamiliar to us. Equation 6.50 of section 6.5 is of similar form, with one major difference – the argument of the exponential is negative rather than positive, which signifies damping rather than growth. What, then, is the difference between wave propagation through a medium as discussed in section 6.5 and wave propagation as discussed in this section? Mathematically, the difference, of course, is fairly trivial, a mere change in the sign of the coefficient α. To achieve it physically is another thing altogether. This is where the initial preparation of the matter is so very important. The basic underlying idea, broadly speaking, is the following. Matter, as it most commonly occurs in nature, is composed of a completely chaotic assembly of randomly jiggling atoms. When a wave passes through this assembly it will invariably lose energy (α is negative) because it sets the atoms in motion and the atoms, through colliding with one another, dissipate this motion in the form of heat (the temperature of the medium rises). On the other hand, when the medium is properly prepared, by which we mean that the atoms are promoted into a more ordered, less chaotic state of motion, there is available some ordered energy (known as free energy in thermodynamics) which can be delivered to the wave to make it grow.

Detailed analysis shows that in any system of waves and particles, there are two competing processes that go on at the same time. And it is these two processes that when added up give the net, observed, coefficient α appearing in Eq. 7.71

$$\underbrace{\alpha}_{\substack{\text{net measured}\\ \text{by observer}}} = \underbrace{\alpha_1}_{\substack{\text{stimulated}\\ \text{emission}}} - \underbrace{\alpha_2}_{\substack{\text{stimulated}\\ \text{absorption}}}. \tag{7.72}$$

The coefficient α_2 represents processes whereby the wave loses energy in the form of heat dissipation in the medium. Coefficient α_1 represents the reverse process in which atoms jiggling in the correct phase relationship with the wave can give up some of their energy to the wave. In an unprepared medium, in which the atoms are in their most chaotic state, α_2 is always found to be greater than α_1; thus, α is negative and there is a net loss of energy from the wave. In a suitably prepared medium, α_1 can exceed α_2, α is positive, and the wave grows: the medium is said to "lase." Note that the exponential growth given by Eq. 7.71 cannot continue indefinitely; if it did, the wave would gain an infinite amount of energy. Exponential growth continues until all the free energy is exhausted.

Some of these concepts can be demonstrated by taking the simple case of a classical harmonic oscillator driven by an electromagnetic wave

$$\frac{d^2x}{dt^2} + \omega_o^2 x = \frac{q}{m} E_o \cos(\omega t). \tag{7.73}$$

Since, as we have said, phase relationships between particle and wave must now be carefully kept track of, we must seek the complete solution of the differential equation. It is given in Appendix 2, Eq. A2.22:

$$x = A\cos(\omega_0 t) + B\sin(\omega_0 t) + \frac{(q/m)}{\omega_0^2 - \omega^2} E_0 \cos(\omega t). \qquad (7.74)$$

Suppose that at time $t = 0$ the system exhibits only free, undriven oscillations, that is, as $t \to 0$, $x \to C\cos(\omega_0 t + \phi)$ where C and ϕ are the amplitude and phase, respectively. From these initial conditions it follows that $A = C\cos\phi - (q/m)E_0/\left(\omega_0^2 - \omega^2\right)$ and $B = -C\sin\phi$, so that

$$x = C\cos(\omega_0 t + \phi) + \frac{(q/m)E_0}{\left(\omega_0^2 - \omega^2\right)}[\cos(\omega t) - \cos(\omega_0 t)]. \qquad (7.75)$$

The instantaneous work done per second by the oscillator is $Fv = [qE_0 \cos\omega t][dx/dt]$. The time-average work done per second is obtained by integrating over a period $T = 2\pi/\omega$ of the driving oscillation. Noting that

$$\int_0^T \cos(\omega t)\sin(\omega t)\,dt = 0 \qquad (7.76)$$

we obtain that

$$\langle P \rangle = \frac{1}{T}\frac{(qE_0)^2}{m}\left(\frac{\omega_0}{\omega_0^2 - \omega^2}\right)\int_0^T \cos(\omega t)\sin(\omega_0 t)\,dt$$

$$- \frac{1}{T}(qE_0 C\omega_0)\int_0^T \cos(\omega t)\sin(\omega_0 t + \phi)\,dt. \qquad (7.77)$$

When we evaluate the integrals we find that the first term of Eq. 7.77 is always positive; however, the second term can be either positive or negative, depending on the value of the phase ϕ. When the two terms combine to give a positive $\langle P \rangle$, the electron does net work on the field, the electron gains energy, and by energy conservation the electromagnetic oscillations lose energy — there is damping. But, for certain phase angles ϕ, $\langle P \rangle$ can be negative and the oscillation will grow. We also observe that the oscillations can only grow when C, the initial amplitude, is nonzero. The reason for this is clear: when $C = 0$ at $t = 0$, the particle has initially neither kinetic nor potential energy and thus there is no reservoir from which the wave can feed itself.

To verify some of these statements, we must evaluate the right-hand side of Eq. 7.77. Using standard tables of integrals we find, after a little tedious algebra, that

$$\langle P \rangle = \frac{1}{T}\frac{(qE_o)^2}{m}\left(\frac{\omega_o}{\omega_o^2 - \omega^2}\right)^2\left[1 - \cos\left(2\pi\frac{\omega_o}{\omega}\right)\right]$$

$$-\frac{1}{T}(qE_o C\omega_o)\left(\frac{\omega_o}{\omega_o^2 - \omega^2}\right)\left[\cos\phi\left\{1 - \cos\left(2\pi\frac{\omega_o}{\omega}\right)\right\}\right.$$

$$\left. + \sin\phi \sin\left(2\pi\frac{\omega_o}{\omega}\right)\right]. \tag{7.78}$$

This result is not very clear, but it simplifies greatly when we allow the driving frequency ω to approach the resonant frequency ω_o, which is where the wave-particle interaction can be expected to be strongest. Thus without further ado we let $\omega \to \omega_o$ on the right-hand side of Eq. 7.78 and obtain that

$$\langle P \rangle = \frac{\pi}{T}\left[\frac{\pi}{2}\frac{(qE_o)^2}{m\omega_o^2} - (qE_o C)\sin\phi\right]. \tag{7.79}$$

We now see that $\langle P \rangle$ is negative and wave growth occurs provided that

$$\left.\begin{aligned} &\sin\phi > 0 \\ \text{and} \quad & \\ &|C \sin\phi| > \left|\frac{\pi}{2}\frac{qE_o}{m\omega_o^2}\right|. \end{aligned}\right\} \tag{7.80}$$

The last two conditions are initial requirements to be imposed on the system; they illustrate what we meant earlier when we talked about suitable "preparation of the medium."

We shall now make a maser from nothing but free electrons. The mechanism by which it operates is not only instructive, but the gadget also works – in fact, a gigawatt (10^9 watts) of microwave power has been generated in it. Here, then, are the ingredients (see Fig. 7.26a): A beam of electrons all traveling in one direction is acted upon by a

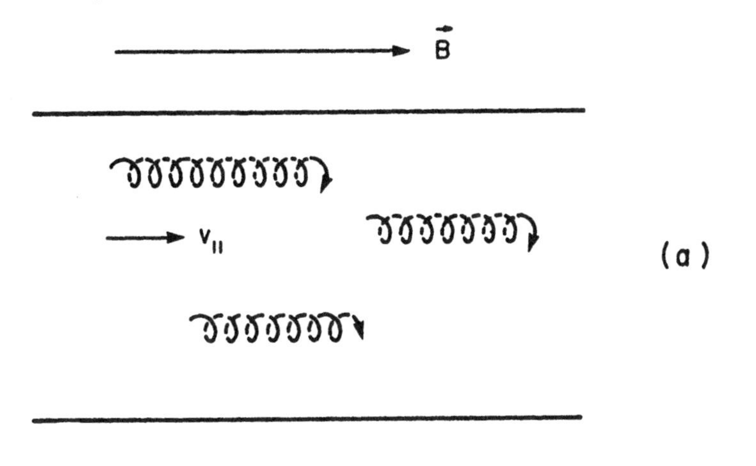

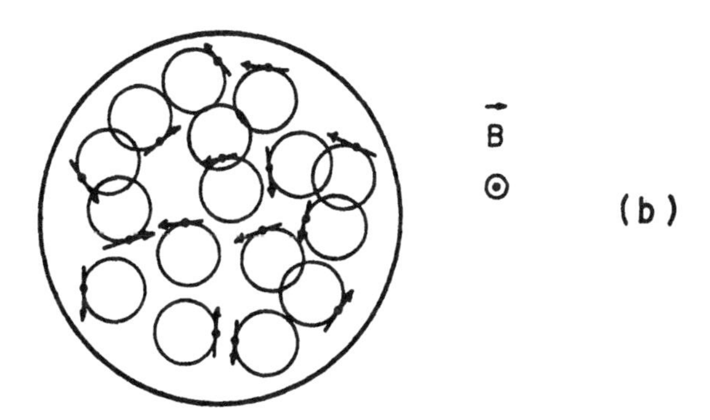

Fig. 7.26 A beam of electrons executing helical motions in a steady, uniform magnetic field $\vec{B}$. (a) Side view of the beam. (b) The parallel electron velocity $v_{\parallel}$ has been set equal to zero; the beam is shown front view. Observe that the electrons, as they move in their orbits, have completely random phases relative to one another.

steady uniform axial magnetic field $\vec{B}$. The electrons also have transverse energy, so they do not go in straight lines but in helices of radius $R = v_\perp/\omega_c$ where $\omega_c = qB/m$ is the familiar electron cyclotron frequency and $v_\perp$ is their velocity perpendicular to the magnetic field. There is an electromagnetic wave permeating the electron gas. The wave is linearly polarized. The parallel energy of the electrons does not enter the picture at all and so we may assume that $v_\parallel$ is zero; all the electrons are now doing is making circular orbits in the magnetic field. The simplified picture one then obtains is like that shown in Fig. 7.26b.

We now pick a class of electrons whose motion is in phase with the applied electric field $E = E_0 \cos \omega t$. That is, we pick those electrons that move up the page when the electric field points up the page. If the frequency ω of the wave is close to the cyclotron frequency $\omega_c = qB/m$, then half a period later, when the field points down the page, the electron velocity is also directed down the page. This is shown in Fig. 7.27. The electrons are being continuously decelerated by the electric field. Therefore particles with approximately the phase shown [call it θ(right)] keep losing energy. Of course, there are electrons with just the opposite phase [θ(wrong)] that move down the page when the field points up the page, and vice versa. This second class of electrons gains energy at the expense of the wave which loses energy.

The last-mentioned class of electrons which cause loss of wave energy are the ones responsible for stimulated absorption (see Eq. 7.72); the former class are responsible for stimulated emission. Thus, if the number of particles with the "right" phase is larger than with the "wrong" phase, the net effect is for the assembly to lose energy, and for the wave to grow.

What kind of mechanism (or is it some sort of a demon) can one envisage which puts particles preferentially into θ(right), and thus makes the medium lase? There are several such mechanisms but the one we shall concentrate on has to do with relativity. It is known that the mass of a relativistic particle varies with speed as $m = m_0 [1 - (v^2/c^2)]^{-1/2}$, that is, the larger the speed, the larger is the mass m

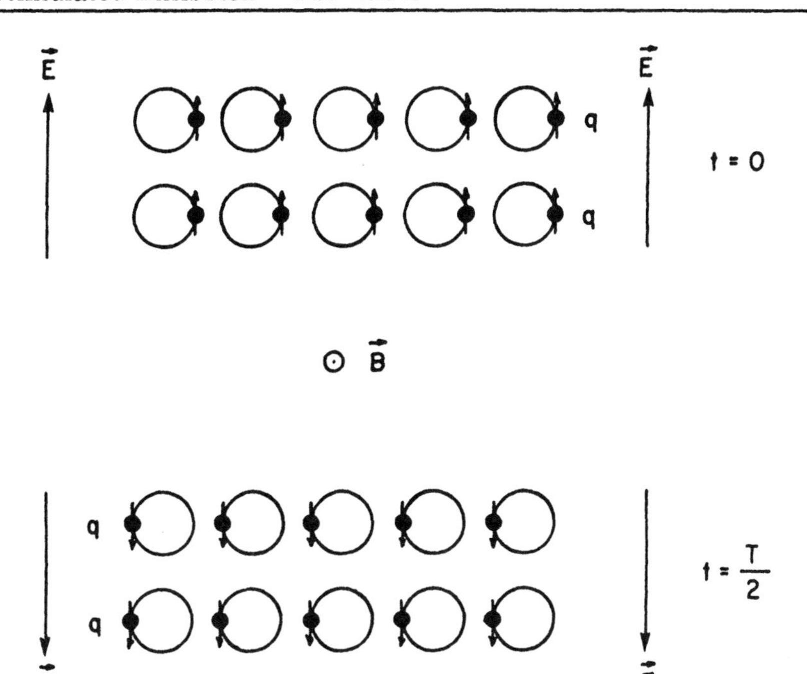

Fig. 7.27 Schematic drawing of a certain class of electrons moving in phase with the oscillating applied electric field $\vec{E} = \vec{E}_0 \cos \omega t$. The rotation frequency $\omega_c = qB/m$ is equal, or close to, the frequency ω of the electric field oscillations. The lower drawing represents conditions one-half period later. Observe that since q for an electron is negative, the force $qE_0 \cos \omega t$ tends to decelerate the electrons in orbits shown.

(m_0 is the rest mass). As a result, the rotation frequency ω_c of a relativistic electron is dependent on the speed, and is given by

$$\omega_c = \frac{qB}{m} = \frac{qB}{m_0}\sqrt{1 - \left(\frac{v_\perp}{c}\right)^2}$$
$$\equiv \frac{\omega_{co}}{\gamma}. \tag{7.81}$$

where we define $\omega_{co} = qB/m_0$ and $\gamma \equiv [1 - (v_\perp/c)^2]^{-1/2}$. From this equation it can be seen that the more energetic the particle, the slower is its rotational frequency.

Consider, then, a particle which enters the interaction region with a phase θ(right); it begins to lose energy, γ decreases and the particle rotation frequency ω_{co}/γ increases. As the rotation frequency increases, it comes ever closer to the wave frequency ω. This is

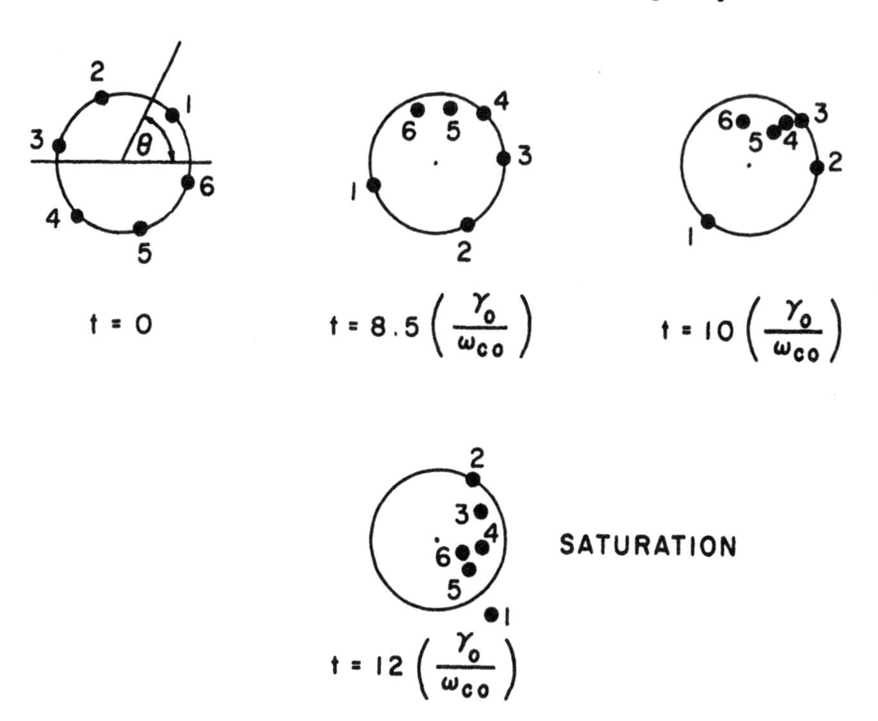

Fig. 7.28 Graphical demonstration of the bunching of electrons, as obtained by computer simulation. Initially (t = 0) six particles are uniformly spaced in their phase θ along the circle shown. The radius of the circle is proportional to γ_o, that is, to the initial energy of the electrons (in these plots $\gamma_o = 2$ and $v_\perp = v_\parallel = 0.61\ c$ for each of the six electrons). At a time 8.5 cyclotron periods later we see a bunching of particles in the upper half plane, $0 \leq \theta \leq \pi$. At a time of 10 cyclotron periods, the bunching is even more pronounced. Observe that some particles no longer lie on the circle, but within it. This means that they have lost energy to the electric field. One finds that at a time of approximately 12 cyclotron orbits, the degree of bunching has reached its maximum, as has the total energy given up to the wave. [After P. Sprangle and W. M. Manheimer, Phys. Fluids 18, 224 (1975).]

necessary since for a strong interaction to occur ω must be in resonance with ω_{co}/γ; in fact, one finds that ω should be just a little greater than ω_{co}/γ. Since the difference $[\omega - (\omega_{co}/\gamma)]$ decreases, the particle tends to remain for a somewhat longer time in the resonant interaction region. In other words, particles tend to bunch in regions where they lose energy (see Fig. 7.28). On the other hand, when the particles enter in the wrong phase, θ(wrong), they gain energy, their rotation frequency decreases, and so they tend to debunch there. This asymmetry caused by the relativistic mass change provides for the sought-for "demon" that throws particles into the favorable phase. We may mention that the effect has been observed even for very mildly relativistic electrons having perpendicular energies of only several kiloelectron volts. For details of the theory see G. Bekefi, J. L. Hirshfield, and S. C. Brown, Phys. Rev. 122, 1037 (1961). Recent experiments yielding tens to hundreds of megawatts of microwaves, and utilizing beams with energies $\sim 10^5$-10^6 eV and pulsed currents of ~15 kA, are reported by V. L. Granatstein, M. Herndon, R. K. Parker, and P. Sprangle, IEEE J. Quant. Electronics, Vol. QE-10, 652 (1974). Earlier experiments with milliampere currents and kilovolt energies were reported by J. L. Hirshfield and J. M. Wachtel, Phys. Rev. Letters 12, 533 (1964).

The final ingredient in the above experiment, and one which constitutes an important step in the "preparation" of the medium has to do with the distribution of transverse energies of the electrons. All electrons were assumed to have initially the same energy (since $v_\perp$ was taken to be the same for each particle). Suppose, instead, the electrons had a random, Gaussian, distribution of momenta* $p_\perp$, as is illustrated in Fig. 7.29a where most electrons have little or no momentum and only a few have lots of momentum. Now, a particle entering the interaction region with virtually no momentum, even if it is initially in the correct phase, has no energy available to give to the wave. It can only absorb energy, and it does. Particles with somewhat more momentum also have little to give. When such electrons

*Since the particles are relativistic, it is more convenient to talk of momentum rather than velocity.

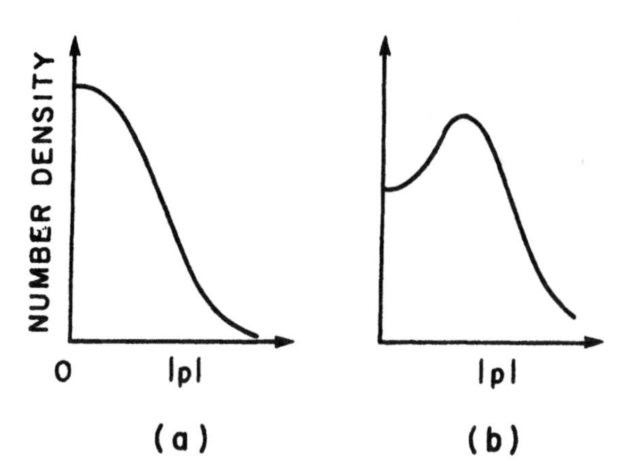

Fig. 7.29 (a) Example of particle distribution in normally occurring matter. This distribution will not lead to laser action, since there are more particles that absorb energy from the wave by stimulated absorption than give energy to the wave by stimulated emission.
(b) Particle distribution function showing inverted populations. Possibility of lasing exists. Note that although such inversion is a necessary condition for lasing, it does not guarantee lasing. Several other criteria must be satisfied in addition.

lose enough energy so that $\omega - (\omega_{co}/\gamma) = 0$, they cannot lose further energy by the mechanism we have just discussed, and from then on they tend to debunch and gain energy in the process, which is counter to what is necessary for lasing. In short, when there are many more interacting particles of low energy (which absorb radiation) than high energy particles (which stimulate the radiation), lasing is impossible. To obtain lasing one must, as it is usually expressed, <u>invert the population</u>. Figure 7.29b shows an example of an inverted population, amongst which the electron beam having electrons with identical $p_{\perp}$'s is but a special case.

We have attempted to explain maser action solely in terms of classical physics, using the classical equations of motion of particles. We have done so purposely in order to emphasize the fact that the concept of stimulated emission is perfectly valid in the context of classical physics. Here recourse to quantum physics is neither necessary nor is it advantageous. To be sure, most masers or lasers do not use free, classical particles* but they rely on quantum transitions of bound states of atoms and molecules. In such cases, of course, a quantum description is mandatory in the final analysis. But, interestingly, even this quantum description does not differ dramatically from the classical formulas, and there are many points of contact. This fact is well brought out in An Introduction to Lasers and Masers by A. E. Siegman (McGraw Hill Book Company, New York, 1971).

Stimulated emission, as the name suggests, is a process which is induced by other electromagnetic radiation being present and interacting with the electrons or atoms. If, in a given region of space, there were nothing but matter around, stimulated emission would not exist. This differs from spontaneous emission which is another process whereby electromagnetic energy is created. Spontaneous emission by a classical oscillator is what Chapter 4 was all about; that is, generation of energy by accelerated charges. In a quantum mechanical system, spontaneous emission is what one sees primarily coming from a fluorescent lamp, an arc lamp, a mercury or neon tube. The photons are created whenever an atom makes a transition (spontaneously) from a higher to a lower state. No background of electromagnetic radiation is needed to keep spontaneous emission going. It is there whether one wants it or not, and in the case of lasers and masers, spontaneous emission is something one does not desire. It represents a background of noiselike radiation that has no temporal or spatial coherence with the amplified, stimulated light.

In summary, then, the interaction of matter and radiation is comprised of three and only three basic processes: spontaneous emission, stimulated emission, and stimulated absorption; they go on simultaneously. The last two combine into one experimentally observable quantity – the damping or growth of a wave shone into the medium at one end and measured at the other end. Spontaneous emission appears to stand apart from the latter two. And although it is a distinct process,

---

*However, for a recent example of a free electron laser see D. A. G. Deacon et al., Phys. Rev. Letters 38, 892 (1977).

all three — spontaneous emission, stimulated emission, and stimulated absorption are connected in a most intimate way, a fact which was first demonstrated by Albert Einstein in a classic paper [Physikalische Zeitschrift, Vol. 18, page 121 (1917)]. It is, in fact, in this paper that the concept of stimulated emission was born.

CHAPTER 8

INTERFERENCE AND DIFFRACTION

We began this book by enunciating the most fundamental property of all wave motion, the Principle of Superposition. And we shall end the book by devoting the present chapter to applications of this principle. When two or more beams of radiation are superposed, the distribution of intensity is, in many situations, not merely the sum of intensities of the individual beams acting alone. Thus, for example, if the beams from two antennas driven from the same rf source are superposed, the intensity in the region of intersection of the beams is found to vary from point to point between maxima, which exceed the sum of the intensities of the beams, and minima which may have zero intensity. This phenomenon is called interference.

When a body is placed between a point source of radiation and an observation screen, the shadow on the screen is not a simple sharp boundary between light and dark, but is comprised of an intricate pattern of bright and dark regions. This phenomenon is called diffraction, and occurs whenever a portion of a wave front is obstructed by material objects, be they opaque or transparent to the incident radiation. At first glance, interference and diffraction appear to contain entirely different physical notions. But this is not so. As we shall show, they are both manifestations of the same principle of superposition of waves. The name "interference" is usually reserved for situations where only a few beams are allowed to interact with one another. The name "diffraction" is usually applied to situations where many, many beams, even a continuous distribution of beams, are made to interact. But the distinction between interference and diffraction is very fuzzy and cannot be defined precisely. Therefore it makes good sense to lump the two into a single chapter. And this is what we have done here.

8.1 INTERFERENCE OF TWO MONOCHROMATIC WAVES

The quantity most accessible to the detection and measurement of electromagnetic radiation is its intensity I (also called the "irradiance" in modern usage). It is this quantity with which we shall be concerned in the present chapter. In section 3.6 the intensity has been defined as the time average ($\langle\ \rangle$) amount of energy which crosses in unit time, a unit area placed perpendicular to the direction of the energy flow. In other words, I is the time-average of the Poynting's vector $\vec{S} = (\vec{E}\times\vec{B})/\mu_o$. For a plane wave in vacuum $\vec{B}$ is perpendicular to $\vec{E}$ and has a magnitude given by E/c. Thus, the intensity can be written as

$$I = \epsilon_o c\langle\vec{E}^2\rangle \text{ watt/m}^2 \tag{8.1}$$

where, in obtaining this result, we made use of the relationship $c^2 = 1/\mu_o\epsilon_o$. We see, therefore, that apart from the constant of proportionality $(\epsilon_o c)$, the quantity $\langle E^2\rangle$ is a measure of the intensity. Later on in the chapter we shall omit $(\epsilon_o c)$ from our equations. This will make the equations a little shorter and neater looking. But, one must remember to insert $(\epsilon_o c)$ in front of $\langle\vec{E}^2\rangle$ whenever I is required in its proper units of watt/m^2.

Suppose we now superpose at some fixed point P in space two monochromatic plane waves whose electric fields at P are $\vec{E}_1$ and $\vec{E}_2$, respectively. By the principle of superposition the total electric field at P is given by

$$\vec{E} = \vec{E}_1 + \vec{E}_2 \tag{8.2}$$

so that

$$\vec{E}^2 = \vec{E}_1^2 + \vec{E}_2^2 + 2\vec{E}_1 \cdot \vec{E}_2. \tag{8.3}$$

We now multiply each term of Eq. 8.3 by $(\epsilon_o c)$ and time-average. This yields the total intensity at P:

$$I = \underbrace{I_1}_{\substack{\text{intensity}\\ \text{of beam 1}\\ \text{alone}}} + \underbrace{I_2}_{\substack{\text{intensity}\\ \text{of beam 2}\\ \text{alone}}} + \underbrace{2\epsilon_o c\langle \vec{E}_1 \cdot \vec{E}_2\rangle}_{\substack{\text{interference}\\ \text{term}}}. \tag{8.4}$$

Here $I_1 = \epsilon_o c\langle \vec{E}_1^2\rangle$ and $I_2 = \epsilon_o c\langle \vec{E}_2^2\rangle$ are the intensities of the two individual waves, and $2\epsilon_o c\langle \vec{E}_1 \cdot \vec{E}_2\rangle$ is the interference term, which can be positive, negative or zero. Hence we see that using little physics other than the principle of superposition, we arrive at the result that the intensity at a point is the sum of the intensities of the individual waves plus (or minus) something extra. It is this "extra" which we shall explore in detail and which provides the basis for the entire discipline called Interference. It is fairly obvious that when more than two waves are superposed, it will result in a sum of cross-terms of the form $\langle \vec{E}_1 \cdot \vec{E}_2\rangle$, $\langle \vec{E}_1 \cdot \vec{E}_3\rangle$, $\langle \vec{E}_2 \cdot \vec{E}_3\rangle$ ... etc. The full result for N waves is actually found to be

$$I = \sum_{i=1}^{N} I_i + 2\epsilon_o c \sum_{j>i}^{N} \sum_{i=1}^{N} \vec{E}_i \cdot \vec{E}_j. \tag{8.5}$$

Before proceeding to special cases, let us note the structure of the interference term $2\epsilon_o c\langle \vec{E}_1 \cdot \vec{E}_2\rangle$. Expanding the dot product yields

$$\text{Interference term} = 2\epsilon_o c\langle |\vec{E}_1||\vec{E}_2| \cos \alpha\rangle \tag{8.6}$$

where α is the angle between the electric vectors of the two waves. The equation tells us that two beams polarized at right angles to one another <u>cannot</u> interfere with each other, a fact which was demonstrated experimentally with polarized light by Arago and Fresnel in the year 1816, long before the advent of Maxwell's electromagnetic theory of light.

Using Eq. 8.6 and Arago and Fresnel's observation one can also prove that light vibrations must be purely transverse, a fact already well known to us. To verify this, suppose two linearly polarized waves are propagating in the z direction. Let the electric vector of wave 1 lie in the xz plane and let the electric vector of wave 2 lie in the yz plane. Written out in full we have

$$\vec{E}_1 = \hat{x}E_{1x} + \hat{y}(0) + \hat{z}E_{1z}$$
$$\vec{E}_2 = \hat{x}(0) + \hat{y}E_{2y} + \hat{z}E_{2z}. \tag{8.7}$$

Inserting Eqs. 8.7 in Eq. 8.4 and working out the dot product gives

$$\text{Interference term} = 2\epsilon_o c\langle E_{1z}E_{2z}\rangle. \tag{8.8}$$

But Arago's experiment says that there is no interference, a result which can only be true if $E_{1z} = E_{2z} = 0$. Thus the electric vectors of the two waves are perpendicular to the direction of propagation z. What beautiful conclusions we have come to on the basis of such simple physics and mathematics!

Interference of waves from two oscillating current elements

There are numerous ways of causing interference, many of which have great practical applications. The most basic is the interference of waves emanating from two neighboring elementary point sources. In electromagnetism the oscillating current element discussed in section 4.1 of Chapter 4 is by far the simplest and most primitive of all sources, and we shall discuss what happens when the radiations from two such sources, oscillating at the same frequency ω, interfere with one another. As we have learned, there is no interference if the waves are polarized at right angles to one another. We shall therefore align the two current elements parallel to each other, along the z axis, as is illustrated in Fig. 8.1. Furthermore, we shall require that the point of observation P lie at all times in the xy plane, such that the angle θ between the radius vector $\vec{r}$ and the current element $I_o\vec{d\ell}$ be 90°. With these simplifications the wave amplitude is angle-independent, which helps in the algebra. Even more important, the electric vectors of the two waves are always

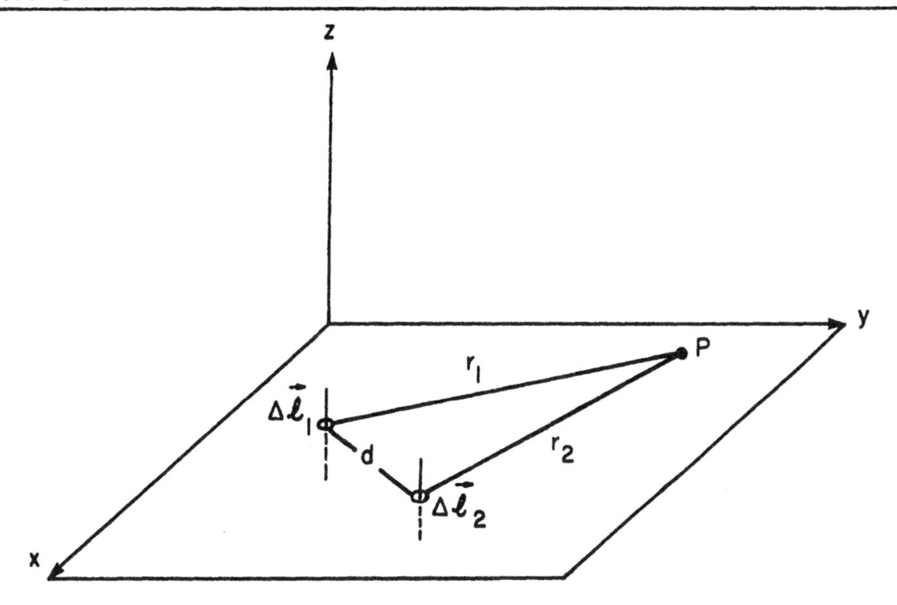

Fig. 8.1 Electromagnetic emission from two parallel current elements $\overrightarrow{\Delta\ell}_1$ and $\overrightarrow{\Delta\ell}_2$ driven at the same frequency ω. In the plane of observation (the x-y plane), the electric field associated with each radiator is oriented parallel to the z axis. In this plane the wave fronts from each radiator are concentric circles, and the amplitude falls off as 1/distance r.

parallel; this removes polarization effects from the problem to be solved. For this situation, then, the electric fields of the individual sources at point P become, in accordance with Eq. 4.16,

$$\vec{E}_1(P) = \hat{z}\,\frac{I_{o1}\Delta\ell_1\omega}{4\pi\epsilon_o r_1 c^2}\cos(\omega t - kr_1 + \phi_1) \qquad \text{V/m}$$

$$\vec{E}_2(P) = \hat{z}\,\frac{I_{o2}\Delta\ell_2\omega}{4\pi\epsilon_o r_2 c^2}\cos(\omega t - kr_2 + \phi_2) \qquad \text{V/m} \tag{8.9}$$

where we changed for convenience from $\sin(\omega t - kr)$ to $\cos(\omega t - kr)$, which is merely a change in phase. We also inserted phase factors ϕ_1 and ϕ_2 to provide for the possibility that the two current elements may oscillate out of phase with one another. Then defining

$$A_1 = \frac{I_{o1}\Delta\ell_1\omega}{4\pi\epsilon_o c^2} \quad \text{and} \quad A_2 = \frac{I_{o2}\Delta\ell_2\omega}{4\pi\epsilon_o c^2} \tag{8.10}$$

gives

$$\begin{aligned} \vec{E}_1 &= \hat{z}\,\frac{A_1}{r_1}\cos(\omega t - kr_1 + \phi_1) \quad \text{V/m} \\ \vec{E}_2 &= \hat{z}\,\frac{A_2}{r_2}\cos(\omega t - kr_2 + \phi_2) \quad \text{V/m} \end{aligned} \tag{8.11}$$

which are seen to be the equations of two spherical waves. The complicated coefficient A_1 can be interpreted as the electric field amplitude at a distance one meter away from the current element 1; similarly A_2 is the electric field amplitude at a distance of 1 meter from the current element 2.

The total electric field $\vec{E}(P) = \vec{E}_1 + \vec{E}_2$ at the observation point P is inserted from Eqs. 8.11 into Eq. 8.1 with the result that

$$I = \epsilon_o c \left\langle \left[\frac{A_1}{r_1}\cos(\omega t - kr_1 + \phi_1) + \frac{A_2}{r_2}\cos(\omega t - kr_2 + \phi_2) \right]^2 \right\rangle . \tag{8.12}$$

Expanding the bracket and time-averaging term by term gives

$$I = \epsilon_o c \left[\frac{A_1^2}{2r_1^2} + \frac{A_2^2}{2r_2^2} + 2\left\langle \frac{A_1 A_2}{r_1 r_2}\cos(\omega t - kr_1 + \phi_1) \cos(\omega t - kr_2 + \phi_2) \right\rangle \right] . \tag{8.13}$$

Use of the trigonometric identity $2\cos(a)\cos(b) = \cos(a+b) + \cos(a-b)$ followed by time averaging of the last term of the above equation leads to the sought-after result

$$I = \frac{\epsilon_o c}{2}\left\{ \left(\frac{A_1}{r_1}\right)^2 + \left(\frac{A_2}{r_2}\right)^2 + 2\left(\frac{A_1}{r_1}\right)\left(\frac{A_2}{r_2}\right)\cos[k(r_1 - r_2) - (\phi_1 - \phi_2)] \right\} . \tag{8.14}$$

Observe that Eq. 8.14 is identical in form to Eqs. A1.8 and A1.10 of Appendix 1. This, of course, is no coincidence, because all we are doing here, as in Appendix 1, is the mathematics of superposing two periodic oscillations. The first term on the right is the intensity I_1 at P due to source 1 radiating alone. The second term is the intensity at P due to source 2 radiating alone, and the last term accounts for the interference between the waves. The last-named can be positive, negative, or zero depending on the value of the argument under the cosine. Having identified the terms on the right-hand side of Eq. 8.14, we can cast the equation into the more elegant form given by

$$I = I_1 + I_2 + 2\sqrt{I_1 I_2} \cos \delta$$

where (8.15)

$$\delta = k(r_1 - r_2) - (\phi_1 - \phi_2).$$

It is evident that there occur maxima in intensity

$$I(\max) = I_1 + I_2 + 2\sqrt{I_1 I_2} \tag{8.16}$$

whenever

$$|\delta| = 0, 2\pi, 4\pi, 6\pi, \ldots \quad \text{(for maxima)} \tag{8.17}$$

and minima in intensity

$$I(\min) = I_1 + I_2 - 2\sqrt{I_1 I_2} \tag{8.18}$$

whenever

$$|\delta| = \pi, 3\pi, 5\pi, 7\pi, \ldots \quad \text{(for minima).} \tag{8.19}$$

Special case of equal intensities

When the intensities are the same, such that $I_1 = I_2 \equiv I_o$, Eq. 8.15 reduces to

$$I = 4I_o \cos^2(\delta/2), \tag{8.20}$$

a formula which is obtained by invoking the identity $\cos 2a = 2 \cos^2 a - 1$. The result is plotted in Fig. 8.2 showing that the intensity varies between $I(\min) = 0$ and $I(\max) = 4I_o$.

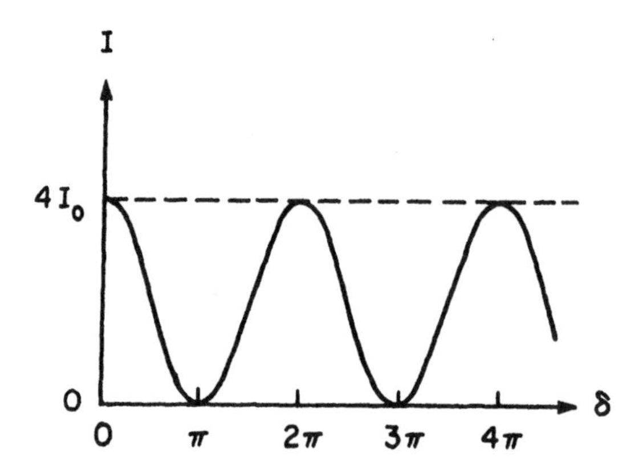

Fig. 8.2 Interference of two waves of equal strength, showing the intensity variation as a function of the phase difference $\delta = k(r_1 - r_2) - (\phi_1 - \phi_2)$.

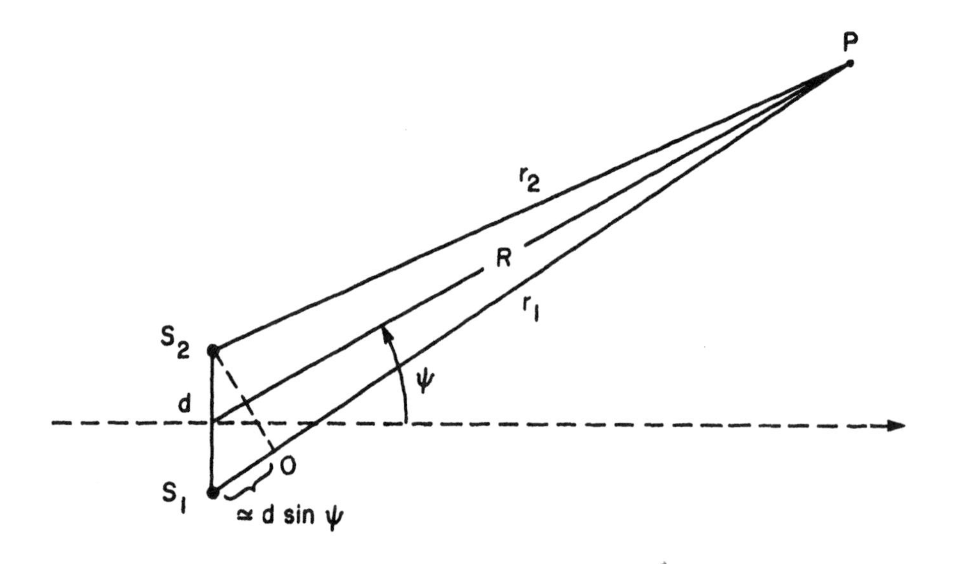

Fig. 8.3 Geometry for calculating the phase difference δ in two-wave interference. When $d \ll R$, the path length difference $(r_1 - r_2) \approx d \sin \psi$ (see text) and the associated phase difference $k(r_1 - r_2) \approx kd \sin \psi$. The total phase difference $\delta \approx kd \sin \psi - (\phi_1 - \phi_2)$.

Of course, even if the two radiating antennas are of identical size and are fed with currents of identical magnitude, I_1 and I_2 at P can never be precisely the same because the distances r_1 and r_2 from the sources to P are not absolutely identical. As a result I(min) will not be exactly zero. However, when the distances r_1 and r_2 are very large in comparison with the separation d between antennas (see Fig. 8.3), our assumption of equal intensities is very good and we can safely write $r_1 \approx r_2 \approx R$ where R is the distance from the midway point between the antennas to the observation point P, as in Fig. 8.3. We cannot, however, treat r_1 and r_2 in this fashion when these lengths appear in the phase factor $\delta = [k(r_1 - r_2) - (\phi_1 - \phi_2)]$. In the phase factor one is concerned with small length differences; and when $(r_1 - r_2)$ changes by as little as one quarter of a wavelength of the radiation, $k(r_1 - r_2)$ changes by $\pi/2$ radians, thus causing a major change in the value of $\cos^2(\delta/2)$ appearing in Eq. 8.20.

Suppose, then, the intensity distribution I(P) is to be measured by an observer moving along the arc of a circle of radius $R \gg d$. From the geometry of Fig. 8.3 we have that

$$r_1^2 = R^2 + (d/2)^2 + Rd \sin \psi$$
$$r_2^2 = R^2 + (d/2)^2 - Rd \sin \psi \tag{8.21}$$

so that

$$r_1^2 - r_2^2 = 2Rd \sin \psi \tag{8.22}$$

But $\left(r_1^2 - r_2^2\right) = (r_1 - r_2)(r_1 + r_2)$ and when r_1 and r_2 are large compared with antenna separation d, it follows that $\left(r_1^2 - r_2^2\right) \approx (r_1 - r_2) \times (2R)$. Hence, it follows from Eq. 8.22 that $(r_1 - r_2) \approx d \sin \psi$ with the result $\delta = kd \sin \psi - (\phi_1 - \phi_2)$ and

$$I = 4I_0 \cos^2\left[\frac{1}{2} kd \sin \psi - \frac{1}{2}(\phi_1 - \phi_2)\right]$$
$$= 4I_0 \cos^2\left[\frac{\pi d}{\lambda} \sin \psi - \frac{1}{2}(\phi_1 - \phi_2)\right] \tag{8.23}$$

where the second form is obtained from the first by writing $k = 2\pi/$

wavelength λ. We have done quite a lot of algebra to derive the explicit value of the phase factor δ in terms of the geometry of the system. A quick derivation results from the following observation: when P is very distant, the vectors $\vec{r}_1$ and $\vec{r}_2$ are virtually parallel; thus $(r_1 - r_2)$ can be found by dropping a perpendicular from source point S_2 onto vector $\vec{r}_1$ (see Fig. 8.3) and noting that the distance $S_1 0 \approx (r_2 - r_1) = d \sin \psi$.

Suppose the two antennas are fed in phase so that $\phi_1 - \phi_2 = 0$. Then the phase factor $\delta/2 = k(r_2 - r_1)/2$ appearing in Eq. 8.20 tells us that intensity maxima occur whenever $(r_2 - r_1) = 0, \lambda, 2\lambda, 3\lambda \ldots,$ that is, when the waves from the two sources arrive so that at P the crest (or trough) of one wave adds to the crest (or trough) of the other wave. Zeros of intensity occur when $(r_2 - r_1) = \lambda/2, 3\lambda/2, 5\lambda/2 \ldots,$ showing, in accord with expectation, that when crest meets trough there is destructive cancellation and thus a minimum in intensity.

We have shown in Fig. 8.2 how the intensity varies with δ and thus with angle ψ. It is instructive to plot the intensity in the form of a radiation pattern, namely, as a polar plot of I versus the angle ψ (see section 4.1). As an example, suppose the two antennas are separated by one-half of a wavelength. Then from Eq. 8.23 it follows that

$$I = 4I_o \cos^2 \left[\frac{\pi}{2} \sin \psi - \frac{1}{2} (\phi_1 - \phi_2)\right]. \tag{8.24}$$

Figure 8.4a shows the radiation pattern when the currents in the two antennas oscillate exactly in phase, that is, when $\phi_1 - \phi_2 = 0$. We see that the radiation is "beamed" symmetrically along the line of intersection between the two sources (the y axis). Figure 8.4b shows a similar plot for the situation $\phi_1 - \phi_2 = \pi/2$, showing that now the maximum has swung over and the pattern is tilted at an angle of 30° to the y axis. Other angles result by choosing other values of $(\phi_1 - \phi_2)$. This is very amusing and has also practical application in fast scanning of a section of the sky, for example. Instead of mechanically rotating the two antennas (which is slow), one can "flood" a piece of sky with radiation by electronically changing the relative phases of the two currents feeding the antennas. Normally, the two

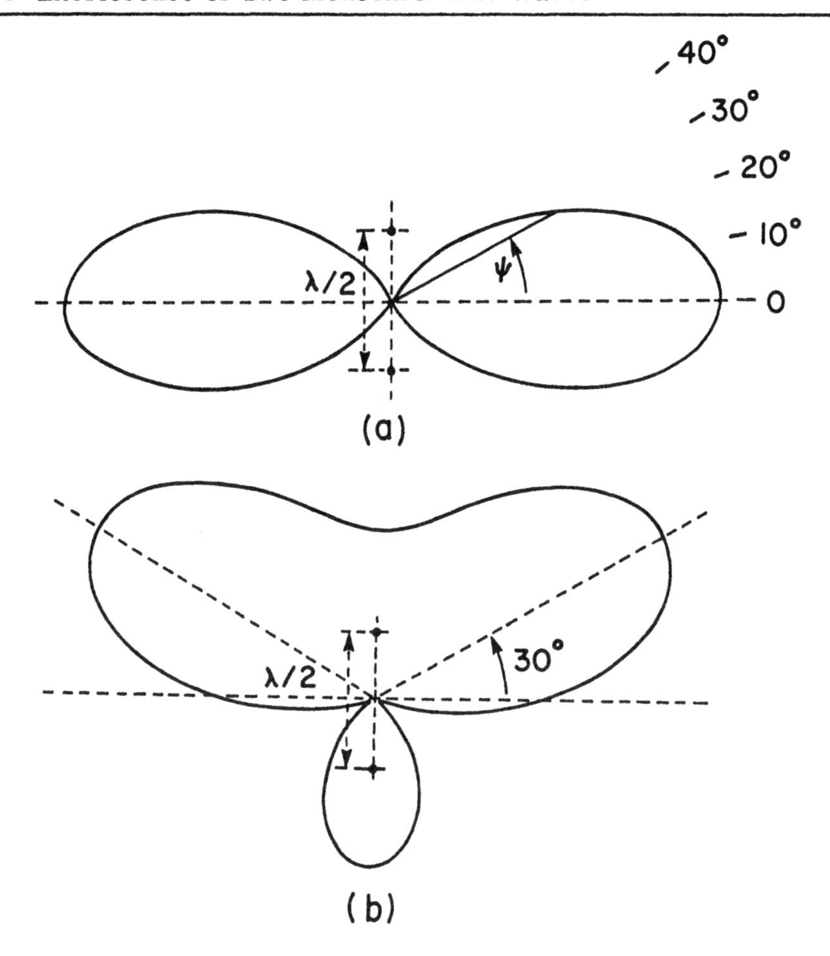

Fig. 8.4 (a) Radiation pattern of two point sources separated by a distance d = λ/2 and driven in phase with respect to one another ($\phi_1 - \phi_2 = 0$); (b) Radiation pattern of the same two sources driven with a phase difference $\phi_1 - \phi_2 = \pi/2$ radians.

antennas are excited from the same high-frequency generator for example, a high-frequency oscillator or a magnetron. The radiation is split into two equal halves, and fed by equally long coaxial cables or by waveguides to the two antennas. In this manner the two antennas are maintained at constant relative phase as long as is desired. The relative phase ($\phi_1 - \phi_2$) can be changed by inserting a so-called "phase-shifter" into one of the feed cables. If one is in no hurry, the desired phase change can be produced by altering the relative lengths of the feed cables. A difference ℓ in their lengths yields a phase difference $(\phi_1 - \phi_2) = 2\pi\ell/\lambda$ radians.

Interference has been known and studied long before the advent of antennas, electromagnetism, and the like. On the following pages we shall describe a few ways of achieving two-beam interference at optical wavelengths.

Example

A television station located in a city on the eastern coast of the United States, such as Boston, with a transmitter site located on the ocean front, does not want to radiate half its transmitted power out over the ocean where it has zero audience. Yet no other transmitter site is economically feasible. Show, by deriving and sketching the radiation pattern, that they can solve their problem by placing two antennas, a quarter wavelength apart, on an east-west line and maintaining a 90° phase difference between the currents in the two antennas. But be careful about that 90° phase difference: Should the western antenna lead or lag the eastern antenna by 90 degrees?

Solution

The problem resembles in many ways that posed earlier in this subsection. The geometry of Fig. 8.1 is appropriate, except we place the antennas along the $\pm x$ axis. Thus the north-south line is the $\pm y$ axis and we assume the ocean to be located somewhere at $x < -\lambda/8$. (See the geometry shown in the figure.)

The fields from the antennas (Eq. 8.11) are

$$\vec{E}_1 = \frac{A_o}{r_1}\hat{z}\cos(\omega t - kr_1 + \phi_1)$$

$$\vec{E}_2 = \frac{A_o}{r_2}\hat{z}\cos(\omega t - kr_2 + \phi_2)$$

where antenna 1 is the western antenna and antenna 2 the eastern.

A_o is the common amplitude of each field. The intensity at point P, lying in the x-y plane is given by Eq. 8.23 for $d = \lambda/4$ and $\phi_1 - \phi_2 = -\pi/2$ as [$\beta = -\psi$ of Eq. 8.23]

$$I = 4I_o \cos^2\left[-\frac{\pi}{4}\sin\beta + \frac{\pi}{4}\right].$$

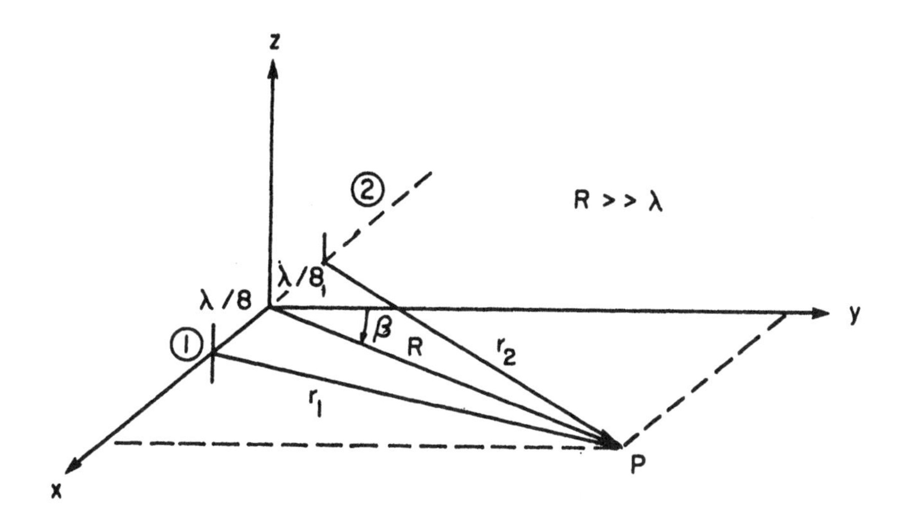

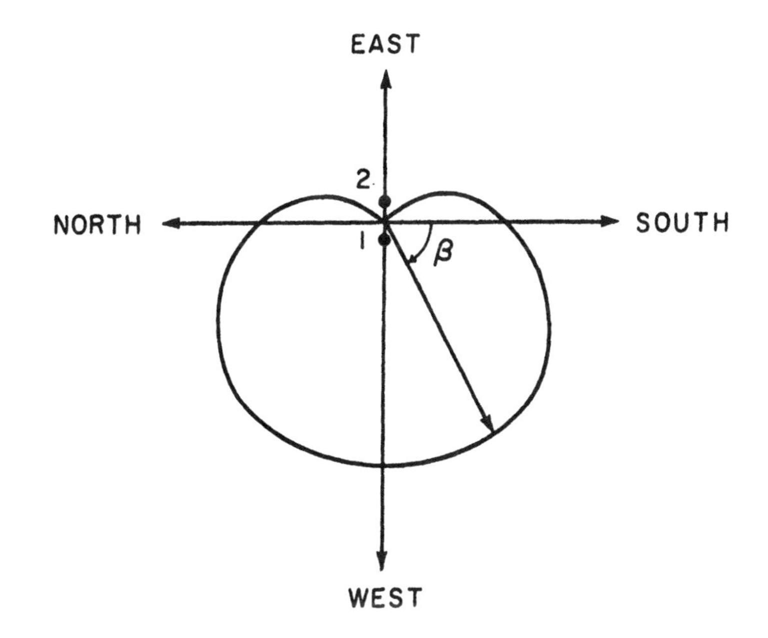

This is the radiation pattern for the case when the phase of the western antenna ϕ_1 lags that of the eastern antenna ϕ_2 since $\phi_1 = \phi_2 - \pi/2$. A plot of the radiation pattern I vs β is as shown.

Comparing this pattern with those of Fig. 8.4 should show you that the radio engineer, with a little ingenuity on his part, has a wide variety of radiation patterns at his disposal. Needless to, say, extension to more than two radiating elements increases immensely the possibilities, an example of which is shown in Fig. 8.10.

8.2 EXAMPLES OF TWO-BEAM INTERFERENCE

Young's experiment

The earliest experimental arrangement used in demonstrating the interference of light was made by Thomas Young (1773-1829). Light from a monochromatic point source S_0 is allowed to impinge on a screen provided with two pinholes S_1 and S_2 separated by a small distance d of the order of a millimeter or less, as is shown in Fig. 8.5. An observation screen is placed several meters beyond the pinholes. The pinholes act as secondary monochromatic sources somewhat akin to the antennas discussed in the previous section. To be sure, there are some important differences. For one, the light used is randomly polarized rather than being linearly polarized as was the case with the antennas. But it turns out that a beam of natural (i.e., randomly) polarized light may be represented as a superposition of two incoherent beams linearly polarized, and at right angles to one another (say, in the x and z directions). The interference between the x components of the two beams and the z components, respectively, can then be considered separately, and the total intensity obtained by addition of the separate intensities. Since the phase factor δ has the same value for each case, the formulas derived in the previous subsection remain valid. Since the two pinholes S_1 and S_2 are placed symmetrically with respect to the point source S_0, the phase difference $\phi_1 - \phi_2 = 0$ and Eq. 8.23 becomes

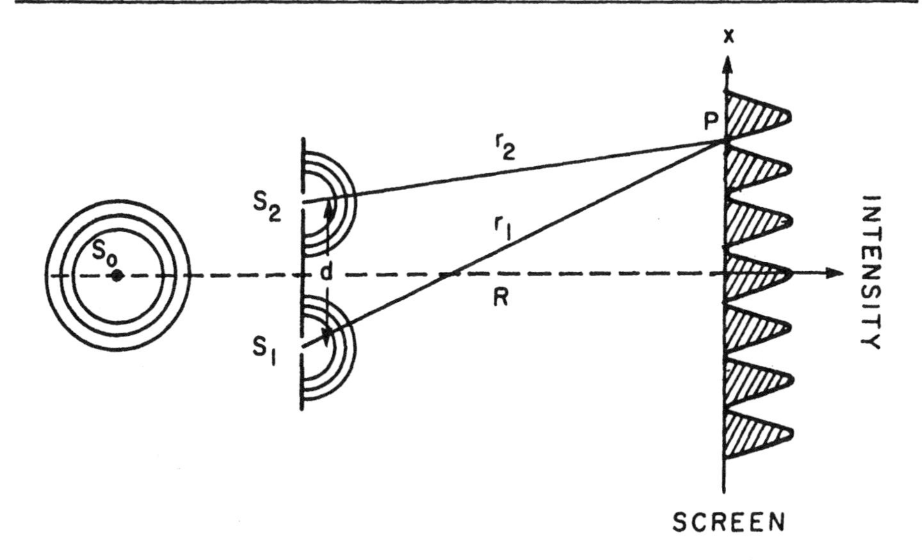

Fig. 8.5 Young's experimental arrangement. The oscillating curve on the screen represents a plot of the intensity as a function of position x. Note that on axis the intensity is maximum. What would you do to shift the pattern so as, for example, to make the intensity minimum on axis?

$$I = 4I_o \cos^2 \left[\frac{\pi d}{\lambda} \sin \psi\right]. \tag{8.25}$$

The interference fringes are observed on a screen stationed at a fixed distance R from the pinholes, and the quantity of interest is not the angle ψ, but rather the distance x of the point of observation P from the symmetry axis. Since x is generally small compared with R, one can write that $\sin \psi \approx \tan \psi = x/R$, with the result that

$$I = 4I_o \cos^2 \left(\frac{\pi d}{\lambda R} x\right). \tag{8.26}$$

Thus it follows that intensity maxima occur when

$$x(\text{max}) = m \frac{R\lambda}{d}, \qquad m = 0, \pm1, \pm2, \pm3 \ldots \tag{8.27}$$

and intensity minima when

$$x(\text{min}) = \left(m + \frac{1}{2}\right) \frac{R\lambda}{d}, \qquad m = 0, \pm1, \pm2, \pm3 \ldots \tag{8.28}$$

The bright (and dark) fringes are equally spaced with a separation

$$\begin{aligned}\Delta x &= (m+1)\frac{R\lambda}{d} - m\frac{R\lambda}{d} \\ &= \frac{R\lambda}{d}.\end{aligned} \tag{8.29}$$

Consider, for example, two pinholes separated by a distance of 0.1 mm and irradiated by yellow light from a sodium lamp of wavelength 5892 Å. The screen is 2 meters from the pinholes. Then, in accordance with Eq. 8.29 the separation between neighboring dark (or neighboring bright) fringes is 1.18 cms, a distance that is easily resolved by the naked eye. Note that if the wavelength is not known, it can be determined from a simple measurement of Δx, and knowledge of R and d. Thus, Young's experimental arrangement can be used as a crude spectrometer. It is a crude instrument because the fringes are not very sharp (the intensity varies as $\cos^2(\ldots . x)$ and therefore Δx cannot be determined with precision. But we shall see subsequently that by increasing the number of sources from two to many, an incredible sharpening of the fringes occurs, although the separation Δx remains unchanged from that given by Eq. 8.29.

Lloyd's mirror

Instead of using one point source and two pinholes, an equally simple arrangement is provided by Lloyd's mirror, illustrated in Fig. 8.6. A point source S_1 is placed some distance away from a plane mirror M. The distance of the source from the plane of the mirror is quite small so that light is reflected at nearly grazing incidence. The interference occurs between the real source S_1 and its virtual image S_2 in the mirror. The perpendicular bisector of S_1S_2 then lies in the plane of the mirror surface.

Fringes are observed on the screen as is indicated in the figure by the plot of intensity as a function of position. The geometric distances from S_1 and S_2 to the far end of the screen (point 0) are equal, and one would therefore expect a maximum of intensity at that point. But, recall that there is a reversal of the electric vector on reflection

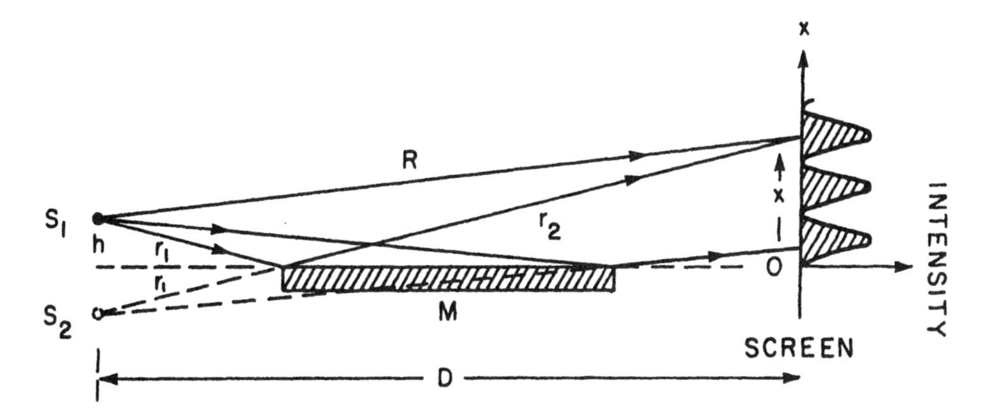

Fig. 8.6 Lloyd's mirror. Observe that the intensity is zero at point 0, that is, in the plane of the mirror.

when a wave propagates from a less dense to a denser medium (see section 7.1). This, in effect, makes the optical path length $S_1 0$ differ from the optical path length $S_2 0$ by one-half of a wavelength, and is equivalent to a phase reversal of 180°. As a result, destructive interference occurs at point 0, that is, a dark fringe occurs right at the mirror surface.

Example

a) Derive the intensity distribution on the observing screen of the Lloyd's mirror experiment shown in Fig. 8.6.

b) If the screen is 1 meter from the source S_1 and the source is 0.25 mm above the plane of the mirror M, what is the linear separation between adjacent maxima of intensity on the screen for light of wavelength 5000 Å?

Solution

Interference at point x on the screen occurs between the direct ray (R) and the ray reflected from the mirror ($R' = r_1 + r_2$). The phase difference $\Delta\phi$ between these two rays is $2\pi/\lambda$ times the path difference δ. From the geometry:

$$R^2 = D^2 + (x-h)^2 \qquad R'^2 = D^2 + (x+h)^2.$$

If we now use $\sqrt{1+x^2} \approx 1 + x^2/2$, $D \gg x$, and $D \gg h$, we can write

$$R = D + (x-h)^2/2D \qquad R' = D + (x+h)^2/2D.$$

Hence we have

$$\delta = R' - R = 2xh/D$$

$$k\delta = (2\pi/\lambda)\delta = 4\pi xh/\lambda D = \Delta\phi.$$

The distances R and R' are nearly equal so we can assume that amplitudes of the fields arriving at the screens are equal, but of course the phases are unequal. Thus the combined field at point x on the screen is

$$E_T = E_o\, e^{j(\omega t - kR)} - E_o\, e^{j(\omega t - kR')}$$

where the minus sign is due to the phase change of π upon reflection at the mirror. The fields can be written as

$$E_T = 2E_o \sin\frac{\Delta\phi}{2} \sin\left(\omega t - kR - \frac{\Delta\phi}{2}\right)$$

and the time-average intensity becomes

$$\langle I \rangle = \epsilon_o c \langle E^2 \rangle = 2\epsilon_o c E_o^2 \sin^2\frac{\Delta\phi}{2}$$

a) $$\langle I \rangle = 4I_o \sin^2\left(\frac{2\pi xh}{\lambda D}\right).$$

Note that I = 0 at x = 0 as shown in Fig. 8.6.

The maxima of the intensity occur whenever

$$\frac{2\pi xh}{\lambda D} = (m + 1/2)\,\pi$$

where m is 0, 1, 2, 3, ... The spacing Δx between maxima is

$$\frac{2\pi h}{\lambda D}(\Delta x) = (\Delta m)\,\pi$$

and for adjacent maxima $\Delta m = 1$. Thus we have

$$\Delta x = \frac{\lambda D}{2h}.$$

Inserting numerical values we find

b) $$\Delta x = \frac{(5\times 10^{-7})(1)}{(2)(2.5\times 10^{-4})} = 10^{-3}\ \text{m} = 0.1\ \text{cm}.$$

Interference by reflection from a thin film

Suppose a thin plate of transparent material such as glass or an oil slick is illuminated by monochromatic light emanating from a point source S as is shown in Fig. 8.7. An observation point P situated on the same side of the plate as the source S will then receive light waves that have traveled by two different paths, one wave reflected from the upper surface, and the other wave reflected from the lower surface, so there is an interference pattern on the same side of the plate as S. The light intensity varies in the usual manner and its value is given by Eq. 8.15. Our task is to determine the explicit form of the phase factor δ defined as

$$\delta = [k \times \text{path traversed}]_{\text{ray }1} - [k \times \text{path traversed}]_{\text{ray }2}\ \text{radians}. \quad (8.30)$$

Note that as rays 1 and 2 travel through different materials the magnitude of the propagation constant $k \equiv \omega/v$ (v is the phase velocity) must be adjusted to take account of these changes.

Let the thickness of the film be d, and let it be made from a lossless dielectric of refractive index n. Then, while ray 1 has traversed the distance

$$AB = AC \sin\theta = 2d \tan\theta' \sin\theta, \quad (8.31)$$

ray 2 has traversed the distance

$$AE + EC = 2AE = 2d/\cos\theta'. \quad (8.32)$$

It follows that the phase difference δ is

$$\delta = \frac{\omega}{c} n[AE + EC] - \frac{\omega}{c}[AB] - \pi \quad (8.33)$$

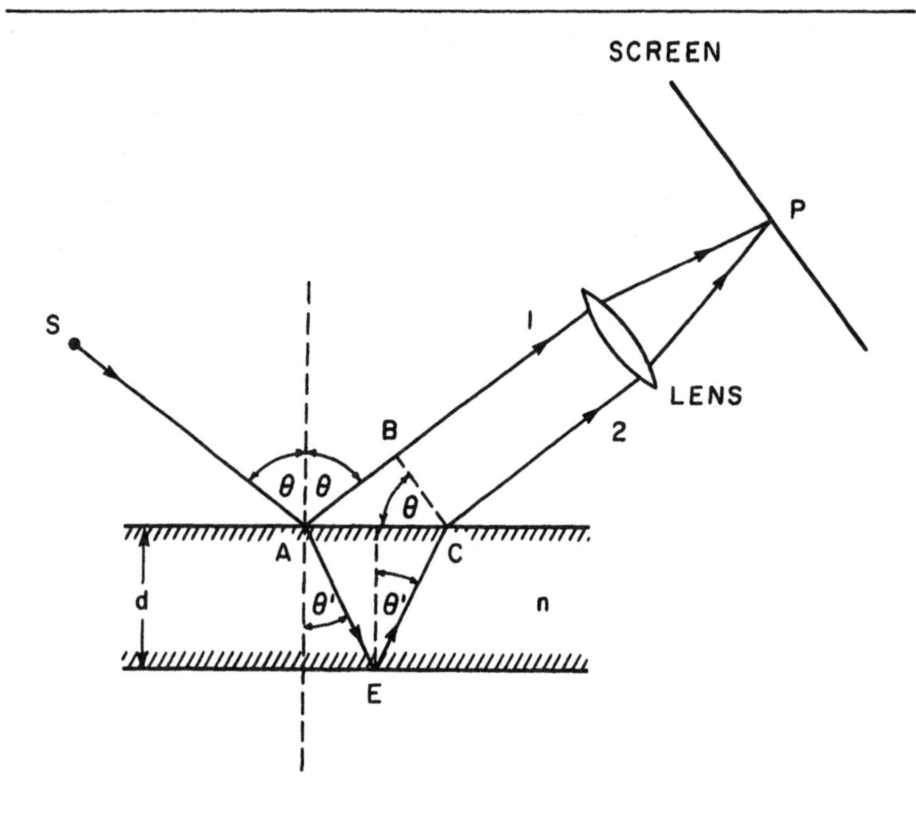

Fig. 8.7 Formation of interference fringes by reflection from a film. A lens is used to focus the fringes on the screen. If a lens is not used, the interference fringes are localized at infinity and that is where they would have to be observed. Instead of a screen, the fringes can be seen with a relaxed eye, focused for infinity.

where the additional factor π in Eq. 8.33 allows for the phase change suffered by the electric vector of ray 1 on being reflected from the upper surface (no such change takes place at the lower surface as is discussed in section 7.1). Substituting Eqs. 8.31 and 8.32 in Eq. 8.33 yields

$$\delta = \frac{2\omega d}{c}\left[(n/\cos\theta') - \tan\theta' \sin\theta\right] + \pi. \tag{8.34}$$

But, from Snell's law of refraction $\sin\theta = n \sin\theta'$ with the result that

$$\delta = \frac{2\omega nd}{c} \cos \theta' + \pi. \tag{8.35}$$

In accordance with Eq. 8.17 constructive interference, and thus intensity maxima, occur when $\delta = 0, 2\pi, 4\pi, 6\pi \ldots$. Use of this fact in conjunction with Eq. 8.35 gives

$$nd \cos \theta' = \left(m - \frac{1}{2}\right) \frac{\lambda}{2}, \qquad m = 1, 2, 3 \ldots \text{ for maxima.} \tag{8.36}$$

Similarly, Eqs. 8.19 and 8.35 yield the result that destructive interference occurs whenever

$$nd \cos \theta' = (m - 1) \frac{\lambda}{2}, \qquad m = 1, 2, 3 \ldots \text{ for minima.} \tag{8.37}$$

It is noteworthy that as the film thickness d goes to zero, the condition for destructive interference can be satisfied for all angles θ', by setting $m = 1$. This says that no fringes will be seen on the observation screen, but only blackness. The reason for this, and the experimental demonstration of the phenomenon are given in section 7.1.

This treatment of the problem is only approximate because we neglected multiple reflections in the film. In reality, a series of waves, in addition to the two considered, interfere with one another and they must be taken into account in calculating the intensity reaching the observer. This will be done in the next section. Our present treatment is quite adequate for films of low reflectivity because for such films successive internally reflected waves, after the first two, carry little energy.

8.3 INTERFERENCE OF WAVES FROM A MULTIPLE ARRAY OF SOURCES

Let us now consider an array of monochromatic point sources in place of the two sources studied in the previous section. Imagine, for the sake of concreteness, a linear array of N identical current elements all aligned parallel to one another and separated by a distance d as is illustrated in Fig. 8.8. The amplitude of the current through each little antenna is the same; and for simplicity, assume further

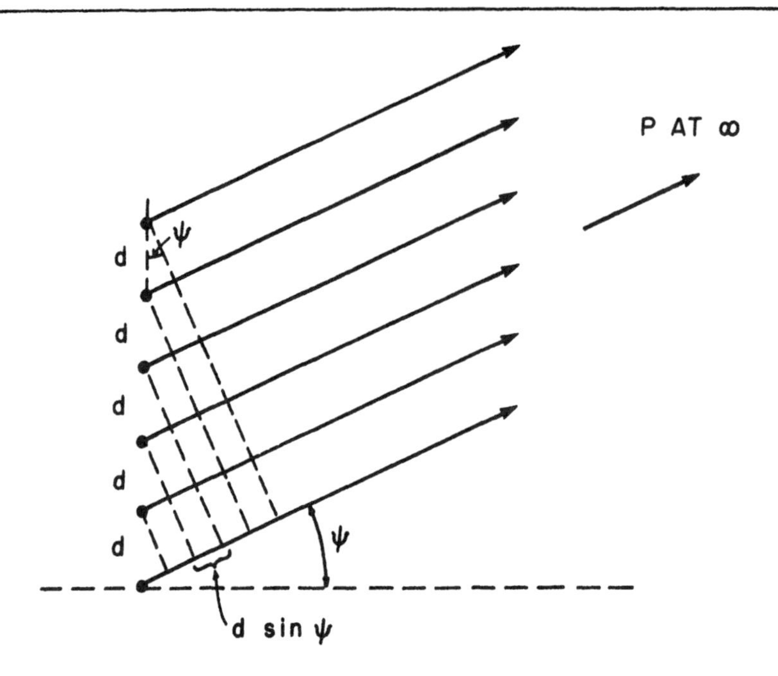

Fig. 8.8 An array of identical point sources driven in phase and at the same frequency ω. The point of observation P is at a large distance from the sources, so that the path length difference at P of the neighboring beams equals approximately d sin ψ.

that the elements are driven in phase with one another. Thus, in place of the two equations (8.11) we shall now have N equations with $A_1 = A_2 = \ldots = A_N \equiv A$ and $\phi_1 = \phi_2 = \ldots = \phi_N \equiv \phi_o$. Also recalling that when the observation point P is at a large distance from the sources, the amplitude terms A_1/r_1, A_2/r_2, etc. can be approximated by A_1/R, A_2/R, ... As a result, the total electric field amplitude at P is

$$\vec{E} = \hat{z}\left[\frac{A}{R}\cos(\omega t - kr_1) + \frac{A}{R}\cos(\omega t - kr_2) + \ldots \frac{A}{R}\cos(\omega t - kr_N)\right]$$

$$(\text{with } \phi_o = 0). \qquad (8.38)$$

A graphical method of summing such a series is given in Appendix 1, and we shall leave it to the reader to perform such a graphical solution.

Here we shall do the same thing analytically, a task made simpler by casting Eq. 8.38 into the customary exponential form,

$$\vec{E} = \hat{z}\frac{A}{R}\left[e^{j(\omega t - kr_1)} + e^{j(\omega t - kr_2)} + \ldots e^{j(\omega t - kr_N)}\right]. \tag{8.39}$$

Factoring out the first term, we obtain

$$\vec{E} = \hat{z}\frac{A}{R}e^{j(\omega t - kr_1)}\left[1 + e^{-jk(r_2 - r_1)} + \ldots e^{-jk(r_N - r_1)}\right]. \tag{8.40}$$

The phase difference δ at P between light emanating from sources 2 and 1 equals $k(r_2 - r_1) = kd \sin \psi$. The phase difference between sources 3 and 1 equals $k(r_3 - r_1) = 2\delta = 2kd \sin \psi$, between 4 and 1 it equals $k(r_4 - r_1) = 3\delta = 3kd \sin \psi$, etc. The reason is that the phase differences between rays from adjacent sources are equal to one another. Hence the field at P may be written as

$$\begin{aligned}\vec{E} &= \hat{z}\frac{A}{R}e^{j(\omega t - kr_1)}\left[1 + e^{-j\delta} + e^{-2j\delta} + \ldots . e^{-(N-1)j\delta}\right] \\ &= \hat{z}\frac{A}{R}e^{j(\omega t - kr_1)}\{1 + (e^{-j\delta}) + (e^{-j\delta})^2 + \ldots . (e^{-j\delta})^{N-1}\}.\end{aligned} \tag{8.41}$$

The term in the braces represents a geometric series whose value equals $[\exp(-jN\delta) - 1]/[\exp(-j\delta) - 1]$. Inserting this expression in the second equation (8.41), and rearranging terms leads to the result

$$\vec{E} = \hat{z}\frac{A}{R}e^{j(\omega t - kr_1)}e^{-j(N-1)\delta/2}\left[\frac{\sin(N\delta/2)}{\sin(\delta/2)}\right]. \tag{8.42}$$

The intensity at P is obtained by taking the real part of the foregoing equation, squaring the result, and time-averaging it in accordance with the prescription given by Eq. 8.1. The resultant intensity is

$$I = \frac{\epsilon_o c A^2}{2R^2}\left[\frac{\sin(N\delta/2)}{\sin(\delta/2)}\right]^2. \tag{8.43}$$

But the intensity due to the emission from a single source radiating alone is given by $I_o = \epsilon_o c A^2/2R^2$, and we can therefore write that

$$I = I_o\left[\frac{\sin(N\delta/2)}{\sin(\delta/2)}\right]^2 \tag{8.44}$$

where

$$\begin{aligned} \delta &= kd \sin \psi \quad \text{radians} \\ &\approx kd\, \psi \quad \text{for small } \psi. \end{aligned} \tag{8.45}$$

As a check on our mathematics, let us take two sources and see if we can recover the equations of the previous section. With N = 2, $\sin(N\delta/2) = \sin \delta = 2 \sin(\delta/2)\cos(\delta/2)$ and Eq. 8.44 reduces to $I = 4I_o \cos^2(\delta/2)$, which is precisely our earlier equation (8.20).

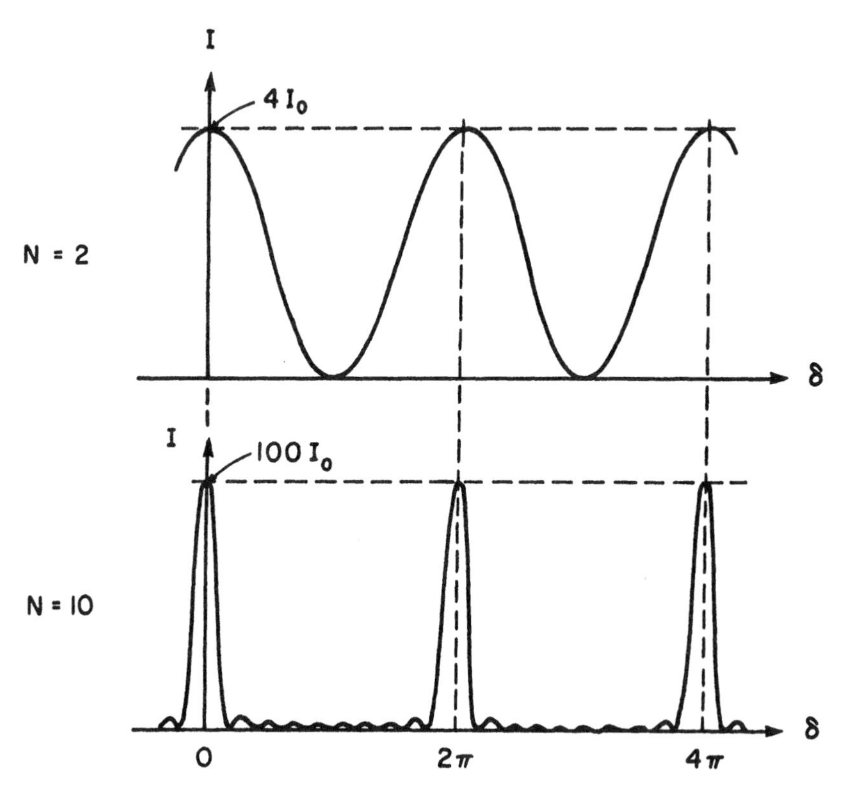

Fig. 8.9 Interference from an array of sources (N = 10), compared with the case of two sources (N = 2). The small secondary maxima are of unequal size, the largest being adjacent to the principal maximum. When N is large, the intensity of these adjacent secondary maxima are approximately (1/22) times the principal maxima [i.e., they are down by ~13.5 dB]. The next secondary maximum is down from the principal maximum by 17.9 dB, etc.

Figure 8.9 shows plots of intensity I vs δ obtained from Eq. 8.44 for the case of two sources (N = 2) and ten sources (N = 10) all driven in phase. The strong maxima, known as principal maxima, occur whenever the denominator of Eq. 8.44 vanishes, that is, when $\delta/2$ is zero or an integral multiple of π; that is, whenever $(kd \sin \psi)/2 = 0, \pm\pi, \pm 2\pi, \ldots$. Expressed in terms of wavelengths, this means that the principal maxima are located at angles ψ given by

$$\sin \psi = \frac{m\lambda}{d} \qquad m = 0, \pm 1, \pm 2, \pm 3 \ldots \qquad \text{for maxima} \tag{8.46}$$

where the integer m is known as the "order of the interference." The maxima are not of infinite intensity (despite the fact that the denominator vanishes) but are of height $N^2 I_0$. The reason is that when the denominator vanishes, so does the numerator, and you can prove to yourselves that this ratio 0/0 has a value equal to N^2. The principal maxima are large because the radiation from all the oscillators arrives exactly in phase at the observation point.

Between the principal maxima there are weak secondary maxima, the first secondary maximum being only a few percent of the principal maximum when N is large. The secondary maxima are small because for them only certain sets of radiators give rise to constructive interference at the observation point P. The maxima are separated by points of zero intensity which occur when $\delta/2 = \pm(\pi/N), \pm(2\pi/N), \pm(3\pi/N) \ldots \pm(n\pi/N)$, but excluding the cases for which the ratio n/N is an integer. Therefore the zeros are located at angular positions given by

$$\sin \psi = \frac{n\lambda}{Nd} \qquad n = \pm 1, \pm 2, \pm 3, \ldots \qquad \text{for minima.} \tag{8.47}$$

Observe that for a given number N of radiators, there are (N-1) zeros between any two neighboring principal maxima; and (N-2) secondary maxima.

The importance of such linear arrays of point sources lies in the fact that the larger N is, the narrower are the widths of the principal maxima. This means that one can pinpoint with great accuracy the region of space one wishes to illuminate (see Fig. 8.10). Indeed, from Eq. 8.47 it can be seen that the full width of a principal maxi-

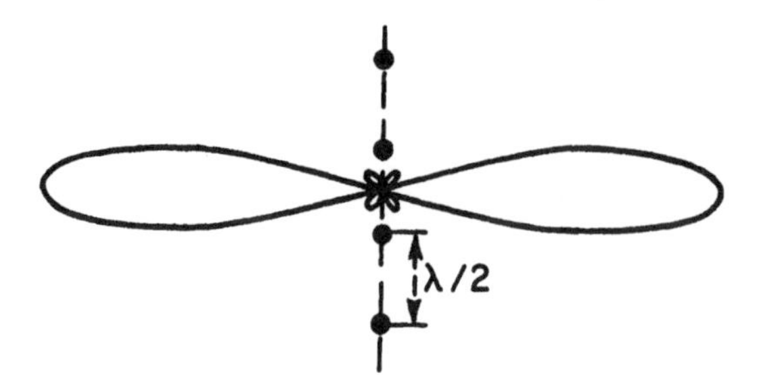

Fig. 8.10 Radiation pattern of four point sources (N = 4) driven in phase and separated by a distance d = λ/2.

mum (as measured by the separation between adjacent zeros) equals

$$\Delta \sin \psi \approx \Delta\psi = \frac{2\lambda}{Nd} . \qquad (8.48)$$

Take, for example, the interferometric radio telescope at Westerbork, Holland, illustrated in Fig. 8.11. It is comprised of 11 elements separated from one another by a distance of d = 1062 meters and operating at a wavelength of 6 cm. With these values, $\Delta\psi = (2 \times 0.06)/(11 \times 1062) = 1.03 \times 10^{-5}$ rad or 2.1 arc seconds! This is about 1/1000 the angular size of the moon.

Spectral resolving power of an array of sources

We have seen that spatial resolution becomes progressively better the larger the number of elements N in the array. Here we wish to show that wavelength resolution when the sources are not purely monochromatic likewise increases with increasing N. This is of great importance for the familiar diffraction grating used so extensively in optics, and we shall use the grating as a basis of our discussion.

One form of grating (see Fig. 8.12) consists of parallel grooves scratched on a piece of optically flat and perfect glass. The grating is then illuminated by a plane beam of radiation whose spectrum is under study. The grooves are virtually opaque to the radiation, and the transparent spaces between the grooves act as antennas or secondary sources of radiation whose beams then interfere with

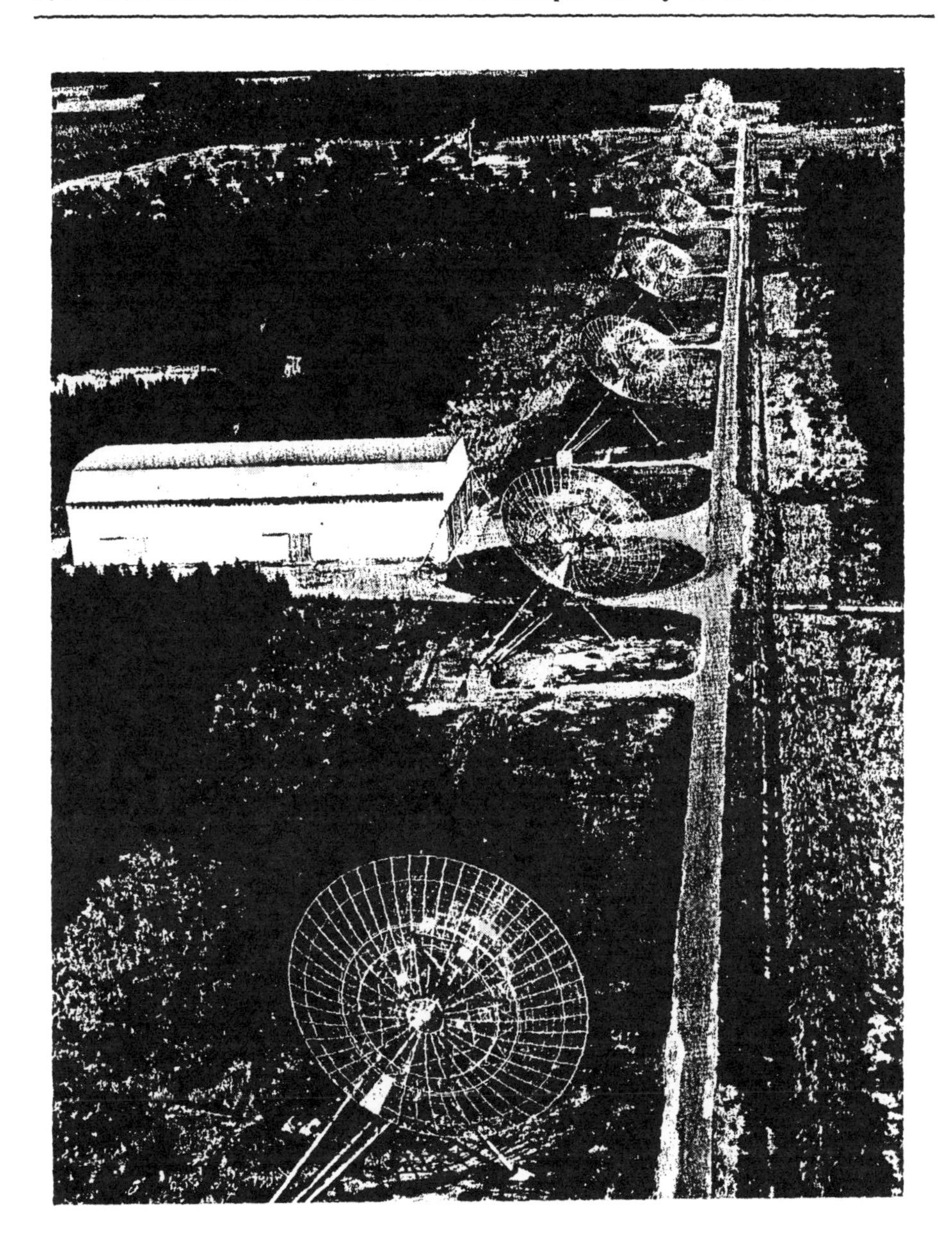

Fig. 8.11 The Westerbork Radio Telescope (courtesy of the Netherlands Foundation for Radio Astronomy. Copyright by Aerophoto Eelde).

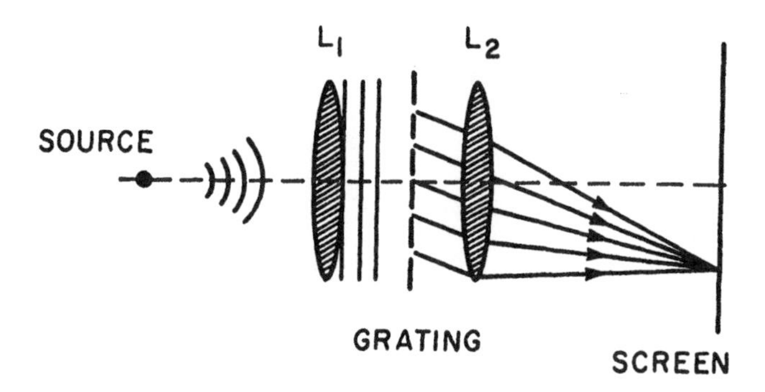

Fig. 8.12 Principle of the diffraction grating. A lens L_1 is interposed between the source and the grating to insure plane-wave illumination of the latter. Since the interference fringes are located at infinity, a lens L_2 is placed behind the grating. The screen is in the focal plane of lens L_2.

one another. If the spaces between the grooves were tiny compared with their separation d, so that they could be considered as point sources, the intensity I given by Eq. 8.44 would in fact be exactly correct. Since the spaces between the grooves are large, and of the same order in size as the opaque portions between them, Eq. 8.44 does not tell the full story. Diffraction associated with each of the sources needs to be considered, as will be done correctly in section 8.9. However, in determining the spectral resolving power, our analysis carried out so far suffices.

Suppose the radiation illuminating the grating consists of two neighboring wavelengths λ_1 and λ_2. This gives rise to two overlapping interference patterns. What is the wavelength resolution that can be attained with such an instrument? Asked differently, what is the minimum wavelength difference $\Delta\lambda \equiv (\lambda_1 - \lambda_2)$ that can be measured? According to Eq. 8.46 the angular position of a principal maximum of order m and wavelength λ_1, equals

$$\sin \psi_1 \approx \psi_1 = m\frac{\lambda_1}{d}. \tag{8.49}$$

And, for wavelength λ_2 the angular position equals

$$\sin\psi_2 \approx \psi_2 = m\frac{\lambda_2}{d}. \tag{8.50}$$

As a result $(\psi_1 - \psi_2) = m(\lambda_1 - \lambda_2)d$. If the angular separation $(\psi_1 - \psi_2)$ is too small, it will be impossible to resolve the two wavelengths.

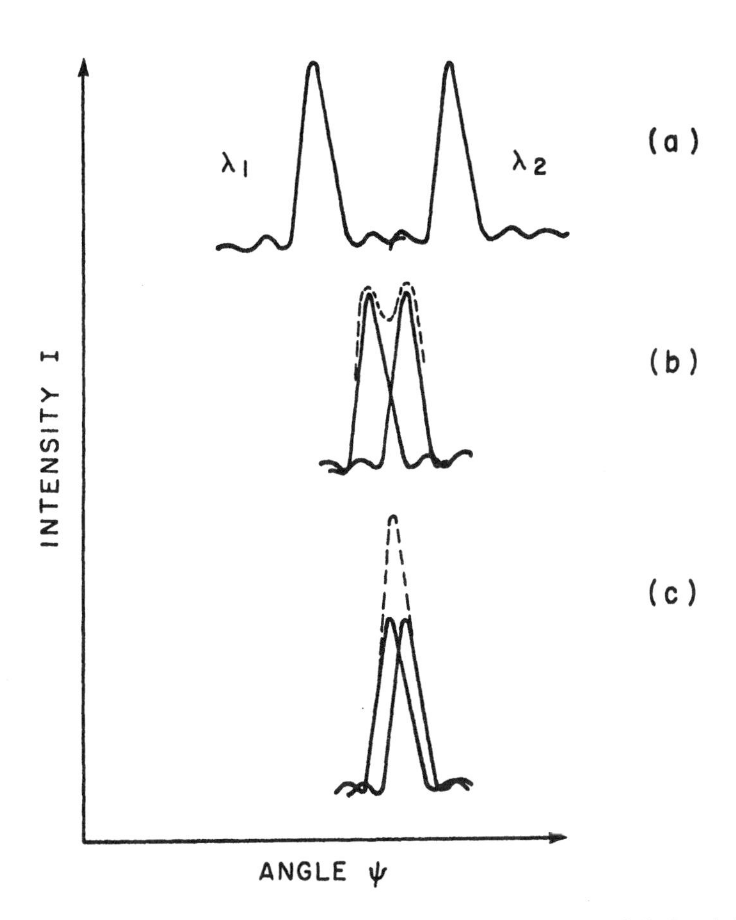

Fig. 8.13 Illustration of the Rayleigh criterion: In (a) the interference patterns of waves of wavelength λ_1 and λ_2 are well resolved. In (b) the intensity maximum of one falls on the zero of the other, and the resultant intensity (shown dashed) suggests that the two principal maxima can still be resolved. In (c) the two maxima are no longer resolved. Note that we obtained the net intensity by summing the intensities of the two interference patterns, rather than summing amplitudes, squaring, time-averaging etc. Is that all right?

But what is too small? Figure 8.13 illustrates what happens as λ_1 and λ_2 approach one another. The total intensity is the sum of the intensities of the individual interference patterns of wavelength λ_1 and λ_2; it is shown dashed in the figure. We see that as λ_1 approaches λ_2 the dimple between the maxima becomes shallower and shallower, and eventually one cannot tell if there are two principal maxima or just one maximum. Lord Rayleigh decided, on the basis of observation, that two interference patterns can just be resolved if the maximum intensity of one coincides with the first minimum of the other. Using this criterion, it then follows from Eq. 8.47 that the minimum angular separation of the two maxima is $\Delta \sin\psi \approx \Delta\psi = \bar{\lambda}/(Nd)$ where $\bar{\lambda}$ is the mean of the two wavelengths, $\bar{\lambda} = (\lambda_1 + \lambda_2)/2$. But from Eqs. 8.49 and 8.50 it follows that $\Delta\psi = \psi_1 - \psi_2 = m(\lambda_1 - \lambda_2)/d$. Thus equating the $\Delta\psi$'s from the two equations just derived, we get

$$\frac{\bar{\lambda}}{(\lambda_1 - \lambda_2)} \equiv \frac{\bar{\lambda}}{\Delta\lambda} = mN. \tag{8.51}$$

The quantity $\bar{\lambda}/\Delta\lambda$ is known as the "resolving power" of the instrument; it is equal to the product of the order number m and the total number of secondary sources (i.e., transparent spaces between grooves) left after engraving the glass plate. Interestingly, the resolving power does not depend on the spacing d, but only on the total number N. Therefore, the larger the grating is physically (for a given spacing d) the better neighboring wavelengths can be resolved.

As an example consider a grating 15 cm wide containing 10,000 lines per cm (not an unusually large number). Then N = 150,000, and $\bar{\lambda}/\Delta\lambda = 1.5 \times 10^5$ m. Suppose $\bar{\lambda} = 6000$ Å, then $\Delta\lambda = (4 \times 10^{-2}/m)$ Å. This says that when such a grating is used in first order, m = 1, we can resolve radiation containing wavelengths which do not differ by more than 0.04 Å. We can do twice as well in second order m = 2, etc.*

*In our simplified grating theory (Eq. 8.44) the intensities of the principal maxima are the same for all orders m, and are given by $I = I_0 N^2$. In practice, as a result of diffraction, the intensity goes down as m goes up, so that m cannot be excessively large. Useful values of m are m = 1, 2, and 3 but not much higher. (Why not m = 0?)

These are small wavelength differences, indeed; they allow us to make detailed studies of the fine structure of atomic spectra.

Interference by multiple reflections – the Fabry-Perot etalon

When a film or plate is illuminated with plane, monochromatic

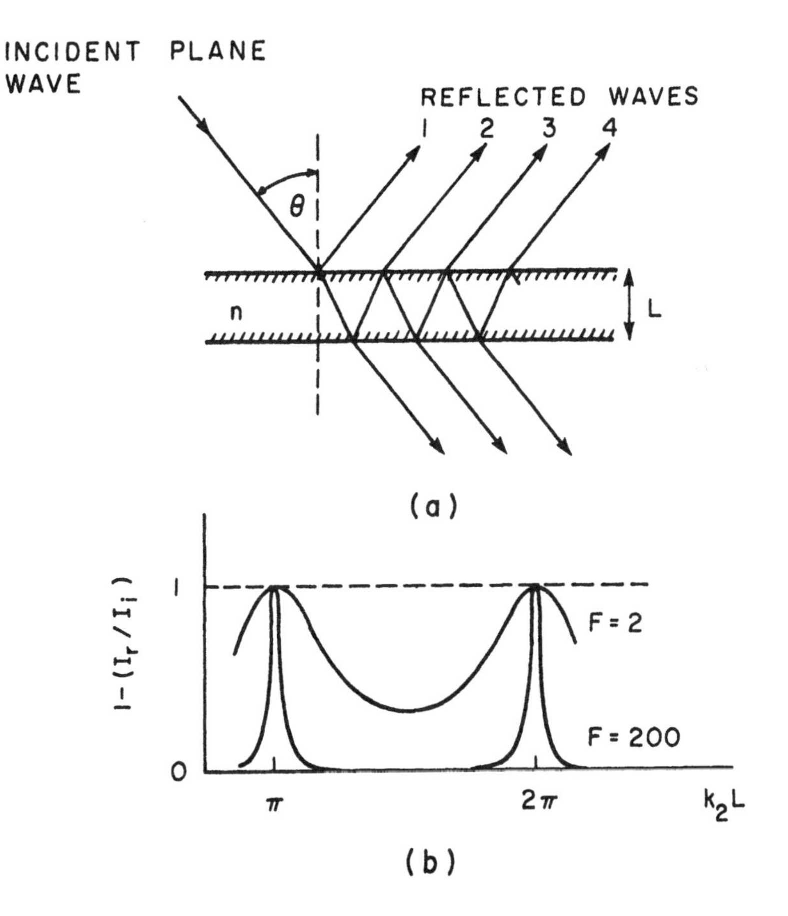

Fig. 8.14 (a) Multiple reflections from a plane parallel dielectric plate of thickness L and refractive index n. Normal incidence $\theta \rightarrow 0$ is assumed. At each reflection the intensity is reduced by $r^2 = [(n-1)/(n+1)]^2$. (b) A plot of the function $1 - (I_r/I_i)$ vs the phase factor k_2L. When F is large, the fringes are very sharp. Note that when the function $1 - (I_r/I_i)$ equals unity, the reflected power I_r is zero.

waves and the plate is made with optically flat, accurately parallel, clean, highly reflecting surfaces, it is insufficient to consider just one reflected wave from each surface as was the case in section 8.2. The radiation keeps bouncing back and forth as is illustrated in Fig. 8.14. Under such conditions the interference between multiply reflected beams must be considered and the contribution from all the waves must be summed.* But, we have already solved this problem, in a different way to be sure, in section 7.2. Although we had barely mentioned the word interference at that stage of the book the phenomenon is implicit in the calculations. Equation 7.11 is the exact result for the more general situation in which the slab is in contact with one medium in front and another medium behind. For the special case when there is vacuum on both sides of the slab, one merely sets $Z_1 = Z_3 = Z_o = \mu_o c$ wherever these quantities appear in Eq. 7.11. Z_o is the wave impedance of the vacuum and is given by Eq. 7.3. On making this substitution one obtains for the reflection coefficient R from a slab illuminated at normal incidence

$$R \equiv \frac{\text{Intensity } I_r \text{ of the reflected wave}}{\text{Intensity } I_i \text{ of the incident wave}}$$

$$= \frac{\left(Z_2^2 - Z_o^2\right)^2 \sin^2(k_2 L)}{\left(Z_2^2 + Z_o^2\right)^2 \sin^2(k_2 L) + 4Z_o^2 Z_2^2 \cos^2(k_2 L)} \tag{8.52}$$

where $Z_2 = \mu_o \omega/k = \mu_o c/n$ is the wave impedance of the slab and L is its thickness. The propagation constant $k_2 = \omega n/c$, with n as the refractive of the material. After a little algebra, the foregoing formula can be written in the following more compact way:

$$I_r = I_i \frac{F \sin^2(k_2 L)}{1 + F \sin^2(k_2 L)} \tag{8.53}$$

---

*The conventional way of summing the multiply reflected beams is done in most text books on optics. See, for example, Optics by B. Rossi (Addison-Wesley Publishing Company, South Reading, Mass., 1957); or Optics by E. Hecht and A. Zajac (Addison-Wesley, 1974).

where the quantity F, known as the "coefficient of finesse," is defined as

$$\left.\begin{aligned} F &= \frac{\left[(Z_2/Z_o)^2 - 1\right]^2}{4(Z_2/Z_o)^2} \\ &= \frac{[(1/n^2) - 1]^2}{4/n^2} \\ &= \left(\frac{2r}{1-r^2}\right)^2 \end{aligned}\right\} \tag{8.54}$$

The finesse is a measure of the amount of impedance mismatch suffered by the incident wave (when $Z_2 = Z_o$ there is no mismatch; $F = 0$, and there is no reflected wave). F may be written in various alternate ways: the second line of Eq. 8.54 gives F in terms of the refractive index n and the third line gives it in terms of the amplitude reflection coefficient r from a single surface,

$$r = \frac{Z_o - Z_2}{Z_o + Z_2} = \frac{n-1}{n+1} \tag{8.55}$$

a result which we have derived in section 7.1 (Eq. 7.6).

The zeros of intensity in the reflected wave occur when k_2L of Eq. 8.53 equals zero or an integral multiple of π. Expressed in terms of the free-space wavelength λ and refractive index n,

$$nL = (m-1)\frac{\lambda}{2} \qquad m = 1, 2, 3, \ldots \tag{8.56}$$

which is exactly Eq. 8.37 for the case of a wave incident normally ($\theta' = 0$) on a thin film in which only two reflections are taken into account.

Thus multiple reflections do not change the position of the minima. What multiple reflections do is to make the minima extremely sharp, a fact which follows (see Fig. 8.14b) by plotting I_r of Eq. 8.53 as a function of (k_2L), for a given fixed value of F. It is this sharpness which causes interference by multiple reflections to be of high interest to the spectroscopist. Note that by silvering the surfaces a typical value of $r^2 \approx 0.92$ and therefore F of Eq. 8.54 is 575.

An important optical instrument known as the Fabry-Perot interferometer is based on multiple-beam interference occurring between two accurately parallel surfaces. It is used in the analysis of spectra in which extremely high wavelength resolution is required. For details see Optics by Hecht and Zajac (Addison-Wesley Publishing Company, Inc., South Reading, Mass., 1974). Laser mirrors placed on either side of the lasing medium, and thus forming a high Q resonant cavity (see section 5.4) operate on the multibeam interference principle just described.

Example

Equation 8.53, which is valid for normal incidence, predicts that the reflected intensity is zero for some combinations of wavelength and plate thickness. These combinations are stated in Eq. 8.56. Referring to Fig. 8.14a, show by direct addition of the amplitudes of the electric fields that the sum of fields of rays 2, 3, 4, ... ∞ just equals that of ray 1.

Solution

Let r be the reflection coefficient for the electric field (i.e., $r = E_r/E_i$) and τ and τ' be the transmission coefficients for the electric fields for transmission from air to a dielectric and from a dielectric to air, respectively. Then the wave amplitudes are:

Incident ray: E_o

Ray 1: rE_o

Ray 2: $\tau\tau' rE_o$

Ray 3: $\tau\tau' r^3 E_o$

Ray 4: $\tau\tau' r^5 E_o$

. .
. .
. .

Thus the sum of all rays excluding the first reflection is

$$E_{2\to\infty} = \sum_{i=2}^{\infty} E_i = \tau\tau' r E_o (1 + r^2 + r^4 + \dots).$$

But

$$\sum_{j=0}^{\infty} r^{2j} = \frac{1}{1 - r^2} \quad \text{for} \quad r < 1.$$

Hence

$$E_{2\to\infty} = \frac{\tau\tau' r}{1 - r^2} E_o.$$

Now from Eqs. 7.7 we have

$$r = \frac{n - 1}{n + 1}, \qquad \tau = \frac{2}{n + 1}, \qquad \tau' = \frac{2n}{n + 1}$$

so $\tau\tau' = 1 - r^2$. This gives as the final result $E_{2\to\infty} = rE_o$ and the sum of all reflected rays 2 to ∞ just cancels the first reflected ray.

8.4 TWO- AND THREE-DIMENSIONAL ARRAYS

The two-dimensional array

A dramatic illustration of the interference caused by a two-dimensional array is to hold a fine wire mesh in front of a laser beam (for example, a helium-neon laser which emits at wavelength of 6328 Å). On a distant screen one observes a rectangular array of bright red dots of varying intensity, with the brightest dots arranged in the pattern of a cross situated symmetrically in the middle of the interference pattern.

Figure 8.15 depicts an idealized two-dimensional array of sources situated in the x-y plane. Let d_x be the spacing between the sources along the x axis and d_y the spacing between the sources along the y axis. Each source emits monochromatic radiation of wavelength λ, and all sources oscillate in phase with each other. (This approximates the wire mesh situated in the x-y plane and irradiated at normal

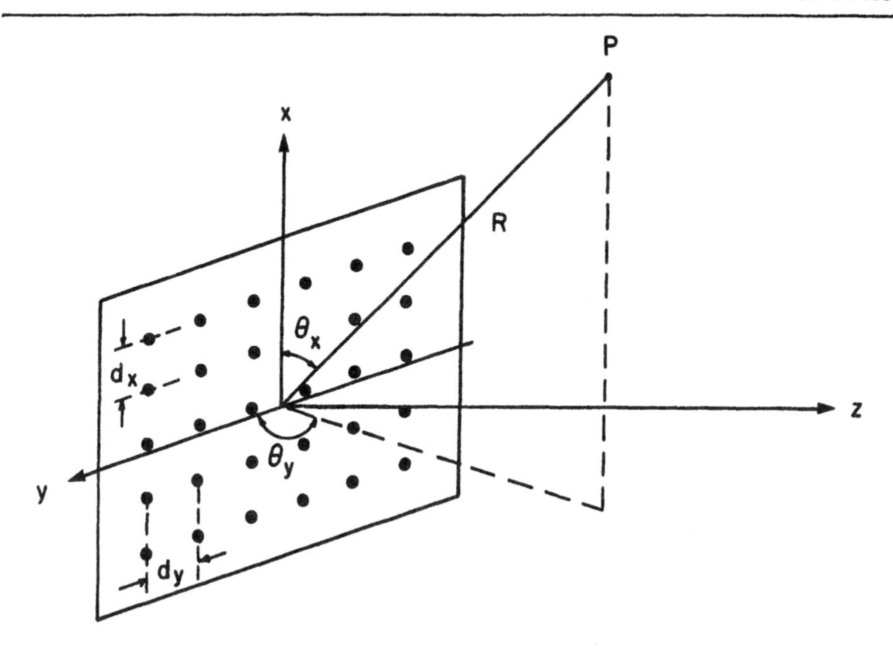

Fig. 8.15 A two-dimensional array of point sources with spacings d_x and d_y between neighboring elements. The angular position of the point of observation P is given in terms of the angle θ_x and θ_y [cos θ_x and cos θ_y are known as the "direction cosines"].

incidence by laser light propagating along z axis.) Let there be N layers of sources counting along the x axis and N layer of sources counting along the y axis, so that the total number of sources is N^2. To calculate the intensity at a distant point P on a screen one proceeds in the same way as was done in the previous subsection for the one-dimensional array. We shall not carry out the algebra but merely state the result, which one could almost guess by examining Eq. 8.44:

$$I = I_0 \left[\frac{\sin(N\delta_x/2)}{\sin(\delta_x/2)}\right]^2 \left[\frac{\sin(N\delta_y/2)}{\sin(\delta_y/2)}\right]^2 \tag{8.57}$$

where

$$\delta_x = kd_x \cos\theta_x$$
$$\delta_y = kd_y \cos\theta_y. \tag{8.58}$$

Observe that we have changed to new angles θ; the angle ψ previously defined the angular position of P relative to the normal to the array. In two and three dimensions this is not a useful quantity. The angles θ are complementary to ψ: θ_x is the angle which the radius vector OP subtends with the x axis, and θ_y is the angle which OP subtends with the y axis (see Fig. 8.15).

It is readily seen from Eq. 8.57 that principal maxima will be obtained at those angular positions for which the conditions

$$\frac{\delta_x}{2} = 0, \pm\pi, \pm 2\pi, \pm 3\pi \ldots$$

$$\frac{\delta_y}{2} = 0, \pm\pi, \pm 2\pi, \pm 3\pi \ldots \tag{8.59}$$

are satisfied simultaneously; this is analogous to the one-dimensional array, where the less-restrictive requirement $\delta/2 = 0, \pm\pi, \pm 2\pi, \ldots$. had to be met. Substituting for δ_x and δ_y from Eqs. 8.59 it follows that principal maxima will occur at those angular positions (θ_x, θ_y) for which the requirements

$$\cos\theta_x = \frac{m\lambda}{d_x}, \qquad m = 0, \pm 1, \pm 2, \ldots$$

$$\cos\theta_y = \frac{n\lambda}{d_y}, \qquad n = 0, \pm 1, \pm 2, \ldots \tag{8.60}$$

are satisfied simultaneously. As a result, a series of intense spots will be seen on a distant screen, each spot corresponding to a different combination of the intergers (m, n). Figure 8.16 sketches the interference pattern with numbers attached to each spot indicating the values of m and n.

If Eq. 8.57 were fully representative of the physics of the problem, the spots would all be of equal intensity given by $I = N^4 I_0$. Such would be the case with infinitely small point sources. For finite-sized sources, diffraction causes the intensity to decrease as the order of the interference (m, n) goes up. Thus, the bright central cross mentioned at the beginning of this subsection is made up of spots with low values of m and n. The brightest dot is of course the one cor-

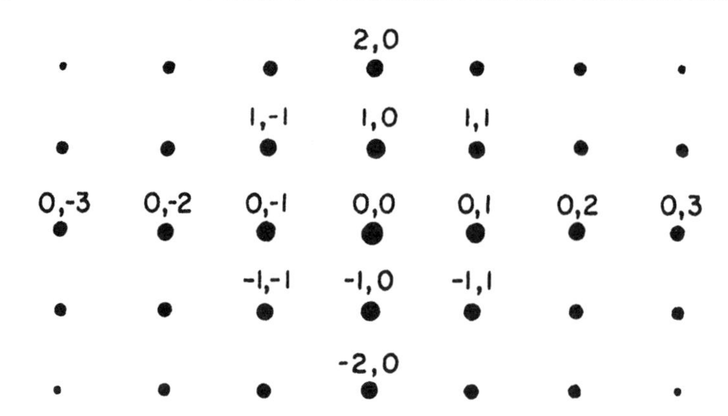

Fig. 8.16 Schematic diagram of the interference pattern of a two-dimensional array. The numbers attached to each bright spot denote the order of the interference (m, n). [After B. Rossi, Optics (Addison-Wesley Publishing Company, Inc., South Reading, Mass., 1957).]

responding to $m = n = 0$.

Two-dimensional arrays with various configurations are of great importance in radio astronomy. The pencil-like radiation pattern allows one to scan the sky in two dimensions. And, if the arrays consist of many elements N, the beam width of the radiation pattern is very narrow, so that any new stellar objects can be pinpointed with great precision.

Three-dimensional arrays — x-ray diffraction

A crystal is a three-dimensional array of atoms. If it is carefully grown and is free from faults and dislocations, the spacing between neighboring atoms is very accurately maintained. When irradiated with electromagnetic waves, the atoms act as secondary sources; each reemits the incident radiation and the scattered waves interfere with one another. From an analysis of the interference pattern the wavelength of the incident radiation can be determined. Thus, a crystal can be used as a three-dimensional diffraction grating. But for what wavelength waves? This is readily answered. For constructive interference to occur, giving rise to strong principal maxima, an equation of the form $\sin \psi = m\lambda/d$ must be satisfied (see Eq. 8.46). Since $\sin \psi$ is always less than or equal to unity, it follows that $\lambda \lesssim d/m$ where d is the spacing between the sources i.e., the atoms. Atomic spacings are typically a few angstroms, and since m is one or greater, λ is of the order of a few angstroms. Therefore, a crystal composed of an array of atoms can act as a spectrometer for x rays.

We shall not go into the complex theory of x-ray diffraction except insofar as to present a very simplified picture of the process given by Bragg. He pointed out that through any crystal one can draw a set of equidistant parallel planes all of which pass through similar groups of atoms which compose the crystal. A few such planes, known as "Bragg planes," are illustrated in Fig. 8.17.

When plane monochromatic radiation is incident on a given Bragg plane, the atoms reradiate (i.e., scatter) the incident wave (see section 7.5). However, strong constructive interference occurs only for one angle which corresponds to the direction of specular reflection.

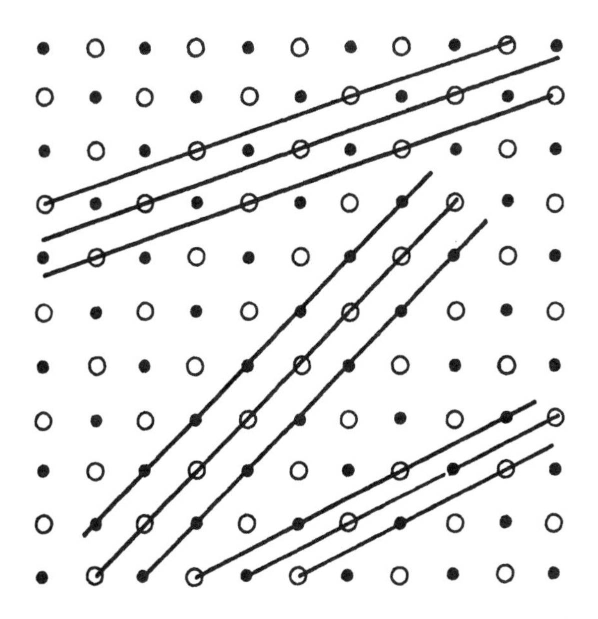

Fig. 8.17 Schematic of a crystal of NaCl showing several typical Bragg planes from which x rays are specularly reflected.

That is, the plane of atoms acts as a mirror, and the intensity is appreciable when the angle of reflection θ' is equal to the angle of incidence θ. But we note that each Bragg plane is but one of many similar, regularly spaced, parallel planes. The beams reflected from the various parallel planes will combine giving rise to an interference pattern, and one can compute where the principal maxima are located. In summary then, the individual atoms are replaced by equidistant parallel planes, and it is these planes that can then be viewed as the secondary sources of the interfering x-ray waves.

The rest of the problem is just a little algebra and geometry. Figure 8.18 illustrates two adjacent Bragg planes. The radiation is incident at an angle θ to the planes. As a result, a beam is reflected specularly from the top plane and another beam is reflected specularly from the bottom plane. For constructive interference to take place, the path length difference $0_1 0_2 B - 0_1 A$ must be an integral number of wavelengths. It is readily seen from the construction of Fig. 8.18 that the path length difference equals $2d \sin \theta$. Hence

$$2d \sin \theta = m\lambda \qquad m = 1, 2, 3 \ldots . \tag{8.61}$$

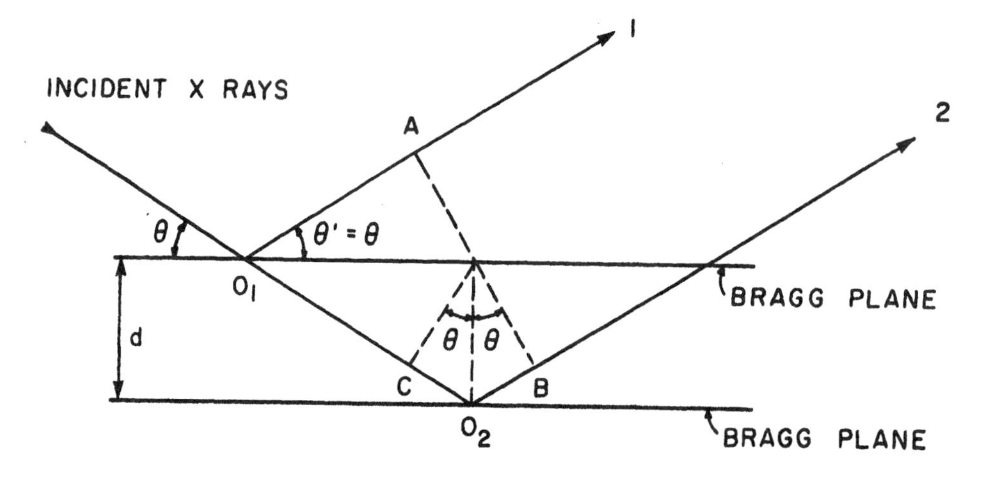

Fig. 8.18 Interference of waves 1 and 2 reflected from two adjacent Bragg planes. For maxima, $0_1 0_2 B - 0_1 A = C0_2 B = 2d \sin \theta$ must equal an integral number of wavelengths; that is, $2d \sin \theta = m\lambda$.

is the condition for constructive interference, a result known as Bragg's law. Remember, however, that Eq. 8.61 must be supplemented by the statement $\theta = \theta'$, which is implicit in the way each Bragg plane reflects the incident radiation.

8.5 DIFFRACTION

When an object is placed in a beam of radiation, it causes a portion of it to "bend," i.e., the radiation deviates from rectilinear propagation. This is known as diffraction. It manifests itself in that the shadow of the object projected onto a screen contains a fine structure of interference fringes in the vicinity of the boundary separating the dark shadow from the rest of the brightly illuminated screen. This may come as a surprise to those readers who have had some experience of geometrical optics, and who are familiar with plotting rays through various optical instruments. To them light travels in straight lines, unless it is bent by reflection or refraction. To be sure, rectilinear propagation is a valid notion if (a) the obstruction placed in the beam is very, very large compared with the wavelength, and (b) if at the same time the observer does not look too closely at what is going on near the boundary of the shadow.

Figure 8.19 illustrates the extent to which rectilinear propagation is violated. It is a plot of the computed ray paths in the immediate vicinity of a perfectly conducting half plane illuminated by plane, linearly polarized monochromatic waves. Experimentally, a razor blade stuck in the beam of a laser would do very well. Observe how some rays are bent into what would normally be the geometrical optics shadow. Also note the reflection of rays from the conducting plate on the side of the incident radiation. But a few wavelengths to the left of the razor-blade edge, the rays are almost straight lines, just as geometrical optics would have it.

Physically, diffraction comes about because the incident radiation induces oscillating currents to flow in the edges of the obstructing body, and as we know, oscillating currents radiate. On the microscopic level it is of course the electrons in the body that are set into

Fig. 8.19 Lines of average energy flow in the neighborhood of a perfectly conducting half plane illuminated at normal incidence with linearly polarized electromagnetic waves. The electric vector is in the plane of the page and the magnetic vector is perpendicular to the plane of the page. The lines correspond to light rays in geometrical optics, and they represent directions of the time-averaged Poynting flux. In some places the radiation is seen to flow in circles. Explain. This exact calculation of the diffraction near the edge comes from W. Braunbek and G. Laukien, Optik, Vol. 9, page 174 (1952).

oscillation causing them to emit electromagnetic waves. The waves then interfere.

The physical picture is very straightforward. The theoretical basis is also well understood. It is Maxwell's equations (Eqs. 3.62) coupled with the standard boundary conditions (Eqs. 5.12) that must be obeyed at every point of the conducting obstacle. But the attainment of an exact solution of this "boundary value problem" is one of the most difficult mathematical tasks of classical electromagnetic theory. Indeed, less than a handful of solutions have been obtained in the 100 years since Clerk Maxwell formulated the electromagnetic theory of light. They are: diffraction by a conducting half-plane, by a circular aperture cut in a conducting screen, and by spherical and spheroidal bodies. To be sure, diffraction has been well known and well studied

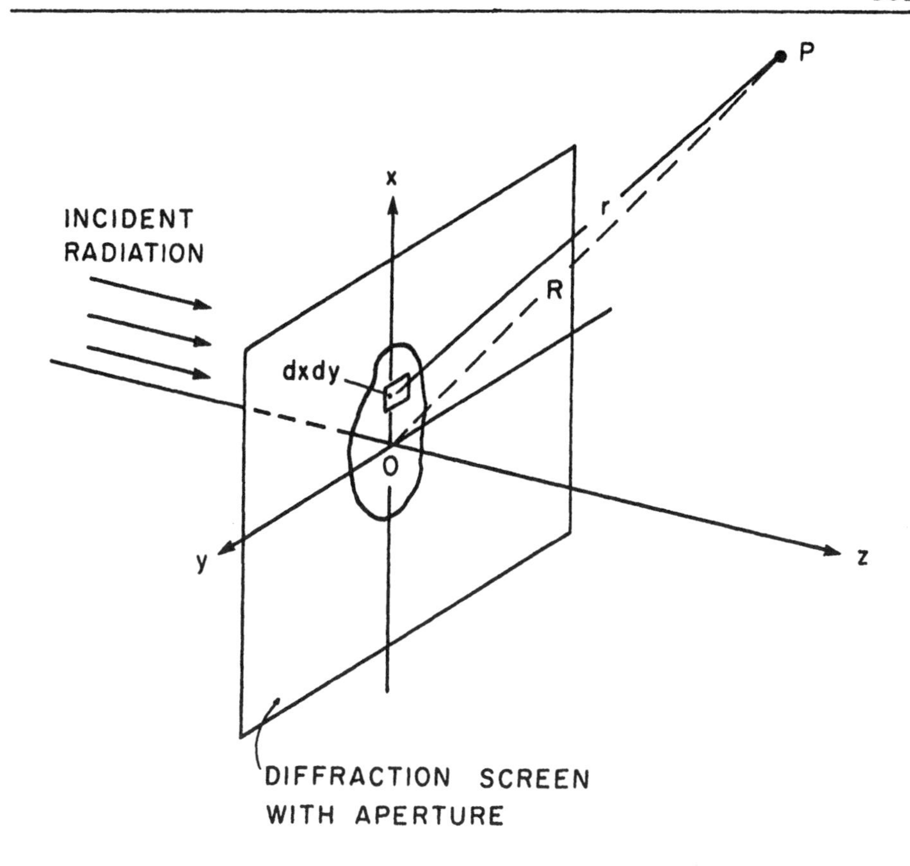

Fig. 8.20 Diffraction by an aperture in a screen.

long before the advent of electromagnetic theory. And all of it is based on what is known as the Huygens-Fresnel principle. It is a model that is easy to apply and gives quite accurate results, as long as the point of observation is at least a dozen or so wavelengths away from the diffracting object.*

Huygens-Fresnel principle

To illustrate the Huygens-Fresnel principle, consider a mono-

*The Huygens-Fresnel principle fails when we wish to calculate the field inside the diffracting aperture itself. It also fails when we wish to compute the total electromagnetic energy flowing through the aperture. But it is fine when one merely requires relative intensities far from the aperture.

chromatic wave incident on a screen which has an aperture cut into it, as is shown in Fig. 8.20. According to this principle, all points in the aperture plane may be thought of as secondary point sources of spherical waves which interfere with one another in the region to the right of the screen. The point sources replace the real source flooding the screen. The screen itself is to be considered as a perfect absorber of all radiation falling upon it — absorbing all and reemitting none, either to the right or to the left of the aperture plane. It is known as the "black screen" and has no equivalent in the real world; herein lies the approximate nature of the problem. This statement, combined with the principle of superposition of waves, yields the amplitude and hence the intensity at different points P behind the screen.

Let us now divide the aperture plane into elementary areas dxdy, so small that each little area can be thought of as a source of a spherical wave varying as $A \exp(j\omega t - jkr)/r$. The strength of the source is expected to increase as the size dxdy of the elementary area increases. Thus, we may write for the electric field dE at P due to one spherical source,

$$dE = \underbrace{\left(\frac{jk}{2\pi}\right)}_{\substack{\text{funny}\\ \text{constant}}} \quad \underbrace{(A\,dx\,dy)}_{\substack{\text{amplitude}\\ \text{of source}}} \quad \underbrace{\frac{e^{j(\omega t - kr)}}{r}}_{\substack{\text{spherical}\\ \text{wave}}} \tag{8.62}$$

where r is the distance from the middle of the little area dxdy to the observation point P.

We have pulled a couple of tricks. First, the electric field is a vector but we shall ignore this fact and just treat it as a scalar quantity. Second, we multiplied the right-hand side by the funny looking constant $(jk/2\pi)$. It makes the amplitudes come out right dimensionally as we shall see. But, since it is merely a constant number, it will change nothing in the magnitudes of the relative intensities at different points P; and it is the relative intensities that concern us here.*

*The constant $(jk/2\pi)$ arises naturally when the scalar wave equation is solved with so-called "Dirichelet" boundary conditions.

The final step in the analysis is to invoke the superposition principle which tells us that in order to obtain the total electric field at P we must sum the amplitudes dE over all the radiating sources lying in the aperture of the screen. Thus, replacing the summation by an integration over a continuum of elementary spherical waves, we obtain

$$E = \frac{jk}{2\pi} \int_{\text{aperture}} A \frac{e^{j(\omega t - kr)}}{r} \, dxdy. \tag{8.63}$$

This is the "master equation" and the starting point of all computations of diffraction. As was mentioned already, it is an approximate result. Many attempts have been made to improve it but with relatively little success. And, apart from the few rigorous solutions based on Maxwell's equations, Eq. 8.63 is as good as any, when it is tested against experiment. For a plane wave incident normally on the screen, the amplitude A is a constant and can be taken from under the integral sign. For any other type of illumination (for example, an incident spherical wave) A becomes a function of x and y.

8.6 DIFFRACTION BY A CIRCULAR APERTURE

An instructive and straightforward calculation is the intensity distribution along the axis of a circular aperture illuminated by waves incident normally on the screen. As is seen from Fig. 8.21 the distance r from a point in the aperture to the observation point P is given by

$$\begin{aligned} r &= \sqrt{x^2 + y^2 + z^2} \\ &= \sqrt{\rho^2 + z^2} \end{aligned} \tag{8.64}$$

where $\rho = \sqrt{x^2 + y^2}$ is the distance from the origin 0 to the point in the aperture in question. The elementary area can be taken as a concentric annulus of width $d\rho$ and circumference of $2\pi\rho$. Thus Eq. 8.63 becomes

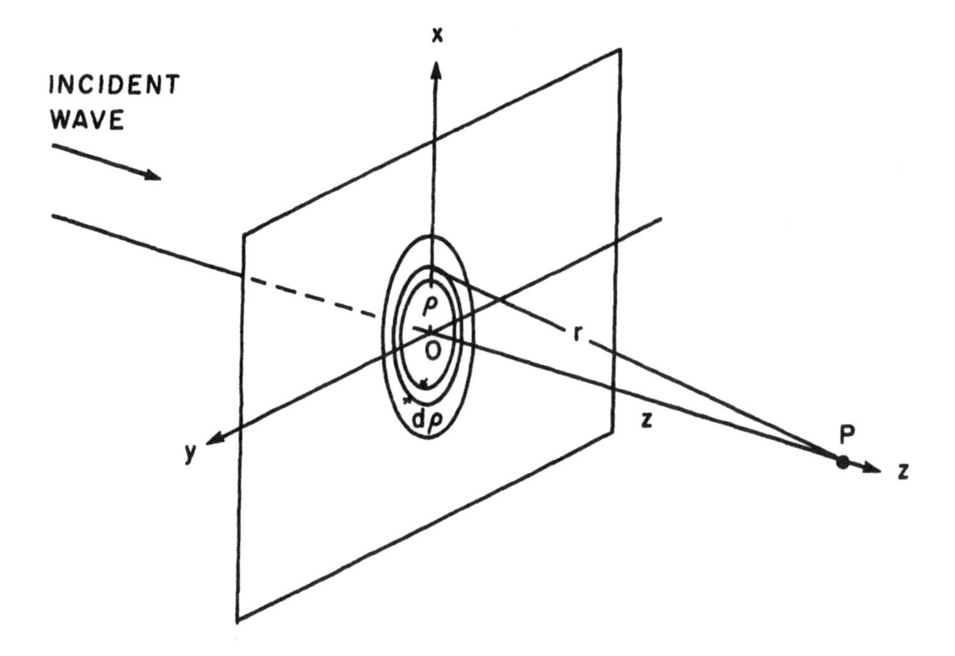

Fig. 8.21 Diffraction on axis of a circular aperture of radius a. P is the observation point.

$$E = \frac{jk}{2\pi} A\, e^{j\omega t} \int_0^a \frac{e^{-jk\sqrt{\rho^2+z^2}}}{\sqrt{\rho^2+z^2}}\, 2\pi\rho d\rho. \tag{8.65}$$

To integrate this, let $\rho^2 + z^2 = r^2$; thus $\rho d\rho = r dr$ with the result that

$$E = jkA\, e^{j\omega t} \int_z^{\sqrt{a^2+z^2}} e^{-jkr}\, dr \tag{8.66}$$

which can be integrated immediately to give

$$E = -A\, e^{j\omega t - jk\sqrt{a^2+z^2}} + A\, e^{j\omega t - jkz}. \tag{8.67}$$

The equation has an interesting structure. The first term on the right represents a plane wave that has traveled the distance $\sqrt{a^2 + z^2}$ from the rim of the aperture to the observation point P; the second term represents a plane wave that has traveled from the origin 0 to P. To get the intensity, we take the real part of Eq. 8.67, we square it and time-average it. Substitution into Eq. 8.1 then yields in a straightforward way

$$I = 2\epsilon_0 cA^2 \sin^2\left[\frac{1}{2}k\left(\sqrt{a^2+z^2}-z\right)\right]. \tag{8.68}$$

Normally, the distance z is large compared with the radius of the aperture a and we can expand the square root in a binomial expansion $\sqrt{a^2+z^2} \approx z + (a^2/2z) + \ldots$. Hence,

$$I \approx 2\epsilon_0 cA^2 \sin^2(ka^2/4z), \qquad z \gg a, \tag{8.69}$$

a result that is plotted in Fig. 8.22. As one proceeds along the z axis the intensity varies between zero and $2\epsilon_0 cA^2$. The violent intensity fluctuations are caused, of course, by the constructive and destructive interference of the spherical waves emanating from the point sources filling the circular aperture. The zeros and maxima occur

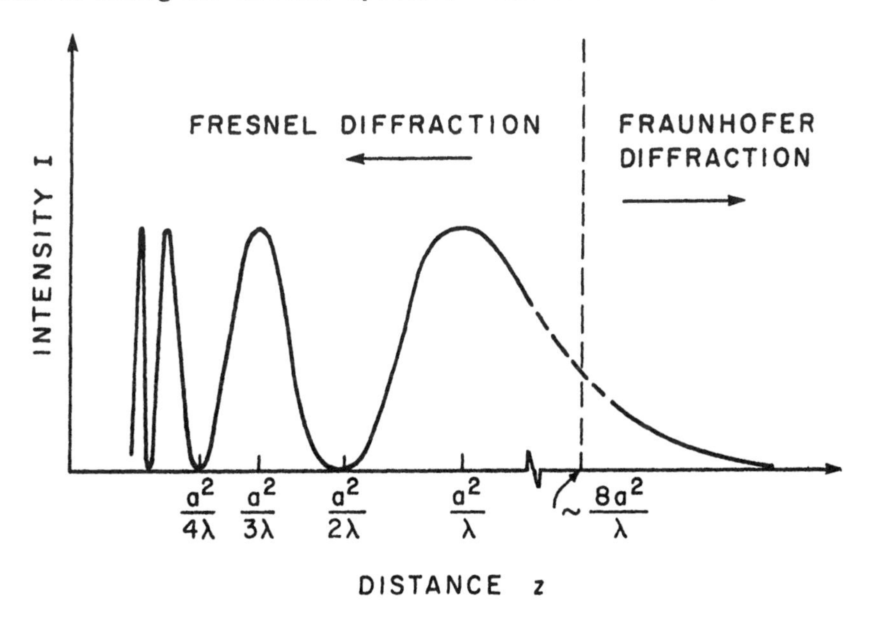

Fig. 8.22 Intensity variation on the axis of a circular aperture illuminated by plane waves.

whenever

$$z = \frac{a^2}{2\lambda}, \frac{a^2}{4\lambda}, \frac{a^2}{6\lambda}, \ldots \quad \text{(minima)}$$
$$z = \frac{a^2}{\lambda}, \frac{a^2}{3\lambda}, \frac{a^2}{5\lambda}, \ldots \quad \text{(maxima).} \tag{8.70}$$

Take, for example, an aperture of 1 mm radius illuminated with radiation of wavelength $\lambda = 5000$ Å. The maxima occur at distances $z = 200$ cm, 66.7 cm, 40 cm, etc., and the minima occur at distances $z = 100$ cm, 50 cm, 33.3 cm, etc. Observe that the spacings between neighboring fringes increase with increasing z.

It is noteworthy that the succession of maxima and minima does not continue indefinitely with ever increasing z; beyond the maximum distance $z = z(\max) = a^2/\lambda = D^2/4\lambda$ (D is the diameter), the radiation intensity falls off monotonically as z becomes larger. This point in space delineates approximately two kinds of diffraction, known as Fraunhofer and Fresnel diffraction. Actually, the demarkation line between the two regimes is usually taken to be $z = 2D^2/\lambda$, so that

$$\text{Fraunhofer diffraction for } z \gtrsim \frac{2D^2}{\lambda}$$
$$\text{Fresnel diffraction for } z \lesssim \frac{2D^2}{\lambda}. \tag{8.71}$$

These two diffraction regimes differ but little from one another as to physics; but they differ a lot as regards mathematical complexity. The diffraction integral (8.63) is easy to solve in the Fraunhofer regime, and generally quite difficult to solve in the Fresnel regime. The reason is the following. When z is sufficiently large, so that $z \gtrsim 2D^2/\lambda$, a wave arriving at P from the aperture plane approaches a plane wave (see Fig. 8.23) so that a plane-wave approximation in the integrand of Eq. 8.63 is valid. But, it is not valid when $z \lesssim 2D^2/\lambda$. For this reason, we shall restrict our discussion to the Fraunhofer regime. For a detailed discussion of Fresnel diffraction, the reader is referred to Optics by B. Rossi (Addison-Wesley, 1957).

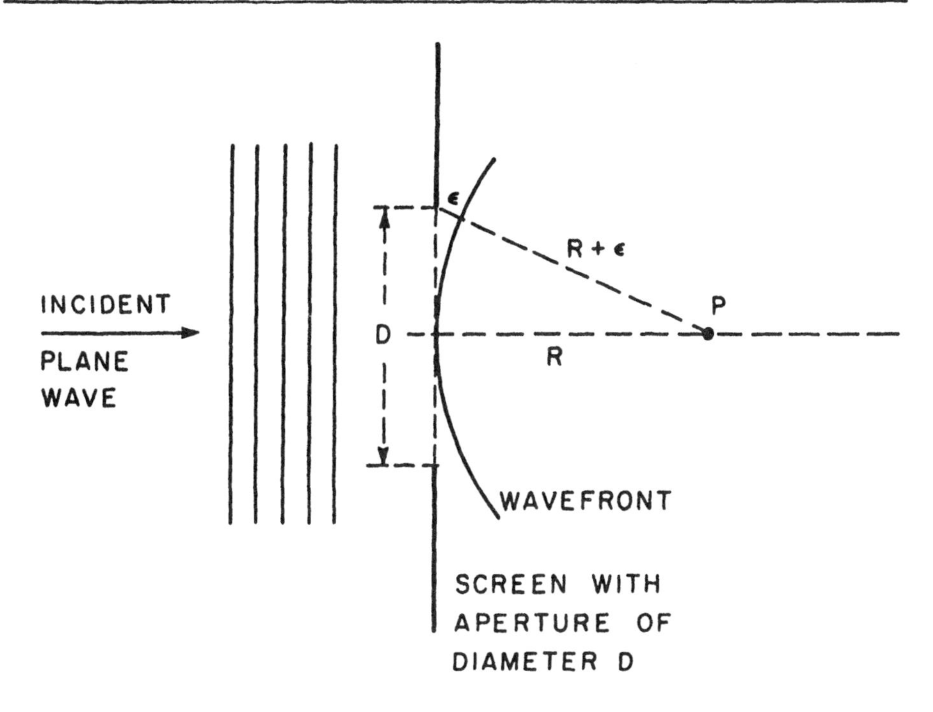

Fig. 8.23 The distance ϵ represents the departure from a plane wave over the aperture D, with P as a hypothetical point source (or point sink). Now, $(R+\epsilon)^2 = R^2 + (D/2)^2$, which on expanding the bracket yields, $R^2 + \epsilon^2 + 2R\epsilon = R^2 + D^2/4$. When ϵ is small so that the wave is almost plane, we can neglect ϵ^2 with the result that $R \approx D^2/8\epsilon$. For a plane wave the path length difference ϵ is exactly zero. When $\epsilon \ll$ wavelength λ of the radiation, the wave is for all practical purposes plane. It has been agreed upon that $\epsilon \approx \lambda/16$ is the breakeven point. Hence setting ϵ equal to this value it follows that $R \approx 2D^2/\lambda$, which delineates the Fraunhofer and Fresnel diffraction regions.

The limit of an infinitely large aperture

What would one expect to see at the observation point P behind the screen as the aperture is made ever larger? In the limit as $a \to \infty$, there is no screen left, there is no diffraction, and what one must see at P is merely the plane wave that flooded the screen in the first place. Does Eq. 8.63, and the mathematics that follows bear out our expectations? It better!

Let us go back to Eq. 8.66 and allow the aperture radius a to go to infinity. Then the electric field at P is

$$E = jkA\, e^{j\omega t} \int_z^\infty e^{-jkr}\, dr. \tag{8.72}$$

But there is trouble: the integral diverges at the upper limit. No matter. We shall fix things up by assuming that the waves damp a little with distance, and introduce an exponential damping term $\exp(-\alpha r)$ in the integrand of the foregoing equation. At the end of the calculations we let $\alpha \to 0$. (Mathematicians love this trick; they say that they are introducing a "Tauberian parameter.") Thus,

$$E = jkA\, e^{j\omega t} \int_z^\infty e^{-(jk+\alpha)r}\, dr \tag{8.73}$$

a result which can now be integrated with the result that

$$\begin{aligned} E &= jkA\, e^{j\omega t} \left[\frac{e^{-(jk+\alpha)r}}{-(jk+\alpha)}\right]_z^\infty \\ &= A\, e^{j(\omega t - kz)} \quad \text{as} \quad \alpha \to 0. \end{aligned} \tag{8.74}$$

We have proved our point: as the aperture is made infinitely large, there is no diffraction, and a regular plane wave is received at the observation point P. Also, we can now appreciate why we have introduced in Eqs. 8.62 and 8.63 the strange looking constant $(jk/2\pi)$. It makes the amplitude of the plane wave come out just right. Without that term, the second equation (8.74) would have the unusual form, $(2\pi/jk)A \exp[j(\omega t - kz)]$.

Example

Calculate the intensity at a point P on the axis of a circular disc of radius a, illuminated by a plane wave.

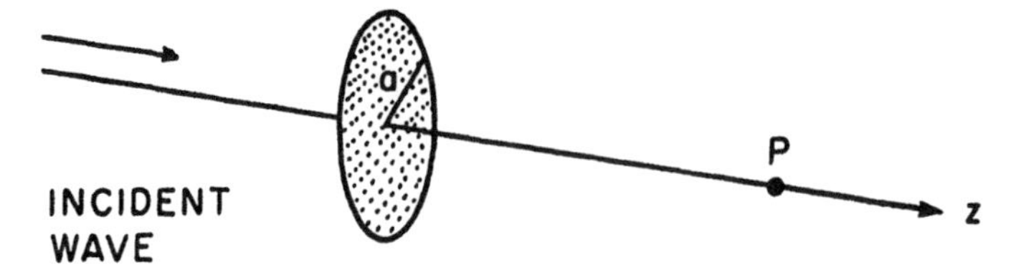

Solution

We could start with Eq. 8.63 and compute the intensity at P from first principles as we did in this section for the case of the circular aperture. However, there is a shorter way known as Babinet's principle. It says that if we know the amplitude behind an aperture in the screen (say, a circular hole) we can determine almost by inspection the amplitude at the same point behind the complementary obstacle (the disc). This principle follows from Eq. 8.63 and the fact that when we add up contributions from the entire plane occupied by the screen we get at P a plane wave, a fact we have just proved above. Breaking up the integral (8.63) into parts we get

$$\underbrace{(jk/2\pi)\int_{\substack{\text{entire}\\\text{plane}}}\text{(integrand of (8.63))}}_{\substack{A\,e^{j(\omega t-kz)}\\\text{unobstructed wave front at P}}} = \underbrace{(jk/2\pi)\int_{\substack{\text{aperture}\\\text{in screen}}}\text{(integrand of (8.63))}}_{\substack{E(\text{aperture})\\\text{field at P due to aperture in screen}}} + \underbrace{(jk/2\pi)\int_{\substack{\text{remainder of}\\\text{aperture plane}}}\text{(integrand of (8.63))}}_{\substack{E(\text{obstacle})\\\text{field at P due to complementary obstacle}}}$$

or

$$A\,e^{j(\omega t-kz)} = E(\text{aperture}) + E(\text{obstacle}) \tag{8.74a}$$

which is the mathematical statement of Babinet's principle.

Now, since E(aperture) has already been calculated, E(disc) follows directly. Substituting Eq. 8.67 in Eq. 8.74a we find that

$$E(\text{disc}) = A\,e^{j\omega t-jk\sqrt{a^2+z^2}}$$

Take the real part of E(disc) and square it:

$$E^2(\text{disc}) = A^2 \cos^2\left(\omega t - k\sqrt{a^2 + z^2}\right).$$

Time averaging then gives

$$\langle E^2(\text{disc})\rangle = \frac{A^2}{2}.$$

Inserting this in Eq. 8.1 for the intensity finally yields

$$I = \frac{1}{2}\epsilon_0 c A^2$$

showing that the intensity along the axis of a disc is constant independent of distance z. This remarkable result that there is a bright spot on axis, right in the middle of the geometrical optics shadow, was proved theoretically and experimentally in 1818 by Fresnel and Arago.

8.7 FRAUNHOFER DIFFRACTION BY A LONG SLIT

Consider a long straight slit of width D and length L ($L \gg D$) cut in an opaque screen illuminated by plane monochromatic light of wavelength λ. The slit is so long in the y direction (see Fig. 8.24) that diffraction from its far edges is negligible. The area of the aperture can now be subdivided into elements, each of length L and width dx'. One such element situated at a distance x' from the origin 0 is shown in the figure. The observation point P is located on a distant screen. Its position coordinates are $(x, 0, z)$. Thus the distance r from element dx' to P is

$$\left.\begin{aligned} r &= \sqrt{(x-x')^2 + z^2} \\ &= \sqrt{x^2 + x'^2 - 2xx' + z^2} \\ &= \sqrt{R^2 + x'^2 - 2xx'} \\ &= R\sqrt{1 + \frac{x'^2}{R^2} - \frac{2xx'}{R^2}} \end{aligned}\right\} \qquad (8.75)$$

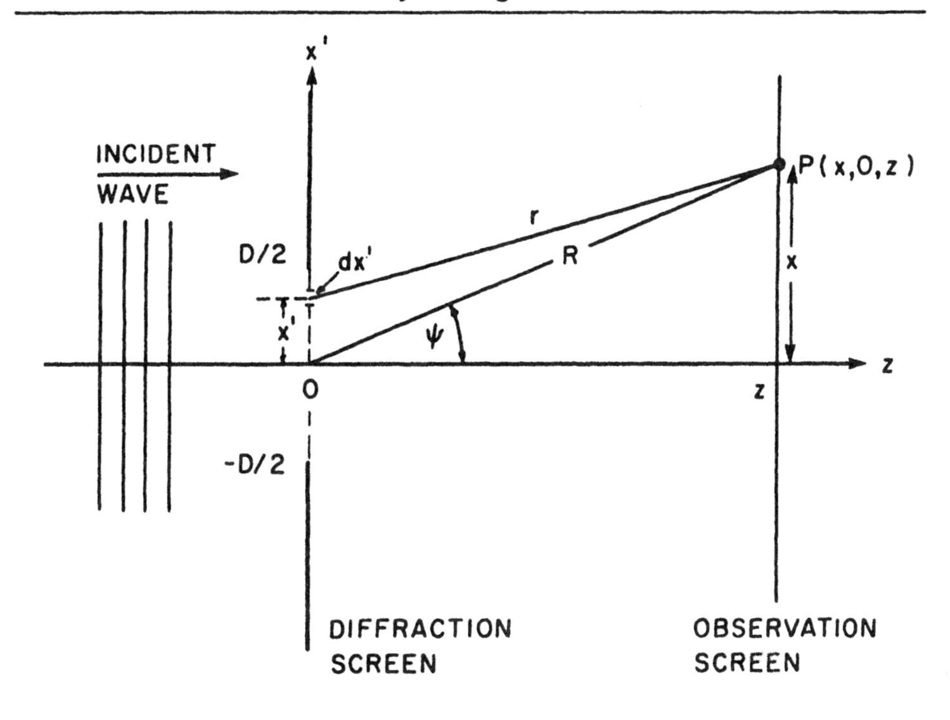

Fig. 8.24 Geometry for computing the diffraction by a long slit of width D.

where R is the distance from P to 0. We now assume that R is large compared with respect to x', and we expand the last expression (8.75) in a binomial expansion

$$r \approx R + \frac{x'^2}{2R} - \frac{xx'}{R}. \tag{8.76}$$

Substituting this value of r in Eq. 8.63 yields the following result for the electric field at P:

$$E \approx \left(\frac{jkA}{2\pi R}\right) e^{j(\omega t - kR)} \int_{-D/2}^{D/2} e^{-j(kx'^2/2R) + j(kxx'/R)} \, L dx'. \tag{8.77}$$

This is a fairly difficult integral to evaluate and can be expressed only in terms of special mathematical functions. But, we recall that in Fraunhofer diffraction the term $(kx'^2/2R)$ (cf. Eqs. 8.71 and the discussion that follows them) can be neglected compared with the other term in the exponential. Therefore,

$$E \approx \left(\frac{jkLA}{2\pi R}\right) e^{j(\omega t-kR)} \int_{-D/2}^{D/2} e^{j(kx/R)x'}\, dx'. \tag{8.78}$$

This integration can be performed immediately, with the result that

$$E \approx \left(\frac{jk}{2\pi}\right) A\left(\frac{LD}{R}\right) e^{j(\omega t-kR)} \left(\frac{\sin\beta}{\beta}\right) \tag{8.79}$$

where the parameter β can be written in several ways:

$$\left.\begin{aligned} \beta &= \left(\frac{kD}{2R}\right) x \\ &= \left(\frac{\pi D}{\lambda R}\right) x \\ &\approx \left(\frac{\pi D}{\lambda}\right) \sin\psi. \end{aligned}\right\} \tag{8.80}$$

Here $\sin\psi \approx x/R$ is the angle between the z axis and the line 0P. As usual, the intensity I at point P is obtained by taking the real part of Eq. 8.79, squaring it, time averaging, and substituting the result in Eq. 8.1:

$$I = \frac{(\epsilon_o c)}{2} \left(\frac{kALD}{2\pi R}\right)^2 \left(\frac{\sin\beta}{\beta}\right)^2. \tag{8.81}$$

Lumping all the constants of Eq. 8.81 into one constant, I_o, yields the sought-after result

$$I = I_o \left(\frac{\sin\beta}{\beta}\right)^2. \tag{8.82}$$

A plot of this equation is illustrated in Fig. 8.25, which shows a strong central peak flanked by a series of weak interference fringes. The principal maximum occurs on axis x = 0 (or $\psi = 0$). It has a value $I = I_o$, because in the limit $\beta \to 0$, $\sin\beta/\beta \to 1$.

The minima in intensity occur when

$$\left.\begin{aligned} &\beta = \pi, \pm 2\pi, \pm 3\pi, \ldots. \\ \text{or} \quad &\sin\psi = \frac{\lambda}{D}, \frac{2\lambda}{D}, \frac{3\lambda}{D}, \ldots \end{aligned}\right\} \text{for minima.} \tag{8.83}$$

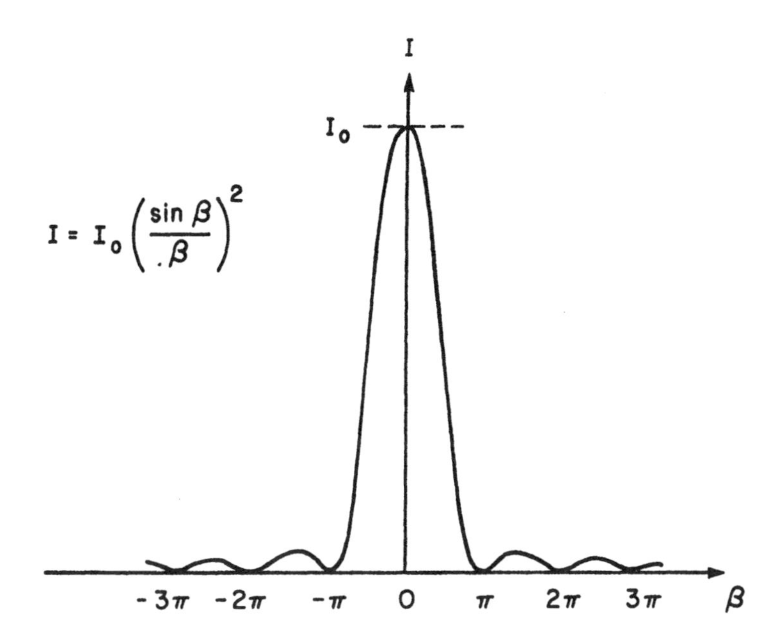

Fig. 8.25 The intensity distribution in the Fraunhofer region of a long slit in a diffraction screen. Note that this figure is identical to Fig. 2.27 for the Fourier spectrum of a wave packet. So, the Fraunhofer diffraction pattern is obviously the spatial Fourier transform of something. But, of what?

The positions of the small subsidiary maxima occur roughly midway between the minima. More precisely, one finds that

$$\left.\begin{array}{l} \beta = 4.49,\ 7.73,\ 10.9,\ \ldots \\ \text{or} \\ \sin\psi = 1.43\,\frac{\lambda}{D},\ 2.46\,\frac{\lambda}{D},\ 3.47\,\frac{\lambda}{D},\ \ldots . \end{array}\right\} \text{for secondary maxima.} \tag{8.84}$$

Observe that as β increases, the peaks of the secondary maxima keep decreasing. The first one has a value equal to $0.047\ I_o$, the second, $0.017\ I_o$, the third $0.0088\ I_o$, etc. By and large, all of the secondary maxima are so small that the bulk of the radiant intensity is concentrated in the central, principal maximum between values of $\beta \approx \pm\pi$. Therefore, the halfwidth $\Delta\beta = \pi$, or written in terms of angles

$$\Delta \sin\psi = \frac{\lambda}{D} \text{ radians} \tag{8.85}$$

which, for small λ/D, can be approximated by $\Delta\psi \approx \lambda/D$ radians. If, for example, $D = 1$ mm and $\lambda = 5000$ Å, then $\Delta\psi = 5 \times 10^{-4}$ rad = 2.9×10^{-2} degrees of arc. Thus, if a screen is placed 5 meters from the aperture, the halfwidth of the central interference fringe will be ~0.25 cm. Its full width is therefore 0.50 cm.

The result that the angular halfwidth $\Delta\psi = \lambda/D$ is not restricted to the infinitely long slit, but is true (except for numerical factors) for diffraction by any shape of the hole. Take, for example, a rectangular slit of width D and length L. Then, in the plane containing the width D, $\Delta\psi_D = \lambda/D$; in the plane containing L, the halfwidth $\Delta\psi_L = \lambda/L$. For the circular aperture, the angular halfwidth, that is, the angular radius of the circular diffraction spot $\Delta\psi = 1.22\ \lambda/D$ where D is the diameter of the aperture, and so on.

Suppose one has a radar antenna in the form of a paraboloidal dish like that shown in Fig. 8.11. Radiation from the feed at the focus illuminates the dish, which reflects the waves and sends out a beam of radiation. If diffraction caused by the rim of the dish did not exist, parallel rays of radiation would emanate from this antenna, just as geometrical ray tracing would predict (see Fig. 8.26a). But diffraction occurs, in exact analogy to the case of a circular aperture in a screen irradiated by plane waves. Therefore, the beam emanating from the dish diverges, and the angular halfwidth (i.e., angular radius) of the central maximum is $\Delta\psi = 1.22\ \lambda/D$. The central maximum is flanked by small secondary maxima, as is shown in Fig. 8.26b. The paraboloidal dish shown in Fig. 8.11 has a diameter of 25 meters. Suppose it emits radiation of wavelength $\lambda = 6$ cm. The angular halfwidth of the emitted beam is $\Delta\psi = 1.22 \times 6/2500$ radians, or 0.17 degrees of arc.

A lens acts in a similar way. Consider, for example, that the lens is illuminated by plane waves. In the geometrical optics approximation, a perfectly sharp, infinitely small image is formed at the focus (provided the lens is perfect and free from aberrations). But diffraction spoils this nice picture. Instead, one has in the focal plane an interference pattern of circular fringes with a sharp central maximum surrounded by weak circular fringes. The angular halfwidth of

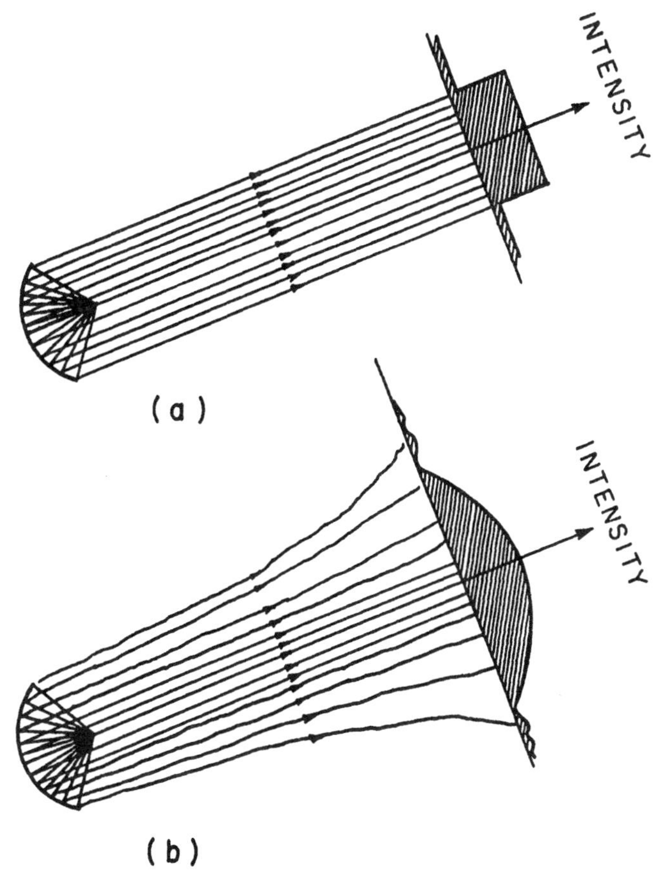

Fig. 8.26 (a) Schematic of the average energy flow in the geometrical optics limit $\lambda \to 0$ emanating from a parabolic antenna illuminated by a point source placed at the focus. It is clear from Eq. 8.85 that for fixed D, the beam divergence $\psi \to 0$ as $\lambda \to 0$. Thus, geometrical optics is an approximation that becomes better and better as $\lambda \to 0$. (b) Schematic of the energy flow in the presence of diffraction.

the principal maximum is $\psi = 1.22\ \lambda/D$. If f is the focal length, then the spot size of the diffraction-limited image has a radius equal to $1.22\ \lambda f/D$, a result that follows from the fact that the linear distance on the observation screen $\approx f\psi$.

Example

The third minimum of a single-slit diffraction pattern occurs

3 mm from the central maximum for light of wavelength 4000 Å. For light of another wavelength λ_o the second minimum occurs at the same place on the screen. The screen is in the focal plane of a lens of focal length 1.5 meter. What is λ_o and what is the slit width D?

Solution

From Figs. 8.24 and 8.25 and Eq. 8.80 we have

$$\beta = \frac{\pi D}{\lambda} \sin \psi$$

and

$$\tan \psi = \frac{3 \times 10^{-3}}{1.5} = 2 \times 10^{-3} \approx \sin \psi.$$

For λ = 4000 Å we have $\beta = 3\pi$, hence

$$3\pi = \frac{\pi D}{4 \times 10^{-7}} (2 \times 10^{-3})$$

$$D = 6 \times 10^{-4} \text{ m} = 0.6 \text{ mm}.$$

For λ_o we have $\beta = 2\pi$, so

$$2\pi = \frac{\pi(6 \times 10^{-4})}{\lambda_o} (2 \times 10^{-3})$$

$$\lambda_o = 6 \times 10^{-7} \text{ m} = 6000 \text{ Å}.$$

Example

Calculate the single-slit diffraction pattern using an array of N discrete radiators in a Huygens construction. Then let $N \to \infty$ and thus proceed to a continuous distribution of radiators.

Solution

For a discrete array, Eq. 8.44 written out in full gives

$$I = I_o \left\{ \frac{\sin[(Nkd/2)\sin\psi]}{\sin[(kd/2)\sin\psi]} \right\}^2$$

where d is the distance between neighboring sources. Therefore the width of the array equals Nd. We call this the slit width D in our Huygens construction, that is,

$$D = Nd.$$

We insert this result in the above equation and obtain

$$I = I_o \left\{ \frac{\sin[(kD/2)\sin\psi]}{\sin[(kD/2N)\sin\psi]} \right\}^2$$

With D kept fixed, let N become very large. As $N \to \infty$

$$\sin[(kD/2N)\sin\psi] \to (kD/2N)\sin\psi$$

and

$$I = N^2 I_o \left\{ \frac{\sin[(kD/2)\sin\psi]}{(kD/2)\sin\psi} \right\}^2$$

$$= N^2 I_o \left(\frac{\sin\beta}{\beta} \right)^2$$

which is the sought-after expression for the single-slit diffraction pattern. Note that as $N \to \infty$, the intensity I_o of an individual slit must be allowed to go to zero in such a way that the product $N^2 I_o$ equals a finite constant.

In this calculation we began with an expression giving the interference pattern between an array of radiators and we ended up with an expression for the diffraction pattern of a slit. This once again reinforces our earlier statement that diffraction is just another manifestation of interference.

8.8 RESOLVING POWER

Thus far we have considered the case of a single monochromatic source illuminating the diffraction screen. Suppose now that there are

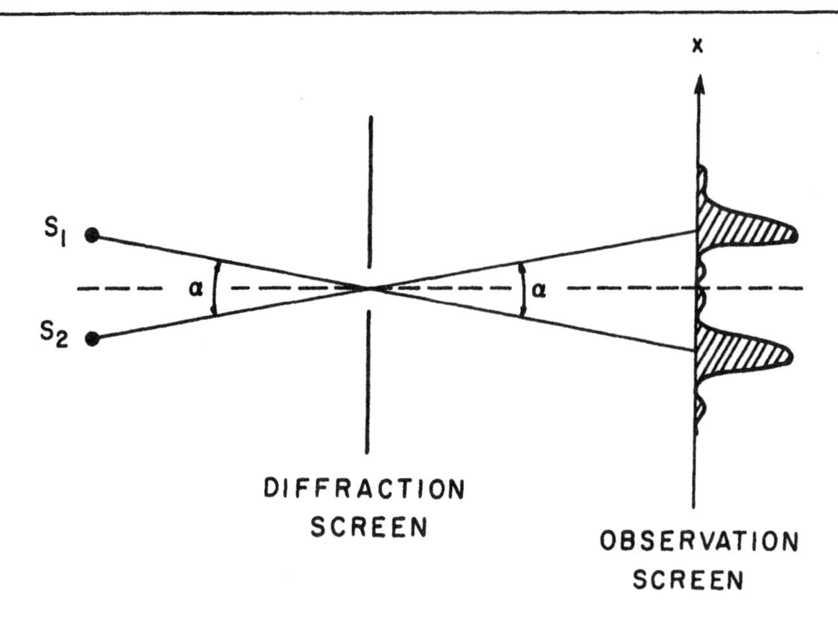

Fig. 8.27 Diffraction by an aperture in a screen illuminated by two independent sources separated spatially, and subtending an angle α with one another.

two independent sources emitting radiation of the same wavelength, separated spatially and subtending an angle α with respect to one another, as is shown in Fig. 8.27. Their diffraction patterns overlap and it is obvious that as α gets smaller, a point is eventually reached where it will be difficult to say whether there is one source or two sources. At what value of $\alpha = \alpha_o$ does this occur? To answer this, let us see when the minimum in the resultant pattern begins to disappear. Figure 8.28 suggests the answer, which is that when the maximum of one is at the first minimum of the other a dimple in the resultant intensity can just be made out. (Recall that we have repeated the Rayleigh criterion already discussed in section 8.3.) Hence

$$\alpha_o = \frac{\lambda}{D}. \tag{8.86}$$

For circular apertures the corresponding result is

$$\alpha_o = 1.22\frac{\lambda}{D} \tag{8.87}$$

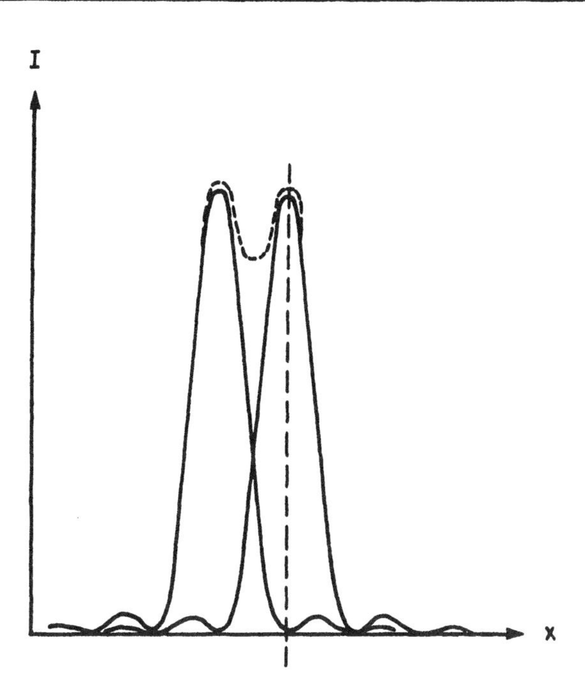

Fig. 8.28 The diffraction patterns from two spatially separated sources are superposed for the case when the maximum of one pattern falls exactly on the first zero of the other pattern (see also Fig. 8.13). The dashed line equals the sum of the two intensities. The two sources can still be resolved (Rayleigh criterion). This corresponds to an angular separation $\alpha = \alpha_0 = \lambda/D$. When α is reduced further, the dimple in the net intensity will quickly disappear. The peak will first become flat-topped and then peaked. One may argue that a flat-topped peak could well be used as an alternate measure of the angular resolution. Such is indeed the case: it is called the "Sparrow criterion," and is sometimes used in place of the Rayleigh criterion.

where D is the diameter of the hole in the diffracting screen.

The quantity α_0 is the so-called "resolving power" or "angular resolution." It determines the minimum angular separation of two point sources, such that an observer watching the images on the distant screen can just tell them apart. The result is applicable to any optical instrument, a lens, binoculars, the eye, as well as to radar transmitters and radio telescopes. The angle α_0 is the best that can be achieved, and no amount of ingenuity can lead to a resolution better than that given by Eqs. 8.86 and 8.87. The culprit is diffraction. For the human eye D ≈ 3 mm, $\lambda \approx 5 \times 10^{-5}$ cm and α_0 = 42 arc-seconds. For the Palomar 200" telescope, $\alpha_0 \approx 1.2 \times 10^{-7}$ radians or 0.025 arc-seconds. But neither of these values can be realized, because of imperfections in the optics, and because atmospheric turbulence destroys phase coherence.

Example

A ruby laser emits a beam of radiation of wavelength λ = 6943 Å. When the beam emerges from the laser, it has a diameter of 2 cm. Assuming that the beam is diffraction limited

a) what is the angular beam divergence?

b) what is the beam diameter 100 meters from the laser?

Solution

a) For a circular aperture the angular position of the first zero of the diffraction pattern is given by $\Delta \sin\psi \approx \Delta\psi = 1.22\ \lambda/D$. Therefore, the angular spread from the symmetry axis is

$$\Delta\psi = \frac{1.22\ \lambda}{D} = \frac{1.22 \times 6943 \times 10^{-8}}{2.0}$$

$$= 4.2 \times 10^{-5} \text{ radians}$$

b) The spatial spread ≈ distance × Δψ

$$= 10^{4} \times 4.2 \times 10^{-5} \text{ cm}$$

$$= 0.42 \text{ cm.}$$

Therefore the beam diameter is ≈2 + 0.42 + 0.42 = 2.84 cm.

Compare this with a flashlight with the same emergent beam diameter that has a small filament placed at the focus of a parabolic mirror. Since the filament cannot be an infinitesimally small point, the different parts of the filament emit as independent (i.e., incoherent) beams. As a result of the finite size of the filament, there is an angular spread which is found to be approximately

$$\Delta\psi \approx \frac{a}{f}$$

where a is the size of the filament and f is the focal length of the mirror. Taking a = 0.01 cm and f = 1 cm, $\Delta\psi \approx 0.01$ radians. At a distance of 100 meters from the flashlight the spatial spread in the beam is $10^4 \times 0.01 = 100$ cm or 1 meter!

Example

The headlights on automobiles are approximately 1.5 meters apart. If the diameter of the pupil of the eye is 3 mm and an average wavelength of visible light is 5000 Å, what is the maximum distance at which the headlights will be resolved (appear as two separate lights) assuming that diffraction effects at the circular aperture of the eye are the limiting factors?

Solution

The angular resolving power of a circular aperture is given by Eq. 8.87 as

$$\alpha_o = 1.22\frac{\lambda}{D} = 1.22\frac{5\times 10^{-7}}{3\times 10^{-3}} = 2.03\times 10^{-4}.$$

Two sources separated by a distance d at a maximum distance R will appear separate if

$$\alpha_o = \frac{d}{R} \qquad R = \frac{d}{\alpha_o} = \frac{1.5}{2.03\times 10^{-4}} \qquad R = 7.38 \text{ km} = 4.6 \text{ miles}.$$

8.9 THE DIFFRACTION GRATING

In section 8.3 we discussed the interference of waves emanating from an array of point or line sources. We said that the results apply also to the diffraction grating; but only approximately. The reason is that a diffraction grating is more nearly an array of wide slits rather than point or line sources. With this realization, and our understanding of the diffraction by a single slit (section 8.7), we are now in a position to present a more accurate picture of a real grating.

The electric field E at a point P on an observation screen placed at a large distance from a single slit of width D is given (Eq. 8.79) by

$$E = A\left(\frac{\sin \beta}{\beta}\right) \frac{e^{j(\omega t - kR)}}{R}. \tag{8.88}$$

The parameter $\beta = (\pi D/\lambda) \sin \psi$ specifies the angle of emergence ψ of the diffracted "rays," R determines the distance of the center of the slit from the observation point P, and A is a constant. When there is more than one slit, the contributions from all slits must be summed . If the slits are identical, then the term $A(\sin \beta/\beta)$ is the same for each, and we have

$$E = A\left(\frac{\sin \beta}{\beta}\right)\left[\frac{e^{j(\omega t - kR_1)}}{R_1} + \frac{e^{j(\omega t - kR_2)}}{R_2} + \ldots . \frac{e^{j(\omega t - kR_N)}}{R_N}\right]. \tag{8.89}$$

The distances R_1, R_2 ... R_N appearing in the denominators are almost equal to one another and we set them all equal to, say, R_o:

$$E = \frac{A}{R_o} \frac{\sin \beta}{\beta}\left[e^{j(\omega t - kR_1)} + e^{j(\omega t - kR_2)} + \ldots . e^{j(\omega t - kR_N)}\right]. \tag{8.90}$$

The series to be summed is identical to the one given by Eq. 8.39 for the array of point radiators. Indeed, Eq. 8.90 differs from Eq. 8.39 only by the term $(\sin \beta/\beta)$ which allows for diffraction caused by the finite extent of the emitting sources. Proceeding with the sum exactly as we did in section 8.3, we obtain

$$E = \frac{A}{R_o}\left(\frac{\sin \beta}{\beta}\right) e^{j(\omega t - kR_1)} e^{-j(N-1)\delta/2}\left[\frac{\sin(N\delta/2)}{\sin(\delta/2)}\right] \tag{8.91}$$

where

$$\delta = \left(\frac{2\pi d}{\lambda}\right) \sin \psi \tag{8.92}$$

with d as the distance separating the center of one slit from the center of the neighboring slit. The intensity, obtained from E in the usual

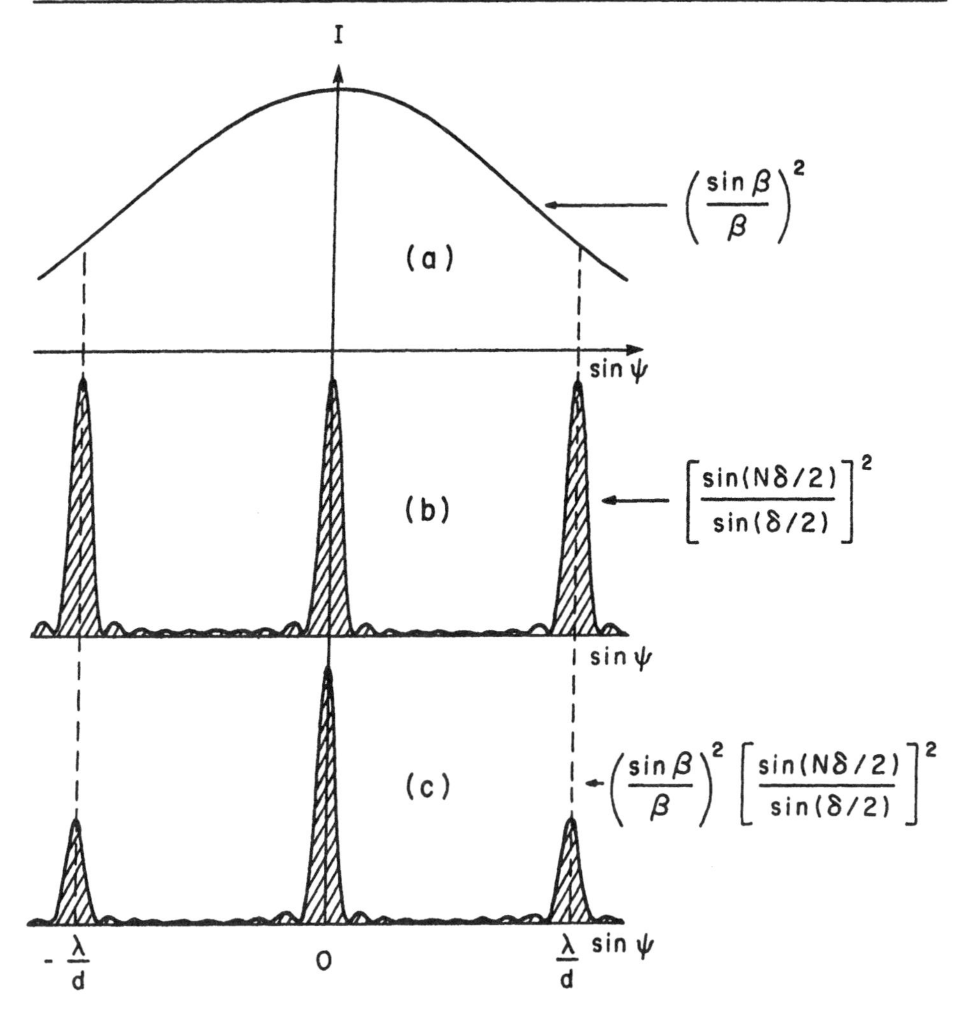

Fig. 8.29 Diffraction pattern of a grating with ten elements ($N = 10$) and a spacing d equal to twice the aperture width D (i.e., $d = 2D$). (a) Diffraction component; (b) interference component; (c) combined pattern.

way, is then given by

$$I = I_o \underbrace{\left(\frac{\sin\beta}{\beta}\right)^2}_{\substack{\text{diffraction}\\ \text{by a single}\\ \text{slit}}} \underbrace{\left[\frac{\sin(N\delta/2)}{\sin(\delta/2)}\right]^2}_{\substack{\text{interference}\\ \text{between slits}}} \tag{8.93}$$

Comparing this result with Eqs. 8.44 and 8.82 allows us to identify those parts responsible for diffraction by a single slit of the array and those parts responsible for the interference between waves emanating from neighboring slits of the array. What role each term plays is illustrated in Fig. 8.29 for the case $N = 10$, $d = 2D$. We see that diffraction causes a slow "modulation" of the intensity so that the principal maxima are no longer of equal intensity for all diffraction orders m. The narrower the slit width D (for a given fixed separation d) the less important is the effect of this modulation. But, observe that under certain circumstances, an interference order m may be missing altogether. This occurs whenever the zero in the term $(\sin\beta/\beta)^2$ falls on a principal maximum. The reader should have no difficulty to show that such "missing orders" occur whenever $d/D = m/n > 1$, m, n integers. Some of these effects are best illustrated when the number of elements N in the array is small. Figure 8.30 shows what happens when $N = 2$ and $d = 3D$.

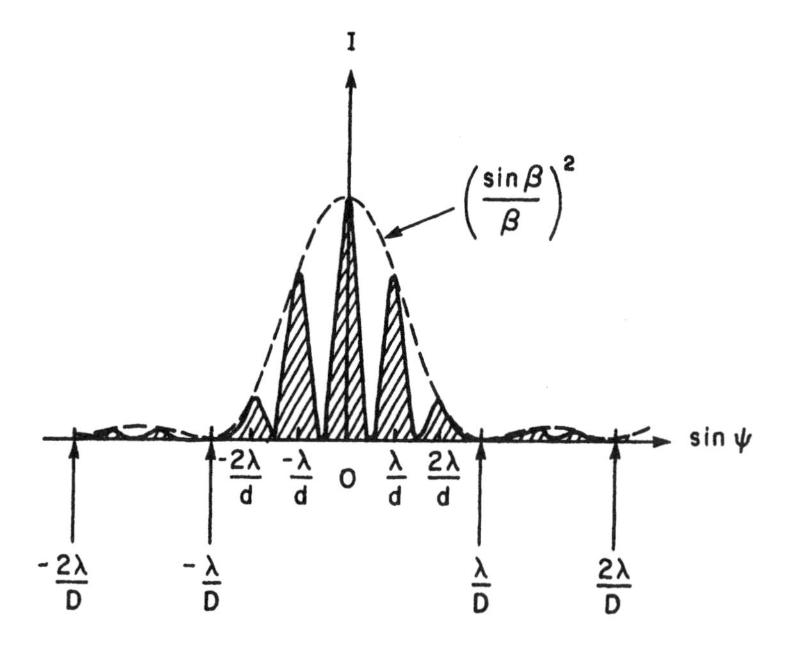

Fig. 8.30 Double slit system is comprised of two equal slits of width D whose centers are separated by a distance $d = 3D$. Note that because of diffraction the $m = 3$ principal maximum is missing.

8.10 HOLOGRAPHY

According to Huygens' principle (section 8.5) each point of a wave front acts as a secondary source and the sum total of the radiation from all such secondary sources reproduces the wave at points beyond this surface. A wave is specified by two quantities, its amplitude and its phase. Thus, if the amplitude A and phase ϕ of the electromagnetic field from any source, however complicated, are recorded at every point of a surface (for example a plane), then we know all about the source; and from this knowledge we can reconstruct the precise features of the object that emitted the waves in the first place. At optical wavelengths, recording the amplitude of a wave is readily accomplished, for example with a photographic plate (strictly speaking one records the square of the amplitude A^2). Unfortunately, this procedure tells nothing about the phase. This suggests that since information has been lost in the recording procedure, the image of the object must also be wanting. We know that this is indeed the case: the image has no depth at all and its three-dimensional properties are missing. This is the fate of most recording procedures such as photos, TV, movies, etc. which record only intensities A^2.

Holography, or "wavefront reconstruction" as it is sometimes called, allows one to record simultaneously both the amplitude and the phase. How is this done? The principle is very simple and has been used for decades at radio and microwave frequencies. A signal of adjustable but known phase is added to the signal whose properties are desired. By the principle of superposition, a minimum in the resulting amplitude indicates a 180° phase difference between the known and unknown signals; a maximum indicates that the two are in phase, and so on. The point is that the addition of a reference signal converts phase variations into amplitude variations (there is a standing wave pattern) which we know how to record.

There are two steps to wave-front reconstruction. The first is to prepare the "hologram" which is the amplitude and phase record of the desired object. This is a complicated standing wave pattern imprinted onto a photographic plate. It was named hologram by the inventor,

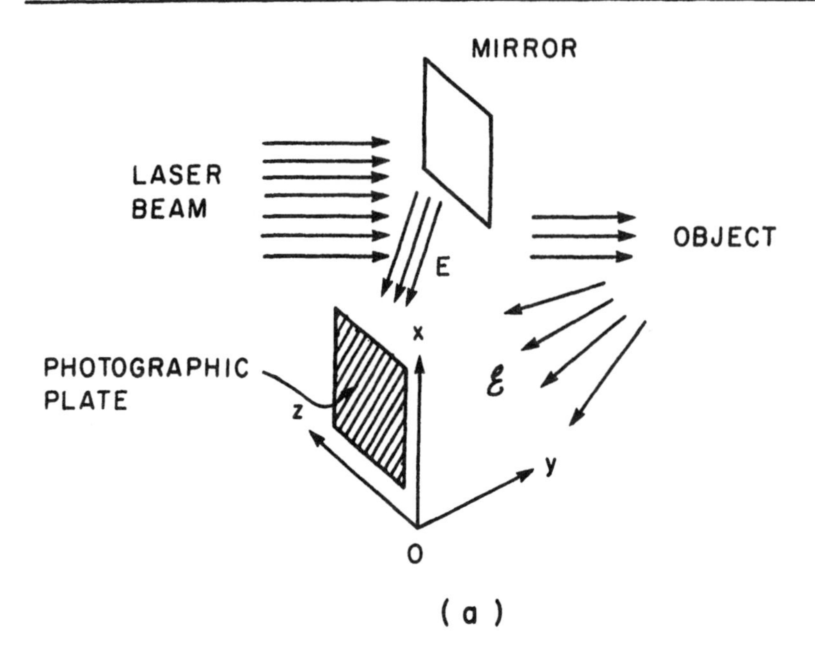

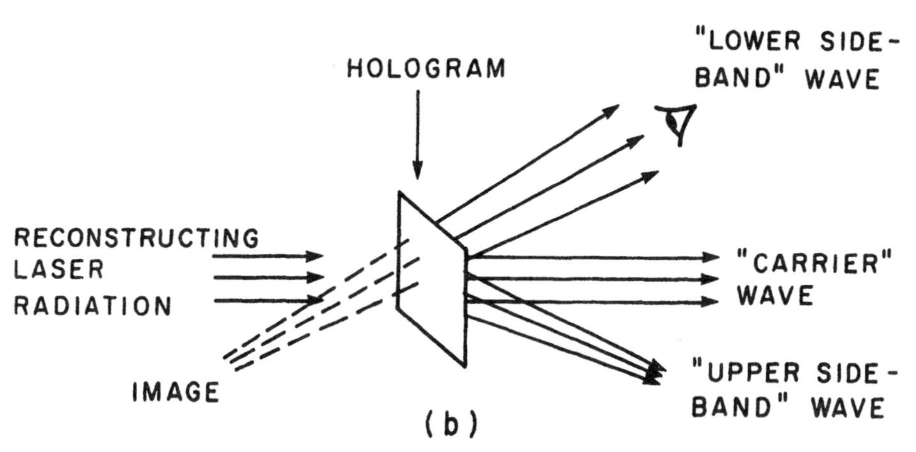

Fig. 8.31 (a) The making of a hologram. A laser beam is split into two parts. One part is allowed to illuminate a mirror and after reflection gives rise to a reference field E (Eq. 8.95) over the surface of the photographic plate. The other part illuminates the object and after scattering from it gives rise to the scattered field $\mathscr{E}$ (Eq. 8.96) over the plate.
(b) Viewing the hologram with laser radiation to reconstruct the object. Observe that the reconstructing wave (Eq. 8.100) is incident on the photographic plate at the same angle as was the reference wave in making the hologram.

D. Gabor [Nature 161, 777 (1948)]; holos in Greek means complete. The second step is to use the hologram and to reconstruct an image of the object originally used.

One of the experimental arrangements used in preparing a hologram is illustrated in Fig. 8.31a. Light from a monochromatic source, preferably a laser, is split in two beams. One beam is the reference beam which is allowed to shine on the photographic plate. The other beam illuminates the object, and the light scattered from the object also floods the photographic emulsion. Suppose the reference wave is incident at an angle θ on the photographic plate situated in the x-z plane of a rectangular coordinate system. Assuming that it is a plane wave, its equation is (cf. Eq. 5.20)

$$E = E_o \cos[\omega t + ky \cos\theta - kz \sin\theta] \tag{8.94}$$

where, for the sake of simplicity, we treat the electric field E as if it were a scalar quantity. Alternately, one can assume that the electric field of this wave and all other waves discussed below have identical fixed polarizations. Over the surface of the photographic plate where $y = 0$ the above equation reduces to

$$E = E_o \cos[\omega t - kz \sin\theta] \qquad \text{(reference wave).} \tag{8.95}$$

The radiation scattered from the object being viewed can be written as

$$\mathscr{E} = \mathscr{E}_o(x, z) \cos[\omega t - \phi(x, z)] \qquad \text{(scattered wave)} \tag{8.96}$$

where the amplitude $\mathscr{E}(x, z)$ and the phase $\phi(x, z)$ are now complicated functions of position corresponding to an irregular wave front. Of course, when the scatterer is a simple object, an electron for example, the $\mathscr{E}_o$ and ϕ are just the amplitude and phase of a simple spherical wave.

The total field at the surface of the photographic plate is equal to $E + \mathscr{E}$. The emulsion responds to the energy rather than the electric field E. Indeed, since the chemical process involved is slow, and takes a long time compared with the period of the wave, the relevant quantity is the time-averaged intensity defined by Eq. 8.1

$$I = \epsilon_o c \langle \vec{E}^2 \rangle. \tag{8.97}$$

Summing the right-hand sides of Eqs. 8.95 and 8.96, squaring, time-averaging and, for the sake of brevity, neglecting the term $\epsilon_o c$ yields

$$I = \frac{1}{2} E_o^2 + \frac{1}{2} \mathscr{E}_o^2 + E_o \mathscr{E}_o \cos(\phi - kz \sin \theta) \tag{8.98}$$

which is the familiar result of two-wave interference (cf. Eqs. 8.4 or 8.14). The key point of this result is that the interference pattern, by way of the fringe configuration distributed over the x-z plane of the film, contains information both about the amplitude distribution $\mathscr{E}_o(x, z)$ and the phase distribution $\phi(x, z)$ across the wave field scattered by the object. This imprint manifests itself as a spatially varying transparency T

$$T = \text{constant} \left[\frac{1}{2} E_o^2 + \frac{1}{2} \mathscr{E}_o^2 + E_o \mathscr{E}_o \cos(\phi - kz \sin \theta) \right] \tag{8.99}$$

when the emulsion is viewed by transmitting light through it.

To reconstruct the original scattered wave field one shines a plane monochromatic (laser) beam obliquely at the hologram (see Fig. 8.31b) at the same angle of incidence θ as was used for the reference beam in making the hologram. Thus

$$E_R = E_{oR} \cos(\omega t + ky \cos \theta - kz \sin \theta) \quad \text{[reconstructing wave]} \tag{8.100}$$

where E_{oR}, ω and k are the amplitude and frequency and wave number of the reconstructing wave. When the reconstructing wave passes through the hologram it emerges modified by the varying transparency T and is of the form $E_R T$. By Huygens' principle this light sets up over the emulsion secondary sources whose strength is likewise proportional to $E_R T$. The sources in turn radiate, giving rise to a wave field which in the plane of the hologram $(y = 0)$ is, to within unimportant constants of proportionality, equal to $E_R T$ or, written out in full,

$$E(\text{reconstructed}) = \left[\left(\frac{1}{2} E_o^2 + \frac{1}{2} \mathscr{E}_o^2 \right) + E_o \mathscr{E}_o \cos(\phi - kz \sin \theta) \right] \times E_{oR} \cos(\omega t - kz \sin \theta). \tag{8.101}$$

Observe that the right-hand side of this equation is reminiscent of an amplitude-modulated sinusoidal disturbance (see Eq. 1.152) in which, however, the modulation is spatial via $\phi(x, z)$ and $\mathscr{E}_o(x, z)$ rather than temporal, as is the case in conventional radio wave modulation. And, just as in conventional modulation, the result can be expressed (using standard trigonometric relations) as a sum of three terms; a carrier and two sidebands (see Eq. 1.153 and Fig. 1.33):

$$E(\text{reconstructed}) = \underbrace{\frac{1}{2} E_{oR}\left(E_o^2 + \mathscr{E}_o^2\right) \cos(\omega t - kz \sin\theta)}_{\text{"carrier"}}$$

$$+ \underbrace{\frac{1}{2} E_{oR} E_o \mathscr{E}_o \cos(\omega t - \phi)}_{\text{"lower sideband"}}$$

$$+ \underbrace{\frac{1}{2} E_{oR} E_o \mathscr{E}_o \cos(\omega t + \phi - 2kz \sin\theta)}_{\text{"upper sideband"}}. \qquad (8.102)$$

The first term corresponds to a wave that passes right through the hologram with no change in direction. It carries no phase information about the radiation scattered from the object and is thus of no further interest here. The second term is the reason for all the foregoing algebra. It represents a wave field which, except for the multiplicative constant $(1/2)\, E_{oR} E_o$, has precisely the form of the original wave scattered from the object (see Eq. 8.96). Therefore, if we peer into the illuminated hologram as if it were a window looking out onto a scene, we see the object just as if it were really sitting there. By a slight movement of the head we can look "around the corner" to see details which were originally hidden by the foreground. In summary, then, this holographic image is not only three-dimensional but it also reproduces parallax effects faithfully.

The third term of Eq. 8.102 represents yet another wave which, because of the phase factor $2kz \sin\theta$, is separated in angle from the foregoing wave, and thus does not interfere with the viewing. This last-named wave produces an image that is all wrong because the wave carries a phase factor $+\phi$ rather than $-\phi$. This has the effect, for example, that a hill on the object appears as a valley and the whole scene is turned around in a strange manner. Consequently this image has little practical utility.

For further details concerning the physics and applications of holography see M. P. Givens, Am. J. Phys. 35, 1056 (1967); G. W. Stroke, An Introduction to Coherent Optics and Holography (Academic Press, Inc., New York, 1969); E. R. Robertson and J. M. Harvey (Eds.), The Engineering Uses of Holography (Cambridge University Press, London, 1970). Acoustical holography is discussed by A. F. Metherell, Scientific American 221, 36 (1969), and by P. Greguss in Physics Today 27, 42 (1974).

8.11 COHERENCE

When one opens an older treatise on optics written before the Second World War, one is likely to find a remark that can be paraphrased as follows: "It is impossible to observe interference effects between light beams emitted by independent sources." We can understand why this is so when we consider the so-called incoherent light sources available at that time, as, for example, the incandescent lamp, the gas discharge tube, the Sun, etc. These sources consist of a large number of uncorrelated, microscopic emitters (electrons, atoms, molecules) each of which radiates light for a short time and is quiescent the rest of the time. Even the most nearly monochromatic light sources of this type are known to radiate an unbroken wave train for periods Δt of approximately 10^{-8} to 10^{-9} sec. (Δt is called the coherence time.) After this interval, another wave train will start but it will bear no phase relationship to the first. If it were possible to photograph with exposure times of this order or less, interference fringes could be obtained from separate sources. Again, with a sufficiently short exposure time we should be able to obtain interference effects (i.e., beats) between beams of different wavelengths [see A. T. Forrester et al., Phys. Rev. 99, 169 (1955)]. Since in practice most experiments require a relatively long time to register these various effects, the interfering sources must be identical in all respects, and this is only possible when the light waves originate from a common source.

In view of these facts, all methods of obtaining interference with incoherent light sources fall into two broad classes. In the first (called the Division of Wave Front) the source is so small that the wave fronts emerging in slightly different directions have similar phases. Examples of this class are Young's two slit experiment (section 8.2), Lloyd's mirror (section 8.2), and the diffraction grating (sections 8.3 and 8.9). In the second class (known as the Division of Amplitude) the beam is divided by partial reflection from a half-silvered mirror; there is a point-to-point correspondence between the wave fronts of the transmitted and reflected beams. Any peculiarity in one is also present in the other. Examples of this class are the interference in thin films (section 8.2), the Fabry-Perot interferometer (section 8.3), and the Michelson interferometer described below.

The situation becomes very different when we consider the new, man-made sources of radiation. These are, for example, the magnetron, the klystron, and the maser, all of which operate in the microwave frequency range; and the various lasers which operate in the infrared, visible, and ultraviolet regions of the spectrum. These sources emit essentially monochromatic, sinusoidal wave trains of very long duration. For example, a stabilized CO_2 laser may have a coherence time Δt of several milliseconds duration. Such sources are said to be coherent. With these, interference effects are readily observed in situations in which beams from different sources are allowed to interact.

We mentioned the high degree of monochromaticity of the coherent sources. Coherence and monochromaticity are closely linked concepts. Recall from section 2.5 that when we Fourier-analyze a wave train of duration Δt, its frequency spectrum exhibits a spread $\Delta\omega$ about a center frequency ω_0 given by

$$\Delta\omega \approx \frac{2\pi}{\Delta t}. \qquad (8.103)$$

The longer the coherence time Δt, the more nearly sinusoidal is the wave train and the narrower is the spectrum. In the ideal case of an emitting source of single frequency ($\Delta\omega = 0$), the coherence time is infinite.

In nature there is no such thing as a completely coherent or a completely incoherent source. There are only degrees of coherence. The Sun's rays composed of short-lived wave trains are very incoherent; the laser with its very pure frequency spectrum is highly coherent. These ideas can be expressed in precise mathematical language but we shall not do so here because that would require a lengthy discussion of correlation functions and such concepts are beyond the scope of this book. The interested reader will find a comprehensive treatment of partial coherence in Principles of Optics by M. Born and E. Wolf (Pergamon Press, New York, 1964).

Temporal coherence

The consequences of short and long temporal coherence can be demonstrated on an instrument known as the Michelson interferometer (Fig. 8.32a). It consists of a half-silvered mirror M which acts as a beam splitter dividing the radiation from the source into two beams of nearly the same intensity. One beam goes to the right and is reflected from a fully silvered mirror M_1. It retraces its path and part of it is deflected by the beam splitter toward the observation screen (the remainder of the beam goes back toward the source but this portion is of no further interest). Similarly, the other beam travels upward toward the fully silvered mirror M_2, is reflected from it, and retraces its path. It strikes the beam splitter and a portion of it travels toward the screen. Thus the two beams, having traveled along different paths, are reunited and interference can be expected.

Now assume that all parts of the optical system are fixed except that mirror M_1 can be moved along a precision track, as is indicated in Fig. 8.32a. In this way the optical path length difference L between the two beams can be changed at will. With a good monochromatic source, intensity maxima and minima will appear on the screen as a function of L, as is illustrated in Fig. 8.32b. For reasonably small path length differences (and we shall see in a moment what that

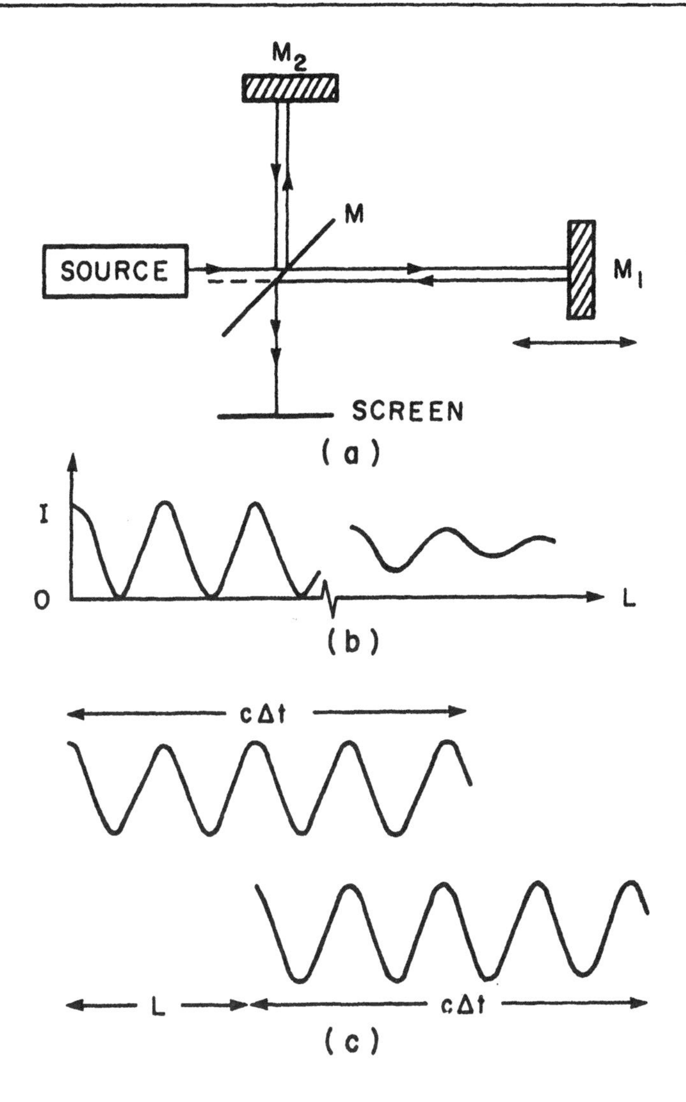

Fig. 8.32 Demonstration of temporal coherence with a Michelson interferometer.

means) the fringe pattern is very sharp; that is, the maxima are bright and the minima are dark. But this does not persist indefinitely. For large L the maxima and minima will become hazier and hazier and eventually all interference will disappear. This happens when the coherence length $c\Delta t$ of the beams begins to be shorter than the path length difference L. This is shown in Fig. 8.32c. For not too large L, the two wave trains, each of length $c\Delta t$, will overlap over a length $c\Delta t - L$. In the region where they overlap interference can occur but no interference effects are possible outside the overlap region. Thus, the condition for clearly discernable fringes is that

$$L < c\Delta t. \tag{8.104}$$

For a largely incoherent source we set $\Delta t = 10^{-9}$ sec and find that $L \lesssim 30$ cm. But for a laser with $\Delta t = 3 \times 10^{-6}$ sec, $L \lesssim 0.9$ km! This illustrates in a quite dramatic way the differences we would see if we illuminated the Michelson first with, say, light from a mercury discharge tube, and then with light from a laser.

Spatial coherence

A beam from a magnetron or from a laser is said to have spatial coherence as well as temporal coherence. We illustrate this in Fig. 8.33 by contrasting the passage of incoherent light through a lens with the passage of coherent light through the same lens. In Fig. 8.33a we show a source comprised of a collection of independent microscopic sources all of which oscillate in a quite uncorrelated way with one another. The light emitted from any one of these sources passes through the lens and is focused in the image plane. And although the source is a point source the image is not. The reason is that the light emitted from this single point source is spatially coherent with itself. This means that the Huygens wavelets emerging from different portions of the lens interfere with one another giving rise to the characteristic diffraction pattern discussed in sections 8.7 and 8.8. The radius of the little diffraction image as obtained from the discussion following Eq. 8.85 is

$$1.22 \frac{\lambda}{D} z$$

where D is the lens diameter and z is the distance of the image plane from the lens. What is true for one of these microscopic emitters is true for all the others and each point source generates a diffraction-broadened image. These images lie next to each other and the image

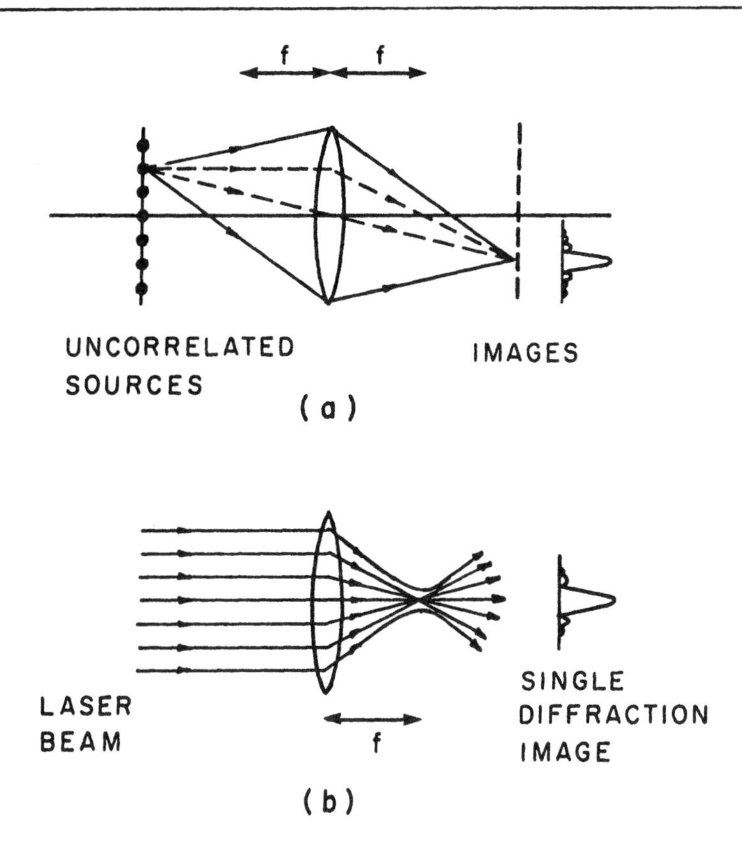

Fig. 8.33 Passage of light through a lens when (a) the source is incoherent, (b) it is coherent.

plane is filled with them. By summing their intensities (not the amplitudes, since they are incoherent with one another!) one arrives at the typical image of the source distribution produced by the lens.

Let us now compare this situation with the one shown in Fig. 8.33b where the lens is illuminated by a coherent beam of light from a laser, say. Here every Huygens wavelet emerging from the lens interferes with every other wavelet. To obtain the net effect in the image plane, which is now also the focal plane of the lens, one sums at any given point of the image plane the amplitudes of all interfering waves, and squares the result to obtain the intensity (cf. section 8.7). The net result of this procedure is that all contributions over the image plane cancel (the waves interfere destructively) except over a very small spot on axis the size of which is approximately 1.22 $\lambda f/D$ where D and f are the lens diameter and focal length, respectively. Thus the entire power from the incoming laser beam can be focused down to a tiny image.

Example

It was implied in the text that the total time averaged intensity at a point due to a superposition of incoherent waves equals the sum of intensities of the individual waves.

(a) Prove this statement using the following idealized model. N waves with identical amplitudes A and frequencies ω interfere at a point. They have the same polarizations and they differ from one another in one respect only, their time-dependent phases $\phi(t)$ bear no relationship to one another. Assume that the phase variations are rapid on the time scale in which the averaged intensity is measured, but slow compared with the period $T = 2\pi/\omega$.

(b) Repeat the calculations for the total average intensity, but now assume that the phases of all the waves are identical.

Solution

The total electric field at a point is

$$E = A\cos(\omega t + \phi_1(t)) + A\cos(\omega t + \phi_2(t)) + A\cos(\omega t + \phi_3(t)) + \ldots\ldots A\cos(\omega t + \phi_N(t)). \quad (8.105)$$

Taking two terms at a time and using standard trigonometric relations as discussed in Appendix 1, it is found that

$$E(t) = E_o(t)\cos(\omega t + \phi(t)) \quad (8.106)$$

where

$$E_o^2(t) = NA^2 + 2A^2 \sum_{j>i}^{N} \sum_{i=1}^{N} \cos(\phi_i(t) - \phi_j(t)) \quad (8.107)$$

and

$$\tan \phi(t) = \frac{\sum_{i=1}^{N} \sin \phi_i(t)}{\sum_{i=1}^{N} \cos \phi_i(t)}. \tag{8.108}$$

Observe that the total amplitude $E_o(t)$ and the net phase $\phi(t)$ are functions of time.

The intensity is obtained by squaring Eq. 8.106, time-averaging and substituting the result in Eq. 8.1. Squaring leads to

$$E^2(t) = E_o^2(t) \cos^2(\omega t + \phi(t)). \tag{8.109}$$

We now invoke the fact that the time variations of $E_o(t)$ and $\phi(t)$ are slow compared with the period $T = 2\pi/\omega$. So that, to good approximation, the cycle average is merely the average over $\cos^2(\omega t + \phi)$ which is 1/2, and

$$\langle E^2(t) \rangle = \frac{1}{2} E_o^2(t)$$

$$= \frac{1}{2} NA^2(t) + A^2(t) \sum_{j>i}^{N} \sum_{i=1}^{N} \cos(\phi_i(t) - \phi_j(t)). \tag{8.110}$$

The intensity $I(t)$ is $\epsilon_o c \langle E^2(t) \rangle$ and thus

$$I(t) = \frac{1}{2} \epsilon_o c N A^2 + \epsilon_o c A^2 \sum_{j>i}^{N} \sum_{i=1}^{N} \cos(\phi_i(t) - \phi_j(t)). \tag{8.111}$$

But, the intensity I_o of a single source acting alone is just

$$I_o = \frac{1}{2} \epsilon_o c A^2 \tag{8.112}$$

and substituting for this value in Eq. 8.111 yields

$$I(t) = NI_o + 2I_o \sum_{j>i}^{N} \sum_{i=1}^{N} \cos(\phi_i(t) - \phi_j(t)). \tag{8.113}$$

It is important to observe that even after averaging over a cycle of the oscillation, the intensity $I(t)$ still varies with time as a result of the fluctuations in the phase. So we must average again, this time over the time it takes to make an average measurement of the intensity with some instrument, say, the eye. Let this time be τ. Then, in accordance with the time-averaging prescription given by Eq. 1.20, we have that

$$\overline{I(t)} = NI_0 + 2I_0 \sum_{j>i}^{N} \sum_{i=1}^{N} \frac{1}{\tau} \int_0^{\tau} \cos(\phi_i(t) - \phi_j(t))\, dt. \tag{8.114}$$

Now, the argument $\phi_i(t) - \phi_j(t)$ can take on any value between 0 and 2π and $\cos(\phi_i(t) - \phi_j(t))$ can take on any value between plus and minus 1. If the fluctuations are completely random (the sources are incoherent), the cosine will, on the average, be positive as often as negative. And if τ is long enough, the second term of Eq. 8.114 vanishes with the result

$$\text{(a)} \quad \overline{I(t)} = NI_0 \tag{8.115}$$

which proves the statement that the total time-averaged intensity equals the sum of the intensities of the superposed, incoherent sources. Observe that the second term of Eq. 8.114 is precisely the term responsible for interference (cf. Eq. 8.4). Since it vanishes, we have, in fact, proved that interference between these independent sources is impossible.

Now assume that all sources are exactly in phase. The term $\cos(\phi_i(t) - \phi_j(t)$ becomes unity and Eq. 8.114 becomes

$$\overline{I(t)} = NI_0 + 2I_0 \sum_{j>i}^{N} \sum_{i=1}^{N} 1_{ij}. \tag{8.116}$$

It is left to the reader to show that this is equivalent to

$$\text{(b)} \quad \overline{I(t)} = N^2 I_0. \tag{8.117}$$

Thus, in the case of coherent sources acting in phase with one another, the total intensity is N^2 times the intensity of a single source acting alone. When $N = 2$, $\overline{I} = 4I_0$, a result that was already derived in section 8.1 (see also Fig. 8.2).

APPENDIX 1

A LITTLE MATHEMATICS ON SUPERPOSITION OF PERIODIC MOTIONS

In a linear system, the resultant of two or more vibrations equals the sum of the individual vibrations. If the vibrations are simple harmonic with amplitudes $A_1, A_2, A_3 \ldots$, angular frequencies $\omega_1, \omega_2, \omega_3 \ldots$ and phases $\phi_1, \phi_2, \phi_3 \ldots$, then the resultant is given by

$$x(t) = A_1 \cos(\omega_1 t + \phi_1) + A_2 \cos(\omega_2 t + \phi_2) + A_3 \cos(\omega_3 t + \phi_3) + \ldots \quad \text{(A1. 1)}$$

We could equally well have written a sum of sines rather than cosines.

This sum is a very general one and is not very tractable. Therefore we consider some special cases.

A1.1 OSCILLATIONS OF THE SAME FREQUENCY ω

Suppose we have two such oscillations,

$$\begin{aligned} x_1 &= A_1 \cos(\omega t + \phi_1) \\ x_2 &= A_2 \cos(\omega t + \phi_2) \end{aligned} \quad \text{(A1. 2)}$$

and we wish to find the resultant

$$\begin{aligned} x &= x_1 + x_2 \\ &= A_1 \cos(\omega t + \phi_1) + A_2 \cos(\omega t + \phi_2). \end{aligned} \quad \text{(A1. 3)}$$

We expand the two cosines, using $\cos(a+b) = \cos a \cos b - \sin a \sin b$ and rearranging terms obtain

$$\begin{aligned} x = {} & [A_1 \cos \phi_1 + A_2 \cos \phi_2] \cos(\omega t) \\ & - [A_1 \sin \phi_1 + A_2 \sin \phi_2] \sin(\omega t) \end{aligned} \quad \text{(A1. 4)}$$

Now, to simplify, we employ a little trick: we write the two terms in square brackets as $A \cos \phi$ and $A \sin \phi$, respectively, with A and ϕ yet to be determined. That is,

$$\begin{aligned} x(t) = {} & A \cos \phi \cos \omega t \\ & - A \sin \phi \sin \omega t \end{aligned} \quad \text{(A1. 5)}$$

where

$$A \cos \phi \equiv A_1 \cos \phi_1 + A_2 \cos \phi_2$$
$$A \sin \phi \equiv A_1 \sin \phi_1 + A_2 \sin \phi_2. \tag{A1.6}$$

And what is the good of this trick? It is simply that Eq. A1.5 for the total disturbance is just

$$x(t) = A \cos(\omega t + \phi). \tag{A1.7}$$

Thus, we have proved that two sinusoidal oscillations of the same frequency combine to yield yet another purely sinusoidal oscillation, albeit it has a new amplitude A and a new phase ϕ. To find A and ϕ we merely solve the two simultaneous equations (A1.6). Squaring each and adding, yields

$$A^2 = A_1^2 + A_2^2 + 2A_1A_2 \cos(\phi_2 - \phi_1). \tag{A1.8}$$

And, dividing the second equation (A1.6) by the first equation (A1.6) yields

$$\tan \phi = \frac{A_1 \sin \phi_1 + A_2 \sin \phi_2}{A_1 \cos \phi_1 + A_2 \cos \phi_2}, \tag{A1.9}$$

which ends the mathematics. Figure A1-1 shows some special cases. By repeated application of the procedure we have just followed it can be shown that a superposition of any number of sinusoidal oscillations with identical frequencies leads to a resultant harmonic oscillation of the same frequency. The calculation of its amplitude and phase, however, is a little difficult. Because of such difficulties, a graphical method is often preferred.

Graphical method

The graphical procedure rests on the rotating vector description of simple harmonic motion. Consider one oscillator, say, $x_1(t) = A_1 \cos(\omega t + \phi_1)$. As shown in Fig. A1-2, the amplitude A_1 is represented by a polar vector of length A_1, making an angle $(\omega t + \phi_1)$ with the x axis. The projection of this vector onto the x axis equals the

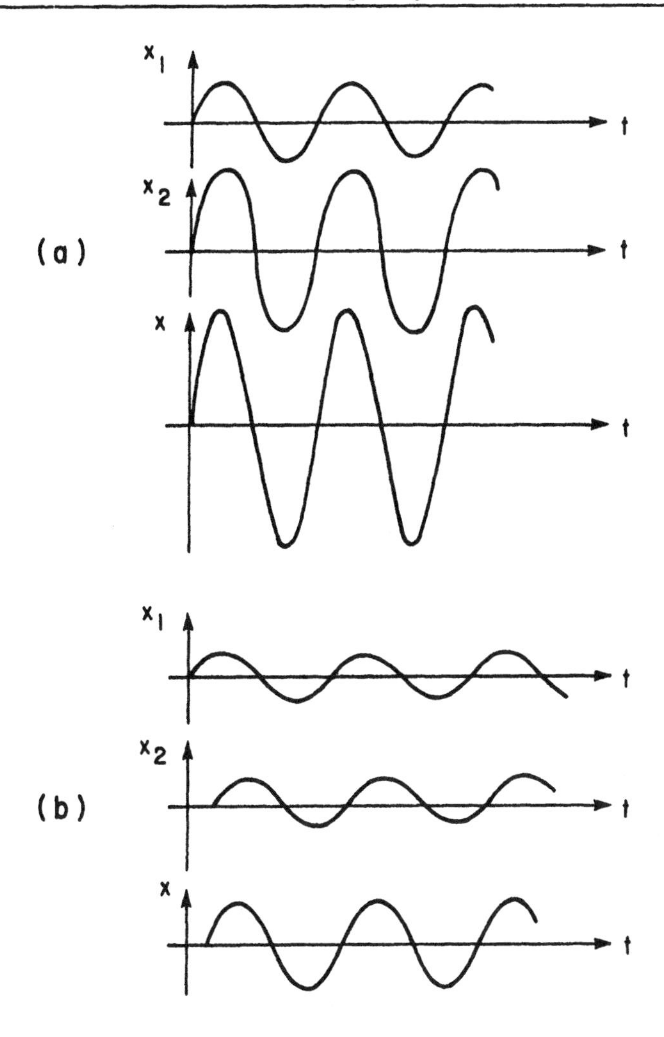

Fig. A1-1. Addition of two oscillations x_1 and x_2 of the same frequency, giving a resultant oscillation x. In (a), x_1 and x_2 are in phase. In (b), x_1 leads x_2 by a small amount.

sought-after value $A_1 \cos(\omega t + \phi_1)$. As time progresses, the vector rotates counterclockwise about the origin 0.

Now, suppose we add to $x_1 = A_1 \cos(\omega t + \phi_1)$ a second oscillation, $x_2 = A_2 \cos(\omega t + \phi_2)$ represented by yet another rotating vector, as is shown in Fig. A1-3. Since ω is the same for both oscillators, the two vectors rotate together and maintain the same relative position.

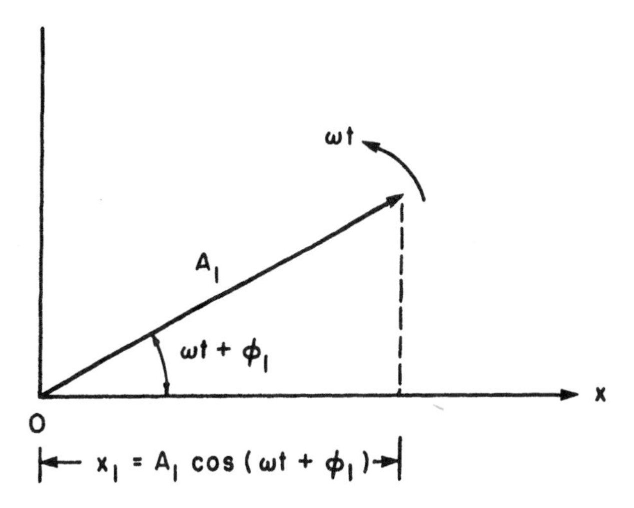

Fig. A1-2. Graphical representation of the simple harmonic vibration, $x_1 = A_1 \cos(\omega t + \phi_1)$, as the projection of a rotating vector of length A_1.

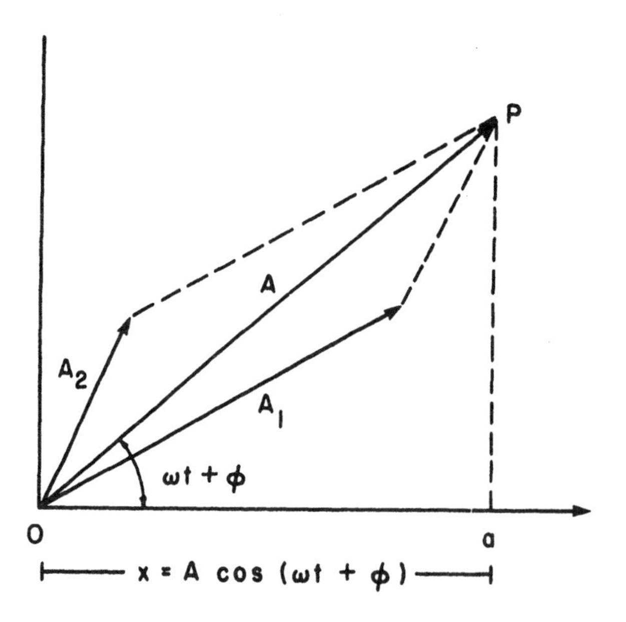

Fig. A1-3. Graphical addition of two simple harmonic vibrations $x_1 = A_1 \cos(\omega t + \phi_1)$, $x_2 = A_2 \cos(\omega t + \phi_2)$ of the same frequency ω. The amplitude A of the resultant vibration $x = A \cos(\omega t + \phi)$ is given by the parallelogram addition of vectors A_1 and A_2.

The resultant $x = x_1 + x_2 = A\cos(\omega t + \phi)$ is given by the projection Oa of the vector OP defined by the parallelogram law of vector addition. The length OP equals the amplitude A of the resultant. To be sure, we are not dealing here with real vectors; x_1 and x_2 are scalar quantities, and the rotating vectors are just geometrical constructs. For that reason they are often called "phasors."

In place of the parallelogram way of adding the two disturbances, it is often more convenient to use the equivalent procedure of the

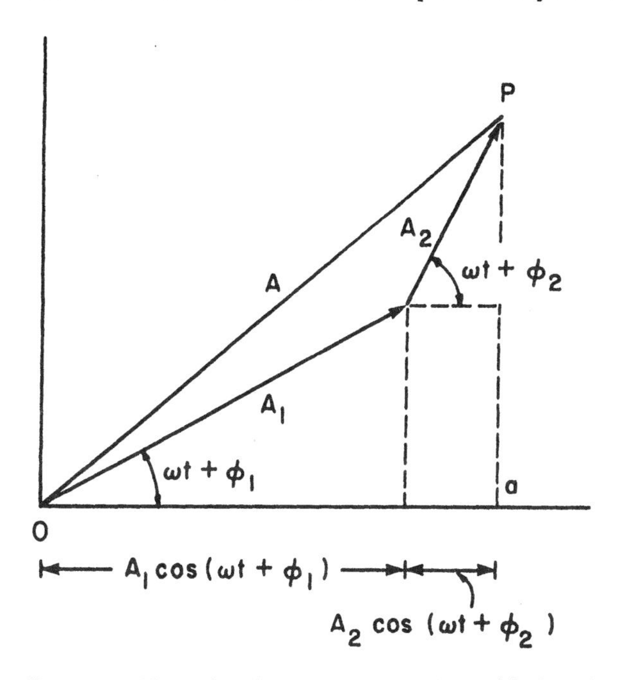

Fig. A1-4. Same as Fig. A1-3 except that the addition is made using the method of vector triangles. The law of cosines applied to the triangle gives the amplitude of the resultant: $A^2 = A_1^2 + A_2^2 + 2A_1A_2\cos(\phi_2 - \phi_1)$. Note that A^2 is proportional to the energy of the resultant oscillation. Thus the total energy equals the sum of the individual energies plus (or minus) something extra $[2A_1A_2\cos(\phi_2 - \phi_1)]$. Is this all right? If ϕ_2 and ϕ_1 vary randomly in time relative to one another, then averaging over all phases gives $\int_0^{2\pi}\cos[\phi_2(t) - \phi_1(t)]\,d\phi = 0$. In this case $A^2 = A_1^2 + A_2^2$; that is, the total energy is the sum of the individual energies with nothing funny left over.

method of triangles shown in Fig. A1-4. We employ the law of cosines applied to the triangle shown and it follows directly that the amplitude A is given by

$$A^2 = A_1^2 + A_2^2 + 2A_1A_2 \cos(\phi_2 - \phi_1), \tag{A1.10}$$

a result which we derived analytically earlier in this section. Observe how quickly the graphical method yields this result.

It is clear that the graphical method of adding periodic disturbances that have the same frequency ω is readily extended to three, four, or more superposed vibrations. In such cases a vector polygon, rather than a vector triangle will result, as is illustrated in Fig. A1-5. Note that in drawing the polygon we set time $t = 0$, a convenience that does not affect the generality of the technique.

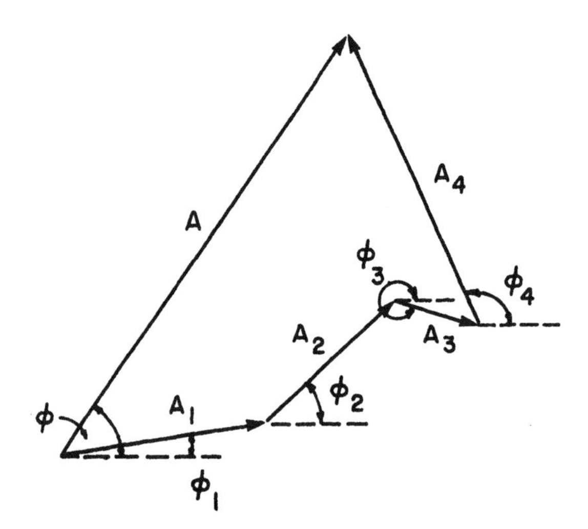

Fig. A1-5. Superposition of 4 oscillators $A_1 \cos(\omega t + \phi_1)$, $A_2 \cos(\omega t + \phi_2)$, $A_3 \cos(\omega t + \phi_3)$, $A_4 \cos(\omega t + \phi_4)$. Without loss of generality, the superpostion is carried out at time $t = 0$. Observe that when the polygon closes such that the resultant amplitude A becomes zero, the oscillations cancel destructively. On the other hand, when the oscillators are all in perfect phase such that $\phi_1 = \phi_2 = \phi_3 = \phi_4$, the polygon becomes a straight line and the resultant amplitude has its maximum value $A = A_1 + A_2 + A_3 + A_4$. Then it is said that the oscillations add constructively.

A1.2 OSCILLATIONS OF DIFFERENT FREQUENCIES

Consider two vibrations of different frequencies, but, for the sake of simplicity, let their amplitudes be the same and the phases zero:

$$x_1(t) = A\cos(\omega_1 t)$$
$$x_2(t) = A\cos(\omega_2 t). \tag{A1.11}$$

Adding the two disturbances, and using the trigonometric relation, $\cos a + \cos b = 2\cos[(a+b)/2]\cos[(a-b)/2]$ gives

$$x(t) = x_1 + x_2$$
$$= 2A\cos\left[\frac{\omega_1 - \omega_2}{2}t\right]\cos\left[\frac{\omega_1 + \omega_2}{2}t\right]. \tag{A1.12}$$

From the example illustrated in Fig. A1-6, it is clear that the resultant displacement $x(t)$ is no longer simple harmonic. Indeed, it may not even repeat itself. The condition for any true periodicity of the resultant motion is that the frequencies be commensurable, that is,

$$\frac{\omega_1}{\omega_2} = \frac{n}{m} \tag{A1.13}$$

where n and m are any two integers. If, for example, the ratio ω_1/ω_2 is some irrational number (say, π or $\sqrt{3}$), there is no time, however long, after which the pattern $x(t)$ repeats itself.

There is one important special case in which the resultant motion is almost, but not quite, simple harmonic. It is the case in which two simple harmonic oscillations of nearly the same frequency act together. Of course, the resultant again obeys Eq. A1.12, but ω_1 and ω_2 have almost the same values. Let us then write

$$\Omega \equiv \frac{\omega_1 - \omega_2}{2} \quad \text{low frequency}$$
$$\omega \equiv \frac{\omega_1 + \omega_2}{2} \quad \text{high frequency} \tag{A1.14}$$

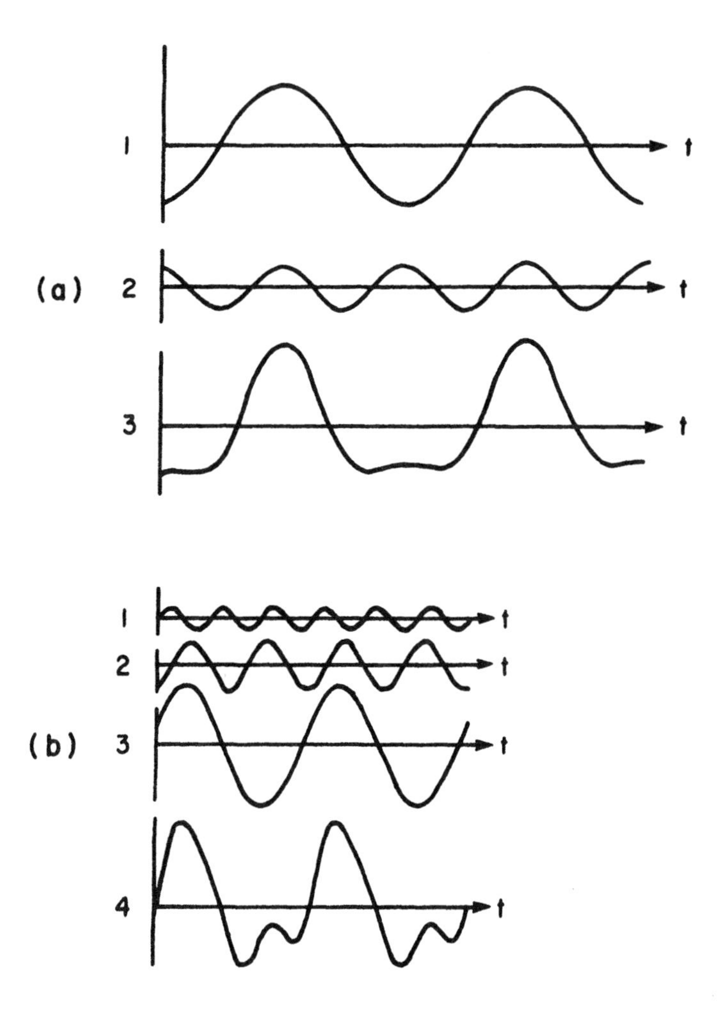

Fig. A1-6. Superposition of oscillations of different amplitudes and frequencies. In (a) we have a superposition of two disturbances (1) and (2) to yield the resultant disturbance (3). In (b) we have a superposition of three oscillations (1), (2) and (3), to yield the resultant (4).

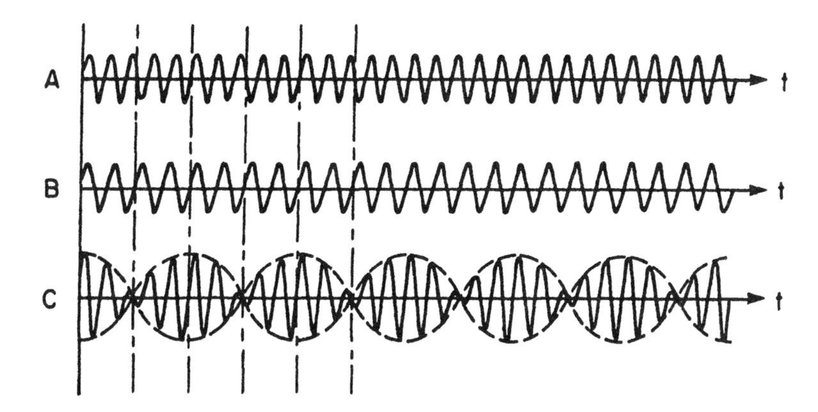

Fig. A1-7. Addition of two simple harmonic vibrations A and B of equal amplitude but somewhat different frequencies. Curve C is the resultant. The envelope shown dashed is the amplitude modulation given by the equations $x = \pm 2A \cos[(\omega_1 - \omega_2)t/2]$.

where Ω is a very low frequency, small compared with ω, and ω itself is nearly equal to ω_1 (or ω_2). With this notation, Eq. A1.12 becomes

$$x(t) = \underbrace{[2A \cos \Omega t]}_{\substack{\text{slowly}\\ \text{varying}\\ \text{amplitude}}} \times \underbrace{\cos(\omega t)}_{\substack{\text{rapidly}\\ \text{oscillating}\\ \text{term}}} \tag{A1.15}$$

which is seen to be composed of a rapidly varying simple harmonic term $\cos(\omega t)$ and a component $2A \cos \Omega t$ which may be viewed as a relatively slow, time-dependent amplitude. A plot of Eq. A1.15 is illustrated in Fig. A1-7. It shows the familiar phenomenon of beats. That is, the net vibration is basically a disturbance that has a frequency equal to the average of the two combined frequencies ($\omega = (\omega_1 + \omega_2)/2$), but with a periodically varying amplitude with a frequency $\Omega = (\omega_1 - \omega_2)/2$; one cycle of the slow "modulation" containing many, many cycles of the basic frequency.

APPENDIX 2

A LITTLE MATHEMATICS ON COMPLEX QUANTITIES

In section 1.1 it was shown that the equation

$$\frac{d^2x}{dt^2} + \omega^2 x = 0 \quad \text{(A2.1)}$$

has a solution which can be expressed in terms of trigonometric functions in one of two ways:

$$x = A\cos(\omega t) + B\sin(\omega t) \quad \text{(A2.2)}$$

or

$$x = C\cos(\omega t + \phi). \quad \text{(A2.3)}$$

Another useful approach is to abandon trignometric functions and attempt a solution in terms of the complex exponential function exp(jp) where p is a real number and $j = \sqrt{-1}$. It is equally as common to write exp[ip] where the letter i defines the quantity $\sqrt{-1}$. The reasoning behind this way of dealing with the solution of Eq. A2.1 is that the complex exponential can be expressed as a combination of sines and cosines as is demanded by Eq. A2.2. To show this let us begin with the regular exponential exp(y), where y is a real quantity. This exponential is defined by the power series,

$$e^y = 1 + y + \frac{y^2}{2} + \frac{y^3}{6} + \frac{y^4}{24} + \frac{y^5}{120} + \frac{y^6}{720} + \ldots . \quad \text{(A2.4)}$$

By setting y = jp, and rearranging terms, it follows that

$$e^{jp} = \underbrace{\left(1 - \frac{p^2}{2} + \frac{p^4}{24} - \frac{p^6}{720} + \ldots\right)}_{\cos(p)} + j\underbrace{\left(p - \frac{p^3}{6} + \frac{p^5}{120} - \ldots\right)}_{\sin(p)} \quad \text{(A2.5)}$$

where the two power series in p will be recognized as defining the sine and cosine functions. Hence,

$$e^{jp} = \cos p + j \sin p \quad \text{(A2.6)}$$

which is a famous result that Euler deduced in the year 1748.

The right-hand side of (A2.6) contains a combination of sines and cosines, as is demanded by Eq. A2.2. We seem to be on the right track! Now, the guess that

$$x = x_0 \, e^{j\omega t} \tag{A2.7}$$

is a solution of Eq. A2.1 is not difficult to make. To check that the guess is the correct one, we substitute Eq. A2.7 in Eq. A2.1. Differentiating twice, we get that $d^2x/dt^2 = -x_0\omega^2 \exp[j\omega t]$. Adding to this the second term $\omega^2 x = \omega^2 x_0 \exp[j\omega t]$ gives zero for the left-hand side. Since this balances the zero on the right-hand side, it follows that Eq. A2.7 is indeed a solution of our differential equation (A2.1). There is one slight problem – the quantity $x_0 \exp[j\omega t]$ is a complex number, but the physical displacement x appearing in the differential equation is a real number, as are all quantities in the physical world. Therefore, if we wish to use complex notation (and it is a most useful notation), we are impelled to find a convention that enables us to express physical quantities in terms of complex numbers. One such convention is to write the solution as $x = x_0 \exp(j\omega t)$ and agree that we use only the real part of this solution as being representative of the physical quantity. This is what we shall do.

Graphical representation

It is common practice to represent a complex number, with its real and imaginary parts, as a point on the "complex plane" whose abscissa is the real part of the function and its ordinate the imaginary part. It can also be represented by a vector drawn from the origin to this point on the complex plane. Thus, for example, Fig. A2-1a gives the graphical representation (known as the Argand diagram) of the complex function $\exp(jp) = \cos(p) + j\sin(p)$. We see that it is a vector of unit length inclined at an angle p to the real axis.

Equation A2.6 shows that any complex number

$$z = a + jb \tag{A2.8}$$

can, in fact, be expressed as a complex exponential having the form

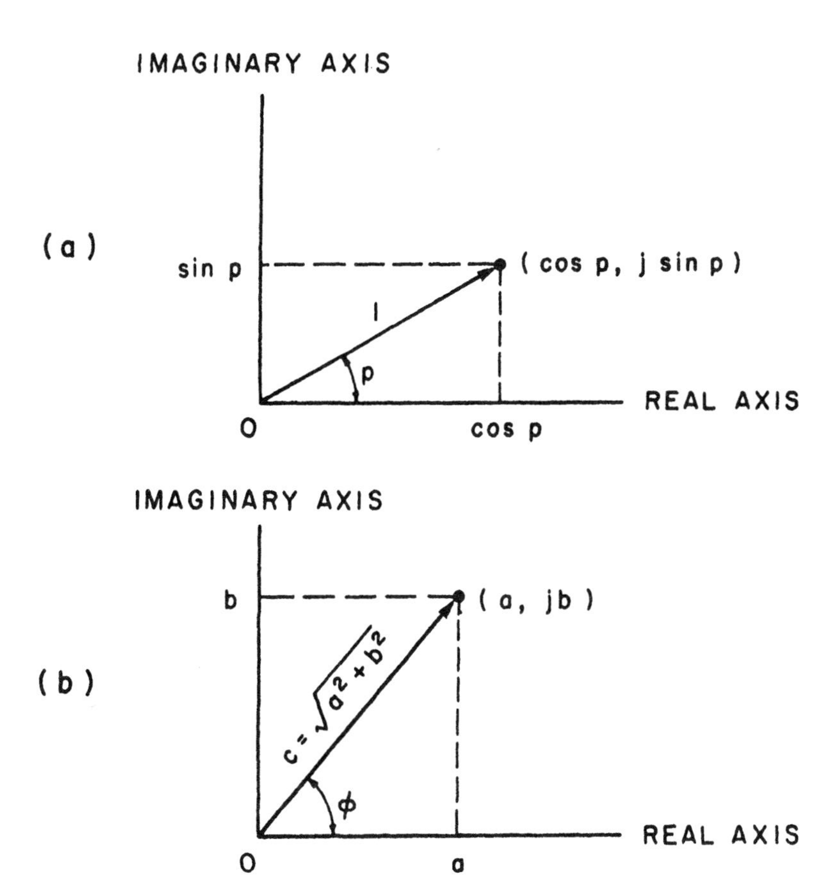

Fig. A2-1. Graphical representation of complex numbers. (a) Representation $\exp(jp) = \cos p + j \sin p$. (b) The complex number $z = (a + jb) = c \exp(j\Phi)$

$z = c \exp(j\Phi)$ with c and Φ as real numbers. Using relation (A2.6) this can be written as

$$z = c \cos \Phi + jc \sin \Phi. \qquad \text{(A2.9)}$$

To make the right-hand sides of Eqs. A2.8 and A2.9 equal, we equate the respective real and imaginary parts of the two equations to one another and obtain that $a = c \cos \Phi$ and $b = c \sin \Phi$. Solving leads to the result that

$$c = \underbrace{|a+jb| = \sqrt{a^2 + b^2}}_{\substack{\text{"magnitude of (a+jb)"} \\ \text{or} \\ \text{"modulus of z"}}} \tag{A2.10}$$

$$\underbrace{\Phi = \tan^{-1}\left(\frac{b}{a}\right)}_{\substack{\text{"phase of (a+jb)"} \\ \text{or} \\ \text{"argument of z"}}}. \tag{A2.11}$$

Figure A2-1b illustrates these results on the complex plane. Note that the factor $\exp(j\Phi)$ of the equation $z = c \exp(j\Phi)$ signifies a rotation of the vector c through an angle Φ.

Let us now go back to the solution (A2.7), $x = x_o \exp(j\omega t)$, of the differential equation (A2.1). We agree that only the real part is meaningful. We write that

$$x = x_o\, e^{j\omega t} \quad \text{with} \quad x_o = A - jB \tag{A2.12}$$

and check that the real part agrees with the known solution (A2.2). Using Eq. A2.6 it follows that

$$\begin{aligned} x &= (A - jB)\, e^{j\omega t} \\ &= A\cos(\omega t) - jB\cos(\omega t) + jA\sin(\omega t) + B\sin(\omega t). \end{aligned} \tag{A2.13}$$

The real part of this is precisely the value given by Eq. (A2.2). Figure A2-2 shows the result graphically.

An alternative but equivalent description of the solution comes from the fact that the complex quantity x_o can be written in terms of the complex exponential (see also Fig. A2-2) as

$$\left.\begin{aligned} x_o &= C\, e^{j\phi} \\ \text{where} \quad C &= \sqrt{A^2 + B^2} \\ \phi &= \tan^{-1}\left(-\frac{B}{A}\right) \end{aligned}\right\} \tag{A2.14}$$

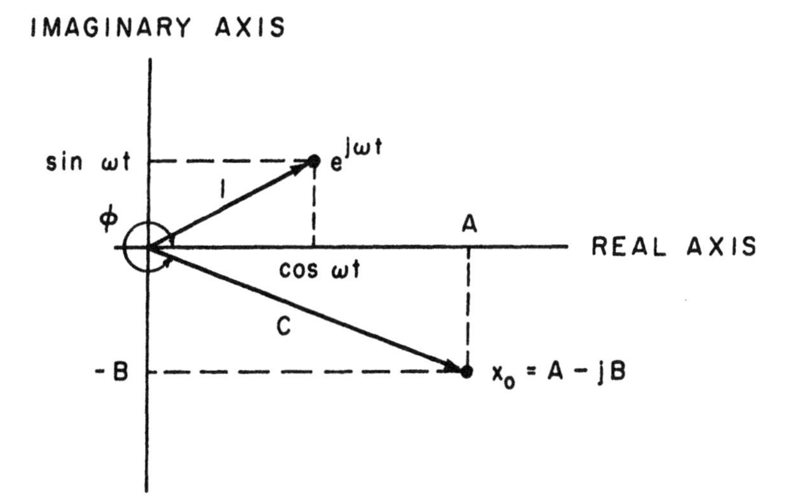

Fig. A2-2. Representation in the complex plane of $x_o = A - jB = C \exp(j\phi)$ and $\exp(j\omega t)$.

so that

$$x = C\, e^{j(\omega t + \phi)}. \tag{A2.15}$$

We take the real part, which yields

$$x = C \cos(\omega t + \phi) \tag{A2.16}$$

which is nothing more than result (A2.3).

In summary, then, we have <u>four</u> alternative ways of representing the solution of the basic differential equation (A2.1). Two of the ways are in terms of real quantities only, and the remaining two are in terms of complex numbers:

$$\left.\begin{array}{lll} x = A\cos(\omega t) + B\sin(\omega t) & & \text{I} \\ x = C\cos(\omega t + \phi) & & \text{II} \\ x = x_o\, e^{j\omega t} & ; \; x_o \text{ complex} = A - jB & \text{III} \\ x = C\, e^{j(\omega t + \phi)} & ; \; C \text{ real} & \text{IV} \end{array}\right\} \tag{A2.17}$$

Recall that our convention requires that we take the real part of the last two expressions.

Solutions of other differential equations

The equation

$$\frac{d^2x}{dt^2} + \omega_o^2 x = p_o\, e^{j\omega t} \quad (p_o \text{ real}) \tag{A2.18}$$

is the equation typical of the driven, undamped oscillator. We know that when $p_o = 0$, the solution has the form $x_o \exp(j\omega_o t)$ (see above). With p_o not equal to zero, we guess that the solution has the form

$$x = x_o\, e^{j\omega_o t} + B\, e^{j\omega t} \tag{A2.19}$$

and attempt to find the value of B in terms of the quantity p_o specified in the problem. We insert Eq. A2.19 in Eq. A2.18, do the differentiation and obtain

$$-\omega_o^2 x_o\, e^{j\omega_o t} - \omega^2 B\, e^{j\omega t} + \omega_o^2 x_o\, e^{j\omega_o t} + \omega_o^2 B\, e^{j\omega t} = p_o\, e^{j\omega t} \tag{A2.20}$$

from which it follows that $B = p_o\left(\omega_o^2 - \omega^2\right)^{-1}$, so that

$$x = x_o\, e^{j\omega_o t} + \frac{p_o}{\omega_o^2 - \omega^2}\, e^{j\omega t}. \tag{A2.21}$$

We take the real part which leads to

$$x = \underbrace{A\cos(\omega_o t) + B\sin(\omega_o t)}_{\substack{\text{called the} \\ \text{"complementary function"}}} + \underbrace{\frac{p_o}{\omega_o^2 - \omega^2}\cos(\omega t)}_{\substack{\text{called the} \\ \text{"particular solution"}}} \tag{A2.22}$$

where the names of the different pieces of the solution come from the theory of differential equations. (See also section 1.4 of the text, The effect of transients.)

Yet another equation is

$$\frac{d^2x}{dt^2} + \beta\frac{dx}{dt} + \omega_o^2 x = p_o\, e^{j\omega t} \qquad \text{(A2. 23)}$$

which is typical of a damped, driven system (see section 1.4, The sinusoidally driven oscillator). We know that when β is finite and the time t is long enough, the part of the solution given by the complementary function falls to zero exponentially with time and may be neglected. Subject to this assumption we guessed in the text (section 1.4, below Equation 1.90) a solution of the form $x = I_o \exp[j(\omega t+\delta)]$ and proceeded to find I_o and δ. Note that I_o was assumed to be a real quantity. We now know that an alternative method of solution is to set $I_o \exp(j\delta)$ equal to some complex quantity, say x_o, and use as our trial function the equation

$$x = x_o\, e^{j\omega t}. \qquad \text{(A2. 24)}$$

↑ complex number (pointing to x_o)

We insert Eq. A2.24 in Eq. A2.23, carry out the differentiation and obtain that

$$-\omega^2 x_o\, e^{j\omega t} + j\beta\omega x_o\, e^{j\omega t} + \omega_o^2 x_o\, e^{j\omega t} = p_o\, e^{j\omega t} \qquad \text{(A2. 25)}$$

from which it follows that $x_o = p_o / \left(\omega_o^2 - \omega^2 + j\beta\omega\right)$ and therefore, that

$$x = \left[\frac{p_o}{\left(\omega_o^2 - \omega^2 + j\beta\omega\right)}\right] e^{j\omega t} \qquad \text{(A2. 26)}$$

We must now take the real part. To do so we rationalize the denominator of the term shown in brackets. If, for example, p_o is real, one obtains

$$x = \frac{p_o\left(\omega_o^2 - \omega^2\right)}{\left(\omega_o^2 - \omega^2\right)^2 + (\beta\omega)^2}\cos(\omega t) + \frac{p_o\beta\omega}{\left(\omega_o^2 - \omega^2\right)^2 + (\beta\omega)^2}\sin(\omega t) \qquad \text{(A2. 27)}$$

showing that damping (i.e., the term β) introduces into the motion an out of phase component which varies as $\sin(\omega t)$ rather than $\cos(\omega t)$. The foregoing result can, of course, be cast into the form $x = A \cos(\omega t + \delta)$ but we shall not pursue this piece of arithmetic any further.

The complex conjugate

If $z = a + jb$, where a and b are real, then the number $a - jb$ is said to be the complex conjugate of z, and is denoted by z^*. For example, if

$$x = C\, e^{j(\omega t + \phi)} \qquad \text{(C real)} \tag{A2.28}$$

then

$$x^* = C\, e^{-j(\omega t + \phi)}. \tag{A2.29}$$

Now, the real part of x, which is the quantity one normally seeks, is simply

$$\left.\begin{aligned} \text{Re}(x) &= \frac{1}{2}(x + x^*) \\ &= \frac{1}{2} C \left[e^{j(\omega t + \phi)} + e^{-j(\omega t + \phi)}\right] \\ &= C \cos(\omega t + \phi) \end{aligned}\right\} \tag{A2.30}$$

Thus, we see that Eq. A2.30 provides a fifth way of expressing simple harmonic motions (the other four ways are given by Eqs. A2.17).

Although we do not use the complex conjugate extensively in this book, the proof of many theorems regarding complex numbers is often greatly simplified by this concept. All one needs to remember are the following easily proved relationships:

$$\text{Re}(z) = \frac{1}{2}(z + z^*) \tag{A2.31}$$

$$\text{Im}(z) = -\frac{1}{2} j(z - z^*) \tag{A2.32}$$

$$|z|^2 = zz^*. \tag{A2.33}$$

Take as an example the product of any two complex numbers z_1 and z_2. Then in accordance with Eq. A2.33

$$\left.\begin{aligned} |z_1 z_2|^2 &= (z_1 z_2)(z_1 z_2)^* \\ &= \left(z_1 z_1^*\right)\left(z_2 z_2^*\right) \\ &= |z_1|^2 \, |z_2|^2 \end{aligned}\right\} \tag{A2. 34}$$

so that

$$|z_1 z_2| = |z_1| \cdot |z_2|. \tag{A2. 35}$$

This says that the modulus of the product of two complex numbers is equal to the product of their moduli.

As a second example, let us calculate the product of the real parts of two complex quantities z_1 and z_2. Using formula (A2. 31) it then follows that

$$\begin{aligned} \mathrm{Re}(z_1)\ \mathrm{Re}(z_2) &= \frac{1}{2}\left(z_1 + z_1^*\right)\frac{1}{2}\left(z_2 + z_2^*\right) \\ &= \frac{1}{4}\left[z_1 z_2 + z_1^* z_2^* + z_1 z_2^* + z_2 z_1^*\right]. \end{aligned} \tag{A2. 36}$$

Such products appear time and again in the physical world, such as in

$$\text{Poynting Flux } \vec{S} = \frac{1}{\mu_o}\left[\mathrm{Re}(\vec{E}) \times \mathrm{Re}(\vec{B})\right] \tag{A2. 37}$$

or in Ohm's law

$$\text{Power } P = \mathrm{Re}(V)\ \mathrm{Re}(I). \tag{A2. 38}$$

Often one desires the time-averaged values of $\vec{S}$ and P. If $\vec{E}$, $\vec{B}$, V, and I are simple harmonic functions of time, their mean values $\langle\vec{E}\rangle$, $\langle\vec{B}\rangle$, $\langle V\rangle$, and $\langle I\rangle$ are, of course, zero, a result which follows from Eq. 1. 20. Moreover, the time-averaged value of such functions as $\cos(\omega t)\sin(\omega t)$ also vanishes. As a consequence, if z_1 and z_2 are simple harmonic functions of time, it follows from Eq. A2. 36 that

$$\begin{aligned} \langle \mathrm{Re}(z_1)\ \mathrm{Re}(z_2)\rangle &= \frac{1}{4}\left[z_1 z_2^* + z_2 z_1^*\right] \\ &= \frac{1}{2}\,\mathrm{Re}\left(z_1 z_2^*\right). \end{aligned} \tag{A2. 39}$$

Applying this result to Eqs. A2. 37 and A2. 38 tells us that

$$\langle \text{Poynting flux } \vec{S} \rangle = \frac{1}{2\mu_o} \text{Re}(\vec{E} \times \vec{B}^*) \tag{A2.40}$$

$$\langle \text{Power } P \rangle = \frac{1}{2} \text{Re}(VI^*). \tag{A2.41}$$

Suppose, then, that in some electrical circuit there is an oscillatory voltage

$$V = V_o \, e^{j(\omega t + \phi)} \quad [V_o \text{ real}] \tag{A2.42}$$

giving rise to an oscillatory current

$$I = I_o \, e^{j(\omega t + \phi)} \quad [I_o \text{ real}]. \tag{A2.43}$$

What is the time-averaged power dissipated? By Eq. A2.41 it is simply

$$\begin{aligned} \langle P \rangle &= \frac{1}{2} \text{Re}\left[V_o \, e^{j(\omega t + \phi)} \, I_o \, e^{-j(\omega t + \phi)}\right] \\ &= \frac{1}{2} V_o I_o. \end{aligned} \tag{A2.44}$$

But note that we could have obtained this result by the following technique without ever mentioning complex conjugate quantities:

$$\begin{aligned} P &= \text{Re}(V) \, \text{Re}(I) \\ &= \text{Re}\left(V_o \, e^{j(\omega t + \phi)}\right) \text{Re}\left(I_o \, e^{j(\omega t + \phi)}\right) \\ &= V_o I_o \cos^2(\omega t + \phi). \end{aligned} \tag{A2.45}$$

Time averaging in accordance with Eq. 1.20 then yields

$$\begin{aligned} \langle P \rangle &= V_o I_o \frac{1}{T} \int_0^T \cos^2(\omega t + \phi) \, dt \\ &= \frac{1}{2} V_o I_o. \end{aligned} \tag{A2.46}$$

APPENDIX 3

ELECTRON SUBJECTED TO AN RF ELECTRIC FIELD ORTHOGONAL TO A STEADY MAGNETIC FIELD

Below is a FORTRAN computer program* that allows one to watch the motion of a charge subjected simultaneously to a steady magnetic field $\vec{B}_o$ and a perpendicular, oscillatory electric field $\vec{E}(t)$ (see Fig. 1.28, page 76). The equations of motion along the three rectangular coordinate axes are

$$\frac{d^2x}{dt^2} = \frac{qE_o}{m} \sin(\omega t) + \omega_c \frac{dy}{dt}$$

$$\frac{d^2y}{dt^2} = -\omega_c \frac{dx}{dt}$$

$$\frac{d^2z}{dt^2} = 0$$

where ω is the angular frequency of the rf electric field and $\omega_c = qB_o/m$ is the cyclotron frequency. These equations were first solved analytically (see pages 75-78, 614-615) to yield the instantaneous positions $x(t)$, $y(t)$ of the charge q. Two separate cases had to be considered: $\omega/\omega_c \neq 1$ and $\omega/\omega_c = 1$. The resulting equations for $x(t)$, $y(t)$ were then programmed. The particle trajectory depends on (i) ω/ω_c, the ratio of rf to cyclotron frequency, and on (ii) $qE_o/m\omega_c^2$, which specifies the relative strengths of the electric and magnetic field amplitudes.

---

*The authors wish to thank Mr. Alan C. Janos for writing the program and providing the photographs displayed in Fig. 1.28b.

$\frac{\omega}{\omega_c} \neq 1$

```
      DIMENSION  X2(1001),X1(1001)
      DIMENSION IX2(1001),IX1(1001)
      BYTE IWR(11)
101   WRITE(6,300)
300   FORMAT('$WR,EP:')
      READ(5,400) WR,EP
400   FORMAT(2F12.5)
      IF (WR.NE.1.0) GOTO 103
      WRITE(6,302)
302   FORMAT(' W/WC =1')
      GOTO 101
103   TR=1.0/WR
      TMAX=1.0+EP
      IF (WR.LT.1.0) TMAX=(1.0+EP)*TR
      TDEL=TMAX*1E-3
      FACTOR=1.0/(1-(1/TR)**2)
      ARG=2*3.14159
      XMAX=0.0
      DO 100 I=1,1001
           ARG1=ARG*TDEL*(I-1)
           ARG2=ARG1/TR
           X1(I)=FACTOR*((1.0/TR)*COS(ARG1)-TR*COS(ARG2))
           X2(I)=(-1.0)*FACTOR*(SIN(ARG2)-(1.0/TR)*SIN(ARG1))
           AX1=ABS(X1(I))
           AX2=ABS(X2(I))
           IF (AX1.GT.XMAX) XMAX=AX1
           IF (AX2.GT.XMAX) XMAX=AX2
100   CONTINUE
      DO 102 J=1,1001
           IX1(J)=X1(J)*600.0/XMAX+600.0
           IX2(J)=X2(J)*600.0/XMAX+600.0
102   CONTINUE
      CALL SET611(6)
      CALL BELL(1)
      IF (WR.LT.10) GOTO 105
      IF (WR.LT.100.AND.WR.GE.10) GOTO 106
      IF (WR.GE.1000) GOTO 107
500   FORMAT('W/WC= ',F5.1)
      ENCODE(11,500,IWR) WR
      GOTO 104
105   FORMAT('W/WC= ',F5.3)
      ENCODE(11,105,IWR) WR
      GOTO 104
106   FORMAT('W/WC= ',F5.2)
      ENCODE(11,106,IWR) WR
104   CALL WRI611(IWR,11,1320,1400,6,0)
107   CALL CPL611(1001,IX1,IX2,200,1400,200,1400,0,1200,0,1200)
      GOTO 101
      END
```

$$\omega = \omega_c$$

```
      DIMENSION X2(1001),X1(1001)
      DIMENSION IX2(1001),IX1(1001)
      BYTE IWR(7)
101   WRITE(6,300)
300   FORMAT('$EP:')
      READ(5,400) EP
400   FORMAT(1F12.5)
      TMAX=1.0+EP
      TDEL=TMAX*1E-3
      FACTOR=1.0*(-1.0/2.0)
      ARG=2*3.14159
      XMAX=0.0
      DO 100 I=1,1001
           ARG0=ARG*TDEL*(I-1)
           X1(I)=(-1.0)*FACTOR*(1-COS(ARG0)-ARG0*0.5*SIN(ARG0))
           X2(I)=FACTOR*0.5*(SIN(ARG0)-ARG0*COS(ARG0))
           AX1=ABS(X1(I))
           AX2=ABS(X2(I))
           IF (AX1.GT.XMAX) XMAX=AX1
           IF (AX2.GT.XMAX) XMAX=AX2
100   CONTINUE
      DO 102 J=1,1001
           IX1(J)=X1(J)*500.0/XMAX+600.0
           IX2(J)=X2(J)*600.0/XMAX+600.0
102   CONTINUE
      CALL SET611(6)
500   FORMAT('W/WC= 1')
      ENCODE(7,500,IWR)
      CALL WRI611(IWR,7,1320,1400,6,0)
      CALL CPL611(1001,IX1,IX2,200,1400,200,1400,0,1200,0,1200)
      GOTO 101
      END
```

PROBLEMS

CHAPTER 1

1.1 A mass of 0.5 kg is hung from a light spring and is found to stretch it 4 cm; it is then pulled down a further 2 cm and released. Find the time of a complete oscillation and calculate the kinetic energy of the mass when passing through the position of equilibrium.

1.2 A particle describing simple harmonic motion executes 100 complete vibrations per minute, and its speed at its mean position is 5 m/sec. What is the amplitude of its oscillations?

What is its velocity (i) when it is half-way between its mean position and an extremity of its path, (ii) at a time after leaving its mean position equal to half the time required to reach an extremity of its path?

1.3 An object of mass M rests on a frictionless horizontal surface. Two identical springs of spring constants k and relaxed length ℓ_0 are attached to M as shown.

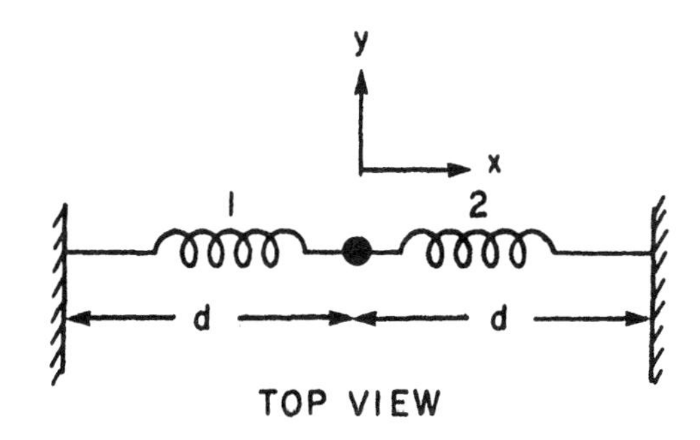

The object is at rest in static equilibrium when each spring is of length d ($d > \ell_0$).

(a) The mass M is given a displacement of x_0 to the right. Give equations for F_1 and F_2, the forces exerted on M by springs 1 and 2, respectively. (Use the sign convention that a positive force acts toward +x.)

The mass is released from its position at $x = x_0$. The initial velocity is zero.

(b) Write the differential equation of motion for the mass M moving in the x direction.

(c) Give a solution of the equation of motion that is consistent with the initial conditions. Express x(t) in terms of d, ℓ_o, x_o, M, and k.

The mass M is brought to rest at its position of equilibrium. It is then given a small displacement $+y_o$ along the y axis. The displacement is so small that the lengths of the springs may be considered to be d, i.e., unchanged.

(d) What is the net restoring force acting on M? Give both magnitude and direction.

(e) What is the period T of the simple harmonic motion along the y axis?

1.4 In a conservative system, the sum of the kinetic and potential energies is a constant, namely,

$$\frac{1}{2}mv^2 + V(x) = E\,(\text{constant}).$$

(a) By direct integration show that

$$t - t_1 = \sqrt{\frac{m}{2}} \int_{x_1}^{x} \frac{dx}{\sqrt{E - V(x)}}$$

where x_1 is the position at time t_1.

(b) If the restoring force is linear, $F(x) = -Kx$, show that the potential energy is

$$V(x) = \frac{1}{2} m\omega_o^2 x^2$$

where $\omega_o = \sqrt{K/m}$.

(c) Now, let position x_1 correspond to the case for which $v = 0$. Show that

$$t - t_1 = \frac{1}{\omega_o} \int_{x_1}^{x} \frac{dx}{\sqrt{x_1^2 - x^2}} = \frac{1}{\omega_o} \cos^{-1}\left(\frac{x}{x_1}\right).$$

(d) Invert the result of part (c). What have you proved?

1.5 A pendulum of length ℓ starts from rest in a position making an angle 2α with the vertical. Show that the time of a complete oscillation is

$$4\sqrt{(\ell/g)}\int_0^{\pi/2}\frac{d\phi}{\sqrt{(1-\sin^2\alpha\,\sin^2\phi)}},$$

and deduce the approximate formula $2\pi\sqrt{(\ell/g)}$.

1.6 A ball of mass m fits snugly (but with negligible friction) inside the neck of a bottle (see figure). The neck has cross-sectional

area A; at equilibrium, the volume of the bottle below the ball is V_0 and is filled with a gas at pressure p_0. The heat capacities of the gas at constant volume and at constant pressure are c_v and c_p, respectively. If the ball is given a small displacement, what is the period of the resulting oscillations about the equilibrium position?

1.7 A mass m = 0.2 kg is hung from a spring, whose spring constant K = 80 newtons/meter. The mass is acted upon by a dissipative force given by (−bv) where v is its velocity (in meters per second).

(a) Set up the equation of motion.

(b) What is the Q of the system, if the observed frequency of oscillation is $(\sqrt{3})/2$ of the oscillatory frequency with no damping present?

(c) The damping b is now increased until the system is critically damped, which occurs when $\sqrt{K/m} = b/2m$. Show, by direct substitution in the differential equation, that the solution is

$$x = e^{-bt/2m}(A + Bt).$$

(d) If the critically damped oscillating mass starts at t = 0 from the origin with a velocity of 3m/sec, what is its maximum displacement, and when does it occur? Draw the displacement and the velocity with time.

1.8 A small mass M slides freely on a frictionless parabolic track under the influence of gravity. The shape of the track is given by

$$y = 0.1\,x^2 \text{ meters.}$$

(a) Derive the equation of motion of the mass M.

(b) Find its angular frequency ω for small amplitude oscillations.

(c) If the frictional forces are given by

$$F_x = -\frac{2M}{3}\frac{dx}{dt} \text{ newtons}$$

$$F_y = 0$$

what is the angular frequency of the damped motion?

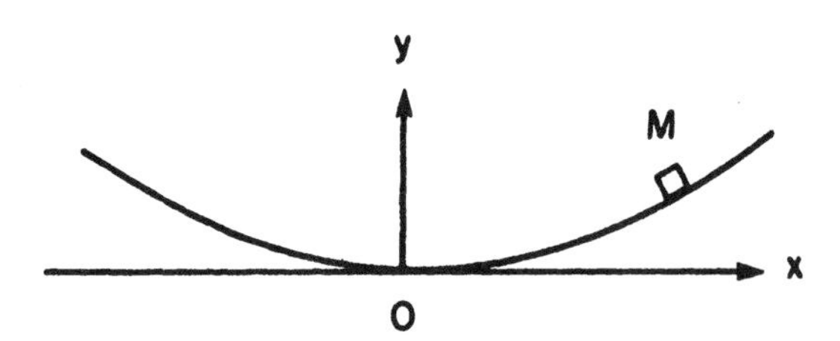

1.9

(a) A generator of constant emf and of variable frequency is connected in series with a resistance R, an inductance L, and a capacitance C. Show that the voltage across the capacitance is a maximum when

$$\omega = \sqrt{\frac{1}{LC} - \frac{R^2}{2L^2}} \, .$$

(b) Show that the voltage across the inductance is a maximum when

$$\frac{1}{\omega} = \sqrt{LC - \frac{R^2C^2}{2}} \, .$$

1.10 A parallel RLC circuit is driven by a current $I = I_o \cos \omega t$.

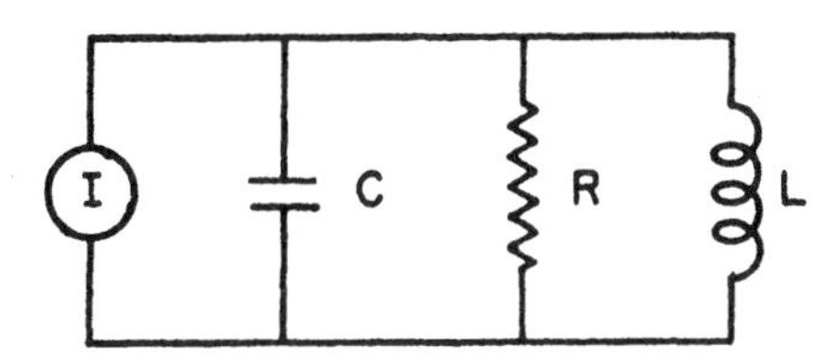

(a) Derive the differential equation for the voltage V across the capacitor.

(b) If the solution to this equation is given by

$$V = \frac{(-\omega I_o/C) \cos(\omega t + \delta)}{\sqrt{(\omega^2 - (1/LC))^2 + \omega^2/(RC)^2}} \, ,$$

determine the values of ω for which the peak voltage V_o is approximately half the maximum.

(c) Find the Q of this system.

1.11 A point on an oscilloscope face has a motion compounded of two simple harmonic motions at right angles to one another, the period of one being twice that of the other. The amplitude of the former is 2 cm, and that of the latter 1.5 cm. The latter reaches its maximum value at the instants when the former reaches $1/\sqrt{2}$ of its maximum value. Draw the path of the point for one complete cycle.

1.12 A radio station transmits at a carrier frequency $\omega_c = 10^6$ rad/sec. The signal is modulated with a sinusoidal oscillation

$\omega_m = 10^5$ rad/sec, that is

$$E(t) = E_o[1 + \alpha \cos \omega_m t] \cos \omega_c t.$$

(a) Taking $\alpha = 0.5$, sketch E(t) as a function of time.

(b) Show that the above equation is equivalent to the superposition of three constant amplitude signals.

(c) Sketch the frequency spectrum of these signals, i.e., intensity vs frequency.

(d) What bandwidth would you need to transmit the complete audible range? (Assume that the audible frequency stretches from 20 c/s to 20 kc/s.)

1.13 A microwave cavity has a resonant frequency of 3000 megahertz (1 hertz = 1 cycle/sec). When it is excited from an external source, the power absorbed by it falls by a factor of 2 over a range of 80 kilohertz on either side of the resonance peak. What is the Q of the cavity? If the cavity is excited and left to itself, how fast will the electromagnetic energy stored in the field within the cavity decay with time?

1.14 The bobs of two simple pendula each have mass m. The pendula are each of length ℓ, and are attached to the ceiling a distance ℓ_o apart as shown. A massless spring (constant k, unstretched length ℓ_o) connects the two bobs. Consider only small displacements and motion in the horizontal direction.

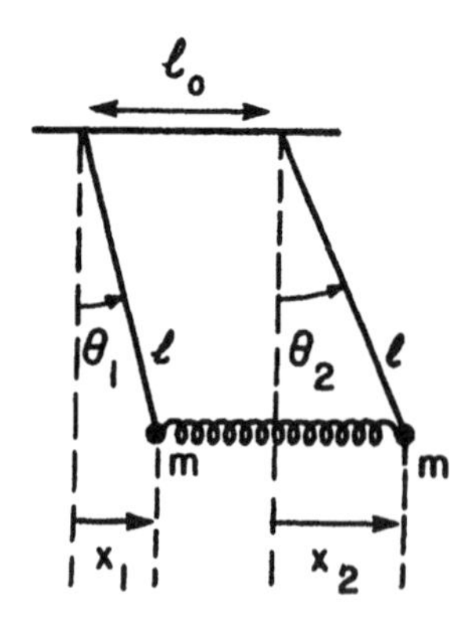

(a) Give the coupled differential equations of motion in terms of $\ddot{x}_1$, $\ddot{x}_2$, x_1, x_2, k, m, g and ℓ. ($\ddot{x}$ means d^2x/dt^2.)

(b) Obtain a single quadratic equation for ω^2 (a fourth-order equation in ω), and solve for the normal frequencies.

(c) Find the ratio of amplitudes of oscillation of the bobs for each mode.

(d) Write the solutions to the differential equations obtained in part (a), under the condition that, at $t = 0$, $x_2 = B$, and $x_1 = 0$ with zero initial velocities for both bobs.

1.15 Two identical pendula are connected by a light coupling spring. Each pendulum has a length of 0.4 m, and they are at a place where $g = 9.8\ \text{m/sec}^2$. With the coupling spring connected, one pendulum is clamped and the period of the other is found to be 1.25 sec exactly.

(a) With neither pendulum clamped, what are the periods of the two normal modes?

(b) What is the time interval between successive maximum possible amplitudes of one pendulum after the other pendulum is drawn aside and released?

1.16 Two springs, each of constant k, support a rigid, massless platform to which a mass m is firmly attached. The position of this mass is $y_1(t)$. A second mass m hangs at the end of another spring (of constant k) from the center of the platform, as shown in the sketch.

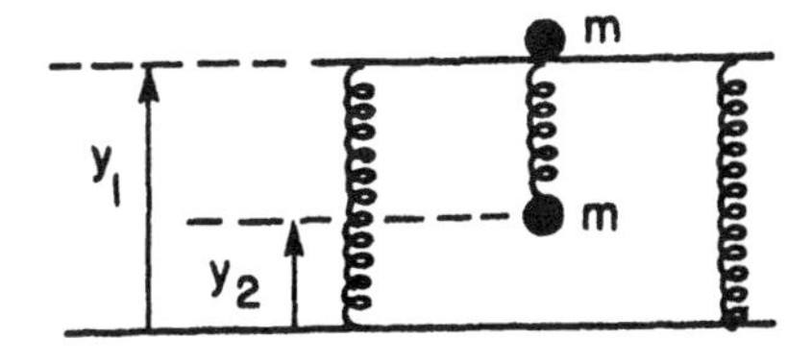

The position of this second mass is $y_2(t)$. Assume that the two longer springs move together with the same frequency and in the same plane.

(a) Write the differential equations of motion for each of the masses.

(b) Solve the equations to find the normal frequencies and find suitable expressions for $y_1(t)$ and $y_2(t)$.

(c) Sketch the configuration of the system for each of the two normal modes. Label the sketches to indicate which configuration corresponds to the normal mode with low frequency ω_1, and which corresponds to the mode with high frequency ω_2.

CHAPTER 2

2.1 Two points on a string are observed as a traveling wave passes them. The points are at $x_1 = 0$ and $x_2 = 1$ m. The transverse motions of the two points are found to be as follows:

$$y_1 = 0.2 \sin 3\pi t$$

$$y_2 = 0.2 \sin(3\pi t + \pi/8)$$

(a) What is the frequency in cycles/sec?

(b) What is the wavelength?

(c) With what speed does the wave travel?

(d) Which way is the wave traveling? Show how you reach this conclusion.

(e) A symmetrical triangular pulse of maximum height 0.4 m and total length 1.0 m is moving to the right with the pulse entirely located between $x = 0$ and $x = 1$ m at $t = 0$. Draw a graph of the transverse velocity vs time at $x = x_2 = +1$ m (assume that the pulse has the speed you calculated in part c).

2.2 The characteristic impedance Z_o of a system is defined as the ratio of the driving force to the displacement produced by the force.

(a) Show that for transverse waves on a string

$$Z_o \equiv \frac{F}{s}$$

$$= \sqrt{\mu T}$$

where μ is the mass per unit length and T is the tension in the string.

(b) Show that when a sinusoidal voltage $V = V_o e^{j\omega t}$ is applied across a capacitor, its characteristic impedance defined as the ratio of driving voltage V to the current I (which is the effective displacement) is given by

$$Z_o \equiv \frac{V}{I} = \frac{1}{j\omega C}$$

where C is the capacitance.

(c) Calculate Z_o for an inductor.

2.3 Strings of tension T and mass density μ_1 and μ_2 are connected together. A traveling wave is incident on the boundary.

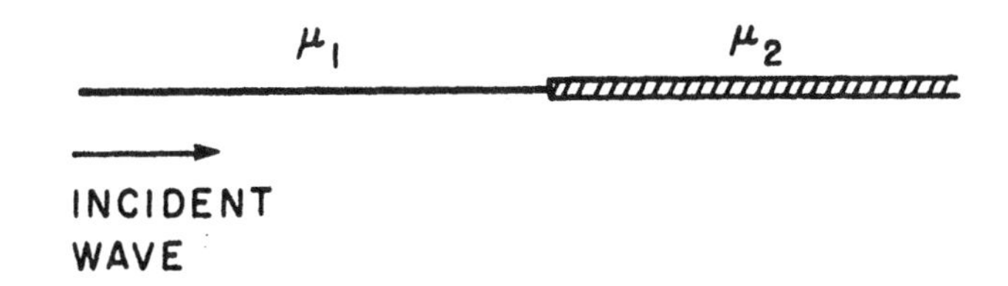

Show that the ratio of reflected amplitude to incident amplitude is given by

$$R = \frac{Z_{o1} - Z_{o2}}{Z_{o1} + Z_{o2}}$$

where Z_{o1} and Z_{o2} are the characteristic impedances given by

$$Z_{o1} = \sqrt{T\mu_1}$$

$$Z_{o2} = \sqrt{T\mu_2}$$

2.4 Two strings with masses per unit length $\mu_1 = 0.1$ kgm/m and $\mu_2 = 0.4$ kgm/m are joined and under tension $T = 10$ n.

A traveling wave of the shape shown in the figure is started moving to the right along the light string. Find the reflected and transmitted

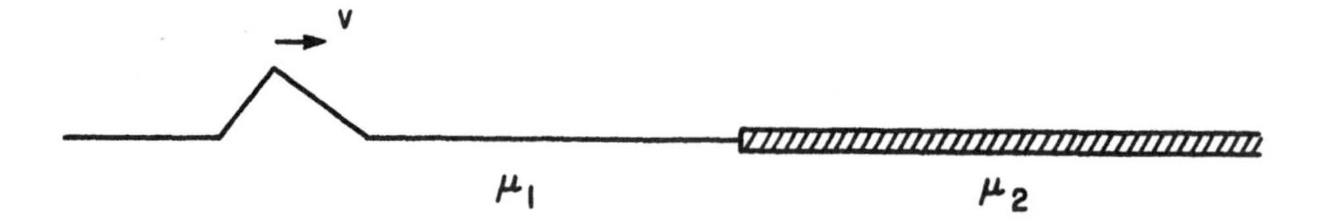

wave pulses (signs and amplitudes) and make sketches for the following cases:

(a) When the incident pulse has its peak at the junction and

(b) When both reflected and transmitted pulses have moved away from the junction.

(c) What is the ratio between the transmitted and incident wave energies?

2.5 Pressure oscillations in a hollow pipe open at both ends are governed by the wave equation

$$\frac{\partial^2 p}{\partial z^2} = \frac{\rho_0}{\kappa} \frac{\partial^2 p}{\partial t^2}$$

where p is the pressure, ρ_0 is the density, and κ is the bulk modulus. Take as a solution

$$p(z,t) = [A \cos kz + B \sin kz] \cos \omega t$$

and require p = 0 at both ends of the pipe. Also take $p = p_0$ at $z = L/2$ and t = 0.

Evaluate A, B, k, and ω for all normal modes meeting the above conditions.

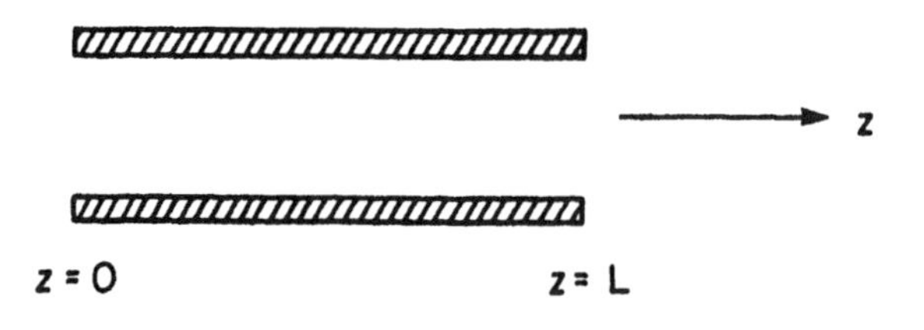

2.6 An isotropic point source of sound radiates spherical waves. The intensity at a distance of one meter from the source is 5 watt/m^2.

What is the intensity, the pressure amplitude, and the particle displacement amplitude in the wave 100 meters from the source?

2.7 The wave equation $\partial^2\psi/dt^2 = v^2(\partial^2\psi/\partial z^2)$ describes wave propagation in a nondispersive medium, that is, a medium in which the phase velocity v is independent of position or time. For a certain class of dispersive media the modified wave equation is found to be

$$\frac{\partial^2\psi}{\partial t^2} = a\frac{\partial^2\psi}{\partial z^2} - \omega_c^2\psi$$

with a and ω_c as constants.

(a) Show that the sinusoidal disturbance

$$\psi = Ae^{j\omega t - jkz}$$

satisfies the modified wave equation.

(b) Find the dispersion relation and plot ω as a function of k.

(c) Discuss the propagation characteristics, and in particular the phase velocity, when $\omega > \omega_c$ and when $\omega = \omega_c$.

(d) What properties does the disturbance have when $\omega < \omega_c$?

2.8 The wave equation for water waves in deep water is given by

$$\frac{\partial^2 y}{\partial x^2} = \frac{2\pi}{g\lambda}\frac{\partial^2 y}{\partial t^2}$$

where y is the vertical displacement of the water, λ the wavelength and g the acceleration due to gravity.

(a) Calculate the velocity of a purely sinusoidal wave of wavelength 62 meters.

(b) Sketch the dispersion diagram, i.e., ω versus k.

(c) Sketch the phase velocity as a function of frequency ω.

2.9 A rectangular room is 6 m long, 4 m wide, 3 m high and has rigid walls. Using appropriate boundary conditions for the excess acoustical pressure, calculate the oscillation frequencies of the three lowest acoustical modes, taking the sound velocity as 330 m/sec. Do the modes form a harmonic series? Discuss.

2.10 A half-wave rectifier removes the negative half-cycles of a pure sinusoidal wave $y = h \sin x$.

(a) Show that the Fourier series of this function is given by

$$\frac{h}{\pi}\left(1 + \frac{\pi}{1.2}\sin x - \frac{2}{1.3}\cos 2x - \frac{2}{3.5}\cos 4x - \frac{2}{5.7}\cos 6x \ldots\right).$$

where $x = \omega t$.

(b) Plot the sum of four terms of the series over the interval 0 to 2π taking $h = \pi$. Compare this with the function itself which you have Fourier analyzed.

2.11 A guitar string of length L is plucked at a distance d from one end. If its initial shape at time $t = 0$ is like that illustrated in the figure, show that its displacement from equilibrium at any subsequent time is given by

$$s(z,t) = \sum_{m=1}^{\infty} A_m \sin\left(\frac{m\pi z}{L}\right)\cos\left(\frac{m\pi vt}{L}\right)$$

where

$$A_m = \frac{2aL^2}{(m\pi)^2 d[L-d]} \sin\frac{m\pi d}{L}.$$

What must you do to suppress the eighth harmonic ($m = 8$)?

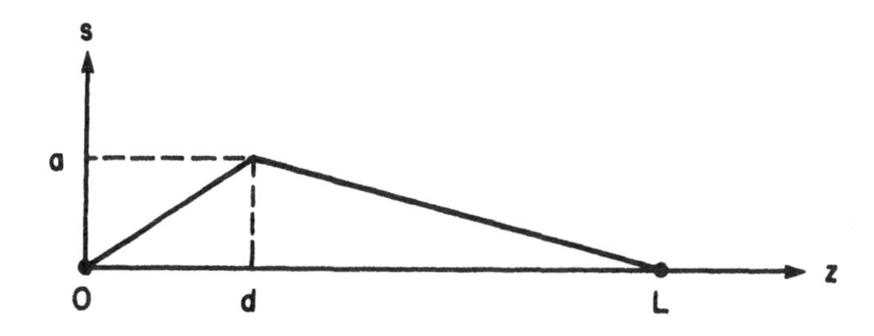

2.12 Find the spectral function F(k) for the following

(i) $f(z) = e^{-\alpha z^2/2}$

(ii) $f(z) = e^{-\beta|z|}$

(iii) $f(z) = 1$ for $-a/2 \leq z \leq a/2$
$\quad = 0$ outside this interval.

Show (in a qualitative way) for the three examples given that: (width of $f(z)$) $\times$ (width of $F(k)$) is a number of order unity.

CHAPTER 3

3.1 The vector function which follows represents a possible electrostatic field which must satisfy $\oint \vec{E} \cdot d\vec{s} = 0$

$$E_x = 6xy; \quad E_y = 3x^2 - 3y^2; \quad E_z = 0.$$

Calculate the line integral of E from the point $(0,0,0)$ to the point $(x_1, y_1, 0)$ along the path which runs straight from $(0,0,0)$ to $(x_1, 0, 0)$ and thence to $(x_1, y_1, 0)$. Make a similar calculation for the path which runs along the other two sides of the rectangle, via the point $(0, y_1, 0)$. Now you have the potential function $\phi(x,y,z)$. Take the gradient of this function and see that you get back the components of the given field. ($\vec{E} = -\nabla\phi$ for an electrostatic field.)

3.2 Calculate the curl and the divergence of each of the following vector fields. If the curl turns out to be zero, try to discover a scalar function ϕ of which the vector field is the gradient.

(a) $F_x = x + y$; $F_y = -x + y$; $F_z = -2z$.

(b) $G_x = 2y$; $G_y = 2x + 3z$; $G_z = 3y$.

(c) $H_x = x^2 - z^2$; $H_y = 2$; $H_z = 2xz$.

3.3 A magnetic field of a uniform plane wave is given by:

$$\vec{B}(\vec{r},t) = B_{oy} f(\vec{k}\cdot\vec{r} \quad \omega t + \phi)\,\hat{y}$$

where B_o is a scalar constant, and $f(\vec{k}\cdot\vec{r} - \omega t)$ is any function of the argument $\vec{k}\cdot\vec{r} - \omega t$. The vectors $\vec{k}$ and $\vec{r}$ are $k_x\hat{x} + k_y\hat{y} + k_z\hat{z}$ and $x\hat{x} + y\hat{y} + z\hat{z}$, respectively.

(a) Give <u>all</u> the conditions imposed on $\vec{E}(\vec{r},t)$ and $\vec{k}$ by Maxwell's equations and the wave equations.

(b) If it is now further specified that $k_z = 0$ give an expression for $\vec{E}(\vec{r},t)$ similar to $\vec{B}(\vec{r},t)$.

3.4 A uniform, transverse, electromagnetic plane wave with the electric vector linearly polarized in the z direction has a flux density, in vacuum, given by

$$|\vec{S}| = S_0 \sin^2(x+y-\omega t)$$

The units of x and y are centimeters and S_0 is a constant.

(a) What is the wavelength, λ?

(b) What is the angular frequency, ω?

(c) What is the direction of the linearly polarized $\vec{B}$ field?

(d) What are the magnitudes of the electric and magnetic fields (expressed in terms of S_0 and other constants, as necessary)?

(e) There are two possible answers to part (c). Why?

3.5 The $\vec{B}$ field of a certain electromagnetic wave in a vacuum is given by

$$\vec{B}(x,y,z,t) = \hat{x}B_0 \sin(\omega t - kz)$$

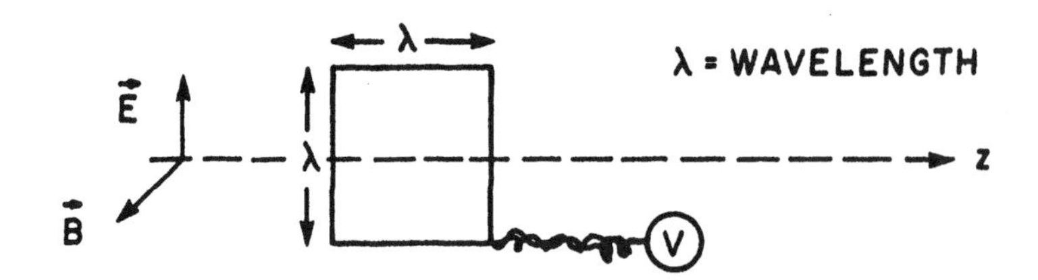

(a) Use Maxwell's equations to deduce the equation for $\vec{E}(x,y,z,t)$ for this wave. A square single-turn loop of wire, with sides of length equal to λ is used to pick up the signal from the wave by detecting the voltage V appearing at the two ends. This will be of the form

$$V = V_0 \sin(\omega t + \phi).$$

(b) The loop is placed as shown, with the two sides parallel to $\vec{E}$ and the other two sides parallel to z. What is the value of V_0 in this situation?

(c) What is the maximum possible value of V_0, and how should the loop be oriented to obtain it?

3.6 A polarizer and an analyzer are oriented so that the maximum amount of light is transmitted.

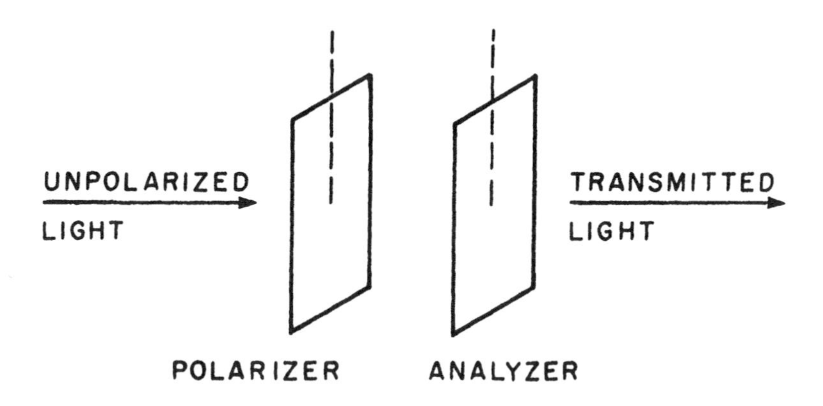

To what fraction of its maximum value is the intensity of the transmitted light reduced when the analyzer is rotated through an angle of 60° ?

3.7 A high-power CO_2 laser (wavelength 10^{-3} cm) emits 10 joules of light in a pulse of 10^{-8} seconds duration. The light is emitted in the form of a parallel beam of 1 cm radius.

(a) Calculate the average energy density and the average Poynting flux.

(b) What are the amplitudes of the electric and magnetic fields?

(c) If the beam falls at right angles on a completely absorbing surface, what is the pressure it exerts on this surface?

(d) An electron is placed in the path of the beam. What maximum velocity does it attain, and what is its maximum displacement?

3.8 A 10 kilowatt light beam from a laser is used to levitate a solid aluminum sphere, as shown.

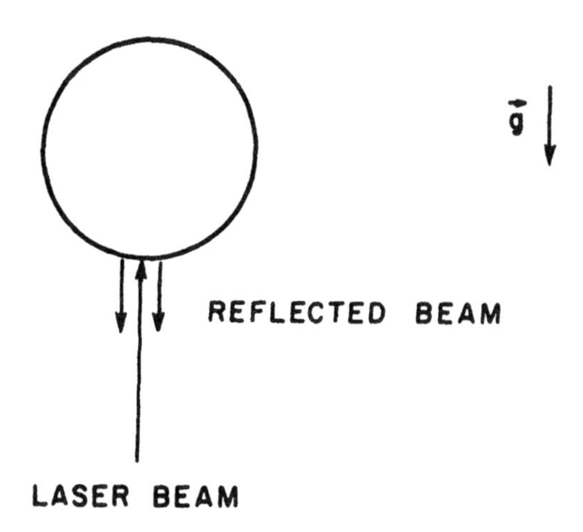

Calculate the mass of the largest sphere that can just be supported. Assume that the laser beam is concentrated onto a small portion of the spherical surface and is totally reflected back from the sphere.

CHAPTER 4

4.1 A charge, q, initially at rest, is given a brief acceleration along the z axis. Let this occur at the origin of the coordinate system at $t = 0$.

(a) Give the arrival times, the relative strengths, and specify the directions of the radiated electric field seen by three observers in the y-z plane at a large distance R from origin. One observer is on the y axis, one on the z axis, and one is on a radius making an angle of 30° to the z axis.

(b) Describe the orientation and magnitude of the associated radiated magnetic fields.

4.2 A point charge +q has been moving with constant velocity w along a straight line until the time $t = t_0$. In the short time interval from $t = t_0$ to $t = t_0 + \Delta t$, a force perpendicular to the trajectory changes the direction without changing the magnitude of the velocity. After the time $t = t + \Delta t$ the charge again moves with velocity w along a straight line forming a small angle $\Delta\alpha$ with the initial trajectory.

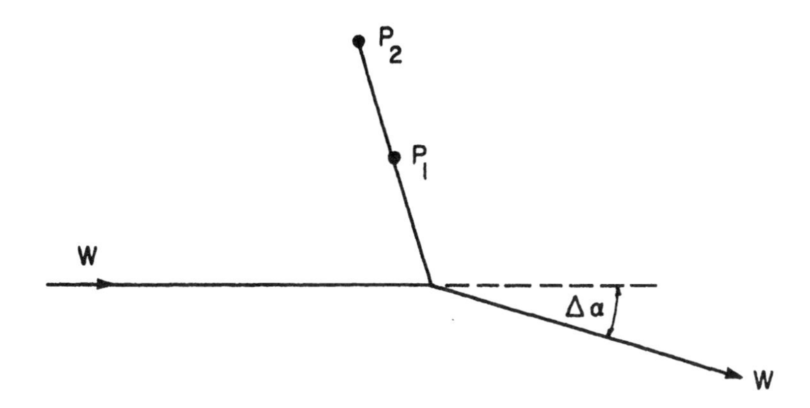

(a) What is the direction of the electric field caused by the acceleration, at the distant point P_1?

(b) In what direction is the radiation intensity of the accelerated charge most intense?

(c) Where is it least intense?

(d) Point P_2 is twice as far from the bend in the trajectory as P_1. By what fraction does the amplitude of the magnetic disturbance decrease as the radiation pulse moved from P_1 to P_2?

(e) What is the total energy radiated?

Make careful sketches in answering parts (a), (b), (c).

4.3 A short vertical transmitting antenna erected at the surface of a perfectly conducting earth transmits a total of 5 kW of power of wavelength λ. What is the value of the electric field strength at a radio receiver 2 km away and stationed at the earth surface? (Assume that there are no houses or other obstacles in the area.)

4.4 In a hydrogen atom an electron of charge −e orbits around a proton of charge +e.

(a) Find the total energy E and the orbital frequency ω as a function of r, the distance between the electron and the proton.

(b) Calculate the energy radiated per unit time as a function of r.

(c) Using $dr/dt = (dr/dE)(dE/dt)$, find the time it takes for a hydrogen atom to collapse from a radius of 10^{-4} cm to a radius of 0.

4.5 A line charge with a total charge Q distributed uniformly over a length L is located along the x axis as shown. The charge is set in motion along the z axis with a motion

$$z(t) = A \sin \omega t$$

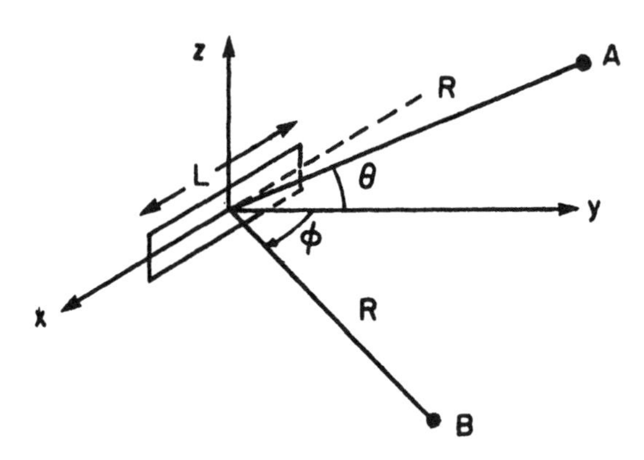

(a) In terms of Q, ω, and A, what is the instantaneous power radiated?

(b) What is the time-average power radiated?

(c) An observer in the y-z plane at point A, a distance R from the source, detects a radiated power P_0 when $\theta = 0$. What power does he detect for any other θ? In other words, what is $P(\theta)$? Assume $R \gg L$ and $R \gg (c/\omega)$.

(d) What polarization does the observer at A detect?

(e) Observer B is also at a distance R from the source but is in the x-y plane. The radius vector makes an angle ϕ with the y axis. He detects power P_0 when $\phi = 0$. Calculate the power he detects for an arbitrary value of ϕ.

(f) If $L = (2\pi c)/\omega$, what power does observer B detect when $\phi = \pi/2$?

Either give your answer as a special case of your answer to part (e) or give a physical reason for your answer.

4.6 Assume that the sun radiates like a black body at 6000°K. Assume that the moon absorbs all the radiation it receives from the sun and reradiates an equal amount of energy like a black body at temperature T. The angular diameter of the sun seen from the moon is about 0.01 radian. What is the equilibrium temperature T of the moon's surface? (Note: You do not need any other data except those contained in the above statement.)

CHAPTER 5

5.1 Elliptically polarized radiation of frequency ω and wave number k propagates in the +z direction and is incident normally on a sheet of perfectly conducting, reflecting metal at $z = 0$. The polarization is as indicated on the sketch. The x component of the incident wave is given by $\vec{E}_{inc,x} = E_o \cos(\omega t - kz)\hat{x}$.

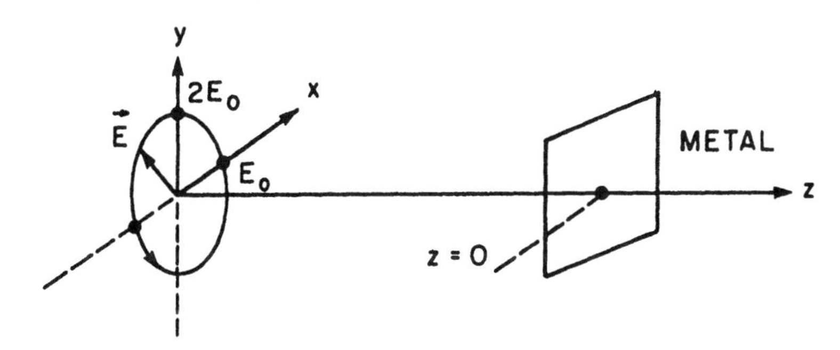

(a) What is $\vec{E}_{ref,x}$ and $\vec{E}_{ref,y}$ in the reflected radiation?

(b) Give a complete, real vector expression for $\vec{E}(z,t)$ in the region $z < 0$. $\vec{E}(z,t)$ is the total field.

(c) What is the radiation pressure on the plate and its time-average value?

5.2 A coaxial cable one wavelength long is terminated by a resistive load equal in magnitude to the characteristic impedance Z_o of the cable. Indicate the magnitude and direction of the Poynting's vector along the line at successive intervals of $(1/8)^{th}$ of a period T, from $t = 0$ to $t = T$.

Make corresponding sketches in the case of a line terminated by a short circuit, rather than by the resistive load.

5.3 A transmission line consists of two parallel wires each of radius a. The distance between the centers of the wires is b.

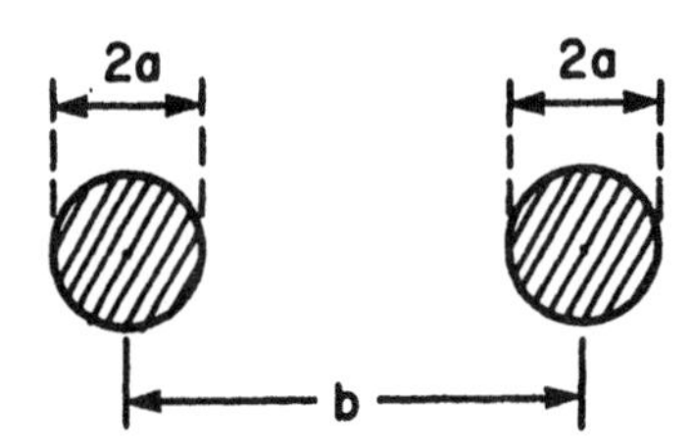

(a) Assuming that $b \gg a$, show that the capacity and inductance per unit length of line are approximately given by

$$C_o \approx \frac{\pi \epsilon_o}{\ln(b/a)} \quad \text{farad/m}$$

$$L_o \approx \frac{\mu_o}{\pi} \ln(b/a) \quad \text{henry/m}$$

(b) Using the results of part (a), compute the phase velocity v of a wave propagating on the line.

(c) Obtain an expression for the characteristic impedance Z_o.

(d) The parallel wire transmission line is made from No. 12 wires (diameter 0.0808 inches) spaced 0.50 inches apart. Calculate C_o, L_o, v and Z_o.

5.4 A waveguide of rectangular cross section operates in the TE_{mn} mode with

$$E_y = E_{oy} \sin(k_x x) \cos(k_y y) \cos(\omega t - k_z z).$$

The field distribution must satisfy the wave equation and boundary conditions at the faces of the guide tube.

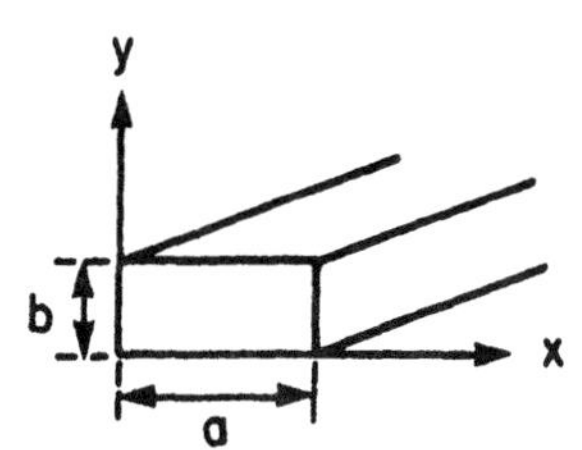

(a) Using the wave equation, develop the necessary relationship between the frequency ω and the various wave numbers.
(b) Using boundary conditions at the faces $x = 0$ and $x = a$, show what restrictions on the wave numbers are required.
(c) Using boundary conditions at the faces $y = 0$ and $y = b$, show what restrictions on the wave numbers are required.
(d) Show that there is a minimum frequency for which propagation will occur and determine this for the TE_{mn} mode.

5.5 Show that the group velocity $v = d\omega/dk$ can be written in the following alternate forms

$$v_g = v - \lambda \frac{dv}{d\lambda}$$

or

$$v_g = \frac{c}{n + \omega(dn/d\omega)}$$

where v is the phase velocity and n the refractive index defined as $n = c/v$.

5.6 The following two waves are superposed:

$$\vec{E}_1 = E_o \cos(5t + 10x)\,\hat{y}$$

$$\vec{E}_2 = E_o \cos(6t + 11x)\,\hat{y}$$

where x is in centimeters and t is in seconds.
(a) Write an equation for the combined disturbance which involves the sum and difference frequencies.

(b) What is the group velocity?

(c) What is the distance between points of zero amplitude in the combined disturbance?

5.7 A copper box with dimensions as shown in the figure acts as a cavity resonator. The electric field

$$E_z = E_o \sin(k_x x) \sin(k_y y) \sin(\omega t)$$

$$E_x = E_y = 0$$

is a possible solution of the wave equation for this case.

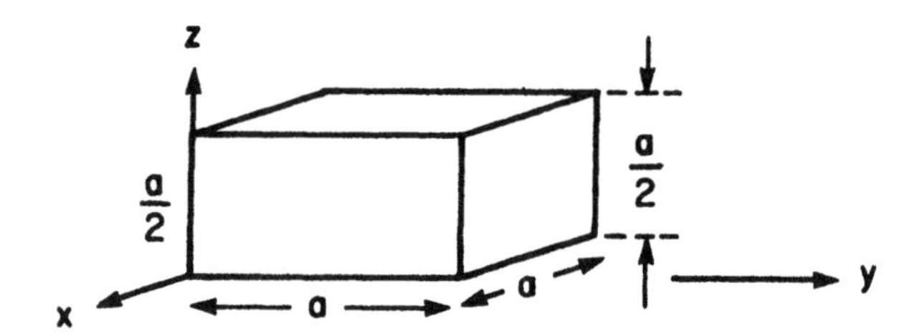

(a) Find the lowest resonance frequency ω_1 and the corresponding free space wavelength λ_1.

(b) Find the next-to-lowest resonance frequency ω_2 and the corresponding free space wavelength λ_2.

CHAPTER 6

6.1 Charge of surface charge density σ is placed on a conductor in contact with a dielectric, as is shown in the figure. The dielectric

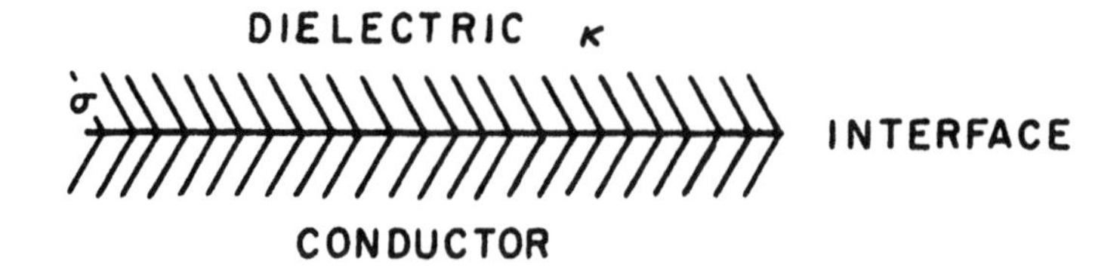

has a dielectric coefficient κ. Determine the polarization charge density σ_p at the interface between the conductor and the dielectric.

6.2 Consider an atom of typical hydrogenic dimension ($r_o = 0.5$ Å). Compute the electric field E_o exerted on the electron by the nucleus at such a distance. (Assume a classical picture.) Now, apply an electric field E of the order E_o. It strongly polarizes the atom producing a dipole moment $p = er_o$. What is the susceptibility of the atom defined as

$$\chi_e = \frac{P}{\epsilon_o E}.$$

P is the total dipole moment per unit volume and is given by Np where N is the atom density. In a solid, $N \simeq 6 \times 10^{23}$ cm^{-3}.

6.3 A wave solution to Maxwell's equations is given by

$$\vec{E}(\vec{r}, t) = \vec{E}_o \cos(\sqrt{6}\, z) \cos(6 \times 10^{10}\, t)$$

where t is in seconds, z is in centimeters, and $\vec{E}_o$ is a constant vector.

(a) What is the index of refraction of the medium?

(b) Give an expression for the associated $B(\vec{r}, t)$ in terms of $\vec{E}_o$, z, and t.

6.4 A plane, monochromatic electromagnetic wave travels in the +z direction within a dielectric medium. The wave is linearly polarized with the $\vec{E}$ field at 45° to the x axis. The frequency of the wave is f Hz, and the dielectric constant of the medium is κ_e. Write expressions for $E_x(z,t)$, $E_y(z,t)$, $B_x(z,t)$, $B_y(z,t)$ in terms of f, c, κ_e and E_o, where E_o is the amplitude of the electric field. What is the rate of energy flow per unit area, per unit time which is transported across a surface perpendicular to the direction of propagation? What is the wave equation for electromagnetic waves traveling within the medium?

6.5 The dielectric constant of hydrogen gas at 0°C and atmospheric pressure is 1.000264.

(a) Compute the polarizability of the hydrogen molecule.

(b) Assuming that the ideal-gas laws are obeyed, compute the dielectric constant of hydrogen at a pressure of 20 atm and a temperature of −200°C.

6.6 A material is considered to be a good conductor when the conduction current exceeds the displacement current. Determine the highest frequency of an electromagnetic oscillation for which the following materials can be considered as good conductors:

	σ $(\text{ohm-m})^{-1}$	κ_e
copper	5.8×10^7	1
sea water	3	81
fresh water	2×10^{-4}	81
earth and rock	1×10^{-5}	6

Here σ is the electrical conductivity of the material and κ_e its dielectric coefficient.

6.7 The ionosphere can be viewed as a dielectric medium of refractive index

$$n = \sqrt{1 - \frac{\omega_p^2}{\omega^2}}$$

where ω is the frequency in rad/sec and ω_p is the plasma frequency, assumed to be a constant.

(a) Make a plot of the magnitude of the propagation vector $\vec{k}$ as a function of the frequency ω for an electromagnetic wave propagating through the ionosphere.

(b) Calculate the phase velocity and group velocity of a radio wave of frequency $\omega = \sqrt{2}\,\omega_p$.

(c) What happens to an electromagnetic disturbance within the ionosphere when $\omega < \omega_p$?

CHAPTER 7

7.1 Two insulators with dielectric coefficients κ_1 and κ_2 are separated by a plane boundary S. Electric fields E_1 and E_2 exist in the two media, having the direction shown.

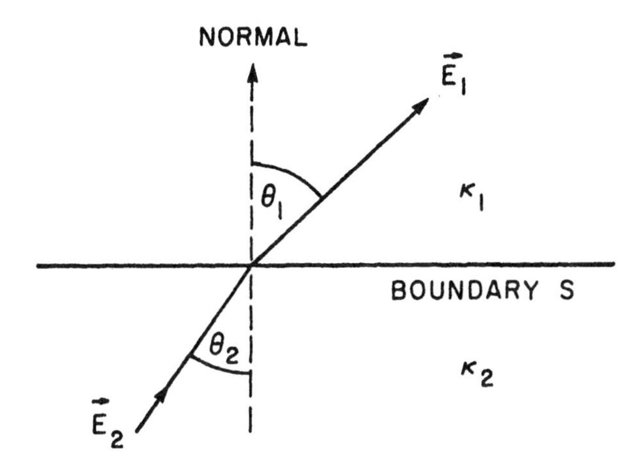

Using appropriate boundary conditions at S, find the relationship between κ_1, κ_2, θ_1 and θ_2. (Your result is the "law of refraction" for electric field lines. Do not confuse this with the true law of refraction, or Snell's law, which tells about the change of direction of a light ray, that is, a change of direction of the Poynting flux vector $\vec{S}$.)

7.2 A point source is viewed through a piece of glass with index n and thickness t. Find the apparent displacement of the source as seen by the observer. Use small angle approximations, that is, assume that the rays strike the glass slab almost at normal incidence.

7.3 Determine the reflection coefficients for normal incidence when an electromagnetic wave of frequency 1×10^9 Hz is incident on the surface of sea water, fresh water, and earth. Use the following values:

	σ $(\text{ohm-m})^{-1}$	κ_e
sea water	3	81
fresh water	2×10^{-4}	81
good earth	10^{-5}	6

7.4 A glass rod of rectangular cross section is bent into the shape shown below. The inner and outer contours of the curved portion of the rod have fixed radii R and R + a, respectively. The index of refraction of the glass is n, equal to 1.5.

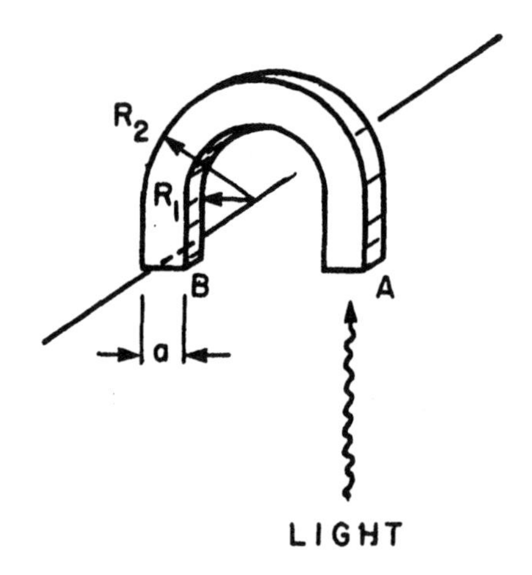

For a given $R \gg \lambda$, what is the maximum value of "a" for which all the light entering at A will emerge at B?

7.5 A plane light wave is incident normally on the face AB of a glass prism as shown in the figure. The index of refraction of glass is 1.50.

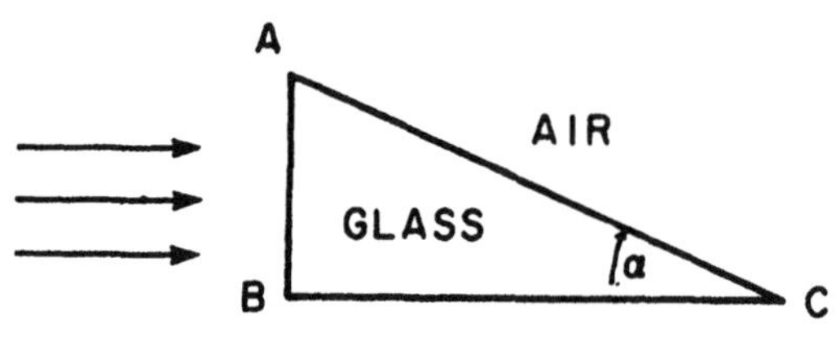

(a) Find the value of the angle α such that the wave will be totally reflected at the surface AC.

(b) Is this the smallest or largest permissible angle for total reflection?

7.6 The nitrogen molecules in the terrestrial atmosphere have an electronic transition in the ultraviolet at a wavelength $\simeq 900$ Å. The total number of N_2 molecules in a column of 1 cm^2 cross section in the atmosphere is 1.3×10^{25}. Compute the fraction of blue sunlight ($\lambda = 4500$ Å) scattered out of the direct path in passing through the atmosphere.

7.7 A beam of light from a star passes through an ionized gas cloud composed of free electrons. The light intensity from the star is diminished as a result of scattering from the electrons. Suppose the electron density is 10^8 electrons per cubic meter. How thick must the gas cloud be in order that the light intensity decrease by 1 percent?

CHAPTER 8

8.1 White light is incident normally on an air film of thickness d formed between two glass plates. What must be the smallest film thickness d if only blue light of wavelength 4000 Å ($= 4 \times 10^{-7}$ m) is to be reflected strongly?

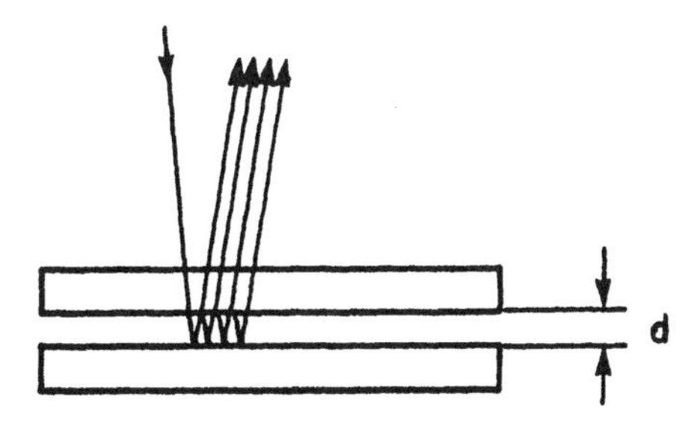

8.2 A radar antenna operating on a wavelength of 0.10 m is located 8 m above the water line of a torpedo boat. Treat the reflected beam from the water as originating in a source 8 m below the water directly under the radar antenna.

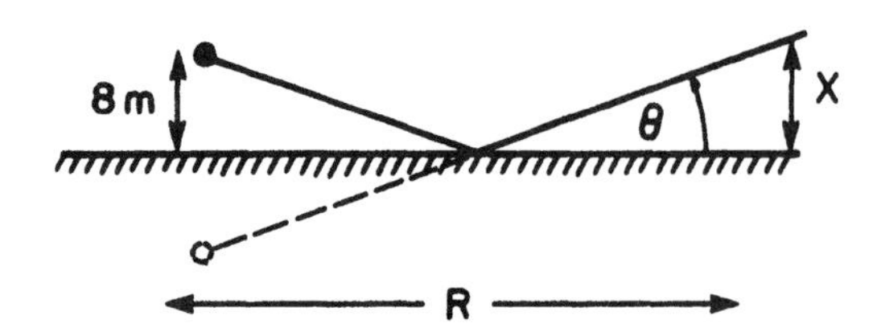

What is the altitude x of an airplane 12 km from the boat if it is to be in the first interference minimum of the radar signal?

What is the total number of minima one observes as one scans the sky in the vertical plane as a function of the angle θ, from $\theta = 0$ to $\theta = \pi$, keeping the distance R fixed?

8.3 Two dipole radiators are separated by a distance $\lambda/2$ along the x axis. The dipoles are oriented along z. Assume $r \gg \lambda$.

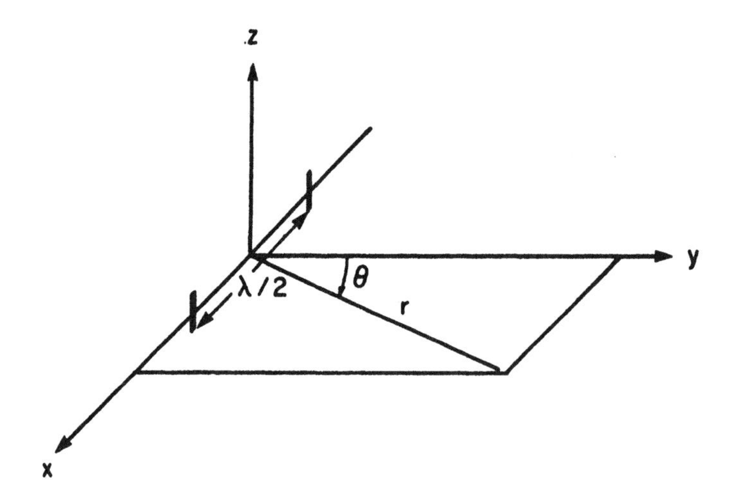

(a) Find the relative intensities of the radiation in the x-y plane at $\theta = 0$, $\pi/3$, $\pi/2$, and π if the oscillators are in phase.

(b) Repeat (a) if the oscillators are 180° out of phase.

(c) The oscillators are spaced by a distance $\lambda/4$ and are 90° out of phase. Find the relative intensities at $\theta = 0, \pi/2, \pi, 3\pi/2$.

8.4 A plano-convex piece of glass (index of refraction n) rests on a plane parallel piece of glass as shown. The radius of the spherical surface is R and is much greater than r_m. Light of wavelength λ is incident normally and reflected at the spherical glass-air interface and at the air-glass interface of the glass plate. The two reflected beams then interfere to produce a series of alternately bright and dark concentric circles when viewed from above.

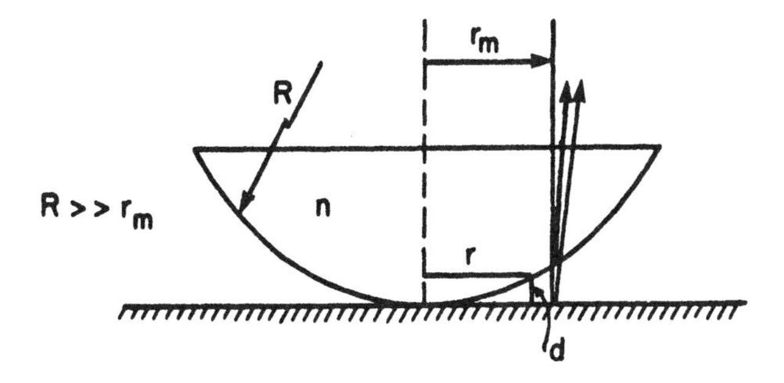

(a) Find the radial distance, r, from the point of contact at which the separation between spherical surface and the plate upon which it rests is d, i.e., find the relation between r and d.

(b) Derive an expression for the radial distances, r_m, at which bright rings will be observed.

8.5 We desire to superpose the oscillations of several simple harmonic oscillators having the same frequency ω and amplitude A, but differing from one another by constant phase increments α; that is,

$$E(t) = A\cos\omega t + A\cos(\omega t+\alpha) + A\cos(\omega t+2\alpha) + A\cos(\omega t+3\alpha) + \dots .$$

(a) Using graphical phasor addition find E(t); that is, writing $E(t) = A_o\cos(\omega t+\phi)$, find A_o and ϕ for the case when there are five oscillators with $A = 3$ units and $\alpha = \pi/9$ rad.

(b) Study the polygon you obtained in part (a) and, using purely geometrical considerations, show that for N oscillators

$$E(t) = (NA)\,\frac{\sin(N\alpha/2)}{N\,\sin(\alpha/2)}\,\cos\left[\omega t + \left(\frac{N-1}{2}\right)\alpha\right].$$

(c) Sketch the amplitude of E(t) as a function of α.

(The above calculation is the basis of finding radiation from antenna arrays and diffraction gratings.)

8.6

(i) Parallel light of wavelength 5461 Å is incident normally on a slit 1 mm wide. If a lens of 100 cm focal length is mounted just behind the slit and the light focused on a screen, what will be the distance in millimeters from the center of the diffraction pattern to (a) the first minimum, (b) the first secondary maximum, (c) the third minimum?

(ii) Plot the intensity curve for the above case, against a scale of distances along the screen in millimeters. Carry as far as the third minimum in either side. Read from this curve the true intensities of the secondary maxima.

8.7 Two arc lights are 1 ft apart and 10 miles away. They are observed in a telescope, the objective of which has a diameter of 4 cm. An adjustable slit is placed in front of the objective and oriented so that its width parallel to the line between the sources can be varied. The aperture is narrowed until the two arc lights are barely resolved. Find its width, assuming the effective wavelength to be 6000 Å.

8.8 A pinhole camera for visible light is made from a cubical box (length of one side = L) by drilling a small circular aperture (diameter = d) in one side and using the opposite inside wall as the screen where the film is placed. Approximately what value of d will provide the sharpest image on the film?

(Hint: Calculate the full size of the geometrical plus diffraction image of a distant source.)

8.9 A plane electromagnetic wave of wavelength λ_o is incident on two long, narrow slits, each having width 2a and separated by a distance 2b, with b ≫ a. One of the slits is covered by a thin dielectric plate of thickness d, and dielectric coefficient κ, with d chosen so that $(\sqrt{\kappa} - 1)d/\lambda_o = 5/2$.

The interference pattern due to the slits is observed in a plane a distance L from the slits, where L is large enough so that the far field approximations may be used, that is, the pattern depends only on the angle θ from the normal to the slits, as shown.

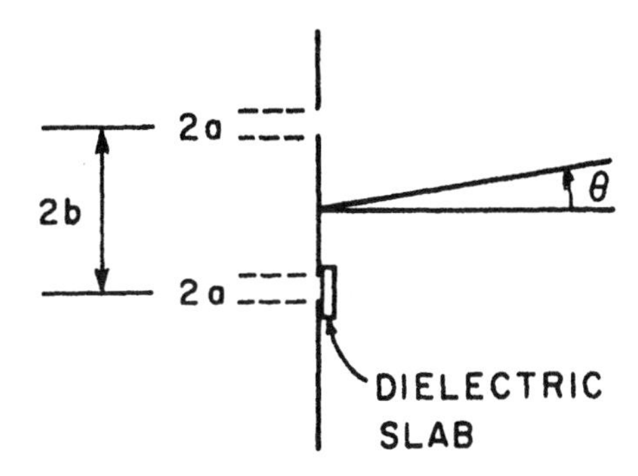

(a) Consider effects due to interference only. What is the condition for a maximum in the pattern? Sketch the interference pattern.

(b) Now include effects due to both interference and diffraction. How is the intensity distribution modified from that obtained in (a)? Let b/a = 10, sketch the resulting interference-diffraction pattern.

(Assume that all angles involved are small enough so that $\cos\theta \simeq 1$, and hence that the optical path through the dielectric is independent of angle.)

UNITS AND DIMENSIONS

To get the value of quantity in Gaussian units, multiply the value in MKS units by the conversion factor.

Quantity	Symbol	Dimensions in MKS	MKS	Conversion Factor	Gaussian
Length	L	L	meter (m)	10^2	centimeter
Mass	M, m	M	kilogram (kg)	10^3	gram
Time	T, t	T	second (sec)	1	second
Charge	Q, q	Q	coulomb (C)	3×10^9	statcoulomb
Energy	U, W	ML^2T^{-2}	joule	10^7	erg
Force	F	MLT^{-2}	newton	10^5	dyne
Power	P	ML^2T^{-3}	watt	10^7	erg/sec
Pressure	p	$ML^{-1}T^{-2}$	newton/m^2	10	dyne/cm^2
Velocity	v, u	LT^{-1}	m/sec	10^2	cm/sec

Capacitance	C	$T^2Q^2M^{-1}L^{-2}$	farad	9×10^{11}	cm
Conductivity	σ	$TQ^2M^{-1}L^{-3}$	mho/m	9×10^9	sec^{-1}
Current	I	QT^{-1}	ampere	3×10^9	statampere
Electric Field	E	$MLT^{-2}Q^{-1}$	volt/m	$(3 \times 10^4)^{-1}$	statvolt/cm
Impedance	Z	$ML^2T^{-1}Q^{-2}$	ohm	$(9 \times 10^{11})^{-1}$	sec/cm
Inductance	L	ML^2Q^{-2}	henry	$(9 \times 10^{11})^{-1}$	sec^2/cm
Magnetic Field	B	$MT^{-1}Q^{-1}$	weber/m^2	10^4	gauss
Magnetization	M	$QL^{-1}T^{-1}$	amp-turn/m	10^{-3}	oersted
Permeability	μ	MLQ^{-2}	henry/m	$10^7/4\pi$	–
Permittivity	ϵ	$T^2Q^2M^{-1}L^{-3}$	farad/m	$36\pi \times 10^9$	–
Polarization	P	QL^{-2}	coulomb/m^2	3×10^5	statcoulomb/cm^2
Potential	ϕ, V	$ML^2T^{-2}Q^{-1}$	volt	$(300)^{-1}$	statvolt
Resistance	R	$ML^2T^{-1}Q^{-2}$	ohm	$(9 \times 10^{11})^{-1}$	sec/cm

SOME CONSTANTS

Velocity of light in vacuo	c	2.998×10^{8} m/sec
Charge on an electron	e	1.602×10^{-19} coulomb
Mass of an electron	m	9.110×10^{-31} kg
Charge to mass ratio	e/m	1.759×10^{11} coulomb per kg
Mass of H atom	M_H	1.673×10^{-27} kg
Ratio of proton mass to electron mass	M_H/m	1836
Planck's constant	h	6.626×10^{-34} joule-sec
Boltzmann's constant	κ	1.381×10^{-23} joule/°K
Stefan-Boltzmann constant	σ	5.670×10^{-8} joule/m^2-sec-°K^4
Classical electron radius	r_o	2.818×10^{-15} m
Permittivity of free space	ϵ_o	8.854×10^{-12} farad/m
Permeability of free space	μ_o	$4\pi \times 10^{-7}$ henry/m
Energy associated with 1 eV		1.602×10^{-19} joule
Temperature corresponding to 1 eV energy		11605°K
Velocity of a 1-eV electron		5.93×10^{5} m/sec
Velocity of a 1-eV H atom		1.38×10^{4} m/sec
Wavelength of a 1-eV photon		1.240×10^{-6} m
Frequency of a 1-eV photon	$\omega/2\pi$	2.418×10^{14} cps (Hz)
Avogadro's number		6.023×10^{23} molecules/mole
Gas density at 1 torr pressure and 0°C		3.53×10^{22} molecules/m^3

ACKNOWLEDGMENTS

We would like to thank our colleagues: William Bertozzi, Judith Bostock, Anthony P. French and Elizabeth S. Hafen. Publication of the preliminary editions was supervised by Peggy Richardson ably supported by Susie Fennelly, Venetia Kaloyanides, Susan Laing and Sari Moeller. The publication of the final camera-ready manuscript was supervised by Alice S. Amdur who had the help of Mary M. Bosco, Inge Calci, Stella J. Foster, Esther D. Grande, Phyllis L. Handel, and Ann G. Serini. The camera-ready manuscript was proofread by Chaia Bekefi.

We also gratefully acknowledge use of the following illustrative material.

Fig. 1.5 from *The Earth's Shape and Gravity*, G. D. Garland, Pergamon Press, 1965.

Fig. 1.11 from *The Sun*, G. P. Kuiper, University of Chicago Press, 1953.

Fig. 1.12 from D. H. Looney and S. C. Brown, Phys. Rev. **93**, 965 (1954).

Fig. 1.23 from *Classical Dynamics of Particles and Systems*, J. B. Marion, Academic Press, 1970.

Fig. 3.20 from J. M. Greenberg, Scientific American, October, 1967.

Fig. 4.3 from film *Electric Fields of Moving Charges* by the Education Development Center, Inc., Newton, Mass.

Fig. 4.24 from M. L. Perlman, E. M. Rowe, and R. E. Watson, Physics Today **27**, 30 (1974).

Fig. 4.25 from J. H. Oort, Scientific American, March 1957.

Fig. 4.29 from P. Thaddeus, Ann. Rev. Astronomy and Astrophys. **10**, 305 (1972).

Fig. 4.30 from R. A. Alphen, G. Gamow, and R. Hartman, Proc. Nat. Acad. Sci. U.S.A. **58**, 2179 (1967).

Fig. 5.25 from *Waves* by F. S. Crawford, McGraw-Hill Book Company, 1965.

Fig. 5.2) from *Optical Waveguides* by N. S. Kapany and J. J. Burke, Academic Press, 1972.

Fig. 7.6 from *Electromagnetic Theory* by J. A. Stratton, McGraw-Hill Book Company, 1941.

Fig. 7.8 from Heliotek, Division of Textron Inc., Sylmar, California.

Fig. 7.21 from the High Altitude Observatory/The National Center for Atmospheric Research.

Fig. 7.28 from P. Sprangle and W. M. Manheimer, Phys. Fluids **18**, 224 (1975).

Fig. 8.11 from the Netherlands Foundation for Radio Astronomy, Copyright by Aerophoto Eelde.

Fig. 8.19 from W. Braunbek and G. Laukien, Optik **9**, 174 (1952).

INDEX

ERRATA

<u>Electromagnetic Vibrations, Waves and Radiation</u>

G. Bekefi and A. H. Barrett

p. 20	17th line from the bottom: . . . suppose $\theta_o = 0.1$ radian . . .
p. 103	Figure 1.39. The (central) coupling spring has a spring constant κ not K.
p. 130	9th line from the top: . . . speed of sound . . .
p. 137	4th line from the top: $\sqrt{\overline{s_o^2}} = 3.44 \times 10^{-4} m$.
	8th line from the top: $\sqrt{\overline{p^2}} = \sqrt{\frac{2\kappa I}{v}} = 29.1 N/m^2$.
p. 167	Lines 15, 19, 20 from the top. All x's should be z's.
p. 175	7th line from the top: . . . which are just two Fourier series. END OF SENTENCE. Since in our problem the string is . . .
p. 240	Line 11: $cos^2(wt + 3x - y - z)$ . . . not $cos(wt + 3x - y - z)$.
p. 368	3rd line from the top: . . . side-band.
p. 432	2nd line from the top: $dn_r/dw < 0$
p. 433	The two equations at the top should be numbered as (6.54); 6th line from the top: $dn_r/dw < 0$
p. 467	5th line from the bottom: . . . the propagation constant k and hence . . .
	4th line from the bottom: Inserting k from Eq. (6.62) in . . .
p. 481	Caption to Fig. 7.15: . . . will be totally internally reflected, provided that the refractive index $n > \sqrt{2}$.
p. 630	3rd line from the bottom: $Z_o \equiv F/v$
	5th line from the bottom: . . . ratio of the driving force to the velocity produced . . .
p. 632	2nd line from the bottom . . . radiates spherical waves . . . AD at a frequency of 1.6 kHz.
p. 634	6th line from the top: in the equation, $1 \cdot 2$ means 1×2; $1 \cdot 3$ means 1×3; $3 \cdot 5$ means 3×5, etc.
p. 635	2nd line from the bottom: $f(\vec{k} \cdot \vec{r} - \omega t + \phi)\hat{y}$
p. 637	5th line from the top: . . . loop be oriented to obtain it? ADD: Assume the loop is rotated about an axis parallel to the electric vector.
p. 646	10th line from the bottom: $n = \sqrt{1 - \frac{\omega_p^2}{\omega^2}}$
p. 650	4th line from the top: . . . under the radar antenna. ADD: the dipole antenna is oriented $\perp$ the plane of the page.